Lecture Notes in Physics

Volume 1033

The series Lecture Notes in Physics (LNP), founded in 1969, reports new developments in physics research and teaching - quickly and informally, but with a high quality and the explicit aim to summarize and communicate current knowledge in an accessible way. Books published in this series are conceived as bridging material between advanced graduate textbooks and the forefront of research and to serve three purposes:

- to be a compact and modern up-to-date source of reference on a well-defined topic;
- to serve as an accessible introduction to the field to postgraduate students and non-specialist researchers from related areas;
- to be a source of advanced teaching material for specialized seminars, courses and schools.

Both monographs and multi-author volumes will be considered for publication. Edited volumes should however consist of a very limited number of contributions only. Proceedings will not be considered for LNP.

Volumes published in LNP are disseminated both in print and in electronic formats, the electronic archive being available at springerlink.com. The series content is indexed, abstracted and referenced by many abstracting and information services, bibliographic networks, subscription agencies, library networks, and consortia.

Proposals should be sent to a member of the Editorial Board, or directly to the responsible editor at Springer:

Dr Lisa Scalone
lisa.scalone@springernature.com

Jana Musilová · Pavla Musilová · Olga Rossi

Calculus of Variations on Fibred Manifolds and Variational Physics

A Geometrical Approach to Variational Problems in Physics

Springer

Jana Musilová
Institute of Theoretical Physics
and Astrophysics
Masaryk University
Brno, Czech Republic

Pavla Musilová
Institute of Theoretical Physics
and Astrophysics
Masaryk University
Brno, Czech Republic

Olga Rossi
Department of Mathematics
University of Ostrava
Ostrava, Czech Republic

ISSN 0075-8450 ISSN 1616-6361 (electronic)
Lecture Notes in Physics
ISBN 978-3-031-77407-2 ISBN 978-3-031-77408-9 (eBook)
https://doi.org/10.1007/978-3-031-77408-9

This Springer imprint is published by the registered company Springer Nature Switzerland AG
The registered company address is: Gewerbestrasse 11, 6330 Cham, Switzerland

We dedicate our joint work and lasting memory to our teacher, colleague and friend, professor of theoretical physics Michal Lenc, long-time director of the Institute of Theoretical Physics and Astrophysics of the Faculty of Science of Masaryk University. His broad knowledge of physics and mathematics, pedagogical mastery and human-friendly approach have become an inspiration not only for us, but also for many of the students he has educated.

And to Tatiana—for future.

Preface

Dear readers,

The text that we would like to offer you is on the border of an advanced study text for students who have decided to study this field of mathematical physics in their professional work, and interested researchers in physics who want to get acquainted with the field at the level of mathematical tools for theories in physics. It is therefore a work on the border between an advanced textbook and a scientific monograph of the basic level.

This was the original idea of our co-author, Prof. Olga Rossi-Krupková, a world-renowned scientist in the field of modern mathematical approaches to the calculus of variations, in which she achieved significant results and successes. We started preparing the book together on the basis of a call from the Springer publishing house, but unfortunately there are only two of us finishing it.

The basic idea of the original concept was to present a mathematical background of fibred manifolds that serves very effectively for the development of most of the fundamental physical theories (classical non-relativistic and relativistic mechanics, quantum mechanics, classical electrodynamics,...). The effectiveness of this approach is given by the possibility to model not only the underlying spaces for physical theories, but also the physical quantities themselves, by geometric, i.e. coordinate free objects (vector fields and differential forms on the mentioned underlying spaces). The expressions representing definitions and theorems of the developed theory are thus very concise and elegant and look practically formally the same at all levels of generality of the theory. Of course, the specific calculations are then done in coordinates.

Our text was originally supposed to be a mathematical one with occasional input into physics in the form of examples. The mathematical background of the calculus of variations at a very general level (the general dimension of all components of the underlying fibred manifold and its prolongations of any order) is developed in a number of works that we cite in Chap. 1 (Introduction). These are mainly the fundamental comprehensive works of D. Krupka, O. Rossi-Krupková and their coauthors, collaborators and students all over the world. However, the difficulty of these texts is quite considerable due to their full generality and complexity.

Although, as already mentioned, the formal expression of the essential ideas of the theory in geometric form is very similar, or almost identical, on both the general and special levels, practical coordinate calculations at a quite general level are technically demanding our idea following the original intention of Olga Rossi is based on simpler situations:

- The one-dimensional bases of a fibred manifold (part of the calculus of variations known as mechanics, although it doesn't always have to be mechanics as a physical discipline).
- In physical theories, extensions of fibred manifolds of the first and second order are mainly used.

Such an approach will not only make it possible to carry out technical calculations in detail, but also, crucially, it will enable the reader to fully understand the geometric nature of the theory, i.e. through carefully and thoroughly executed coordinate calculations at the level of lower dimensions and prolongations of underlying spaces. (This is also the approach that characterizes our text, either directly or through problems and their solutions.)

Originally, the text was supposed to be mathematical. With regard to the conviction that it is necessary to directly introduce a solid, consistent and in all respects correct mathematical background into physics, we decided not to abandon the concept of our colleague Olga Rossi, but to extend it to other physical theories than just those that use the mathematics for mechanics (one-dimensional base of underlying manifolds). These theories represent the majority—they are theories of physical fields (hydrostatics and hydrodynamics, electrodynamics, quantum mechanics, string theory,...) That is why we have included a chapter expanding the mathematical background of fibred manifolds and geometric objects defined on them. This extending Chap. 8 is thus formulated for multidimensional bases of fibred manifolds and in general for arbitrary finite order prolongations. On the other hand, practical calculations in coordinates continue to be performed on first- and second-order prolongations of fibred manifolds, as is consistent with the physical theories included in Chap. 9.

We would like to express our conviction that our co-author Olga Rossi would agree with the (not essential) changes made with respect to the original focus of the text on predominantly mathematical one, also due to her close relationship to physics.

Brno, Czech Republic
July 2024

Jana Musilová
Pavla Musilová

Acknowledgements

The authors are indebted to Prof. Demeter Krupka for his introduction to modern calculus of variations and global analysis as effective methods for the study of physical theories. The first two authors would like to thank Prof. Olga Rossi-Krupková (*1960–†2019) for her long-term cooperation in the given scientific field and helpful and long-term collegial and personal friendship.

We would also like to thank Ing. Jakub Zlámal for the professional editing of the typesetting and Mgr. Jana Hoderová for redrawing the illustrative pictures. Our big thanks go to editor Dr. Lisa Scalone for her recommendations and advice throughout the entire contact with the publisher.

Last but not least, we would like to thank our family for their continued support and patience.

Contents

1 Introduction 1
1.1 Organization of the Book 2
1.2 Notations 5
References 8

2 Fibred Manifolds 11
2.1 Fibred Manifolds and Charts 12
2.2 Sections 17
2.3 Local Isomorphisms 20
2.4 Jets and Prolongations 22
2.5 Problems 30
References 33

3 Vector Fields and Differential Forms 35
3.1 Vector Fields 36
3.2 Differential Forms 48
3.3 Contact Ideal 52
3.4 Distributions 63
3.5 Problems 72
References 77

4 Calculus of Variations 79
4.1 Lagrange Structure and Variational Integrals 80
4.2 First Variational Formula, Lepagean Forms 83
4.3 Extremals, Euler-Lagrange Equations 87
4.4 Symmetries and Conservation Laws, Noether Theorem 97
4.5 Generalized Theorem Emmy Noether 106
4.6 Problems 111
References 113

5 Dynamical Forms and the Inverse Problem 115
5.1 A Note to Higher Order Lagrange Structures 116
5.2 Variational Dynamical Forms 119

5.3 Helmholtz Conditions of Variationality 122
5.4 Lagrangians of Variational Dynamical Forms 128
5.5 Variational Forces in Mechanics 132
5.6 Problems 133
References 135

6 Hamiltonian Systems and the Hamilton–Jacobi Theory 137
6.1 Regular Variational Problem in the First-Order Mechanics and Hamilton Equations 138
6.2 Hamilton Extremals 142
6.3 The Concept of Regularity Revisited 147
6.4 A Note to Higher-Order Hamilton Theory 163
6.5 A Note to Hamilton-Jacobi Theory 170
6.6 Problems 176
References 179

7 Elements of Variational Sequences 181
7.1 Preliminary Considerations 182
7.2 Variational Sequence 187
7.3 Representation of the Variational Sequence by Forms 192
7.4 Variational Sequence in Calculus of Variations 201
7.5 A Note to Local and Global Aspects 207
7.6 Problems 208
References 209

8 Extension: Geometrical Structures for Field Theories 211
8.1 Fibred Manifolds with n-Dimensional Bases, Sections, Jet Prolongations 212
8.2 Vector Fields and Differential Forms 218
8.3 Lagrange Structures, First Variational Formula, Euler-Lagrange Equations 236
8.4 Symmetries, Noether Theorem, Generalized Noether Theorem 245
8.5 Inverse Problem 249
8.6 Problems 252
References 256

9 Variational Physics 257
9.1 Mechanics and Special Relativity 258
9.2 Mechanics of Continuous Media and Waves 269
9.3 Classical Electrodynamics 273

9.4 Quantum Mechanics 277
9.5 A Note to String Theories 281
9.6 Problems 284
References 286

Solutions and Hints 287

Index 325

Introduction

1

The oldest tasks, which are variational in nature, date back to antiquity. For example, the *isoperimetric problem*—finding the shape of a closed curve of a fixed length that should encircle a planar area of the largest possible area content. Everyone probably knows that it is a circle, but also that the shortest line connecting two given points in the Euclidean plane or space is a line segment, on a spherical surface the principal circle, etc. A typical variational problem is also the propagation of light through a homogeneous or inhomogeneous medium—light propagates in such a way that it travels between two specified points in the shortest time (Fermat principle). The first problem, recognized as variational, is that of the *brachistochrone problem*, [1]; see also in the abstract of Chap. 4. In short, the question is: What is the shape of the path along which a body slides from point A to point B in the homogeneous gravitational field of the Earth (without friction) in the shortest time?

A mathematical theory that makes it possible to solve such problems is the *calculus of variations*. Since the days of the brachistochrone, it has evolved to its current modern form, which is based on purely geometric concepts. That includes, in particular, fibred manifolds and their jet prolongations as underlying spaces, as well as geometric structures defined on these manifolds, as vector and tensor fields, especially differential forms (antisymmetric covariant tensor fields of various orders), adapted to the fibred structure of the mentioned underlying spaces. There are a number of founders of this geometric approach. As the pioneering works in the field of modern geometric theory can be considered [2] (among others the introduction of the Cartan form for the formulation of the variational principle), or [3,4] (variational field theory with help of differential forms and their exterior derivatives modulo contact forms). Key works that led to the discussion of global aspects, especially through the gradually built concept of variational sequence are, e.g. [5–19], and other.

An appropriate mathematical framework for calculus of variations is fibred manifolds and their jet prolongations which form very natural and convenient underlying

J. Musilová et al., *Calculus of Variations on Fibred Manifolds and Variational Physics*, Lecture Notes in Physics 1033, https://doi.org/10.1007/978-3-031-77408-9_1

geometrical structures. The reader can find very good presentation and explanation of these structures, e.g. in [20]. The approach in our book is based namely on works of D. Krupka and O. Rossi (Krupková) and her collaborators: [21] (extensive work concerning among other geometry of fibred manifolds), [22,23] (introduction of the concept of Lepage forms), [24] (the first introduction of the concept of *finite order* variational sequence as an appropriate tool for solving global aspects connected, e.g. with the problem of trivial Lagrangians or the inverse problem of calculus of variations), [25,26] (complete general solution of the problem of representation of the finite order variational sequence by forms), [27,28] (latest summarizing works) and many more, [29–33] (Hamilton theory, variational equations) and [34,35] (second-order variational equations).

At present, various problems of the calculus of variations and especially the variational sequence and its applications in physics are studied by R. Vitolo and M. Palese and their collaborators, [36–44]. Of course, the list of references of works that may relate to our book is not and cannot be complete. However, these are the most important links from our point of view.

Note: The classical approach to calculus of variations can be found in the famous book [46]. A modern mathematically correct treatise concerning differentiable manifolds in general and differential forms on these structures gives, e.g. [45].

1.1 Organization of the Book

The structure of this book is chosen to bring the reader to a deeper understanding of the geometric structures needed for the calculus of variations. These structures enable efficient and elegant coordinate free expression of basic concepts and results of the calculus of variations as such, especially its physical applications.

The formal structure of chapters is practically the same, each chapter is divided into sections, mostly six. At the beginning of the chapter, after a short abstract, there are presented prerequisites, i.e. requirements for the reader's knowledge and skills, which will allow him to follow the text without any problems. Then follow eventual motivational considerations, problems or examples. The other parts are structured in a standard way, as is customary in mathematics: definitions, propositions with proofs, key theorems, again with proofs, and examples. Each chapter is concluded with problems to solve, often with hints or complete solutions placed at the end of the book. For some problems, the results are given for checking. The section Problems is always followed by the list of references relevant for the chapter, i.e. that either been used in the text or are recommended to the reader for further study. For a deeper study in a broader context, we recommend a suitable selection from the links given at the end of this chapter.

Chapters 2 and 3 present definitions and properties of underlying geometric concepts for calculus of variations in mechanics, with regard to later applications in physics. The basis for building a geometric approach to the variational mechanics lies in the concept of a differentiable manifold, especially a fibred manifold (Y, π, X) with the one-dimensional base X. A concrete realization of a fibred manifold that we often use in examples is the trinity $(Y, \pi, X) = (\mathbf{R} \times \mathbf{R}^m, p, \mathbf{R})$, where $\mathbf{R}$ is the real axis, $\mathbf{R}^m = \mathbf{R} \times \cdots \times \mathbf{R}$ is the m-dimensional Euclidean space and p is the Cartesian projection on the first factor of the Cartesian product $\mathbf{R} \times \mathbf{R}^m$, $p : \mathbf{R} \times \mathbf{R}^m \to \mathbf{R}$. E.g. in classical mechanics, the coordinate t on the base X is interpreted as time, the space $\mathbf{R}^m$ realizes the configuration space over the time point $t \in \mathbf{R}$. Smooth mappings, or mappings at least differentiable up to a needed order, assigning to a point t a point in the configuration space (so-called sections) represent graphs of time development of the given mechanical system. Prolongations of a fibred manifold with the help of sections allow to enrich the fibred structure with a phase space. A point of the phase space over a certain time point t represents the instant state of the studied mechanical system at the time t. The concept of fibred manifolds and its prolongations for mechanics and its properties is presented in Chap. 2.

Mappings assigning to points of a fibred manifold or its prolongations vector or tensor quantities are vector and tensor fields. Smooth or differentiable antisymmetric covariant tensor fields of various orders, called the differential forms, have the extreme meaning for modern calculus of variations. These concepts and their relations to the fibred structure of a given manifold, especially projectable and vertical vector fields and their prolongations, or horizontal and contact differential forms are introduced and studied in detail in Chap. 3.

A separate area of classical mechanics is Lagrangian mechanics, based on the search for trajectories of mechanical systems as stationary points (called extremals) of a certain functional (in short, a mapping assigning real values to the graphs of the evolution of a system); a stationary point of this mapping, most often a minimum, then determines the trajectory of the mechanical system. Stationary points are determined by the well-known Euler-Lagrange equations. Relevant functionals for physical theories as well as geometric conditions and equations for finding their stationary points are determined by appropriately selected differential forms—Lagrangians and Euler-Lagrange forms. Conservation laws in mechanics are closely related to the symmetries of the Lagrangians or their Euler-Lagrange forms. These symmetries are characterized by theorem Emmy Noether (Noether equation) and generalized theorem Emmy Noether (Noether-Bessel-Hagen equation). It is well known that Lagrangian mechanics cannot be used for any mechanical system. This is especially true for systems in which non-conservative forces are present, so, the potential energy is not defined for such systems. In connection with this, the question arises of how to recognize whether the given equations of motion (equivalently the so-called dynamical form) correspond to a variational mechanical system (system subjected to Lagrangian mechanics). This question representing the so-called inverse problem is important especially for applications in physics. Geometrical concepts (symmetries of Lagrange structures) and related theoretical considerations (conservation laws, inverse problem) are presented in Chaps. 4 and 5.

Another, but fully equivalent approach to solving tasks of calculus of variations, or especially variational physics, is the Hamiltonian theory. As for the practical aspects, the difference between the Lagrangian and Hamiltonian theory lies in the following: While, e.g. in the case of Lagrangian first-order mechanics of a system with m degrees of freedom one must solve m ordinary second-order differential equations (Euler-Lagrange equations), the Hamiltonian theory leads to $2m$ first-order differential equations (Hamilton equations). It is well known that the condition allowing to use the Hamiltonian approach is regularity of the basic structure of the Lagrangian approach—the Lagrangian. In Chap. 6 the concept of regularity is generalized compared to its standard definition. (So, some previously singular problems become regular from the point of view of the new definition.) Also in this chapter, the first-order mechanics is developed in detail, with the exception of a brief extension to higher order mechanics, which gives a better insight into the essence of the generalized definition of regularity. A very brief note is then devoted to the Hamilton-Jacobi theory.

An integral part of the modern calculus of variations is the so-called variational sequence. In our case, it is a special case of a well-known variational bicomplex—a finite order variational sequence. It arises as the factor sequence of the de Rham exact sequence of exterior derivative operators which assign to k-forms its exterior derivatives ($(k+1)$-forms). The classes of forms arise by factorization of spaces of k-forms with respect to their subspaces of forms of a certain type of contactness forming the differential ideal. The mentioned sequence of mappings is called the *finite order variational sequence*. Its first (nontrivial) term is the Euler-Lagrange mapping discussed in Chap. 5. For the calculus of variations also the second term of this sequence, called the *Helmholtz-Sonin mapping*, is important. It assigns to dynamical forms special 3-forms (called *Helmholtz-Sonin forms*) and the set of variational dynamical forms is its kernel. This, among others, allows to solve the inverse problem (to find out whether the given equations of motion are variational, i.e. whether they originate from a certain Lagrangian).

Although our specific considerations and the use of the variational sequence in solving the inverse problem in physics relate to the first order (with the representation by forms lifted in general up to the third jet prolongation of the underlying fibred manifold), definitions and propositions are valid generally.

With regard to the fact that our important goal is to use the mathematical tools of calculus of variations on fibred manifolds in physical theories that include field theory to a greater extent, our text includes Chap. 8 as extension of previous considerations to the case of field theory.

The last chapter is devoted to the use of geometric structures of calculus of variations on fibred manifolds to formulate physical theories and solve their specific problems. It deals with symmetries and conservation laws and solves the inverse problem in classical relativistic and non-relativistic mechanics, continuum mechanics, classical electrodynamics and quantum mechanics (a short note is devoted to the possibility of the use of the presented mathematical background in string theories). It shows, among others, that these key physical theories are variational, i.e. their equations of motion are always derived from the corresponding Lagrange structure on an appropriately chosen fibred manifold.

1.2 Notations

In this section the standard notation used in papers and books is devoted to calculus of variations, especially the notation frequently used and universally accepted by D. Krupka and O. Rossi (Krupková) and their coauthors. In summations within the range of predetermined indices, we use Einstein summation rules, with some exceptions concerning low-dimensional examples. To express the two-sided implication $\Longleftrightarrow$, we use both standard possibilities "iff" and "if and only if".

Underlying geometric structures

$\mathbf{R}$, $\mathbf{R}^m$	Euclidean space of dimension 1 or m, respectively (alternatively $\pi : Y \to X$)
(Y, π, X)	fibred manifold with the base X, total space Y
or $(J^0 Y, \pi_0, X)$	and projection (surjective submersion) π mechanics: $\dim X = 1$, field theory $\dim X = n > 1$
$(\mathbf{R} \times \mathbf{R}^m, p, \mathbf{R})$	realization of a fibred manifold, p is the Cartesian projection on the first factor
$(\mathbf{R}^n \times \mathbf{R}^m, p, \mathbf{R}^n)$	
$(J^1 Y, \pi_1, X)$	first jet prolongations of (Y, π, X)
$(J^r Y, \pi_r, X)$,	r-th jet prolongations of (Y, π, X)
$U \subset X$	open subset of X
$\Omega \subset X$	compact subset of X
$W = W_0, W_1, W_2, W_r$	open subsets of $Y = J^0 Y, J^1 Y, J^2 Y, J^r Y$
$\pi_{s,r} : J^s Y \to J^r Y$	projection of s-th prolongation of (Y, π, X)
$0 \leq r \leq s$	on its r-th prolongation
(V, ψ), $\psi = (t, q^\sigma)$	fibred chart on Y in mechanics, V open set
$\psi = (x^i, y^\sigma)$, $1 \leq i \leq n$	$1 \leq \sigma \leq m$, fibred chart on Y in field theory
(U, φ), $\varphi = (t)$, $U = \pi(V)$	associated chart on X
(V_1, ψ_1), $V_1 = \pi_{1,0}^{-1}(V)$,	associated fibred chart on $J^1 Y$
$\psi_1 = (t, q^\sigma, \dot{q}^\sigma)$	alternatively $q^\sigma = q^\sigma_0$, $\dot{q}^\sigma = q^\sigma_1$
$\psi_1 = (x^i, y^\sigma, y^\sigma_j)$	$1 \leq i, j \leq n$, field theory

(V_2, ψ_2), $V_2 = \pi_{2,0}^{-1}(V)$	associated fibred chart on J^2Y
$\psi_2 = (t, q^\sigma, \dot{q}^\sigma, \ddot{q}^\sigma)$	alternatively $q^\sigma = q_0^\sigma, \dot{q}^\sigma = q_1^\sigma, \ddot{q}^\sigma = q_2^\sigma$
$\psi_2 = (x^i, y^\sigma, y_j^\sigma, y_{j\ell}^\sigma)$	$1 \le i, j, \ell \le n$, field theory
(V_r, ψ_r), $V_r = \pi_{r,0}^{-1}(V)$	associated fibred chart on J^rY
$\psi_r = (t, q^\sigma, \dots q_r^\sigma)$	$1 \le \sigma \le m$, mechanics
$\psi_r = (t, q^\sigma, q_{j_1}^\sigma, \dots q_{j_1 \dots j_r}^\sigma)$	$1 \le j_1, \dots, j_r \le n$, field theory
$TY, TJ^1Y, \dots, TJ^rY$	tangent bundles on $Y, J^1Y, \dots, J^rY$
$\Lambda^k(TJ^rY)$	spaces of differential k-forms on J^rY, $r \ge 0$
$\Lambda_X^k(TJ^rY)$, $\Lambda_Y^k(TJ^rY)$	π_r-horizontal, $\pi_{r,0}$-horizontal forms
Ω_k^r, Θ_k^r	alternative notation of spaces and subspaces of differential forms
$\Omega_{k,c}^r$, $\Theta_{k,c}^r$	spaces of strongly contact forms
$\mathcal{I}, \mathcal{I}_C$	ideal, contact ideal

Most frequent mappings

$T_x f$, $f : x \to f(x)$	tangent mapping to mapping f at point x
$f \circ g$, f^*	compositions of mappings, pullback
$\gamma, J^1\gamma, \dots, J^r\gamma$	section of π and its prolongations
$\Gamma_U(\pi)$, $\Gamma_\Omega(\pi)$, $\Gamma(\pi)$	set of sections on U, Ω, of all local sections
$q^\sigma\gamma$ or $q^\sigma \circ \gamma$, ..., $q_s^\sigma J^r\gamma$ or $q_s^\sigma J^r\gamma$, etc.	equivalent notations
(α, α_0), (α_u, α_{0u})	local diffeomorphisms on Y, $u \in (-\varepsilon, \varepsilon)$
$(J^1\alpha, \alpha_0)$, $(J^1\alpha_u, \alpha_{0u})$	first prolongation of local diffeomorphisms
$(J^r\alpha, \alpha_0)$, $(J^r\alpha_u, \alpha_{0u})$	r-th prolongation of local diffeomorphisms
$\chi : (u, t, q^\sigma, \dot{q}^\sigma, \dots) = (t, uq^\sigma, u\dot{q}^\sigma, \dots)$	special mapping
$\mathcal{A}, \mathcal{I}, \mathcal{R}, \dots$	special operators
$\mathcal{L}$	Legendre transformation mapping

Vector fields and differential forms

$\xi, \tilde{\xi}, \hat{\xi}, \Xi, \tilde{\Xi}, \hat{\Xi}$	vector fields on fibred manifolds
$\zeta, \tilde{\zeta}, \hat{\zeta}$	vector fields on fibred manifolds
$\left(\frac{\partial}{\partial t}, \frac{\partial}{\partial q^\sigma}, \frac{\partial}{\partial \dot{q}^\sigma}, \dots\right)$	bases of vector fields
$\xi^0, \xi^\sigma, \Xi^\sigma, \Xi_1^\sigma, \dots$	components of vector fields and their prolongations in charts
$\zeta^0, \zeta^\sigma, Z^\sigma, Z_1^\sigma, \dots$	components of vector fields and their prolongations in charts
$(\alpha_u, \alpha_{0u}), (J^1\alpha_u, \alpha_{0u})$	one-parameter group of a vector field and his first prolongation
$J^1\xi, J^2\xi, \dots, J^r\xi$	jet prolongations of a vector field
$\omega, \eta, \varrho, \dots$	differential forms (general notation)
$\wedge \dots \omega \wedge \eta$	exterior product
$d_t = \frac{d}{dt}$	total derivative operator
$(dt, dq^\sigma, d\dot{q}^\sigma, d\ddot{q}^\sigma, \dots)$	bases of 1-forms
$(dt, dq^\sigma, dq_1^\sigma, dq_2^\sigma, \dots)$	alternative notation
$(dt, \omega^\sigma, d\dot{q}^\sigma)$	bases of forms on J^1Y adapted to contact structure
$(dt, \omega^\sigma, \dot{\omega}^\sigma, d\ddot{q}^\sigma)$	bases of forms on J^2Y adapted to contact structure
$(dx^i, dy^\sigma, \dots, dy^\sigma_{j_1 \dots j_r})$	base of 1-forms for $\dim X = n > 1$
d	exterior derivative of a form,
∂_ξ	Lie derivative with respect to ξ
h, p	horizontalization, contactization
p_q	q-contactization
$h\,d, p\,d$	horizontal derivative, contact derivative
$\frac{d}{dt}, \frac{d'}{dt}$	total derivative (of a function)
i_ξ	contraction (of form) by vector field
ω_0	volume element dt, $dx^1 \wedge \dots \wedge dx^n$

Calculus of variations—structures

$\lambda, \tilde{\lambda}, L, \tilde{L}$	Lagrangians, Lagrange functions
(π, λ)	Lagrange structure
θ_λ	Lepage equivalent of λ
E, E_λ	dynamical form, Euler-Lagrange form
$[\alpha], \alpha_E$	dynamical (mechanical) system, representative associated with E (Lepage equivalent of E)
$\Delta_\alpha, \mathcal{D}_\alpha, \Delta_\alpha^0, \mathcal{D}_\alpha^0$	dynamical distribution, characteristic distribution, their annihilators
$\Delta_\lambda, \Delta_\lambda^0$	Euler-Lagrange distributions and its annihilator
$p_\sigma = p_\sigma^0, p_\sigma^j$	generalized momenta
H, H_λ	Hamilton function, Hamilton form
$g = (g_{\sigma\nu}), \Gamma_{\sigma\nu\mu}$	metric tensor, Christoffel symbols

References

1. J. Bernoulli, Acta Eruditorum. **MaJi** (1697), 263–270
2. E. Cartan, *Lecons sur les invariants intègraux* (Hermann, Paris, 1922)
3. T.H.J. Lepage, Sur les champ geodesiques du calcul de variations I. Bull. Acad. Roy. Belg. Cl. Sci. **22**, 716–729 (1936)
4. T.H.J. Lepage, Sur les champ geodesiques du calcul de variations II. Bull. Acad. Roy. Belg. Cl. Sci. **22**, 1036–1046 (1936)
5. P. Dedecker, *Systèmes différentiels extérieurs, invariants intégraux et suites spectrales, in Convegno Internazionale di Geometria Differenziale (Italia, 1953)* (Roma, Edizioni Cremonese, 1954), pp.247–262
6. P. Dedecker, Quelques applications de la suite spectrale aux intégrales multiples du calcul des variations et aux invariants intégraux. I. Suites spectrales et généralisations, Bull. Soc. Roy. Sci. Liège **24**, 276–295 (1955)
7. P. Dedecker, Quelques applications de la suite spectrale aux intégrales multiples du calcul des variations et aux invariants intégraux. II. Espaces de jets locaux et faisceaux, Bull. Soc. Roy. Sci. Liège **25**, 387–399 (1956)
8. P. Dedecker, Calcul des variations et topologie algébrique. Mém. Soc. Roy. Sci. Liège **19**, 216 (1957)
9. P. Dedecker, On the generalization of symplectic geometry to multiple integrals in the calculus of variations, in *Differential Geometrical Methods in Mathematical Physics.(Proc. Sympos., Univ. Bonn, Bonn, 1975)*. Lecture Notes in Mathematics, vol. 570 (Springer, Berlin, 1977), pp. 395–456
10. P. Dedecker, W.M. Tulczyjew, Spectral sequences and the inverse problem of the calculus of variations, in *Differential Geometrical Methods in Mathematical Physics (Proc. Conf., Aix-en-Provence/Salamanca)*. Lecture Notes in Mathematics, vol. 836 (Springer, Berlin 1980), pp. 498–503 (1979)
11. W.M. Tulczyjew, The Lagrange complex. Bull. Soc. Math. France **105**, 419–431 (1977)
12. W.M. Tulczyjew, The Euler-Lagrange resolution, in *Differential Geometrical Methods in Mathematical Physics (Proc. Conf., Aix-en-Provence/Salamanca, 1979)*, Lecture Notes in Mathematics, vol. 836 (Springer, Berlin, 1980), pp. 22–48
13. W.M. Tulczyjew, Cohomology of the Lagrange complex. Ann. Scuola Norm. Sup. Pisa Cl. Sci. **14**, 217–227 (1987)
14. F. Takens, A global version of the inverse problem of the calculus of variations. J. Differential Geom. **14**, 543–562 (1979)
15. A.M. Vinogradov, On the algebro-geometric foundations of Lagrangian field theory. Soviet Math. Dokl. **18**, 1200–1204 (1977)
16. A.M. Vinogradov, The C-spectral sequence, Lagrangian formalism, and conservation laws. I. The linear theory. J. Math. Anal. Appl. **100**, 1–40 (1984)
17. A.M. Vinogradov, The C-spectral sequence, Lagrangian formalism, and conservation laws. II. The nonlinear theory. J. Math. Anal. Appl. **100**, 41–129 (1984)
18. I.M. Anderson, The variational bicomplex, Technical report of the Utah State University (1989)
19. I.M. Anderson, T. Duchamp, On the existence of global variational principles. Am. J. Math. **102**, 781–868 (1980)
20. D.J. Saunders, *The Geometry of Jet Bundles*. London Mathematical Society Lecture Note Series, vol. 142 (Cambridge University Press, Cambridge, 1989)
21. D. Krupka, Some geometric aspects of variational problems in fibred manifolds, Folia Fac. Sci. Nat. UJEP Brunensis Physica **14**, 1–65 (1973). ((math-ph/0110005))
22. D. Krupka, A map associated to the Lepagian forms on the calculus of variations in fibred manifolds. Czechoslovak Math. J. **27**, 114–118 (1977)
23. D. Krupka, Lepagean forms in higher order variational theory, Atti Accad. Sci. Torino Cl. Sci. Fis. Mat. Natur. **117**(suppl. 1), 197–238 (1983)

24. D. Krupka, Variational sequences on finite order jet spaces, in *Differential Geometry and its Applications (Brno, 1989)*. ed. by J. Janyška, D. Krupka (World Sci. Publ, Teaneck, NJ, 1990), pp.236–254
25. M. Krbek, J. Musilová, Representation of the variational sequence by differential forms. Rep. Math. Phys. **51**, 251–258 (2003)
26. M. Krbek, J. Musilová, Representation of the variational sequence by differential forms. Acta Appl. Math. **88**, 177–199 (2005)
27. D. Krupka, *Global variational theory in fibred spaces, in Handbook of Global Analysis* (Elsevier Sci. B. V, Amsterdam, 2008), pp.773–836
28. D. Krupka, *Introduction to Global Variational Geometry, Atlantis Studies in Variational Geometry*, vol. 1 (Atlantis Press, Paris, 2015)
29. O. Krupková, *The Geometry of Ordinary Variational Equations*. Lecture Notes in Mathematics, vol. 1678 (Springer, Berlin, 1997)
30. O. Krupková, Lepagean 2-forms in higher order Hamiltonian mechanics. I. Regularity, Arch. Math. (Brno) **22** (1986), 97–120
31. O. Krupková, Hamiltonian field theory. J. Geom. Phys. **43**, 93–132 (2002)
32. O. Krupková, *Lepage forms in the calculus of variations, in Variations, Geometry and Physics* (Nova Sci. Publ, New York, 2009), pp.27–55
33. O. Krupková, Variational equations on manifolds, in *Advances in Mathematics Research*, Adv. Math. Res. vol. 9. (Nova Sci. Publ., New York, 2009), pp. 201–274
34. O. Krupková, G.E. Prince, Lepagean forms, closed 2-forms, and second-order ordinary differential equations. Russian Math. **51**(12), 1–16 (2007)
35. O. Krupková, G.E. Prince G.E., Second order ordinary differential equations in jet bundles and the inverse problem of the calculus of variations, in *Handbook of Global Analysis* (Elsevier Sci. B. V., Amsterdam, 2008), pp. 837–904
36. R. Vitolo, *Variational sequences, in Handbook of Global Analysis* (Elsevier Sci. B. V, Amsterdam, 2008), pp.1115–1163
37. M. Palese, R. Vitolo, On a class of polynomial Lagrangians. Rend. Circ. Mat. Palermo suppl. **66**, 147–159 (2001). (math-ph/0111019)
38. M. Palese, E. Winterroth, On the relation between the Jacobi morphism and the Hessian in gauge-natural field theories. Theoret. Math. Phys. **152**, 1191–1200 (2007)
39. M. Palese, E. Winterroth, Lagrangian reductive structures on gauge-natural bundles. Rep. Math. Phys. **62**, 229–239 (2008). ((arXiv:0712.0925))
40. M. Palese, E. Winterroth, A variational perspective on classical Higgs fields in gauge-natural theories. Theoret. Math. Phys. **168**, 1002–1008 (2011). ((arXiv:1110.5426))
41. M. Palese, E. Winterroth, Symmetries of Helmholtz forms and globally variational dynamical forms, J. Phys. Conf. Ser. **343**, 012129, 4 pages (2012). arXiv:1110.5764
42. M. Palese, E. Winterroth, Higgs fields on spinor gauge-natural bundles. J. Phys. Conf. Ser. **411**, 012025, 5 pages (2013)
43. M. Palese, E. Winterroth, Generalized symmetries generating Noether currents and canonical conserved quantities. J. Phys. Conf. Ser. **563**, 012023, 4 pages (2014)
44. M. Palese, O. Rossi, E. Winterroth, J. Musilová, Variational sequences, representation sequences, and applications in physics. SIGMA **12**, 045, 45 pages (2016)
45. M. Spivak, *Calculus on Manifolds. A Modern Approach to Classical Theorems of Advancer Calculus,* 27 edn. (Perseus Books Publishing, L. L. C., Massachusetts, 1998)
46. I.M. Gelfand, S.V. Fomin, *Calculus of variations* (Dover Publications Inc, Mineola, New York, 2000)

Fibred Manifolds

2

Mechanics is one of the key disciplines of classical physics. It studies the time evolution of systems of classical mass particles or bodies. Trajectories and other physical characteristics of particles and their systems are thus functions of time. The *configuration* of a particle at time t is standardly given by its position $\mathbf{r}(t)$ understood as a point in three-dimensional Euclidean space $\mathbf{R}^3$, the configuration of a system of N particles without constraints is given by the set of positions of all particles $\{\mathbf{r}_1(t), \ldots, \mathbf{r}_N(t)\}$, interpreted as the point in $3N$-dimensional Euclidean space $\mathbf{R}^{3N}$. The configuration of a system of N particles with k constraints restricting their positions, i.e. a system with $m = N - k$ *degrees of freedom* at time t is given by m *generalized coordinates* $\{q^1(t), \ldots, q^m(t)\}$, i.e. as a point in m-dimensional space $\mathbf{R}^m$. Analogously, the *mechanical state* of a system of particles with m degrees of freedom at time t is given by m generalized coordinates and m corresponding *generalized velocities* $\{\dot{q}^1(t), \ldots, \dot{q}^m(t)\}$. So, with every time point t we can connect the m-dimensional *configuration space* $\mathbf{R}^m$ with points $\{q^1(t), \ldots, q^m(t)\}$, and the $2m$-dimensional *phase space* $\mathbf{R}^m \times \mathbf{R}^m$ with points $\{q^1(t), \ldots, q^m(t), \dot{q}^1(t), \ldots, \dot{q}^m(t)\}$. Such an interpretation leads to the quite obvious idea of *fibred spaces*, denoted as $Y = \mathbf{R} \times \mathbf{R}^m$, or $J^1Y = \mathbf{R} \times \mathbf{R}^m \times \mathbf{R}^m$, points of which $\{t, q^1, \ldots, q^m\}$, or $\{t, q^1, \ldots, q^m, \dot{q}^1, \ldots, \dot{q}^m\}$, respectively, are projected onto the corresponding time point t in a natural way. Spaces $\mathbf{R} \times \mathbf{R}^m$ and $\mathbf{R} \times \mathbf{R}^m \times \mathbf{R}^m$ with the Cartesian projection onto the first factor $\mathbf{R}$ have the natural structure of a *fibred manifold*. Such a concept enables us to study trajectories and phase trajectories of particle systems (parameterized by time) as specific mappings of the type $\mathbf{R} \ni t \to \gamma(t) \in \mathbf{R} \times \mathbf{R}^m$ called *sections of a fibred manifold* and $\mathbf{R} \ni t \to J^1\gamma(t) \in \mathbf{R} \times \mathbf{R}^m \times \mathbf{R}^m$, called *jet prolongations* of sections. The concept of prolongations of sections enables us to construct higher order prolongations of fibred manifolds. Sets of coordinates of such spaces contain higher order derivatives of mappings $\gamma(t)$, especially second-order derivatives (accelerations).

J. Musilová et al., *Calculus of Variations on Fibred Manifolds and Variational Physics*, Lecture Notes in Physics 1033, https://doi.org/10.1007/978-3-031-77408-9_2

Both the configuration and phase space are geometrical structures commonly used as domains of functions representing physical quantities depending on time and on coordinates and velocities.

The generalization of the above mentioned idea to spaces with nontrivial topological structure leads to the concept of a *fibred manifold with an one-dimensional base* as the fundamental geometric structure for mechanics developed on the basis of a correct mathematical background. (Note that in some other physical theories the use of fibred manifolds with multidimensional bases is appropriate. Such a typical example is electrodynamic: Quantities characterizing the electromagnetic field, intensities **E**, **B** or potentials **A**, φ are functions of time and coordinates, and the base of the corresponding fibred manifold is four-dimensional.)

As mentioned, the equations of motion of particles and their systems in classical Newtonian and Lagrangian mechanics are of the second order. So, Lagrangians in mechanics are mostly of the first order. Therefore, in this chapter the mathematical background of fibred manifolds is developed in detail for the first- and second-order situations. Of course, generalization of basic concepts and their properties to an arbitrary order is very straightforward. A more detailed explanation of the concept of fibred manifolds and their prolongations can be found in the textbooks [1,2]. A highly advanced approach is contained, e.g. in the works of Krupka [3] or Saunders [4,5].

Prerequisites Smooth (differentiable) manifold, coordinate system (a map) on a manifold, coordinate transitions, differentiable atlas, maps compatible with an atlas, differentiable structure, tangent and cotangent bundle of a smooth manifold, smooth (differentiable) mappings of manifolds, diffeomorphisms, tangent mappings, local and global concepts □

2.1 Fibred Manifolds and Charts

In this section the concept of a fibred manifold with one-dimensional base is introduced, as well as fibred charts for coordinate calculations.

Definition 2.1 (*Fibred manifolds*) Let X be an one-dimensional manifold, Y a manifold of the dimension $m+1$ and $\pi : Y \to X$ a surjective submersion. The triad (Y, π, X) is called a *fibred manifold with one-dimensional base*, Y and X are the *total space* and *base* of the fibred manifold, respectively, and π is the *projection*. The set $\pi^{-1}(x)$ for an arbitrary point $x \in X$ has the structure of m-dimensional submanifold of Y. It is called the *fibre over the point* x.

Fibres are disjoint each to other. Regardless of the fact that all fibres have the same dimension m, they may have different topological structures—they need not be diffeomorphic.

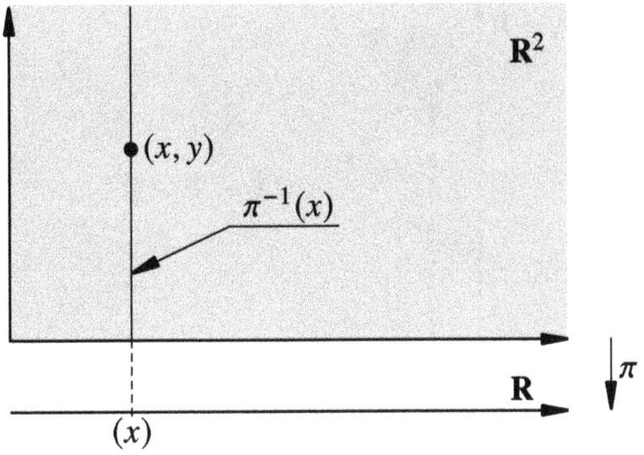

Fig. 2.1 Projection of a Cartesian space as a fibred manifold

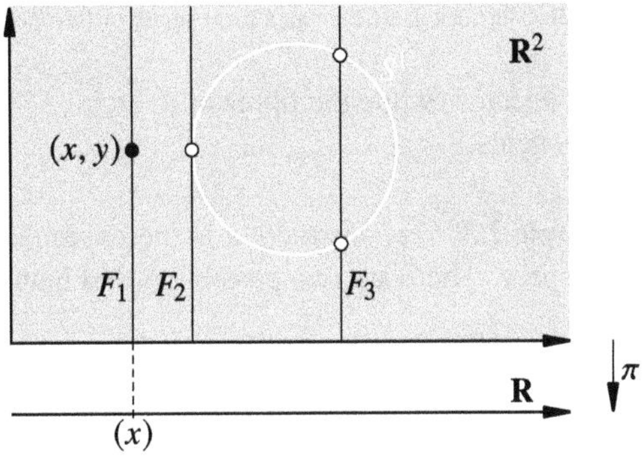

Fig. 2.2 A counterexample of the fibration

Definition 2.2 (*Fibrations*) A fibred manifold is called the *fibration*, if it has a connected base and their fibres are diffeomorphic to each other. A m-dimensional manifold F diffeomorphic with a (arbitrary) fibre of a given fibration is called the *type fibre*.

The concept of a fibred manifold is clarified by some simple examples.

Example 2.1 Cartesian spaces. Let $X = \mathbf{R}$, $Y = \mathbf{R}^2$, $\pi : \mathbf{R}^2 \ni (x, y) \to x \in \mathbf{R}$ (see Fig. 2.1).
This fibred manifold is an example of fibration with the type fibre $\mathbf{R}$.

Example 2.2 Let $X = \mathbf{R}$, $Y = \mathbf{R}^2 \backslash S^1$, where $S^1 = \{(x, y \in \mathbf{R}^2 \mid x^2 + y^2 = 1)\}$, is the unit circle (see Fig. 2.2). Define the projection as

$$\pi : \mathbf{R}^2 \ni (x, y) \to x \in \mathbf{R} .$$

The base of this fibred manifold is connected, but its fibres are not diffeomorphic. There are three types of fibres: $F_1 = \mathbf{R}$ for $x < 1$, $F_2 = \mathbf{R} \backslash \{(\pm 1)\}$ for $x = \pm 1$ and $F_3 = \mathbf{R} \backslash \{(y \in \mathbf{R} \mid -\sqrt{1 - x^2} < y < -\sqrt{1 - x^2}\}$ for $\{x \in \mathbf{R} \mid -1 < x < 1\}$. These types of fibres are not diffeomorphic to each other. The fibres of the type F_1

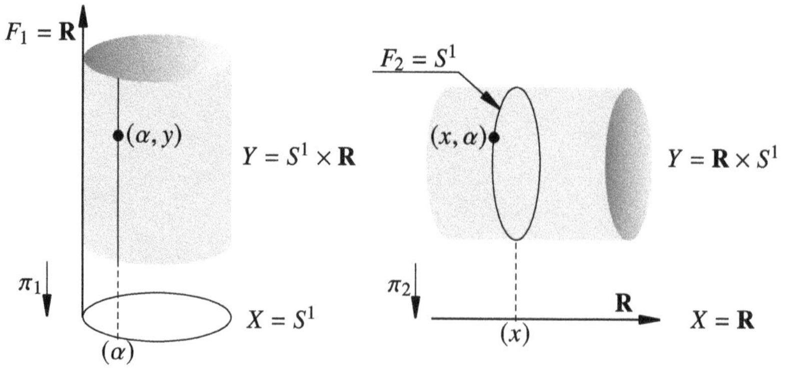

Fig. 2.3 Cylindrical surface as a total space of two possible fibrations

are connected, while the fibres of the type F_2 and F_3 have two and three connected components.

Example 2.3 Very instructive is the example of the manifold $Y = S^1 \times \mathbf{R}$ as the total space. There are two possible fibred manifolds with different bases: S^1 or $\mathbf{R}$,

$$\pi_1 : S^1 \times \mathbf{R} \longrightarrow S^1, \quad \text{or} \quad \pi_2 : \mathbf{R} \times S^1 \longrightarrow \mathbf{R},$$

(see Fig. 2.3). The fibred manifold $(S^1 \times \mathbf{R}, \pi_1, S^1)$ is the cylindrical surface fibred over the circle. It is the fibration with the type fibre $\pi_1^{-1}(\alpha) = \mathbf{R}$ for every point $\alpha \in S^1$. (The coordinate α can have very illustrative meaning as e. g. the angle position of a point on the circle S^1.) The fibred manifold $(\mathbf{R} \times S^1, \pi_2, \mathbf{R})$ is the cylindrical surface fibred over the line $\mathbf{R}$. It is also the fibration. The corresponding type fibre is $\pi_2^{-1}(x) = S^1$ for every point $x \in \mathbf{R}$.

The base X and the total space Y of a $(m+1)$-dimensional fibred manifold (Y, π, X) are interrelated by the projection π. The mapping π is a surjective submersion. Thus, its rank is constant and equals to the dimension of the base, i.e. rank $\pi = 1$. This means that we can choose maps (coordinate systems) on Y and X in such a way that they are adapted to the fibred structure. Such a choice is schematically shown in Fig. 2.4 with the help of a commutative diagram. Let (V, ψ) be a map on Y. Then we can choose the map (U, φ) on X such that $U = \pi(V)$ and

$$p \circ \psi = \varphi \circ \pi ,$$

where p is the Cartesian projection of $\mathbf{R} \times \mathbf{R}^m$ on the first factor. Let us express this relation in coordinates. Denote the coordinate functions of ψ as $\psi = (u, q^\sigma)$, i.e. for $y \in V$ it holds $\psi(y) = \big(u\psi(y), q^\sigma \psi(y)\big)$. Denote $U = \pi(V)$. Then we can choose the map (U, φ), $\varphi = (t)$, on X, $\varphi(x) = \big(t\varphi(x)\big)$, in such a way that

$$u\psi(y) = t\varphi\big(\pi(y)\big) .$$

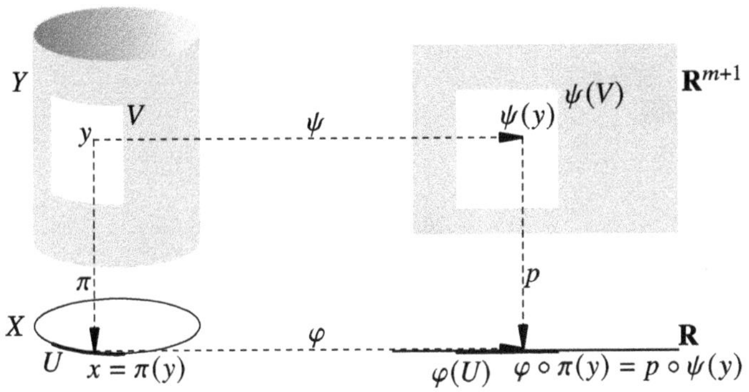

Fig. 2.4 Adapted charts

(V, ψ) and $U, \varphi)$ are called the *fibred chart* and *associated chart*, respectively. If there is not a risk of misunderstanding concerning the chosen charts we can use the shortened notation

$$y \to \psi(y) = \big(t(y), q^\sigma(y)\big), \quad \pi(y) \to \varphi \circ \pi(y) = \big(t(y)\big), \quad 1 \le \sigma \le m\ .$$

The corresponding chart expression of the surjective submersion π is given by the chart expression of the projection $p = \varphi \circ \pi \circ \psi^{-1}$ in Cartesian coordinates, i.e.

$$\varphi \circ \pi \circ \psi^{-1} : \mathbf{R}^{m+1} \ni \psi(V) \longrightarrow \varphi(U) \in \mathbf{R}, \quad \varphi \circ \pi \circ \psi^{-1}(t, q^\sigma) = (t)\ .$$

Let (V, ψ) and $(\bar{V}, \bar{\psi})$ be two fibred charts on Y and (U, φ) and $(\bar{U}, \bar{\varphi})$ the corresponding associated charts, respectively. Recall that coordinate mappings are one-to-one mappings. The transformation mappings on intersections $\bar{V} \cap V$ and $\bar{U} \cap U$ are $\bar{\psi} \circ \psi^{-1}$ and $\bar{\varphi} \circ \varphi^{-1}$, respectively, or vice versa $\psi \circ \bar{\psi}^{-1}$ and $\varphi \circ \bar{\varphi}^{-1}$. There *transformation relations* are of the type

$$\bar{t} = \bar{t}(\bar{\psi} \circ \psi^{-1})(t, q^\nu) = \bar{t}(\bar{\varphi} \circ \varphi^{-1})(t, q^\nu), \bar{q}^\sigma = \bar{q}^\sigma(\bar{\psi} \circ \psi^{-1})(t, q^1, \dots, q^m)\ ,$$

or shortly

$$\bar{t} = \bar{t}(t), \quad \bar{q}^\sigma = \bar{q}^\sigma(t, q^1, \dots, q^m), \quad 1 \le \sigma \le m \tag{2.1}$$

and vice versa. The situation is illustrated by Fig. 2.5.

The Jacobi matrices of these transformation mappings are

$$D(\bar{\psi} \circ \psi^{-1})(t, q^\nu) = \begin{pmatrix} \frac{d\bar{t}(t)}{dt} & \frac{\partial \bar{q}^1(t,q^\nu)}{\partial t} & \dots & \frac{\partial \bar{q}^m(t,q^\nu)}{\partial t} \\ 0 & \frac{\partial \bar{q}^1(t,q^\nu)}{\partial q^1} & \dots & \frac{\partial \bar{q}^m(t,q^\nu)}{\partial q^1} \\ \dots & \dots & \dots & \dots \\ 0 & \frac{\partial \bar{q}^1(t,q^\nu)}{\partial q^m} & \dots & \frac{\partial \bar{q}^m(t,q^\nu)}{\partial q^m} \end{pmatrix}, \quad D(\bar{\varphi} \circ \varphi^{-1})(t) = \left(\frac{d\bar{t}}{dt}\right)\ .$$

The existence of adapted charts on a fibred manifold confirms that a $(m + 1)$-dimensional fibred manifold (Y, π, X) is *locally diffeomorphic* with the Cartesian product

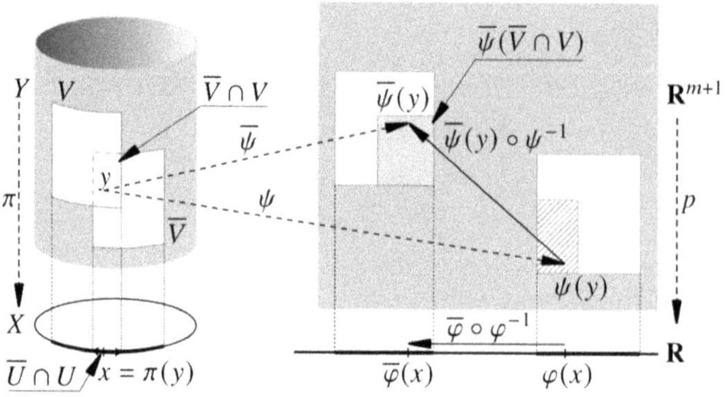

Fig. 2.5 Chart transformations

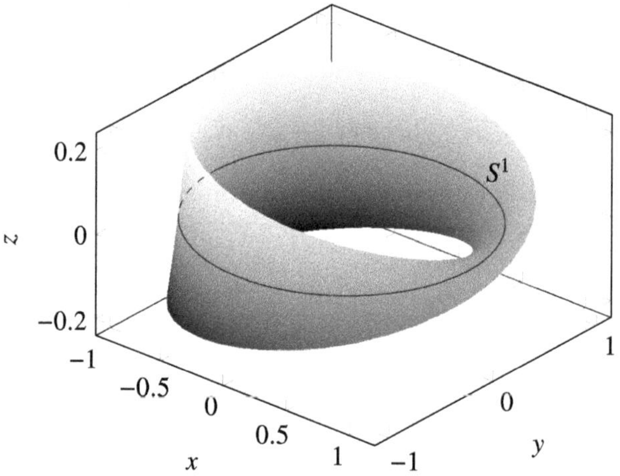

Fig. 2.6 Möbius strip

$\mathbf{R} \times \mathbf{R}^m$ of Euclidean spaces. The following example shows that the above mentioned property of fibred manifolds need not be global.

Example 2.4 The well-known *Möbius strip* (here considered as the manifold without boundary) is a surface in $\mathbf{R}^3$ parametrized as

$$x = \cos t + p \cos \frac{t}{2} \cos t, \quad y = \sin t + p \cos \frac{t}{2} \sin t, \quad z = p \sin \frac{t}{2},$$

where $0 \leq t \leq 2\pi$, $p_{\min} < p < p_{\max}$ and it is shown in Fig. 2.6 for $p_{\min} = -0.2$ and $p_{\max} = 0.2$. A Möbius strip is a two-dimensional manifold $\pi : \mathcal{M} \to S^1$ fibred over the unit circle S^1 (the intersection of the Möbius strip with the plane $z = 0$). Every point of the strip is projected onto this circle. This is also fibration with type fibre $\mathbf{R}$. (The open interval $I = (p_{\min}, p_{\max})$ is diffeomorphic with $\mathbf{R}$.) However, this fibred manifold cannot be expressed as the Cartesian product of $S^1 \times I$. On the other hand, for every point there exists a neighbourhood homeomorphic with a Cartesian product of open intervals, or equivalently with $\mathbf{R} \times \mathbf{R}$.

The general definition of a fibred manifold admits also a non-connected total spaces, fibres or bases. However, in the following text, we consider only fibred manifolds with connected bases and total spaces. (Situations with non-connected fibres are still permissible.) Emphasize that there exist only two types of (globally) non-diffeomorphic one-dimensional connected manifolds, the line $\mathbf{R}$ and the circle S^1. So, in the following we are dealing with two types of fibred manifolds only,

$$\pi : Y \longrightarrow \mathbf{R} \quad \text{or} \quad \pi : Y \longrightarrow S^1 .$$

Note that the second type of fibred manifolds can be useful for the description of periodical motions.

2.2 Sections

A typical task in the classical mechanics is to solve equations of motion and to find trajectories of a particle or all particles in a mechanical system. A trajectory is in fact a "continuous time sequence" of configurations of the system. We show that parametrized trajectories of a system can be described as so-called *sections* of a fibred manifold representing all thinkable configurations at all times.

Definition 2.3 (*Sections*) Let (Y, π, X) be a fibred manifold and $U \subset X$ an open set (most often an open interval I). A smooth mapping

$$\gamma : U \ni x \longrightarrow \gamma(x) \in Y, \quad \text{such that} \quad \pi \circ \gamma = \mathrm{id}_U , \tag{2.2}$$

is called the *(local) section* or simply *section* of the projection π, or alternatively the (local) section of the fibred manifold (Y, π, X). For $U = X$ the section γ is *global.*

Let us express a section γ in coordinates. Let (V, ψ), $\psi = (t, q^\sigma)$, be a fibred chart on Y and (U, φ), $\varphi = (t)$, the corresponding associated chart on X such that U is a subset of the domain of γ. It holds for a point $x \in U$

$$\psi \circ \gamma(x) = \psi \circ \gamma \circ \varphi^{-1}(\varphi(x)) = \left(t(\psi \circ \gamma \circ \varphi^{-1})(\varphi(x)), q^\sigma(\psi \circ \gamma \circ \varphi^{-1})(\varphi(x))\right) .$$

Denote as $\tilde{\gamma} = \psi \circ \gamma \circ \varphi^{-1}$ the chart mapping of γ. We can write

$$\tilde{\gamma}(t) = \psi \circ \gamma \circ \varphi^{-1}(t) = \left(t\tilde{\gamma}(t), q^\sigma \tilde{\gamma}(t)\right), \quad t \in U , \tag{2.3}$$

or, in shortened notation

$$\gamma = (t\gamma, q^\sigma \gamma), \quad 1 \leq \sigma \leq m . \tag{2.4}$$

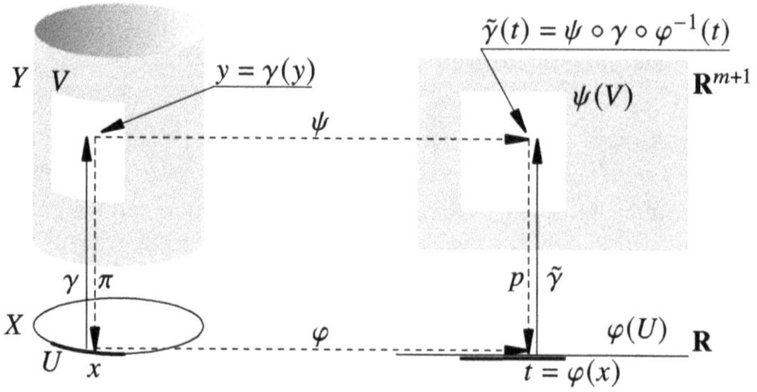

Fig. 2.7 Chart expressions of sections

The mappings appearing in relation (2.3) are schematically shown in Fig. 2.7. Local sections can be interpreted as graphs of parametrized trajectories of mechanical systems. The trajectories are standardly understood as curves in the configuration space, e.g.

$$C : (\alpha, \beta) \ni t \to C(t) = \left(q^1C(t), \dots, q^mC(t)\right) .$$

As a simple example consider two special types of motion of the spherical pendulum, the *simple pendulum* and the *conical pendulum*.

Example 2.5 The simple pendulum is a particle moving in $\mathbf{R}^3$ subjected to two constraints

$$x^2 + y^2 + z^2 = l^2, \quad z = 0 .$$

The solution of its equation of motion for small angular displacements φ represents its trajectory lying in the coordinate plane xy and parametrized by time as follows:

$$C(t) = \left(xC(t), yC(t)\right) = \left(l \cos\{\varphi_0 \cos(\omega t + \alpha)\}, l \sin\{\varphi_0 \cos(\omega t + \alpha)\}\right), \quad \omega = \sqrt{\frac{g}{l}} ,$$

where $g = 9,8\,\mathrm{m\,s^{-2}}$ is the free-fall acceleration and φ_0 and α are integration constants given by initial conditions. The constant ω is the *angular frequency* of the pendulum.

The conical pendulum is also a particle in $\mathbf{R}^3$. Its motion is limited by the constraint $x = l \cos \varphi_0$. The solution of its equations of motion is

$$C(t) = \left(yC(t), zC(t)\right) = \left((l \sin \varphi_0) \cos(\omega t + \alpha), (l \sin \varphi_0) \cos(\omega t + \alpha)\right) ,$$

$$\omega = \sqrt{\frac{g}{l \cos \varphi_0}} ,$$

where $0 < \varphi_0 < \frac{\pi}{2}$ is the given constant and α is the integration constant. The constant ω is the angular frequency (and simultaneously the angular speed). Figure 2.8

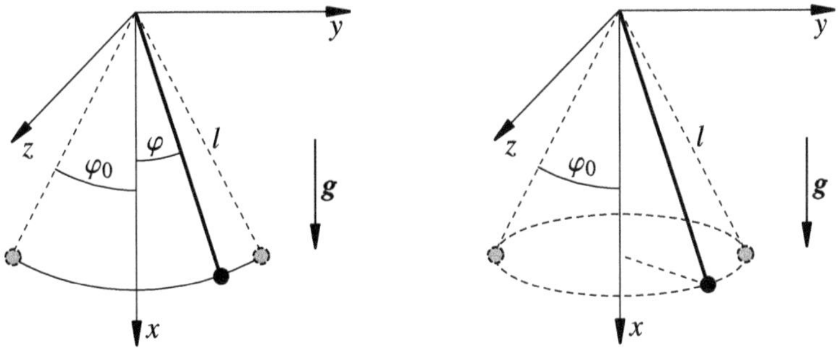

Fig. 2.8 Simple pendulum (left) and conical pendulum (right)

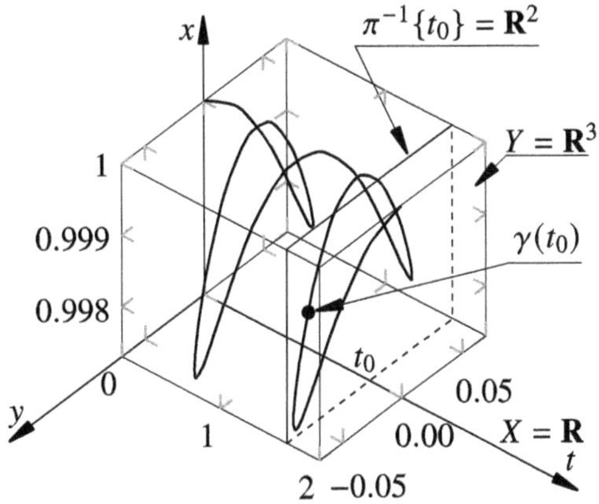

Fig. 2.9 Trajectory of the simple pendulum in xy plane as the section of a fibred manifold $\mathbf{R}^3 \to \mathbf{R}$

presents both situations and Figs. 2.9 and 2.10 show the solutions as sections of the corresponding fibred manifolds. The motion of a simple pendulum in the configuration space is constrained in such a way that it has only one degree of freedom (functions $xC(t)$ and $yC(t)$ are not independent). We can consider the angular displacement $\varphi(t)$ as the generalized coordinate. In such interpretation the configuration space of the pendulum is one-dimensional and the corresponding fibred manifold is $(\mathbf{R} \times \mathbf{R}, p, \mathbf{R})$. The trajectory of the pendulum can be then considered as the section $\gamma : \mathbf{R} \ni t \to \gamma(t) = \big(t, \varphi_0 \cos(\omega t + \alpha)\big) \in \mathbf{R} \times \mathbf{R}$. The situation is depicted in the left Fig. 2.11 for $\alpha = -\pi/2$, i.e. $\gamma(t) = (t, \varphi_0 \sin \omega t)$. The trajectory of the simple pendulum can be alternatively interpreted as the section of the fibred manifold $\pi : S^1 \times \mathbf{R} \to S^1$ (see the right Fig. 2.11). The situation is similar also for the conical pendulum. Its trajectory is described by the angular position $\alpha(t) = l \sin \varphi_0 \sin(\omega t + \alpha)$.

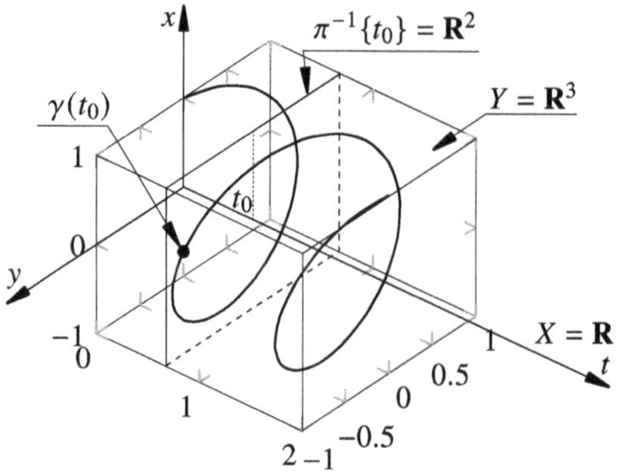

Fig. 2.10 Trajectory of the conical pendulum in xy plane as the section of a fibred manifold $\mathbf{R}^3 \to \mathbf{R}$

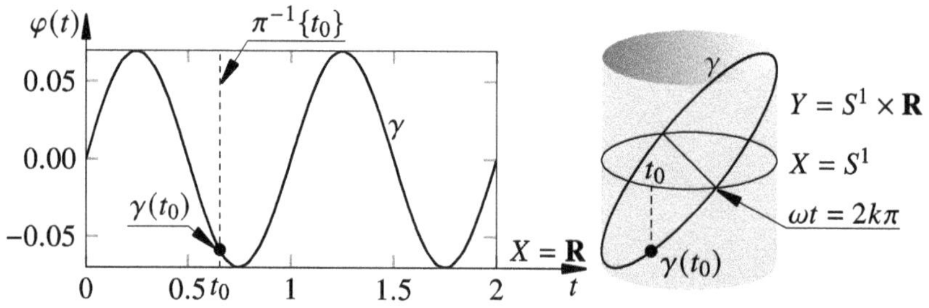

Fig. 2.11 Trajectory of the simple pendulum as the section of fibred manifolds $\mathbf{R} \times \mathbf{R} \to \mathbf{R}$ and $S^1 \times \mathbf{R} \to S^1$

2.3 Local Isomorphisms

This section concerns mappings of fibred manifolds preserving fibres.

Definition 2.4 (*Homomorphisms, isomorphisms*) Let $\pi : Y \to X$ and $\bar{\pi} : \bar{Y} \to \bar{X}$ be fibred manifolds. A mapping $\alpha : \bar{Y} \to Y$ is called a *homomorphism of fibred manifolds* π and $\bar{\pi}$, if there exists a mapping $\alpha_0 : X \to \bar{X}$ such that

$$\alpha_0 \circ \pi = \bar{\pi} \circ \alpha \,, \tag{2.5}$$

i.e. the diagram in Fig. 2.12 commutes. The mapping α_0 is called *projection* of the homomorphism α. A homomorphism of fibred manifolds is called the *isomorphism*, if both mappings α and α_0 are local diffeomorphisms.

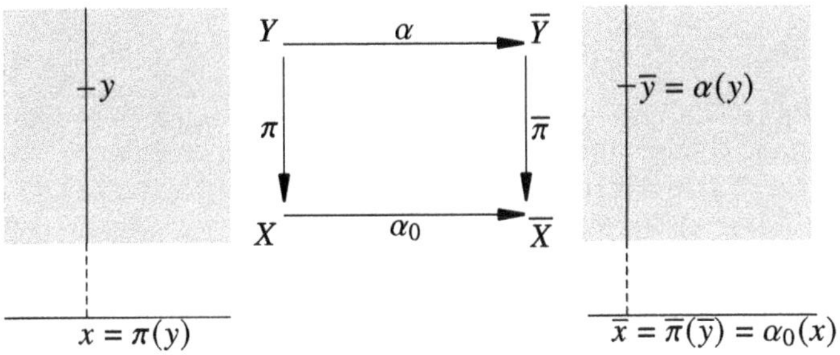

Fig. 2.12 Homomorphism of fibred manifolds—schematic diagram

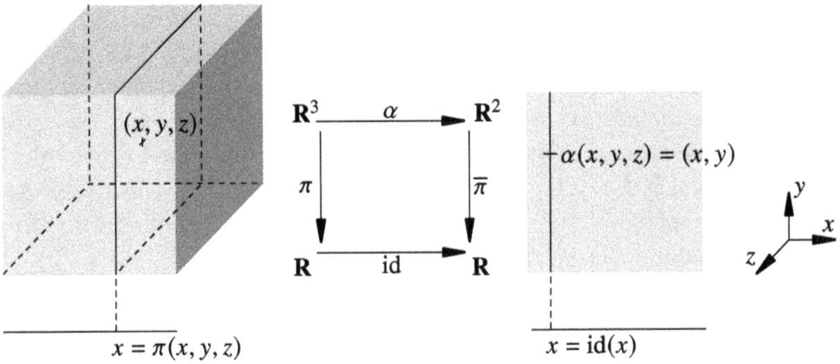

Fig. 2.13 Example 2.6

Proposition 2.1 *Let α be a homomorphism of fibred manifolds π and $\bar{\pi}$. Then its projection is unique.*

Proof Suppose that α_0 and α_0' are two projections of the homomorphism α, i.e.

$$\alpha_0 \circ \pi = \alpha_0' \circ \pi = \bar{\pi} \circ \alpha \ .$$

Suppose that there exists a point $x \in X$ such that $\alpha_0(x) \neq \alpha_0'(x)$. Then for every point $y \in Y$ such that $x = \pi(y)$ we obtain

$$(\alpha_0 \circ \pi)(y) \neq (\alpha_0' \circ \pi)(y),$$

which is the contradiction with the relation $\alpha_0 \circ \pi = \alpha_0' \circ \pi$. □

Example 2.6 Let $\pi : \mathbf{R}^3 \to \mathbf{R}$ and $\bar{\pi} : \mathbf{R}^2 \to \mathbf{R}$. On fibred manifolds $(\mathbf{R}^3, \pi, \mathbf{R})$ and $(\mathbf{R}^2, \pi, \mathbf{R})$ we have global fibred charts $(\mathbf{R}^3, \mathrm{id}_{\mathbf{R}^3})$ and $(\mathbf{R}^2, \mathrm{id}_{\mathbf{R}^2})$, respectively, and corresponding associated charts $(\mathbf{R}, \mathrm{id}_{\mathbf{R}})$. Define $\alpha : \mathbf{R}^3 \ni (x, y, z) \to \alpha(x, y, z) = (x, y) \in \mathbf{R}^2$. Supposing that α is a homomorphism of fibred manifolds π and $\bar{\pi}$, then α_0 is the identity. The situation is illustrated by Fig. 2.13.

2.4 Jets and Prolongations

In the previous section we have described the thinkable trajectories of a mechanical system in the configuration space as sections of an appropriately chosen fibred manifold (Y, π, X). In this section we define the phase trajectories as *jet prolongations* of sections of (Y, π, X) and show that these prolonged sections can generate a fibred manifold with fibres formed by phase spaces of the mechanical system. In such a picture the phase trajectories of a concrete mechanical system are special type sections of this *prolonged* manifold.

Using sections we can obtain another interpretation of points of a fibred manifold. Denote as $\Gamma(\pi)$ the set of all local sections of π. Let $U \subset X$. Denote $\Gamma_U(\pi)$ a set of local sections of π such that the domain of each of them contains the set U. Suppose that U is an open set. We say that the sections $\gamma_1 \in \Gamma_U(\pi)$ and $\gamma_2 \in \Gamma_U(\pi)$ have the *0-th-order contact* at the point $x \in U$ if it holds $\gamma_1(x) = \gamma_2(x)$. Such sections are also called *0-equivalent*. (Verify that the relation "to be 0-equivalent" on $\Gamma_U(\pi)$ is the equivalence.) Denote as $J^0_x\gamma = [\gamma_x]$ the class of all sections having the 0-th-order contact at the point x. Let $y \in Y$ be an arbitrarily chosen point. Then there exists a unique class $J^0_x\gamma$ such that $\pi(y) = x$ and $\gamma(x) = y$. We can identify the point $y \in Y$ with the corresponding class $J^0_x\gamma$. Denote

$$J^0Y = \{J^0_x\gamma \mid \gamma \in \Gamma(\pi), x \in X\} \quad \text{and} \quad \pi_0 : J^0Y \ni J^0_x\gamma \to \pi_0(J^0_x\gamma) = x \in X\,.$$

Then (J^0Y, π_0, X) is a fibred manifold of the dimension $m + 1$. It can be identified with (Y, π, X), i.e. $(J^0Y, \pi_0, X) = (Y, \pi, X)$.

The idea of equivalent sections can be expanded to higher orders of contact and it serves for prolongation procedure of fibred manifolds.

Definition 2.5 (*Equivalent sections*) Let $\Gamma(\pi)$ be the set of all local sections of π. Let U be an open set. We say that the sections $\gamma_1, \gamma_2 \in \Gamma_U(\pi)$ have the *r-th-order contact* at the point $x \in U$ if it holds $\gamma_1(x) = \gamma_2(x)$ and if there exist such a fibred chart (V, ψ) $\psi = (t, q^\sigma)$, $1 \le \sigma \le m$, on Y with the associated chart (U, φ), $U = \pi(V)$, $\varphi = (t)$, on X such that $x \in U$ and, in the shortened notation (2.4), $t = t\varphi(x)$,

$$q^\sigma\gamma_1(t) = q^\sigma\gamma_2(t), \quad \frac{dq^\sigma\gamma_1(t)}{dt} = \frac{dq^\sigma\gamma_2(t)}{dt}, \quad \ldots \quad , \frac{d^rq^\sigma\gamma_1(t)}{dt^r} = \frac{d^rq^\sigma\gamma_2(t)}{dt^r}\,. \tag{2.6}$$

The sections $\gamma_1, \gamma_2 \in \Gamma_U(\pi)$ are called *r-equivalent* at the point $x \in U$ if they have the r-th-order contact at this point. The class of all sections r-equivalent at the point x with the section $\gamma \in \Gamma_U(\pi)$ is called *r-jet* of the section γ and it is denoted as $J^r_x\gamma$.

Proposition 2.2 *Let $\gamma_1, \gamma_2 \in \Gamma_U(\pi)$ be the r-equivalent sections. Then conditions (2.6) holds in every fibred chart $(\bar{V}, \bar{\psi})$, $\pi(\bar{V})$, $\pi(V) \subset U$, $\pi(\bar{V}) \cap \pi(V) \neq \emptyset$, $\bar{\psi} = (\bar{t}, \bar{q}^\sigma)$, $(\pi(\bar{V}), \bar{\varphi})$.*

Proof We prove the proposition for $r = 1$. Suppose that sections $\gamma_1, \gamma_2 \in \Gamma_U(\pi)$ are 1-equivalent. Then there exists a fibred chart (V, ψ), $\pi(V) \subset U$, $\psi = (t, q^\sigma)$, on Y and the corresponding associated chart $(\pi(V), \varphi)$, $\varphi = (t)$, on X such that relations (2.6) hold. Let $V \cap \bar{V} \neq \emptyset$ and let $x \in \pi(V) \cap \pi(\bar{V})$. Then it holds trivially $\bar{q}^\sigma \gamma_1(t) = \bar{q}^\sigma \gamma_2(t)$, $1 \leq \sigma \leq m$. For the coordinate representations on $V \cap \bar{V}$ and $\pi(\bar{V}) \cap \pi(V)$, $\psi \circ \gamma \circ \varphi^{-1}$ and $\bar{\psi} \circ \gamma \circ \bar{\varphi}^{-1}$ of the section γ, we can write

$$D(\bar{\psi} \circ \gamma \circ \bar{\varphi}^{-1}) = D(\bar{\psi} \circ \psi^{-1}) \cdot D(\psi \circ \gamma \circ \varphi^{-1}) \cdot D(\varphi \circ \bar{\varphi}^{-1}) .$$

Then we have

$$D(\bar{\psi} \circ \gamma_1 \circ \bar{\varphi}^{-1}) = D(\bar{\psi} \circ \gamma_2 \circ \bar{\varphi}^{-1}) \iff D(\psi \circ \gamma_1 \circ \varphi^{-1}) = D(\psi \circ \gamma_2 \circ \varphi^{-1}) .$$

Taking into account that (notation $\psi \circ \gamma \circ \varphi^{-1} = \tilde{\gamma}$, $\bar{\psi} \circ \gamma \circ \bar{\varphi}^{-1} = \bar{\tilde{\gamma}}$)

$$D(\bar{\psi} \circ \gamma \circ \bar{\varphi}^{-1})(\bar{t}) = D\bar{\tilde{\gamma}}(t) = \begin{pmatrix} 1 \\ \frac{d\bar{q}^1 \bar{\tilde{\gamma}}(\bar{t})}{d\bar{t}} \\ \cdots \\ \frac{d\bar{q}^m \bar{\tilde{\gamma}}(\bar{t})}{d\bar{t}} \end{pmatrix}, \quad D(\psi \circ \gamma \circ \varphi^{-1})(t) = D\tilde{\gamma}(t) = \begin{pmatrix} 1 \\ \frac{dq^1 \tilde{\gamma}(t)}{dt} \\ \cdots \\ \frac{dq^m \tilde{\gamma}(t)}{dt}, \end{pmatrix},$$

we can see immediately that for 1-equivalent sections condition (2.6) for $r = 1$ is fulfilled in all fibred charts.

Note that in the shortened notation (2.4) and (2.1) we can write

$$\begin{aligned} \dot{\bar{q}}^\sigma J^1_x \gamma(\bar{t}) &= \frac{d\bar{q}^\sigma \gamma(\bar{t})}{d\bar{t}} = \frac{dt}{d\bar{t}} \cdot \frac{d\bar{q}^\sigma \big(t, q^\nu \gamma(t)\big)}{dt} \\ &= \frac{dt}{d\bar{t}} \left(\frac{\partial \bar{q}^\sigma (t, q^\lambda)}{\partial t} + \frac{\partial \bar{q}^\sigma (t, q^\lambda)}{\partial q^\nu} \dot{q}^\nu \right) \left(J^1_x \gamma(t) \right) , \end{aligned} \tag{2.7}$$

where we used the *total derivative operator*

$$\frac{d}{dt} = \frac{\partial}{\partial t} + \dot{q}^\sigma \frac{\partial}{\partial q^\sigma} ,$$

i.e.

$$\frac{d}{dt} : f(t, q^\nu) \longrightarrow \frac{df(t, q^\nu)}{dt} = \frac{\partial f(t, q^\nu)}{\partial t} + \dot{q}^\sigma \frac{\partial f(t, q^\nu)}{\partial q^\sigma} .$$

and analogously, for a function $f(t, q^\nu, \dot{q}^\nu)$,

$$\frac{d}{dt} = \frac{\partial}{\partial t} + \dot{q}^\sigma \frac{\partial}{\partial q^\sigma} + \ddot{q}^\sigma \frac{\partial}{\partial \dot{q}^\sigma}, \quad \text{etc.}$$

Taking into account that $\dot{q}^\sigma J^1_x \gamma_1 = \dot{q}^\sigma J^1_x \gamma_2$ we can conclude that also $\dot{\bar{q}}^\sigma J^1_x \gamma_1 = \dot{\bar{q}}^\sigma J^1_x \gamma_2$. □

We will see that using r-jets the new fibred manifold can be generated as follows:

$$J^rY = \{J^r_x\gamma \mid \gamma \in \Gamma(\pi), x \in X\}, \quad \pi_r : J^rY \ni J^r_x\gamma \to \pi_r(J^r_x\gamma) = x \in X\,.$$

For our purposes it sufficient to consider $r = 1, 2$. Define mappings

$$\begin{aligned}
\pi_{1,0} &: J^1Y \ni J^1_x\gamma \longrightarrow \pi_{1,0}(J^1_x\gamma) = J^0_x\gamma \in Y\,,\\
\pi_{2,0} &: J^2Y \ni J^2_x\gamma \longrightarrow \pi_{2,0}(J^2_x\gamma) = J^0_x\gamma \in Y\,,\\
\pi_{2,1} &: J^2Y \ni J^2_x\gamma \longrightarrow \pi_{2,1}(J^2_x\gamma) = J^1_x\gamma \in J^1Y\,.
\end{aligned}$$

Suppose that $r = 1$. The space J^1Y has the structure of a $(1 + 2m)$-dimensional smooth manifold, generated by the smooth structure of Y as follows: Let $\{(V, \psi)\}$ be an atlas on Y, formed by fibred charts. (The corresponding associated charts on X define the atlas on X.) Then for every (V, ψ) the pair $(V_1, \psi_1) = (\pi^{-1}_{1,0}(V), \psi_1)$, $\psi_1 = (t, q^\sigma, \dot{q}^\sigma)$, where

$$\begin{aligned}
\psi_1 \;:\; J^1Y \ni J^1_x\gamma \;&\longrightarrow\; \psi_1(J^1_x\gamma) = (tJ^1_x\gamma, q^\sigma J^1_x\gamma, \dot{q}^\sigma J^1_x\gamma)\\
&= \left(t\varphi(x), q^\sigma(\psi\gamma\varphi^{-1})(\varphi(x)), \left.\frac{\mathrm{d}q^\sigma(\psi\circ\gamma\circ\varphi^{-1})(t)}{\mathrm{d}t}\right|_{t=\varphi(x)}\right), \quad \gamma \in \Gamma(\pi)\,,
\end{aligned}$$

is a coordinate system (the map) on J^1Y. Indeed, the system of all subsets $\pi^{-1}_{1,0}(V)$ defines a topology on J^1Y. Since mappings $\bar{\psi}\circ\psi^{-1} : \psi(\bar{V}\cap V) \to \bar{\psi}(\bar{V}\cap V)$ and $\psi\circ\bar{\psi}^{-1} : \bar{\psi}(\bar{V}\cap V) \to \psi(\bar{V}\cap V)$ are diffeomorphisms, the mappings

$$\begin{aligned}
\bar{\psi}_1\circ\psi_1^{-1} &: \psi_1(\bar{V}_1\cap V_1) \longrightarrow \bar{\psi}_1(\bar{V}_1\cap V_1) \quad \text{and}\\
\psi_1\circ\bar{\psi}_1^{-1} &: \bar{\psi}_1(\bar{V}_1\cap V_1) \longrightarrow \psi_1(\bar{V}_1\cap V_1)
\end{aligned}$$

are diffeomorphisms as well. The situation is quite analogous for the set J^2Y of all 2-jets $J^2_x\gamma$, $x \in X$, $\gamma \in \Gamma(\pi)$. It has the structure of the smooth manifold, again induced by the structure of Y. The pairs $\{(V_2, \psi_2)\}$, $(V_2, \psi_2) = (\pi^{-1}_{2,0}(V), \psi_2)$, $\psi_2 = (t, q^\sigma, \dot{q}^\sigma, \ddot{q}^\sigma)$, where

$$\begin{aligned}
\psi_2 : J^2Y \ni J^2_x\gamma \;&\longrightarrow\; \psi_2(J^2_x\gamma) = (tJ^2_x\gamma, q^\sigma J^2_x\gamma, \dot{q}^\sigma J^2_x\gamma, \ddot{q}^\sigma J^2_x\gamma)\\
&= \Bigg(t\varphi(x), q^\sigma(\psi\gamma\varphi^{-1})(\varphi(x)), \left.\frac{\mathrm{d}q^\sigma(\psi\circ\gamma\circ\varphi^{-1})(t)}{\mathrm{d}t}\right|_{t=\varphi(x)},\\
&\qquad \left.\frac{\mathrm{d}^2q^\sigma(\psi\circ\gamma\circ\varphi^{-1})(t)}{\mathrm{d}t^2}\right|_{t=\varphi(x)}\Bigg), \quad \gamma \in \Gamma(\pi)\,.
\end{aligned}$$

It is evident that mappings π_1, π_2 (and of course π_r in general) are surjective submersions.

Note that there arise another fibred manifolds with multidimensional bases,

$$\pi_{1,0} : J^1Y \longrightarrow Y, \quad \pi_{2,0} : J^2Y \longrightarrow Y, \quad \pi_{2,1} : J^2Y \longrightarrow J^1Y\,.$$

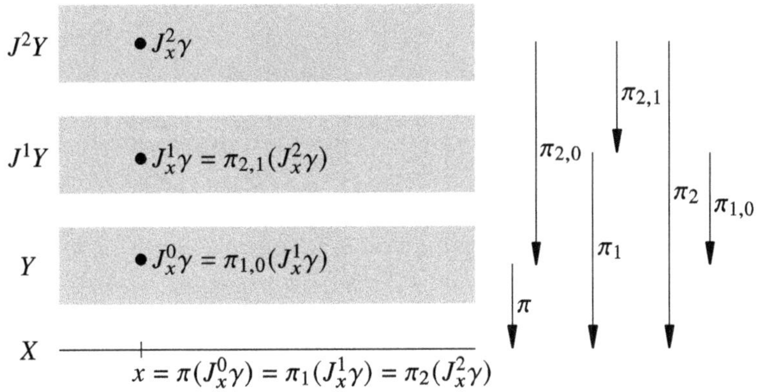

Fig. 2.14 Jet prolongations of a fibred manifold (Y, π, X)

The hierarchic system of fibred manifolds derived from (Y, π, X) by the prolongation procedure is schematically described by Fig. 2.14. (Recall that the fibred manifolds with multidimensional bases are not the subject of our study in this chapter.)

Definition 2.6 (*Jet prolongations of fibred manifolds*) The fibred manifold (J^1Y, π_1, X), $\pi_1 : J^1Y \ni J^1_x\gamma \to \pi_1(J^1_x\gamma) = x \in X$, is called the *first jet prolongation* of the fibred manifold (Y, π, X). The fibred manifold (J^2Y, π_2, X), $\pi_2 : J^2Y \ni J^2_x\gamma \to \pi_2(J^2_x\gamma) = x \in X$, is called the *second jet prolongation* of the fibred manifold (Y, π, X). Let (V, ψ) be a fibred chart on (Y, π, X) $\big((\pi(V), \varphi)$ being the associated chart on $X\big)$. The fibred charts (V_1, ψ_1) and (V_2, ψ_2), on J^1Y and J^2Y, respectively, defined above are called the *associated fibred charts*.

Let $\gamma \in \Gamma_W(\pi)$ be a section of the projection π. The mapping

$$J^1\gamma : W \ni x \longrightarrow J^1\gamma(x) = J^1_x\gamma \in J^1Y$$

is a section of the projection π_1. Indeed, it holds

$$\pi_1 \circ J^1\gamma(x) = \pi_1(J^1_x\gamma) = x \implies \pi_1 \circ J^1\gamma = \mathrm{id}_W\ .$$

Analogously the mapping

$$J^2\gamma : W \ni x \longrightarrow J^2\gamma(x) = J^2_x\gamma \in J^2Y$$

is a section of the projection π_2. Note that for a section $\delta : W \ni x \to \delta(x) \in J^1Y$ need not exist a section $\gamma \in \Gamma_W(\pi)$ such that $\delta = J^1\gamma$.

Definition 2.7 (*Jet prolongations of sections*) The mappings $J^1\gamma$ and $J^2\gamma$ defined above are called the *first* and *second prolongation of the section* γ , respectively. A section δ of the projection π_1 for which there exists a section $\gamma \in \Gamma_W(\pi)$ such that $\delta = J^1\gamma$ is called *holonomic*.

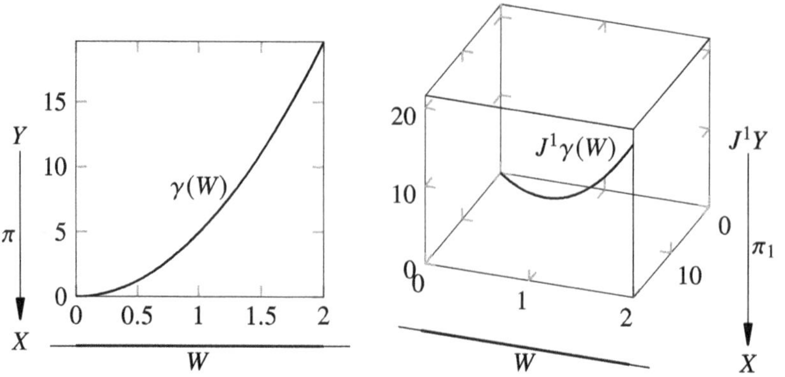

Fig. 2.15 Parametrized phase trajectory of a particle as first jet prolongation of the trajectory

Example 2.7 Consider the free fall of a mass particle in the homogeneous Earth gravitational field (the free-fall acceleration is $g = 9.81\,\mathrm{m\,s^{-2}}$). The motion of the particle is one-dimensional. The corresponding fibred manifold is $(Y, \pi, X) = (\mathbf{R} \times \mathbf{R}, p, \mathbf{R})$ and its first and second prolongations are $(J^1Y, \pi_1, X) = (\mathbf{R} \times \mathbf{R}^2, p_1, \mathbf{R})$ and $(J^2Y, \pi_2, X) = (\mathbf{R} \times \mathbf{R}^3, p_2, \mathbf{R})$. The trajectory of the particle parametrized by time is the section

$$\gamma : \mathbf{R} \ni t \longrightarrow \gamma(t) = \big(t\gamma(t), q^1\gamma(t)\big) = \left(t, \frac{1}{2}gt^2\right) \in \mathbf{R}^2 .$$

The first jet prolongation of this section represents the phase trajectory of the particle parametrized by time,

$$\begin{aligned} J^1\gamma : \mathbf{R} \ni t \longrightarrow J^1\gamma(t) &= \big(tJ^1\gamma(t), q^1J^1\gamma(t), \dot{q}^1J^1\gamma(t)\big) \\ &= \left(t, q^1\gamma(t), \frac{\mathrm{d}q^1\gamma(t)}{\mathrm{d}t}\right) = \left(t, \frac{1}{2}gt^2, gt\right) \in \mathbf{R}^3 . \end{aligned}$$

The situation is presented in Fig. 2.15.

Using jet prolongations of sections of a fibred manifold (Y, π, X) we can define prolongations of its (local) isomorphisms. The situation is shown in Fig. 2.16. Let $W \subset Y$ be an open set and $\pi(W) \subset X$ its projection. Let $\alpha : W \to Y$ be a local isomorphism defined on W and $\alpha_0 : \pi(W) \to X$ its projection. (For the simplicity of coordinate calculations suppose that (V, ψ) is a fibred chart on Y such that both sets W and $\alpha(W)$ are the subsets of V.) Let $\gamma \in \Gamma_{\pi(W)}(\pi)$ be a section. It is evident that the mapping

$$\alpha \circ \gamma \circ \alpha_0^{-1} : \alpha_0\big(\pi(W)\big) \ni \alpha_0(x) \longrightarrow \alpha \circ \gamma \circ \alpha_0^{-1}\big(\alpha_0(x)\big) \in \alpha(W)$$

is a section of π as well (see Problem 2.5). In agreement with the diagram in Fig. 2.16 we now define the first jet prolongation of the isomorphism α.

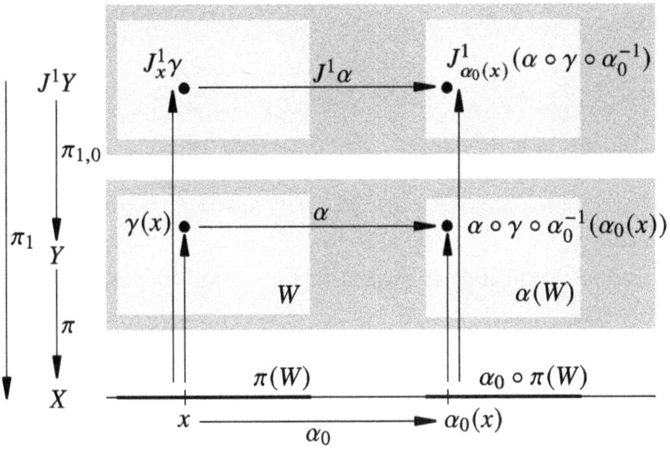

Fig. 2.16 Prolongation of local isomorphisms

Proposition 2.3 *Let $W \subset Y$ be an open set and $\pi(W) \subset X$ its projection. Let $\alpha : W \to Y$ be a local isomorphism defined on W and $\alpha_0 : \pi(W) \to X$ its projection. Then the mapping*

$$J^1\alpha : \pi_{1,0}^{-1}(W) \ni J^1_x\gamma \longrightarrow J^1\alpha(J^1_x\gamma) = J^1_{\alpha_0(x)}(\alpha \circ \gamma \circ \alpha_0^{-1}) \in J^1Y \tag{2.8}$$

is a local isomorphism of the fibred manifold (J^1Y, π_1, X) and its projection is α_0.

Proof Suppose that there exists a fibred chart (V, ψ), $\psi = (t, q^\sigma)$, on Y and corresponding associated chart (U, φ), $\varphi = (t)$, on X such that $W \subset V$. Let (V_1, ψ_1) be the associated fibred chart on J^1Y, i.e. $V_1 = \pi_{1,0}^{-1}(V)$, $\psi_1 = (t, q^\sigma, \dot{q}^\sigma)$. The mapping α is a local isomorphism of the fibred manifold (Y, π, X). That means that the coordinate mappings $\psi \circ \alpha \circ \psi^{-1}$ and $\varphi \circ \alpha_0 \circ \varphi^{-1}$ defined on $\psi(W) \subset \mathbf{R}^{m+1}$ and $\varphi(\pi(W)) \subset \mathbf{R}$, respectively,

$$\begin{aligned}
\psi \circ \alpha \circ \psi^{-1} &: \psi(W) \ni (t, q^\sigma) \longrightarrow \psi \circ \alpha \circ \psi^{-1}(t, q^\sigma) \\
&= \big(t(\psi \circ \alpha \circ \psi^{-1})(t, q^\sigma), q^\nu(\psi \circ \alpha \circ \psi^{-1})(t, q^\sigma)\big) \in \psi\big(\alpha(W)\big) \subset \mathbf{R}^{m+1}, \\
\varphi \circ \alpha_0 \circ \varphi^{-1} &: \varphi\big(\pi(W)\big) \ni (t) \longrightarrow \varphi \circ \alpha_0 \circ \varphi^{-1}(t) \in (\varphi \circ \alpha_0)\big(\pi(W)\big) \subset \mathbf{R}
\end{aligned}$$

are diffeomorphisms. Moreover, the sections $\gamma \in \Gamma_{\pi(W)}(\pi)$ are smooth mappings on $\pi(W)$, i.e.

$$\psi \circ \gamma \circ \varphi^{-1} : \varphi\big(\pi(W)\big) \ni t \longrightarrow \psi \circ \gamma \circ \varphi^{-1}(t) \in \psi(W) \subset \mathbf{R}^{m+1}.$$

So, the chart mapping of $J^1\alpha$ defined on $\psi_1(\pi_{1,0}^{-1}(W)) \subset \mathbf{R}^{2m+1}$,

$$\psi_1 \circ J^1\alpha \circ \psi_1^{-1} : \psi_1(\pi_{1,0}^{-1} W)) \ni (t, q^\sigma, \dot{q}^\sigma) \longrightarrow \psi_1 \circ J^1\alpha \circ \psi_1^{-1}(t, q^\sigma, \dot{q}^\sigma)$$
$$= \Big(t(\psi_1 \circ J^1\alpha \circ \psi_1^{-1})(t, q^\sigma, \dot{q}^\sigma), q^\nu(\psi_1 \circ J^1\alpha \circ \psi_1^{-1})(t, q^\sigma, \dot{q}^\sigma),$$
$$\dot{q}^\nu(\psi_1 \circ J^1\alpha \circ \psi_1^{-1})(t, q^\sigma, \dot{q}^\sigma)\Big) \in \psi_1\Big(J^1\alpha\big(\pi_{1,0}^{-1}(W)\big)\Big) \subset \mathbf{R}^{2m+1},$$

is also the diffeomorphism. It is evident that α_0 is the π_1-projection of $J^1\alpha$ (see Problem 2.6). □

Definition 2.8 (*Prolongations of isomorphisms*) Let α be a local isomorphism of the fibred manifold (Y, π, X) defined on an open set $W \subset Y$ and let α_0 be its projection. The local isomorphism $J^1\alpha$ is called the *first prolongation* of α.

The definition of the second prolongation of a local isomorphism of a fibred manifold (Y, π, X) is based on the same idea as the definition of its first prolongation.

Proposition 2.4 *Let $W \subset Y$ be an open set and $\pi(W) \subset X$ its projection. Let $\alpha : W \to Y$ be a local isomorphism defined on W and $\alpha_0 : \pi(W) \to X$ its projection. Then the mapping*

$$J^2\alpha : \pi_{2,0}^{-1}(W) \ni J_x^2\gamma \longrightarrow J^2\alpha(J_x^2\gamma) = J_{\alpha_0(x)}^2(\alpha \circ \gamma \circ \alpha_0^{-1}) \in J^2Y \tag{2.9}$$

is a local isomorphism of the fibred manifold (J^2Y, π_2, X), and its projection is α_0.

The proof of Proposition 2.4 is quite analogous as that of Proposition 2.3.

Definition 2.9 (*Prolongations of isomorphisms*) Let $W \subset Y$ be an open set and $\pi(W) \subset X$ its projection. Let $\alpha : W \to Y$ be a local isomorphism defined on W and $\alpha_0 : \pi(W) \to X$ its projection. The local isomorphism $J^2\alpha$ is called the *second prolongation* of the local isomorphism α.

Proposition 2.5 *Let α be a local isomorphism of the fibred manifold (Y, π, X) and let α_0 be its π-projection. Then*

a) $J^1\alpha$ is a local isomorphism of the fibred manifold $(J^1Y, \pi_{1,0}, Y)$ and α is its projection,
b) $J^2\alpha$ is a local isomorphism of the fibred manifold $(J^2Y, \pi_{2,0}, Y)$ and α is its projection,
c) $J^2\alpha$ is a local isomorphism of the fibred manifold $(J^2Y, \pi_{2,1}, J^1Y)$ and $J^1\alpha$ is its projection.

A graphic representation of Proposition 2.5 is given in Fig. 2.17.

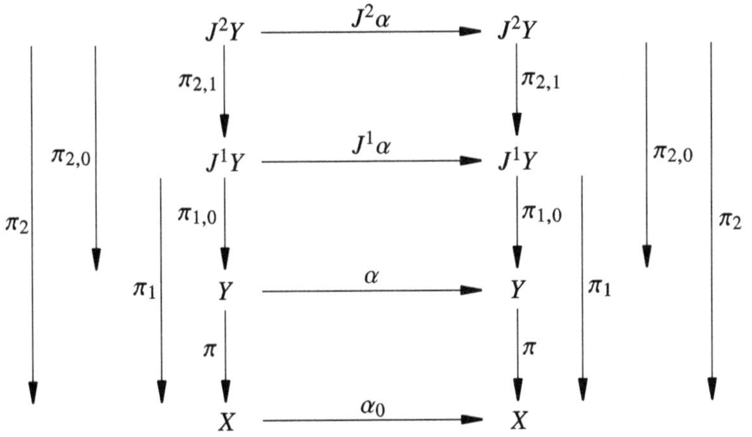

Fig. 2.17 Local isomorphisms and their projections

Example 2.8 Consider the fibred manifold $(Y, \pi, X) = (\mathbf{R} \times \mathbf{R}^2, p, \mathbf{R})$ and its first prolongation $(J^1Y, \pi_1, X) = (\mathbf{R} \times \mathbf{R}^4, p_1, \mathbf{R})$, where p, p_1 denote the Cartesian projections onto the first factor, as usual. Furthermore, consider the local isomorphism of Y as follows:

$$\begin{aligned} \alpha : \mathbf{R} \times \mathbf{R}^2 \ni (t, q^1, q^2) &\longrightarrow \alpha(t, q^1, q^2) \\ &= (t + 1, q^1 \cos\vartheta + q^2 \sin\vartheta, \; -q^1 \sin\vartheta + q^2 \cos\vartheta) \in \mathbf{R} \times \mathbf{R}^2 \, . \end{aligned}$$

Its projection is

$$\alpha_0 : \mathbf{R} \ni (t) \longrightarrow \alpha_0(t) = t + 1 \in \mathbf{R} \, .$$

We find the first prolongation of this local isomorphism. Let γ be a section of the projection p, $\gamma(t) = \big(t, q^1\gamma(t), q^2\gamma(t)\big)$. Then

$$J^1\gamma(t) = \left(t, q^1\gamma(t), q^2\gamma(t), \frac{dq^1\gamma(t)}{dt}, \frac{dq^2\gamma(t)}{dt}\right) .$$

Denote $\bar{\gamma} = \alpha \circ \gamma \circ \alpha_0^{-1}$. Then

$$\bar{\gamma}\big(\alpha_0(t)\big) = \bar{\gamma}(t+1) = \Big(t + 1, q^1\gamma(t) \cos\vartheta + q^2\gamma(t) \sin\vartheta, \; -q^1\gamma(t) \sin\vartheta + q^2\gamma(t) \cos\vartheta\Big),$$

$$\begin{aligned} J^1\bar{\gamma}\big(\alpha_0(t)\big) &= J^1\bar{\gamma}(t+1) \\ &= \Big(t + 1, q^1\gamma(t) \cos\vartheta + q^2\gamma(t) \sin\vartheta, \; -q^1\gamma(t) \sin\vartheta + q^2\gamma(t) \cos\vartheta, \\ &\quad \frac{dq^1\gamma(t)}{dt} \cos\vartheta + \frac{dq^2(t)}{dt} \sin\vartheta, \; -\frac{dq^1\gamma(t)}{dt} \sin\vartheta + \frac{dq^2(t)}{dt} \cos\vartheta\Big) , \end{aligned}$$

and

$$J^1\alpha : \mathbf{R} \times \mathbf{R}^4 \ni (t, q^1, q^2, \dot{q}^1, \dot{q}^2) \longrightarrow J^1\alpha(t, q^1, q^2, \dot{q}^1, \dot{q}^2)$$
$$= \big(t+1, q^1 \cos\vartheta + q^2 \sin\vartheta, q^1 \sin\vartheta + q^2 \cos\vartheta, \dot{q}^1 \cos\vartheta + \dot{q}^2 \sin\vartheta,$$
$$-\dot{q}^1 \sin\vartheta + \dot{q}^2 \cos\vartheta\big) .$$

2.5 Problems

Problem 2.1 Let $\langle O;\ x, y, z\rangle$ be a Cartesian coordinate system in $\mathbf{R}^3$. Consider the surface Y of the toroid arising by rotation of the circle $S^1 : (x-R)^2 + z^2 = r^2$, $y = 0$, $0 < r < R$, around the z-axis (see Fig. 2.18). Define the mapping

$$\pi : Y \ni (\bar{x}, \bar{y}, \bar{z}) \longrightarrow \pi(\bar{x}, \bar{y}, \bar{z}) = (x, y, z) \in S^1$$

assigning to every point $(\bar{x}, \bar{y}, \bar{z}) = Y$ the corresponding point (x, y, z) on the circle S^1, i.e. the intersection of the circle going through the point $(\bar{x}, \bar{y}, \bar{z})$ and lying in the plane $z = \bar{z}$.

(a) Is the mapping π the surjective submersion?
(b) If the answer (a) is positive and thus (Y, π, S^1) is the fibred manifold, describe its fibres. Is this fibred manifold a fibration?
(c) If the answer (a) is positive, i.e. (Y, π, S^1) is the fibred manifold, choose appropriate fibred charts on Y and corresponding associated charts on X.

Problem 2.2 Let X and M be smooth manifolds, $\dim X = 1$, $\dim M = m$. Let $\pi : X \times M \to X$ be the projection of the Cartesian product of manifolds X and M on the first factor. The corresponding fibred manifold $(X \times M, \pi, X)$ is called *trivial*. Supposing that the manifold X is connected, we can conclude that the fibred manifold $(X \times M, \pi, X)$ is the fibration with the type fibre M. (All fibres are identical and thus

Fig. 2.18 Problem 2.1

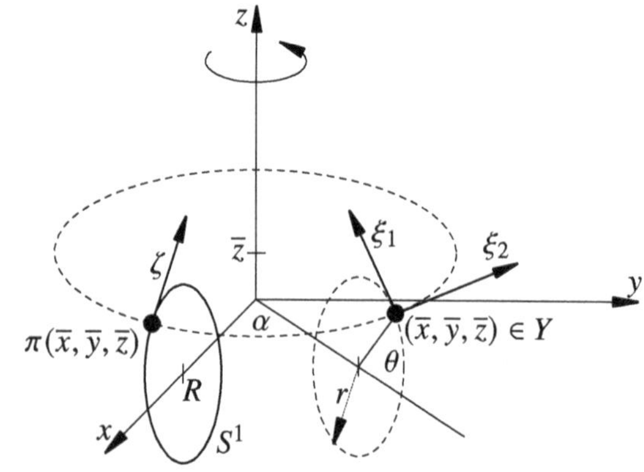

diffeomorphic.) Prove that on the trivial fibred manifold, there exist special charts *adapted to the Cartesian product*, such that transformation rules between them have the form

$$\bar{t} = \bar{t}(t), \quad \bar{q}^\sigma = \bar{q}^\sigma(q^1, \dots, q^m),$$

i.e. the coordinates on the base and on the fibre are not mixed by these transformation rules.

Problem 2.3 Let X be a one-dimensional manifold with the atlas $\mathcal{A} = \{(U_\iota, \varphi_\iota)\}$, $\varphi_\iota = (t\varphi_\iota)$, $\iota \in I$, where I is an index set. Let $\mathcal{U} \subset X$ be an open set, $x \in \mathcal{U}$ a point and $\varepsilon > 0$. The smooth mapping

$$C_x : (-\varepsilon, \varepsilon) \ni s \longrightarrow C(s) \in X$$

such that $C_x(0) = x$ is called a *parametrized curve* in X going through the point x. Two curves $C_{x,1}$ and $C_{x,2}$ are called *1-equivalent at the point* x, if $C_{x,1}(0) = C_{x,2}(0) = x$, and if there exists a coordinate system $(U_\iota, \varphi_\iota) \in \mathcal{A}$ such that $x \in U_\iota \cap \mathcal{U}$ and

$$\left.\frac{\mathrm{d}t\varphi_\iota C_{x,1}(s)}{\mathrm{d}s}\right|_{s=0} = \left.\frac{\mathrm{d}t\varphi_\iota C_{x,2}(s)}{\mathrm{d}s}\right|_{s=0}.$$

It is evident that the above defined relation on the set of curves is the equivalence. The equivalence class $\xi(x) = [C_x]$ is called the *tangent vector at* x to the manifold X.

a) Prove that the property of the equivalence of curves is independent of coordinates.
b) Denote $T_x X = \{\xi(x) \mid x \in X\}$, where $\xi(x)$ is a tangent vector at x to X. Introduce on $T_x X$ the structure of the vector space (called the *tangent space* at the point x to the manifold X). What is the dimension of this space?
c) Denote $TX = \cup_{x \in X} T_x X$, i.e. TX is formed by all pairs $(x, [C_x])$. Prove that TX is the two-dimensional smooth manifold. (Construct the atlas on TX induced by the atlas $\mathcal{A}$ on X).
d) Prove that the mapping $\pi : TX \ni \xi(x) \to \pi\big(\xi(x)\big) = x \in X$ is the surjective submersion and thus (TX, π, X) is the fibred manifold. This fibred manifold is called the *tangent bundle* of the manifold X.
e) Express the components of a vector $\xi(x)$ in coordinates and derive transformation rules between various coordinate systems.

Problem 2.4 Prove Proposition 2.2 for $r = 2$ using relations (2.7).

Problem 2.5 Let $\alpha : W \to Y$ be a local isomorphism of the fibred manifold (Y, π, X) defined on an open set W, $\alpha_0 : \pi(W) \to X$ being its projection. Let $\gamma : \pi(W) \to Y$ be a section. Prove that the mapping $\bar{\gamma} = \alpha \circ \gamma \circ \alpha_0^{-1} : \alpha_0\big(\pi(W)\big) \to \alpha(W)$ is the section of π a well.

Problem 2.6 Let (α, α_0) be a local isomorphism of the fibred manifold (Y, π, X), α being defined on an open set $W \subset Y$. Prove that the local diffeomorphism $J^1\alpha$ introduced by Definition 2.8 is a local isomorphism of the fibred manifold (J^1Y, π_1, X) with the projection α_0.

Problem 2.7 The projection

$$p : \mathbf{R}^3 \backslash \{(0, 0, 0)\} \ni (x, y, z) \longrightarrow p(x, y, z) = (x) \in \mathbf{R}$$

is evidently the surjective submersion and thus it defines the fibred manifold $(\mathbf{R}^3 \backslash (0, 0, 0), p, \mathbf{R}^2)$. Describe all types of fibres of this fibred manifold and decide if this manifold is the fibration.

Problem 2.8 Consider the fibred manifold

$$p : \mathbf{R} \times \mathbf{R}^2 \ni (t, x, y) \longrightarrow p(t, x, y) = (t) \in \mathbf{R} .$$

We have the global fibred chart $(\mathbf{R} \times \mathbf{R}^2, \mathrm{id}_{\mathbf{R}\times\mathbf{R}^2})$ on the total space and the associated chart $(\mathbf{R}, \mathrm{id}_{\mathbf{R}})$ on the base. Let $(\mathbf{R} \times \mathbf{R}^2, \psi)$ be another global fibred chart and $(\mathbf{R}, \varphi)$ the associated chart, where

$$\bar{t}\psi(t) = at + b, \quad \bar{x}\psi(t, x, y) = (at + b, x \cos\alpha + y \sin\alpha, -x \sin\alpha + y \cos\alpha) ,$$

where a, b and α are constants. Derive the transformations rules on the first and second jet prolongations of the given fibred manifold.

Problem 2.9 Let $(Y, \pi, X) = (\mathbf{R}^3, p, \mathbf{R})$ be the fibred manifold defined in Problem 2.8, with the global fibred chart $(\mathbf{R}^3, \mathrm{id}_{\mathbf{R}^3})$ and the associated chart $(\mathbf{R}, \mathrm{id}_{\mathbf{R}})$. Consider the global section

$$\gamma : \mathbf{R} \ni t \longrightarrow \gamma(t) = \big(t, R(\omega t - \sin\omega t), R(1 - \cos\omega t)\big) \in \mathbf{R}^3 ,$$

where ω and R are positive constants. In the associated fibred charts on J^1Y and J^2Y express the first and second jet prolongations of the sections γ.

Problem 2.10 Let (Y, π, X) be the fibred manifold defined in Problems 2.8 and 2.9. Suppose that

$$\delta : \mathbf{R} \ni t \longrightarrow \delta(t) = \big(t, x\delta(t), y\delta(t), \dot{x}\delta(t), \dot{y}\delta(t), 0, -g\big) \in J^2Y ,$$

where $g = \mathrm{const}$. Suppose that the section $\pi_{2,1} \circ \delta$ is holonomic. Determine all sections γ of the projection p such that $\pi_{2,1} \circ \delta = J^1\gamma$.

Problem 2.11 Construct the first and second prolongations of sections discussed in Example 2.5 (Figs. 2.9, 2.10 and 2.11).

Problem 2.12 Consider the fibred manifold $(Y, \pi, X) = (\mathbf{R} \times \mathbf{R}^3, p, \mathbf{R})$, where p is the projection on the first factor. Let

$$\delta : \mathbf{R} \ni t \longrightarrow \delta(t)$$
$$= \left(t, t^2 \cos t, \frac{1}{t} \sin t, 2t^3, 2t \cos t - t^2 \sin t, -\frac{\sin t}{t^2} + \frac{\cos t}{t}, 3t^2\right) \in J^1(\mathbf{R} \times \mathbf{R}^3) .$$

Is this section δ of the projection π_1 holonomic? If yes, determine all sections γ of the projection π such that $\delta = J^1\gamma$. If not, change the coordinate expression of the section δ in such a way that δ will be the first jet prolongation of the section

$$\gamma : \mathbf{R} \ni t \longrightarrow \gamma(t) = \left(t, t^2 \cos t, \frac{1}{t} \sin t, 2t^3\right) \in \mathbf{R} \times \mathbf{R}^3 .$$

Problem 2.13 Let $\delta : X \ni x \to \delta(x) \in J^1Y$ be a section of the projection π_1 of the first prolongation (J^1Y, π_1, X) of the fibred manifold (Y, π, X). Suppose that this section is holonomic, i.e. there exists a section γ of the projection π such that $\delta = J^1\gamma$. Are the sections γ unique? If yes, prove this assertion. If not, describe all such sections γ.

Problem 2.14 Consider the fibred manifold defined in Problem 2.12. Let δ be a section of the projection p_1 such that it holds

$$t\delta(t) = t, \quad \dot{q}^1\delta(t) = R \cos \omega t, \quad \dot{q}^2\delta(t) = R \sin \omega t, \quad \dot{q}^3\delta(t) = b ,$$

where R, b and ω are positive constants. Determine all sections γ of the projection p such that $\delta = J^1\gamma$. Calculate the second jet prolongations of these sections.

Problem 2.15 Write down all possible composed mappings resulting from Fig. 2.14, e.g. $\pi_2 = \pi_1 \circ \pi_{2,1}$, $\pi_2 = \pi \circ \pi_{1,0} \circ \pi_{2,1}$, etc.

Problem 2.16 Prove Proposition 2.5.

Problem 2.17 Consider the fibred manifold and its local isomorphism given in Example 2.8. Find the second prolongation of this local isomorphism.

References

1. O. Krupková, *The Geometry of Ordinary Variational Equations*. Lecture Notes in Mathematics, vol. 1678 (Springer, Berlin, 1997)
2. J. Musilová, P. Musilová, *Mathematics for Understanding and Praxis III* (VUTIUM, Brno, 2017). ((In Czech.))

3. D. Krupka, Global variational theory in fibred spaces. in *Handbook of Global Analysis* (Elsevier Sci. B. V., Amsterdam, 2008), pp. 773–836
4. D.J. Saunders, *The Geometry of Jet Bundles*. London Mathematical Society Lecture Note Series, vol. 142 (Cambridge University Press, Cambridge, 1989)
5. D.J. Saunders, Jet manifolds and natural bundles, in *Handbook of Global Analysis* (Elsevier Sci. B. V., Amsterdam, 2008), pp. 1035–1068

3 Vector Fields and Differential Forms

Mechanical problems are often variational. In such situations a mechanical system moves along the trajectory in the configuration space which can be interpreted as a stationary point of a functional appropriately defined on the set of all possible trajectories (sections of the fibred manifold corresponding to the considered mechanical system). The functional is usually defined as an integral of a special type differential 1-form defined on J^1Y (i.e. depending on time and generalized coordinates and generalized velocities of the system), so-called *variational integral*. The stationary point (stationary section) of this functional is defined by the requirement of zero first variation of the functional when varying the stationary trajectory. Variations of sections are standardly interpreted by means of vector fields on Y and their prolongations. The calculus of variations based on the use of vector fields and differential forms having special properties with respect to the fibred structure of an underlying manifold has a great advantage compared with the "classical" calculus of variations: it works with invariant (geometrical) objects—vector fields and differential forms. Thus the key results can be expressed in a free-coordinate form, even though the practical calculations are made in coordinates, i.e. with the use of fibred charts. The last part of this chapter is devoted to distributions on fibred manifolds and their jet prolongations. The attention is paid to distributions connected with so-called *dynamical forms* which are extremely important for calculus of variations.

Although the definitions and propositions are explicitly formulated for manifolds (J^rY, π_r, X), where $r = 0, 1, 2$, $\dim X = 1$, the generalization for general order r is straightforward and very simple (for a quick overview including key points of proofs, see, e.g. [1], and for details see, e.g. [2]). One of the key theorems concerning differential forms adapted to the fibred structure of manifolds is the Poincaré lemma. An advanced and very detailed (although relatively difficult to study) treatise in full generality can be found in [3]. For multidimensional base see the extension given

J. Musilová et al., *Calculus of Variations on Fibred Manifolds and Variational Physics*, Lecture Notes in Physics 1033, https://doi.org/10.1007/978-3-031-77408-9_3

in the presented text in Chap. 8. (A correct mathematical explanation concerning differential forms on differentiable manifolds in general can be found in [4].)

Prerequisites Covariant tensors on a real vector space, structure of a vector space on tensors, antisymmetric tensors, other algebraic operations (tensor product, alternation, exterior product), tensor bundles of smooth manifolds, vector fields on smooth manifolds, tangent mappings, local one-parameter group of a vector field, components of a vector field expressed via the local one-parameter group, coordinate expressions and transformation rules for vector fields, differential forms on smooth manifolds, coordinate expressions and transformation rules for forms, operations with forms (exterior derivative, pullback by a mapping, contraction of a form by a vector field, Lie derivative). □

3.1 Vector Fields

The well-known concept of a vector field on a manifold can be simply applied to fibred manifolds. Let (Y, π, X) be a fibred manifold, $(\dim X = 1, \dim Y = 1 + m)$ and let (TY, Π, Y) be its tangent bundle (see Problem 2.3), where $\Pi : TY \to Y$ is the projection assigning to every vector $\xi(y) \in T_y Y \subset TY$ its "initial" point y, i.e. $\Pi\big(\xi(y)\big) = y$. Let $W \subset Y$ be an open set. The mapping

$$\xi : W \ni y \longrightarrow \xi(y) \in T_y Y \subset TY$$

is called the (local) *vector field* on Y. For $W = Y$ the vector field is global. The definitions of a vector field on X, J^1Y and J^2Y are quite similar. We will denote sets of all (smooth) vector fields on X, Y, J^1Y and J^2Y as $\mathcal{V}(X)$, $\mathcal{V}(Y)$, $\mathcal{V}(J^1Y)$ and $\mathcal{V}(J^2Y)$, respectively. Let (V, ψ), $\psi = (t, q^\sigma)$, be a fibred chart on Y, (U, φ), $\varphi = (t)$, the corresponding associated chart on X and (V_1, ψ_1), $\psi_1 = (t, q^\sigma, \dot{q}^\sigma)$, and (V_2, ψ_2), $\psi_2 = (t, q^\sigma, \dot{q}^\sigma, \ddot{q}^\sigma)$, the associated fibred charts on J^1Y and J^2Y, respectively. Then in tangent spaces $T_x X$, $T_y Y$, $T_{\tilde{y}} J^1 Y$ and $T_{\hat{y}} J^2 Y$ of manifolds X, Y, J^1Y, J^2Y at points $x \in U$, $y \in V$, $\tilde{y} \in V_1$ and $\hat{y} \in V_2$, respectively, we have the bases connected with the corresponding fibred charts,

$$\left(\frac{\partial}{\partial t}\right)_x, \quad \left(\frac{\partial}{\partial t}, \frac{\partial}{\partial q^\sigma}\right)_y, \quad \left(\frac{\partial}{\partial t}, \frac{\partial}{\partial q^\sigma}, \frac{\partial}{\partial \dot{q}^\sigma}\right)_{\tilde{y}}, \quad \left(\frac{\partial}{\partial t}, \frac{\partial}{\partial q^\sigma}, \frac{\partial}{\partial \dot{q}^\sigma}, \frac{\partial}{\partial \ddot{q}^\sigma}\right)_{\hat{y}},$$

$1 \le \sigma \le m$. The chart expressions of the corresponding vector fields are

$$\xi_0(x) = \xi^0\big(t\varphi(x)\big) \left.\frac{\partial}{\partial t}\right|_x,$$

$$\xi(y) = \xi^0\big(t\psi(y), q^\nu\psi(y)\big) \left.\frac{\partial}{\partial t}\right|_y + \xi^\sigma\big(t\psi(y), q^\nu\psi(y)\big) \left.\frac{\partial}{\partial q^\sigma}\right|_y,$$

$$\begin{aligned}
\tilde{\xi}(\tilde{y}) = {} & \xi^0\big(t\psi_1(\tilde{y}), q^\nu\psi_1(\tilde{y}), \dot{q}^\nu\psi_1(\tilde{y})\big) \left.\frac{\partial}{\partial t}\right|_{\tilde{y}} \\
& + \xi^\sigma\big(t\psi_1(\tilde{y}), q^\nu\psi_1(\tilde{y}), \dot{q}^\nu\psi_1(\tilde{y})\big) \left.\frac{\partial}{\partial q^\sigma}\right|_{\tilde{y}} \\
& + \tilde{\xi}^\sigma\big(t\psi_1(\tilde{y}), q^\nu\psi_1(\tilde{y}), \dot{q}^\nu\psi_1(\tilde{y})\big) \left.\frac{\partial}{\partial \dot{q}^\sigma}\right|_{\tilde{y}} , \\
\hat{\xi}(\hat{y}) = {} & \xi^0\big(t\psi_2(\hat{y}), q^\nu\psi_2(\hat{y}), \dot{q}^\nu\psi_2(\hat{y}), \ddot{q}^\nu\psi_2(\hat{y})\big) \left.\frac{\partial}{\partial t}\right|_{\hat{y}} \\
& + \xi^\sigma\big(t\psi_1(\hat{y}), q^\nu\psi_2(\hat{y}), \dot{q}^\nu\psi_2(\hat{y}), \ddot{q}^\nu\psi_2(\hat{y})\big) \left.\frac{\partial}{\partial q^\sigma}\right|_{\hat{y}} \\
& + \tilde{\xi}^\sigma\big(t\psi_2(\hat{y}), q^\nu\psi_2(\hat{y}), \dot{q}^\nu\psi_2(\hat{y}), \ddot{q}^\nu\psi_2(\hat{y})\big) \left.\frac{\partial}{\partial \dot{q}^\sigma}\right|_{\hat{y}} \\
& + \hat{\xi}^\sigma\big(t\psi_2(\hat{y}), q^\nu\psi_2(\hat{y}), \dot{q}^\nu\psi_2(\hat{y}), \ddot{q}^\nu\psi_2(\hat{y})\big) \left.\frac{\partial}{\partial \ddot{q}^\sigma}\right|_{\hat{y}} ,
\end{aligned}$$

or in shortened notation without explicit writing of the points x, y, $\tilde{y}$, $\hat{y}$ and the coordinate mappings, for simplicity, $1 \leq \sigma, \nu \leq m$,

$$\begin{aligned}
\xi_0 &= \xi^0(t)\,\frac{\partial}{\partial t} , \\
\xi &= \xi^0(t, q^\nu)\,\frac{\partial}{\partial t} + \xi^\sigma(t, q^\nu)\,\frac{\partial}{\partial q^\sigma} , \\
\tilde{\xi} &= \xi^0(t, q^\nu, \dot{q}^\nu)\,\frac{\partial}{\partial t} + \xi^\sigma(t, q^\nu, \dot{q}^\nu)\,\frac{\partial}{\partial q^\sigma} \\
&\quad + \tilde{\xi}^\sigma(t, q^\nu, \dot{q}^\nu)\,\frac{\partial}{\partial \dot{q}^\sigma} , \\
\hat{\xi} &= \xi^0(t, q^\nu, \dot{q}^\nu, \ddot{q}^\nu)\,\frac{\partial}{\partial t} + \xi^\sigma(t, q^\nu, \dot{q}^\nu, \ddot{q}^\nu)\,\frac{\partial}{\partial q^\sigma} \\
&\quad + \tilde{\xi}^\sigma(t, q^\nu, \dot{q}^\nu, \ddot{q}^\nu)\,\frac{\partial}{\partial \dot{q}^\sigma} + \hat{\xi}^\sigma(t, q^\nu, \dot{q}^\nu, \ddot{q}^\nu)\,\frac{\partial}{\partial \ddot{q}^\sigma} .
\end{aligned} \tag{3.1}$$

Definition 3.1 (*Lifts*) Let f be a function on J^sY, and $s = 0, 1$, $r > s$. Pullbacks $F = \pi_r^* f = f \circ \pi_r$ and $F = \pi_{r,s}^* f = f \circ \pi_{r,s}$ are called *lifts of the function* f to J^rY. Let (V, ψ), $\psi = (t, q^\sigma)$, be a fibred chart on Y and (U, φ), $\varphi = (t)$, and (V_1, ψ_1), $\psi_1 = (t, q^\sigma, \dot{q}^\sigma)$, (V_2, ψ_2), $\psi_2 = (t, q^\sigma, \dot{q}^\sigma, \ddot{q}^\sigma)$, the associated chart on X and the associated fibred charts on J^1Y and J^2Y, respectively. Let ξ_0, ξ and $\tilde{\xi}$ be vector fields on X, Y and J^1Y, respectively, i.e.

$$\xi_0 = \xi^0(t)\frac{\partial}{\partial t}, \quad \xi = \xi^0(t, q^\nu)\frac{\partial}{\partial t} + \xi^\sigma(t, q^\nu)\frac{\partial}{\partial q^\sigma},$$

$$\tilde{\xi} = \xi^0(t, q^\nu, \dot{q}^\nu)\frac{\partial}{\partial t} + \xi^\sigma(t, q^\nu, \dot{q}^\nu)\frac{\partial}{\partial q^\sigma} + \tilde{\xi}(t, q^\nu, \dot{q}^\nu)\frac{\partial}{\partial \dot{q}^\sigma}.$$

The vector fields Ξ_0, Ξ, $\tilde{\Xi}$ on $J^r Y$, $r = 0, 1, 2$, are called *lifts* of vector fields ξ_0, ξ and $\tilde{\xi}$ on $J^r Y$, $r = 0, 1, 2$, if

$$\Xi_0 = \Xi^0\frac{\partial}{\partial t}, \quad \Xi = \Xi^0\frac{\partial}{\partial t} + \Xi^\sigma\frac{\partial}{\partial q^\sigma}, \quad \tilde{\Xi} = \Xi^0\frac{\partial}{\partial t} + \Xi^\sigma\frac{\partial}{\partial q^\sigma} + \tilde{\Xi}^\sigma\frac{\partial}{\partial \dot{q}^\sigma}, \tag{3.2}$$

where Ξ^0, Ξ^σ and $\tilde{\Xi}^\sigma$, $1 \leq \sigma \leq m$, are the corresponding lifts of components ξ^0, ξ and $\tilde{\xi}$, respectively.

For example, for a function f on Y the pullbacks $\pi_{1,0}^* f$ and $\pi_{2,0}^* f$ are its lifts to $J^1 Y$ and $J^2 Y$, respectively. In the following we will use for lifts of functions or vector fields the same notation as for the functions and vector fields themselves, because the expression of a lift is the same as for the initial object.

Vector fields on a fibred manifold can have special properties with respect to the fibred structure of the underlying manifold. Suppose that vector fields entering into our considerations are smooth. In the following, by a vector field on X, Y, $J^1 Y$, $J^2 Y$, we mean a local vector field in general, i.e. a vector field defined on an open subset of the corresponding manifold. This admits, of course, global vector fields as well.

Definition 3.2 (*Projectable vector fields*) Let (Y, π, X) be a fibred manifold, $\dim Y = m + 1$, $(J^1 Y, \pi_1, X)$ and $(J^2 Y, \pi_2, X)$ its first and second jet prolongations, respectively. Denote $Y = J^0 Y$, $\pi = \pi_0$ as in Sect. 2.4. Let $r = 0, 1, 2$, $s = 0, 1$, and $s < r$.

(a) A vector field ξ_r on $J^r Y$ is called *π_r-projectable* if there exists a vector field ξ_0 on X, called the *π_r-projection* of the vector field ξ, such that

$$T\pi_r \circ \xi = \xi_0 \circ \pi_r . \tag{3.3}$$

(b) A vector field ξ_r on $J^r Y$ is called *$\pi_{r,s}$-projectable* if there exists a vector field $\xi_{r,s}$ on X, called the *$\pi_{r,s}$-projection* of the vector field ξ_r, such that

$$T\pi_{r,s} \circ \xi_r = \xi_{r,s} \circ \pi_{r,s} .$$

Projectability of vector fields (various types) is schematically illustrated by Fig. 3.1.

Proposition 3.1 *Let (Y, π, X) be a fibred manifold and $(J^1 Y, \pi_1, X)$ and $(J^2 Y, \pi_2, X)$ its prolongations. Denote $J^0 Y = Y$ and $\pi = \pi_0$. Let $r = 0, 1, 2$, $s = 0, 1$ and $s < r$.*

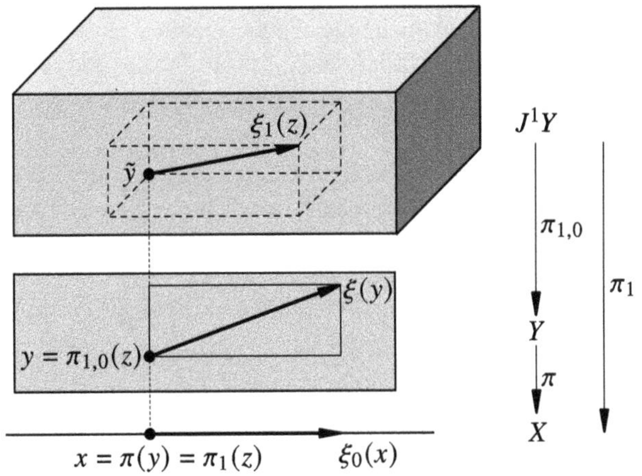

Fig. 3.1 Projectable vector fields

(a) Let the vector field ξ_r on J^rY be π_r-projectable. Then its π_r-projection ξ_0 is unique and

$$\xi_0 = \xi^0(t)\,\frac{\partial}{\partial t}$$

in an arbitrarily chosen fibred chart and corresponding associated charts.

(b) Let the vector field ξ_r on J^rY be $\pi_{r,s}$-projectable. Then its $\pi_{r,s}$-projection $\xi_{r,s}$ is unique and its components in an arbitrarily chosen fibred chart and corresponding associated charts depend only on coordinates on J^sY.

Proof We prove the proposition for a vector field on J^1Y. Let ξ_1 be a vector field on an open set $W_1 \subset J^1Y$. Its chart expression is, in general,

$$\xi_1 = \xi_0(t, q^\nu, \dot{q}^\nu)\,\frac{\partial}{\partial t} + \xi^\sigma(t, q^\nu, \dot{q}^\nu)\,\frac{\partial}{\partial q^\sigma} + \tilde{\xi}^\sigma(t, q^\nu, \dot{q}^\nu)\,\frac{\partial}{\partial \dot{q}^\sigma}\,.$$

Let $(V\,\psi)$, $\psi = (t, q^\sigma)$, be a fibred chart on Y, (U, φ), $\varphi = (t)$, the associated chart on X and (V_1, ψ_1), $\psi_1 = (t, q^\sigma, \dot{q}^\sigma)$, the associated fibred chart on J^1Y such that $W_1 \subset V_1$. The chart representation of the projection π is

$$\varphi \circ \pi_1 \circ \psi_1^{-1} : \psi_1(V_1) \ni (t, q^\sigma, \dot{q}^\sigma) \longrightarrow (t) \in \varphi(U)\,.$$

Its Jacobi matrix representing the tangent mapping $T\pi_1$ is

$$D(\varphi \circ \pi_1 \circ \psi_1^{-1})(t, q^\sigma, \dot{q}^\sigma) = \begin{pmatrix} 1 \\ 0 \\ \vdots \\ 0 \end{pmatrix}.$$

Then the chart representation of the image of the vector $\xi_1(z)$ by the tangent mapping $T\pi_1$ is $\xi^0(\pi_1(z))\,\frac{\partial}{\partial t}$. Thus, the vector field ξ_1 is projectable if and only if there exists a vector field ξ_0 on X such that $\xi_0 \circ \pi_1(z) = \xi^0(\pi_1(z))\,\frac{\partial}{\partial t}$. As ξ_0 should be a vector field on X it is evident that $\xi^0 = \xi^0(t)$, i.e. the component ξ^0 of the vector field ξ_1 depends only on the coordinate along the base. The unique π_1-projection of the vector field ξ_1 is then $\xi_0 = \xi^0(t)\,\frac{\partial}{\partial t}$. Analogously, we express the Jacobi matrix of the coordinate mapping of the projection $\pi_{1,0}$,

$$\psi \circ \pi_{1,0} \circ \psi_1^{-1} : \psi_1(V_1) \ni (t, q^\sigma, \dot{q}^\sigma) \longrightarrow (t, q^\sigma) \in \psi(V)\,,$$

$$D(\psi \circ \pi_{1,0} \circ \psi_1^{-1})(t, q^\sigma, \dot{q}^\sigma) = \begin{pmatrix} 1 & 0 \\ 0 & E_m \\ 0 & 0_m \end{pmatrix},$$

where E_m is the unit matrix of the order m and 0_m is the zero matrix of the same order. □

There are special cases of vector fields corresponding to zero components of certain types.

Definition 3.3 (*Vertical vector fields*) Let (Y, π, X) be a fibred manifold, $\dim Y = 1 + m$, (J^1Y, π_1, X) and (J^2Y, π_2, X) its first and second jet prolongations, respectively. Denote $J^0Y = Y$ and $\pi_0 = \pi$. Let $r = 0, 1, 2$, $s = 0, 1$ and $s < r$.

(a) A vector field ξ on an open set $W_r \subset J^rY$ is called π_r*-vertical*, if $T\pi_r(\xi(y)) = 0$ for every $y \in W_r$.
(b) A vector field ξ on an open set $W_r \subset J^rY$ is called $\pi_{r,s}$*-vertical*, if $T\pi_{r,s}(\xi(y)) = 0$ for every $y \in W_r$.

In Sect. 2.4 we have presented the construction of prolongations of a fibred manifold with the help of jets of sections of π. The prolongation of sections allowed us to define prolongations of local isomorphisms of a fibred manifold (Y, π, X). Now, we will see that projectable vector fields on (Y, π, X) can be prolonged as well, using prolongations of local isomorphisms of their one-parameter groups. Let ξ be a π-projectable vector field on Y (in general this vector field need not be defined globally, but on an open set $W \subset Y$) and ξ_0 its projection (defined on $\pi(W)$). Denote as $\mathcal{G}_W = \{\alpha_u\}$ the local one-parameter group of transformations of the manifold Y belonging to the vector field ξ, and $\{\alpha_{0u}\}$ the local one-parameter group of transformations of the manifold X belonging to the vector field ξ_0. The concept of the local one-parameter group applied to a fibred manifold is as follows. Let ξ be a vector field on $W \subset Y$. Then for or each point $y_0 \in W$ there exists a fibred chart (V, ψ), $\psi = (t, q^\sigma)$, a real number $\varepsilon > 0$ and the uniquely determined group $\mathcal{G}_V$ of local diffeomorphisms

$$\alpha_u : V \ni y \longrightarrow \alpha_u(y) \in \alpha_u(V) \subset V$$

Fig. 3.2 Projectable vector fields and their one-parameter groups

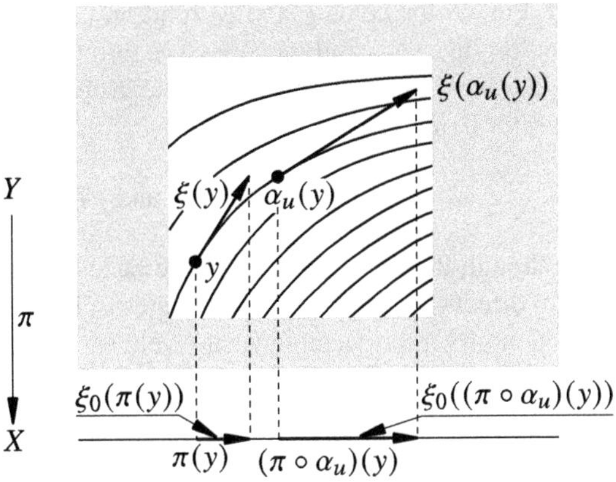

such that for each point $y \in V$ and each value $u \in (-\varepsilon, \varepsilon)$ the vector $\xi\big(\alpha_u(y)\big)$ is the equivalence class

$$\xi\big(\alpha_u(y)\big) = \big[\alpha_u(y), \alpha_{\alpha_u(y)}\big] ,$$

where the (smooth) mapping

$$\alpha_z : (-\varepsilon, \varepsilon) \ni s \longrightarrow \alpha_z(s) = \alpha_s(z) \in V$$

defines the curve α_z going through the point $z \in V$. This means that it holds

$$\xi^0\big(\alpha_u(y)\big) = \left.\frac{\mathrm{d}t(\psi \circ \alpha_{u+s})(y)}{\mathrm{d}s}\right|_{s=0} , \quad \xi^\sigma\big(\alpha_u(y)\big) \left.\frac{\mathrm{d}q^\sigma(\psi \circ \alpha_{u+s})(y)}{\mathrm{d}s}\right|_{s=0} . \tag{3.4}$$

Similarly, if ξ_0 is a vector field on an open set $W_X \subset X$ then there exists a chart (U, φ), $\varphi = (t)$, the real number $\varepsilon > 0$ and the unique local one-parameter group $\mathcal{G}_{W_X}\{\alpha_{0u}\}$ such that for each point $x_0 \in W_X$ and each value $u \in (-\varepsilon, \varepsilon)$ the vector $\xi_0\big(\alpha_{0u}(x)\big)$ is the equivalence class

$$\xi_0\big(\alpha_{0u}(x)\big) = \big[\alpha_{0u}(x), \alpha_{\alpha_{0u}(x)}\big], \quad \text{i.e.}$$

$$\left.\frac{\mathrm{d}t(\varphi \circ \alpha_{0(u+s)})(x)}{\mathrm{d}s}\right|_{s=0} = \xi^0(x) .$$

Now we will show the connection of local one-parameter groups of a π-projectable vector field ξ and its projection ξ_0. The situation is shown in Fig. 3.2.

Proposition 3.2 *Let ξ be a π-projectable vector field on an open set $W \subset Y$, and ξ_0 its π-projection defined on $\pi(W) \subset X$. Then for every value $u \in (-\varepsilon, \varepsilon)$ the mapping $\alpha_u \in \mathcal{G}_W$ is the local isomorphism of the manifold Y and the mapping $\alpha_{0u} \in \mathcal{G}_{\pi(W)}$ is its projection.*

Proof For vector fields ξ and ξ_0 relation (3.3) holds. It is necessary to prove that diffeomorphisms α_u and α_{0u} obey relation (2.5) for every value $u \in (\varepsilon, \varepsilon)$, i.e. $\pi \circ \alpha_u = \alpha_{0u} \circ \pi$. Taking into account the properties of the local one-parameter group of a vector field, we can state that

$$T\alpha_u \circ \xi = \xi \circ \alpha_u \quad \text{and} \quad T\alpha_{0u} \circ \xi_0 = \xi_0 \circ \alpha_{0u} \,. \tag{3.5}$$

Even though these relations are more or less evident, we will prove them with the help of direct calculations in coordinates in Problem 3.1. Following Fig. 3.2 and the definition of a π-projectable vector field we can see that

$$T\pi\big(\xi(y)\big) = \xi_0\big(\pi(y)\big), \quad T\pi\Big(\xi\big(\alpha_u(y)\big)\Big) = \xi_0\Big(\pi\big(\alpha_u(y)\big)\Big),$$

$$\text{and} \quad \xi\big(\alpha_0(y)\big) = \xi(y), \quad \xi_0\Big(\alpha_{00}\big(\pi(y)\big)\Big) = \xi_0\big(\pi(y)\big),$$

because $\alpha_0 = \alpha_u|_{u=0} = \mathrm{id}_W$, $\alpha_{00} = \alpha_{0u}|_{u=0} = \mathrm{id}_{W_X}$. Taking into account relations (3.5) we can see that for some value s of the parameter it must hold

$$T\alpha_{0s}\Big(\xi_0\big(\pi(y)\big)\Big) = \xi_0\Big(\pi\big(\alpha_u(y)\big)\Big) \quad \text{for an arbitrary point} \quad y \in W, \quad \text{i.e.}$$

$$T\alpha_{0s} \circ \xi_0 \circ \pi = \xi_0 \circ \pi \circ \alpha_u \,.$$

Note that we have yet no relation between local one-parameter groups of vector fields ξ and ξ_0. For this reason, we cannot be sure if $s = u$. Taking into account the second relation (3.5) for s we obtain

$$\xi_0 \circ \alpha_{0s} \circ \pi = \xi_0 \circ \pi \circ \alpha_u \implies \alpha_{0s} \circ \pi = \pi \circ \alpha_u \,.$$

Denoting $s = u + v$ we have $\xi_0 \circ \alpha_{0(u+v)} \circ \pi = \xi_0 \circ \pi \circ \alpha_u$, and for $u = 0$

$$\xi_0\Big(\alpha_{0v}\big(\pi(y)\big)\Big) = \xi_0\big(\pi(y)\big) \quad \text{for every point } y \in W \implies v = 0.$$

Thus, we finally obtain

$$\alpha_{0u} \circ \pi = \pi \circ \alpha_u.$$

So, condition (2.5) is fulfilled. This means that α_u is the local isomorphism on Y and α_{0u} is its π-projection. □

Using the concept of jet prolongations of isomorphisms of a fibred manifold introduced in Sect. 2.4 we now define the jet prolongations of π-projectable vector fields. Consider the one-parameter group $\mathcal{G}_W = \{\alpha_u\}$, $u \in (-\varepsilon, \varepsilon)$, of local isomorphisms (α_u, α_{0u}) of a fibred manifold (Y, π, X), where α_{0u} is the π-projection of α_u. It is easy to show (see Problem 3.3) that the families of prolonged isomorphisms $\big\{(J^1\alpha_u, \alpha_{0u})\big\}$ and $\big\{(J^2\alpha_u, \alpha_{0u})\big\}$ have the properties of a local one-parameter group of isomorphisms on the fibred manifolds (J^1Y, π_1, X) and (J^2Y, π_2, X), respectively.

Definition 3.4 (*Prolongations of vector fields*) Let ξ be a π-projectable vector field defined on an open set $W \subset Y$ with the projection ξ_0 defined on $\pi(W)$. Let $\mathcal{G}_W = \{\alpha_u\}$ be the local one-parameter group of the vector field ξ and $\mathcal{G}_{\pi(W)} = \{\alpha_{0u}\}$ the corresponding local one-parameter group of the vector field ξ_0, (α_{0u} is the projection of α_u, $u \in (-\varepsilon, \varepsilon)$). The vector field $J^1\xi$ corresponding to the one-parameter group of isomorphisms $\{J^1\alpha_u\}$ of the fibred manifold (J^1Y, π_1, X) is called the *first jet prolongation* of the vector field ξ. The vector field $J^2\xi$ corresponding to the one-parameter group of isomorphisms $\{J^2\alpha_u\}$ of the fibred manifold (J^2Y, π_2, X) is called the *second jet prolongation* of the vector field ξ.

Definition 3.4 means that the following relations hold for $\tilde{y} \in \pi_{1,0}^{-1}(W)$:

$$J^1\xi(\tilde{y}) = \left.\frac{\mathrm{d}J^1\alpha_u(\tilde{y})}{du}\right|_{u=0}, \quad J^2\xi(\tilde{y}) = \left.\frac{\mathrm{d}J^2\alpha_u(\tilde{y})}{du}\right|_{u=0}.$$

The alternative notation is possible, using jets of sections. Let γ be a section defined on $\pi(W)$. Suppose that $\alpha_{0u}\big(\pi(W)\big) \subset \pi(W)$. (Recall that $\alpha_u \circ \gamma \circ \alpha_{0u}^{-1}$ is again a section of the projection π defined on $\pi(W)$.) For $x \in \pi(W)$ we can write

$$J^1\xi(J^1_x\gamma) = \left.\frac{\mathrm{d}J^1\alpha_u(J^1_x\gamma)}{du}\right|_{u=0}, \quad J^2\xi(J^2_x\gamma) = \left.\frac{\mathrm{d}J^2\alpha_u(J^2_x\gamma)}{du}\right|_{u=0}. \tag{3.6}$$

Now we express the components of prolongations of a vector field ξ in fibred charts. Let $\tilde{y} \in \pi_{1,0}^{-1}W$, $y = \pi_{1,0}(\tilde{y})$, $x = \pi_1(\tilde{y})$, i.e. $y \in W$, $x \in \pi(W)$. Let (V, ψ), $\psi = (t, q^\sigma)$, be a fibred chart on Y. Denote (V_1, ψ_1), $V_1 = \pi^{-1}(V)$, $\psi_1 = (t, q^\sigma, \dot{q}^\sigma)$, the associated fibred chart on J^1Y and (U, φ), $U = \pi(V)$, $\varphi = (t)$, the associated chart on X. Choose a neighbourhood W_y of the point y such that $W_y \subset V$, and that also $\alpha_u(W_y) \subset V$, for certain interval $(-\varepsilon, \varepsilon)$ of parameters u (see Fig. 3.3). The chart expressions of the vector fields ξ and ξ_0 are

$$\xi(y) = \xi^0(t)\frac{\partial}{\partial t} + \xi^\sigma(t, q^\nu)\frac{\partial}{\partial q^\nu}, \quad \xi_0(x) = \xi^0(t)\frac{\partial}{\partial t}.$$

It holds

$$\xi^0(t) = \left.\frac{dt(\psi \circ \alpha_u \circ \psi^{-1})(t, q^\nu)}{du}\right|_{u=0} = \left.\frac{dt(\varphi \circ \alpha_{0u} \circ \varphi^{-1})(t)}{du}\right|_{u=0},$$

$$\xi^\sigma(t, q^\nu) = \left.\frac{dq^\sigma(\psi \circ \alpha_u \circ \psi^{-1})(t, q^\nu)}{du}\right|_{u=0}.$$

The chart expression of the vector field $J^1\xi$ is

$$J^1\xi = \xi^0(t)\frac{\partial}{\partial t} + \xi^\sigma(t, q^\nu)\frac{\partial}{\partial q^\sigma} + \tilde{\xi}(t, q^\nu, \dot{q}^\nu)\frac{\partial}{\partial \dot{q}^\nu},$$

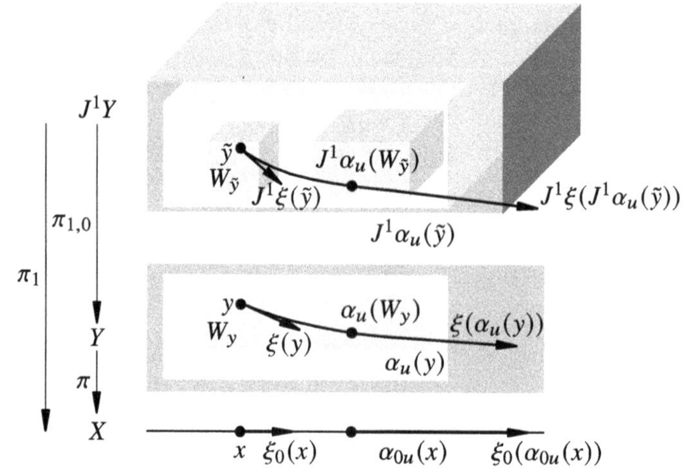

Fig. 3.3 Prolongations of vector fields

where

$$\begin{aligned}
\xi^0(t) &= \left.\frac{dt(\psi_1 \circ J^1\alpha_u \circ \psi_1^{-1})(t, q^\nu, \dot{q}^\nu)}{du}\right|_{u=0} = \left.\frac{dt(\varphi \circ \alpha_{0u} \circ \varphi^{-1})(t)}{du}\right|_{u=0}, \\
\xi^\sigma(t, q^\nu) &= \left.\frac{dq^\sigma(\psi_1 \circ J^1\alpha_u \circ \psi_1^{-1})(t, q^\nu, \dot{q}^\nu)}{du}\right|_{u=0} = \left.\frac{dq^\sigma(\psi \circ \alpha_u \circ \psi^{-1})(t, q^\nu)}{du}\right|_{u=0}, \\
\tilde{\xi}^\sigma(t, q^\nu, \dot{q}^\nu) &= \left.\frac{d\dot{q}^\sigma(\psi_1 \circ J^1\alpha_u \circ \psi_1^{-1})(t, q^\nu, \dot{q}^\nu)}{du}\right|_{u=0}.
\end{aligned} \tag{3.7}$$

The final expression of components $\tilde{\xi}^\sigma$ as functions of coordinates $(t,\ q^\nu, \dot{q}^\nu)$, $1 \leq \sigma,\ \nu \leq m$, is technically somewhat laborious, although we need nothing more than the chain rule for the derivative of functions. We will therefore take it in more detail. We use expression (3.6), i.e. for a point $\tilde{y} \in V_1$, $\tilde{y} = J_x^1\gamma$, $x \in U$,

$$J^1\alpha_u(\tilde{y}) = J^1_{\alpha_{0u}(\pi_1(\tilde{y}))}(\alpha_u \circ \gamma \circ \alpha_{0u}^{-1}) = J^1(\alpha \circ \gamma \circ \alpha_{0u}^{-1})\Big(\alpha_{0u}\big(\pi_1(\tilde{y})\big)\Big).$$

For $\tilde{\xi}^\sigma$, $1 \leq \sigma \leq m$, it holds, following the last equation in set (3.7) and in exact notation with all mappings involved,

$$\tilde{\xi}^\sigma(t,\ q^\nu,\ \dot{q}^\nu) = \left[\frac{d}{du}\Big(\dot{q}^\sigma(\psi_1 \circ J^1(\alpha_u \circ \gamma \circ \alpha_{0u}^{-1}) \circ \varphi^{-1})\big(t\alpha_{0u}(t)\big)\Big)\right]_{u=0}.$$

For shortening and better clarity of calculations we leave out the writing of coordinate mappings ψ_1, ψ, φ given in advance. Then we can write

$$\tilde{\xi}^\sigma(t, q^\nu, \dot{q}^\nu) = \left[\frac{\mathrm{d}}{\mathrm{d}u}\left(\dot{q}^\sigma(J^1\alpha_u \circ \gamma \circ \alpha_{0u}^{-1})(\alpha_{0u}(t))\right)\right]_{u=0}$$
$$= \left[\frac{\mathrm{d}}{\mathrm{d}u}\left(\dot{q}^\sigma(J^1\alpha_u \circ J^1\gamma \circ \alpha_{0u}^{-1})(\alpha_{0u}(t))\right)\right]_{u=0} .$$

Moreover, denote $\tau = \tau(u, t) = \alpha_{0u}(t)$ (τ is a function of the parameter u through the mapping α_{0u} and of the coordinate t, for illustration see Fig. 3.3—the variable t is the t-coordinate of the point x, i.e. $t = \varphi(x)$). These two variables are independent of each other. It holds

$$\dot{q}^\sigma\left(J^1\alpha_u \circ J^1\gamma \circ \alpha_{0u}^{-1}\right)(\tau) = \left.\frac{\mathrm{d}q^\sigma(J^1\alpha_u \circ J^1\gamma)(t)}{\mathrm{d}t}\right|_t \cdot \left.\frac{\mathrm{d}\alpha_{0u}^{-1}(\tau)}{\mathrm{d}\tau}\right|_\tau$$
$$= \left.\frac{\mathrm{d}q^\sigma(\alpha_u \circ \gamma)(t)}{\mathrm{d}t}\right|_t \cdot \left.\frac{\mathrm{d}\alpha_{0u}^{-1}(\tau)}{\mathrm{d}\tau}\right|_\tau ,$$

because $q^\sigma(J^1\alpha_u \circ J^1\gamma) = q^\sigma(\alpha_u \circ \gamma)$. Then

$$\tilde{\xi}^\sigma = \left[\frac{\mathrm{d}}{\mathrm{d}u}\left(\left.\frac{\mathrm{d}q^\sigma(\alpha_u \circ \gamma)(t)}{\mathrm{d}t}\right|_t \cdot \left.\frac{\mathrm{d}\alpha_{0u}^{-1}(\tau)}{\mathrm{d}\tau}\right|_\tau\right)\right]_{u=0}$$
$$= \left[\frac{\mathrm{d}}{\mathrm{d}u}\left.\frac{\mathrm{d}q^\sigma(\alpha_u \circ \gamma)(t)}{\mathrm{d}t}\right|_t\right]_{u=0} \cdot \left[\left.\frac{\mathrm{d}\alpha_{0u}^{-1}(\tau)}{\mathrm{d}\tau}\right|_\tau\right]_{u=0}$$
$$+ \left[\left.\frac{\mathrm{d}q^\sigma(\alpha_u \circ \gamma)(t)}{\mathrm{d}t}\right|_t\right]_{u=0} \cdot \left[\frac{\mathrm{d}}{\mathrm{d}u}\left.\frac{\mathrm{d}\alpha_{0u}^{-1}(\tau)}{\mathrm{d}\tau}\right|_\tau\right]_{u=0}$$
$$= \frac{\mathrm{d}}{\mathrm{d}t}\left[\frac{\mathrm{d}q^\sigma(\alpha_u \circ \gamma)(t)}{\mathrm{d}u}\right]_{u=0} - \dot{q}^\sigma\gamma(t) \cdot \frac{\mathrm{d}}{\mathrm{d}t}\left[\frac{\mathrm{d}\tau(u, t)}{\mathrm{d}u}\right]_{u=0} ,$$

(reminding that variables t and u are independent in our calculations). Taking into account that

$$\left.\frac{\mathrm{d}q^\sigma(\alpha_u \circ \gamma)(t)}{\mathrm{d}u}\right|_{u=0} = \xi^\sigma(\gamma(t)) ,$$
$$\left.\frac{\mathrm{d}\alpha_{0u}^{-1}}{\mathrm{d}u}\right|_{u=0} = \left.\frac{\mathrm{d}\alpha_{(0,-u)}}{\mathrm{d}u}\right|_{u=0} = -\left.\frac{\mathrm{d}\alpha_{0u}}{\mathrm{d}u}\right|_{u=0} = -\xi^0(t) ,$$
$$\left[\left.\frac{\mathrm{d}\alpha_{0u}^{-1}(\tau)}{\mathrm{d}\tau}\right|_\tau\right]_{u=0} = 1 ,$$

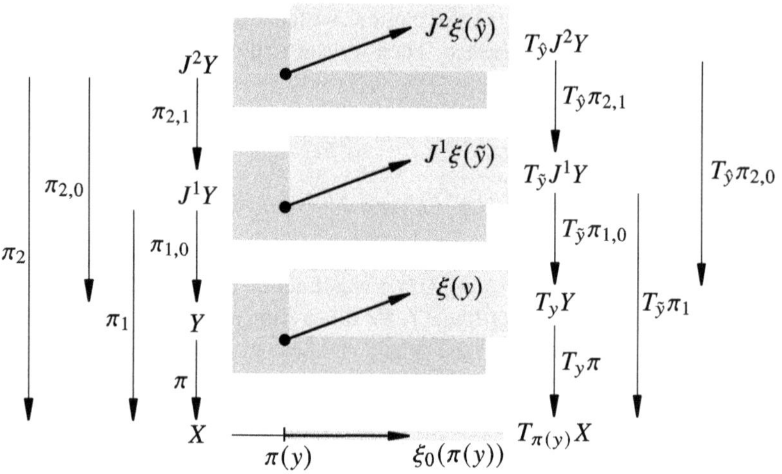

Fig. 3.4 Prolongations of vector fields are projectable

and using the assumption that the obtained relation is valid for all section γ we have the resulting expression

$$\tilde{\xi}^\sigma = \frac{\mathrm{d}\xi^\sigma}{\mathrm{d}t} - \dot{q}^\sigma \frac{\mathrm{d}\xi^0}{\mathrm{d}t} \,.$$

The similar calculation can be made for $J^2\xi$ (components on J^2Y are denoted as $\hat{\xi}^\sigma$). Then we have

$$\tilde{\xi}^\sigma = \frac{\mathrm{d}\xi^\sigma}{\mathrm{d}t} - \dot{q}^\sigma \frac{\mathrm{d}\xi^0}{\mathrm{d}t}, \quad \hat{\xi}^\sigma = \frac{\mathrm{d}\tilde{\xi}^\sigma}{\mathrm{d}t} - \ddot{q}^\sigma \frac{\mathrm{d}\xi^0}{\mathrm{d}t} \,, \tag{3.8}$$

and for a π-vertical vector field ξ

$$\tilde{\xi}^\sigma = \frac{\mathrm{d}\xi^\sigma}{\mathrm{d}t}, \quad \hat{\xi}^\sigma = \frac{\mathrm{d}\tilde{\xi}^\sigma}{\mathrm{d}t} \,. \tag{3.9}$$

It is to be noted that both the first and second prolongations of a (π-projectable) ξ are projectable too. The vector field $J^1\xi$ is π_1-projectable as well as $\pi_{1,0}$ projectable, the vector field $J^2\xi$ is π_2-projectable, $\pi_{2,0}$-projectable and $\pi_{2,1}$-projectable. These properties are shown in Fig. 3.4. While the figure shows vectors, i.e. values of vector fields in concrete points, Fig. 3.5 shows the commutative diagrams of mappings.

Example 3.1 Consider the fibred manifold $(Y, \pi, X) = (\mathbf{R} \times \mathbf{R}^2, p, \mathbf{R})$ and the π-vertical vector field on Y,

$$\xi = \frac{\partial}{\partial q^1} + 3q^1 \frac{\partial}{\partial q^2} \,.$$

This vector field is schematically drawn on Fig. 3.6 in a fibre $\pi^{-1}(t)$ for arbitrary t. (Note that the "arrow representation" of the vector field ξ is the same in all fibres,

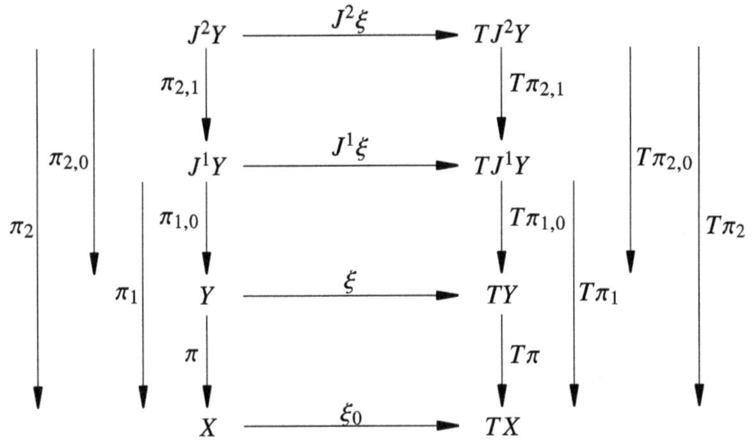

Fig. 3.5 Prolongations of vector fields are projectable

because ξ is vertical.) Let us calculate the prolongations $J^1\xi$ and $J^2\xi$ of this vector field as well as its one-parameter group and its first and second jet prolongations. For obtaining the components of $J^1\xi$ and $J^2\xi$ we use relations (3.9):

$$\tilde{\xi}^1 = 0, \quad \tilde{\xi}^2 = 3\dot{q}^1, \quad \hat{\xi}^1 = 0, \quad \hat{\xi}^2 = 3\ddot{q}^1, \quad \text{i.e.}$$

$$\tilde{\xi} = \frac{\partial}{\partial q^1} + 3q^1 \frac{\partial}{\partial q^2} + 3\dot{q}^1 \frac{\partial}{\partial \dot{q}^2}, \quad \hat{\xi} = \frac{\partial}{\partial q^1} + 3q^1 \frac{\partial}{\partial q^2} + 3\dot{q}^1 \frac{\partial}{\partial \dot{q}^2} + 3\ddot{q}^1 \frac{\partial}{\partial \ddot{q}^2} .$$

Equations for the one-parameter group of the vector field ξ (in the presented example global) read

$$\frac{dt\alpha_u}{du} = 0, \quad \frac{dq^1\alpha_u}{du} = 1, \quad \frac{dq^2\alpha_u}{du} = 3q^1, \quad \frac{dt\alpha_{0u}}{du} = 0 .$$

Integrating these equations and taking into account the conditions

$$t\alpha_{00}(t) = t, \quad t\alpha_0(t) = t, \quad q^1\alpha_0(t, q^1, q^2) = q^1, \quad q^2\alpha_0(t, q^1, q^2) = q^2 ,$$

we finally obtain

$$t\alpha_{0u}(t) = t ,$$
$$t\alpha_u(t) = t, \quad q^1\alpha_u(t, q^1, q^2) = q^1 + u, \quad q^2\alpha_u(t, q^1, q^2) = \frac{3}{2}u^2 + 3q^1 u + q^2 .$$

It is evident that the integral curves of the vector field ξ are parabolas, as it is also indicated by Fig. 3.6. The jet prolongations $\{J^1\alpha_u\}$ and $\{J^2\alpha_u\}$ of the one-parameter group $\{\alpha_u\}$ are as follows:

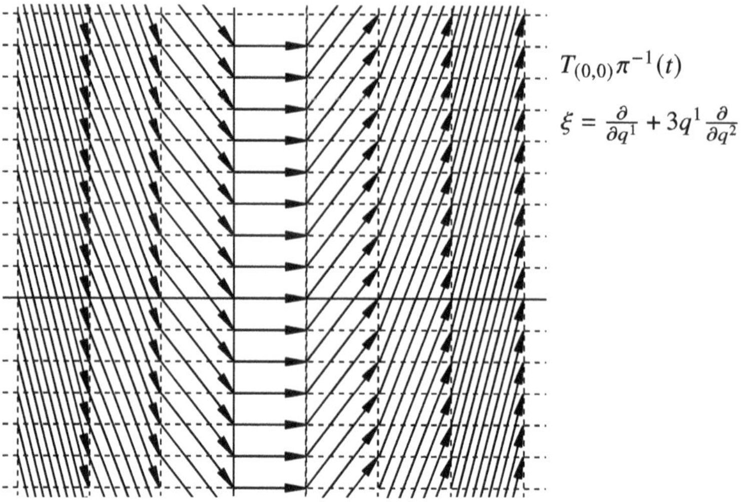

Fig. 3.6 Example 3.1

$$J^1\alpha_u(t, q^1, q^2, \dot{q}^1, \dot{q}^2) = \left(t, q^1 + u, q^2 + 3q^1 u + \frac{3}{2}u^2, \dot{q}^1, \dot{q}^2 + 3\dot{q}^1 u\right) ,$$

$$J^2\alpha_u(t, q^1, q^2, \dot{q}^1, \dot{q}^2, \ddot{q}^1, \ddot{q}^2)$$
$$= \left(t, q^1 + u, q^2 + 3q^1 u + \frac{3}{2}u^2, \dot{q}^1, \dot{q}^2 + 3\dot{q}^1 u, \ddot{q}^1, \ddot{q}^2 + 3\ddot{q}^1 u\right) .$$

3.2 Differential Forms

Differential forms on a smooth manifold are covariant C^r-differentiable (or smooth) antisymmetric tensor fields, i.e. C^r-differentiable (or smooth) mappings assigning to every point of the manifold (or of some its open subset in the local case) a covariant antisymmetric tensors. The concept of differential forms on fibred manifolds is introduced in a quite similar way. Suppose that (Y, π, X) is a fibred manifold. Consider the fibred manifold $(\Lambda^k(TY), \Pi, Y)$, where $\Lambda^k(TY)$ is the bundle of covariant antisymmetric k-tensors on Y, and $\Pi : \Lambda^k(TY) \to Y$ is the projection assigning to every k-tensor $\omega(y) \in \Lambda^k(T_yY) \subset \Lambda^k(TY)$ its "initial" point y. Let $W \subset Y$ be an open set. By a (local) *differential k-form*, $1 \leq k$, we mean a C^r-differentiable (or smooth) mapping

$$\omega : W \ni y \longrightarrow \omega(y) \in \Lambda^k(T_yY) \subset \Lambda^k(TY) .$$

For $W = Y$ the k-form is global. The concepts of a k-form on X, J^1Y and J^2Y are quite similar. A function on W is considered as a *0-form*. Recall that the set of all

k-forms on a given manifold is the module over the ring of functions. We will denote the modules of (smooth) k-forms on X, Y, J^1Y and J^2Y as $\Lambda^k(TX)$, $\Lambda^k(TY)$, $\Lambda^k(TJ^1Y)$ and $\Lambda^k(TJ^2Y)$, respectively, in general $\Lambda^k(TJ^rY)$ – on J^rY. Let (V, ψ), $\psi = (t, q^\sigma)$, be a fibred chart on Y, (U, φ), $\varphi = (t)$, the corresponding associated chart on X and (V_1, ψ_1), $\psi_1 = (t, q^\sigma, \dot{q}^\sigma)$, and (V_2, ψ_2), $\psi_2 = (t, q^\sigma, \dot{q}^\sigma, \ddot{q}^\sigma)$ the associated fibred charts on J^1Y and J^2Y, respectively. Then in spaces of 1-forms $\Lambda^1(T_xX)$, $\Lambda^1(T_yY)$, $\Lambda^1(T_{\tilde{y}}J^1Y)$ and $\Lambda^1(T_{\hat{y}}J^2Y)$ on manifolds X, Y, J^1Y, J^2Y at points $x \in U$, $y \in V$, $\tilde{y} \in V_1$ and $\hat{y} \in V_2$, respectively, we have the bases of 1-forms connected with the corresponding fibred charts,

$$(\mathrm{d}t)_x, \quad (\mathrm{d}t, \mathrm{d}q^\sigma)_y, \quad (\mathrm{d}t, \mathrm{d}q^\sigma, \mathrm{d}\dot{q}^\sigma)_{\tilde{y}}, \quad (\mathrm{d}t, \mathrm{d}q^\sigma, \mathrm{d}\dot{q}^\sigma, \mathrm{d}\ddot{q}^\sigma)_{\hat{y}} \ ,$$

$1 \leq \sigma \leq m$. In shortened notation without explicit writing of the points x, y, $\tilde{y}$, $\hat{y}$ and the coordinate mappings, we can write the chart expressions of 1-forms as follows:

$$\begin{aligned}
\omega_X &= \omega_0(t)\,\mathrm{d}t \ , \\
\omega &= \omega_0(t, q^\nu)\,\mathrm{d}t + \omega_\sigma(t, q^\nu)\,\mathrm{d}q^\sigma \ , \\
\tilde{\omega} &= \omega_0(t, q^\nu, \dot{q}^\nu)\,\mathrm{d}t + \omega_\sigma(t, q^\nu, \dot{q}^\nu)\,\mathrm{d}q^\sigma \ + \ \tilde{\omega}_\sigma(t, q^\nu, \dot{q}^\nu)\,\mathrm{d}\dot{q}^\sigma \ , \\
\hat{\omega} &= \omega_0(t, q^\nu, \dot{q}^\nu, \ddot{q}^\nu)\,\mathrm{d}t + \omega_\sigma(t, q^\nu, \dot{q}^\nu, \ddot{q}^\nu)\,\mathrm{d}q^\sigma \\
&\quad + \tilde{\omega}_\sigma(t, q^\nu, \dot{q}^\nu, \ddot{q}^\nu)\,\mathrm{d}\dot{q}^\sigma + \hat{\omega}_\sigma(t, q^\nu, \dot{q}^\nu, \ddot{q}^\nu)\,\mathrm{d}\ddot{q}^\sigma \ .
\end{aligned} \tag{3.10}$$

The chart expression of a differential k-form, $k \geq 1$, is a linear combination of exterior products ("wedge" product) of k factors of the base 1-forms. For example a k-form on Y can be written as

$$\omega = A_{\sigma_1 \ldots \sigma_{k-1}}(t, q^\nu)\,\mathrm{d}q^{\sigma_1} \wedge \ldots \wedge \mathrm{d}q^{\sigma_{k-1}} \wedge \mathrm{d}t + B_{\sigma_1 \ldots \sigma_k}(t, q^\nu)\,\mathrm{d}q^{\sigma_1} \wedge \ldots \wedge \mathrm{d}q^{\sigma_k} \ ,$$

where $A_{\sigma_1 \ldots \sigma_{k-1}}(t, q^\nu)$ and $B_{\sigma_1 \ldots \sigma_k}(t, q^\nu)$ are functions on Y, antisymmetric in indices $(\sigma_1, \ldots, \sigma_{k-1})$ and $(\sigma_1, \ldots, \sigma_k)$, respectively, $1 \leq \sigma_1, \ldots, \sigma_k \leq m$. Note that the only nontrivial form on X is a 1-form, i.e. $\omega_X = \omega_0(t)\,\mathrm{d}t$.

Similarly as in the case of vector fields there are some special types of differential forms adapted to the fibred structure of underlying manifolds. (Note that the fact that we work only with fibred manifolds with one-dimensional bases makes these concepts comprehensible very easily. The situation is technically not so simple for fibred manifolds with multidimensional bases—see Chap. 8.)

Definition 3.5 (*Projectable forms*) Let (Y, π, X) be a fibred manifold, $\dim Y = 1 + m$, (J^1Y, π_1, X) and (J^2Y, π_2, X) its first and second jet prolongations, respectively. Denote $Y = J^0Y$, $\pi = \pi_0$, as in Sect. 2.4. Let $r = 0, 1, 2$, $s = 0, 1$ and $s < r$. Let $k \geq 1$.

(a) A k-form ω on J^rY is called π_r*-projectable* if there exists a k-form η on X such that

$$\omega = \pi_r^* \eta \ .$$

(b) A k-form ω on $J^r Y$ is called $\pi_{r,s}$*-projectable* if there exists a form η on $J^s Y$ such that

$$\omega = \pi_{r,s}^* \eta \, .$$

Definition 3.6 (*Horizontal forms*) Let (Y, π, X) be a fibred manifold, $\dim Y = 1 + m$, $(J^1 Y, \pi_1, X)$ and $(J^2 Y, \pi_2, X)$ its first and second jet prolongations, respectively. Denote $Y = J^0 Y$, $\pi = \pi_0$, as in Sect. 2.4. Let $r, s = 0, 1, 2$ and $s < r$. Let $k \geq 1$.

(a) A k-form ω on $J^r Y$ is called π_r*-horizontal* if it vanishes whenever any of its vector arguments is π_r-vertical vector, i.e.

$$i_\xi \omega = 0 \quad \text{for every } \pi_r \text{ vertical vertical vector field } \xi .$$

(b) A k-form ω on $J^r Y$ is called $\pi_{r,s}$*-horizontal* if it vanishes whenever any of its vector arguments is $\pi_{r,s}$-vertical vector.

$$i_\xi \omega = 0 \quad \text{for every } \pi_{r,s}\text{-vertical vector field } \xi .$$

π_r-horizontal forms on $J^r Y$, $r = 0, 1, 2$, are also called r*-th-order Lagrangians*, $\pi_{r,0}$-horizontal 2-forms on $J^r Y$, $r = 1, 2$, are called r*-th-order dynamical forms.*

It is easy to verify that sets of all π_r-horizontal and $\pi_{r,s}$-horizontal k-forms are submodules of the module $\Lambda^k(TJ^r Y)$. We denote them as $\Lambda^k_X(TJ^r Y)$ and $\Lambda^k_{J^s Y}(TJ^r Y)$. By direct calculations in charts we can prove the following proposition:

Proposition 3.3 *Let (V, ψ) be a fibred chart on Y, (U, φ) the associated chart on X and (V_1, ψ_1), (V_2, ψ_2) the associated fibred charts on $J^1 Y$ and $J^2 Y$, as usual, $k \geq 1$.*

(a) The chart expression of a π_r horizontal 1-form on $J^r Y$, $r = 0, 1, 2$, is

$$\omega = \omega_0 \, \mathrm{d}t \, ,$$

where ω_0 is a function on $J^r Y$. All π_r-horizontal k-forms are trivial (zero) for $k > 1$.

(b) The chart expression of a $\pi_{r,s}$ horizontal 1-form on $J^r Y$, $r = 1, 2$, $s = 0, 1$, is

$$\omega = \omega_0 \, \mathrm{d}t + \omega_\sigma \, \mathrm{d}q^\sigma \quad \textit{for} \quad s = 0 \, ,$$

where ω_0 *and* ω_σ, $1 \le \sigma \le m$, *are functions on* $J^r Y$, *and*

$$\omega = \omega_0 \,\mathrm{d}t + \omega_\sigma \,\mathrm{d}q^\sigma + \tilde{\omega}_\sigma \,\mathrm{d}\dot{q}^\sigma \quad \textit{for} \quad s = 1 \;,$$

where ω_0, ω_σ *and* $\tilde{\omega}_\sigma$, $1 \le \sigma \le m$, *are functions on* $J^r Y$. *All* $\pi_{r,0}$ *horizontal* k*-forms are trivial (zero) for* $k > m+1$. *All* $\pi_{r,1}$ *horizontal* k*-forms are trivial (zero) for* $k > 2m+1$.

Proposition 3.4 *Let* $r = 0, 1$ *and* $k \ge 1$. *Then there exists a unique* **R***-linear mapping of modules of forms over the field* **R** *preserving the exterior product*

$$h : \Lambda^k(T J^r Y) \ni \omega \longrightarrow h\omega \in \Lambda^k_X(T J^{r+1} Y) \;, \tag{3.11}$$

such that

$$J^r \gamma^* \omega = J^{r+1} \gamma^* h\omega \;. \tag{3.12}$$

Proof Recall that the **R**-linearity and preservation of the exterior product mean the following relations:

$$h(\alpha\omega + \beta\eta) = \alpha h\omega + \beta h\eta, \quad h(\omega \wedge \chi) = h\omega \wedge h\chi$$

for arbitrary $\alpha, \beta \in \mathbf{R}$ and arbitrary $\omega, \eta \in \Lambda^k(J^r Y)$ and $\chi \in \Lambda^l(J^r Y)$. First, we prove the proposition for 1-forms. Let $\omega \in \Lambda^1(TY)$ be defined on an open set $W \subset Y$ and let $\gamma \in \Gamma_{\pi(W)}(\pi)$. Let (V, ψ), $\psi = (t, q^\sigma)$, be a fibred chart on Y, $V \cap W \neq \emptyset$, (U, φ), $\varphi = (t)$, the associated chart on X and (V_1, ψ_1), $\psi_1 = (t, q^\sigma, \dot{q}^\sigma)$, the associated fibred chart on $J^1 Y$. The chart expressions of ω and γ are (in the shortened notation)

$$\omega = \omega_0 \,\mathrm{d}t + \omega_\sigma \,\mathrm{d}q^\sigma, \quad \gamma(t) = \big(t\gamma(t), q^\sigma \gamma(t)\big) \;,$$

where ω_0 and ω_σ, $1 \le \sigma \le m$, are functions on Y. It holds

$$\begin{aligned} J^0\gamma^*\omega = \gamma^*\omega &= (\omega_0 \circ \gamma)\,\mathrm{d}t + (\omega_\sigma \circ \gamma)\,\gamma^*\mathrm{d}q^\sigma \\ &= \left((\omega_0 \circ \gamma) + (\omega_\sigma \circ \gamma) \frac{\mathrm{d}q^\sigma \gamma(t)}{\mathrm{d}t} \right) \mathrm{d}t \;. \end{aligned}$$

We wish to find a horizontal form on $J^1 Y$, i.e. $\eta \in \Lambda^1_X(T J^1 Y)$, $\eta = \eta_0(t, q^\nu, \dot{q}^\nu)\,\mathrm{d}t$, such that relation (3.12) holds, i.e. the requirement is

$$\eta_0 \circ J^1\gamma = \left((\omega_0 \circ \gamma) + (\omega_\sigma \circ \gamma) \frac{\mathrm{d}q^\sigma \gamma(t)}{\mathrm{d}t} \right)$$

for every section γ. This leads to the unique expression of the component η_0, $\eta_0 = \omega_0 + \omega_\sigma \dot{q}^\sigma$. We have obtained the mapping

$$h : \Lambda^1(TY) \ni \omega \longrightarrow h\omega = \eta \in \Lambda^1_X(T J^1 Y), \quad \eta = (\omega_0 + \omega_\sigma \dot{q}^\sigma)\,\mathrm{d}t \;. \tag{3.13}$$

For a form $\omega \in \Lambda^k(TY), k > 1$, we define $h\omega = 0$. It remains to verify the properties of the mapping h. The **R**-linearity is evident from the construction of the form $h\omega$. As for the exterior product the conclusion is quite trivial, because every horizontal k-form for $k > 1$ is zero. □

Definition 3.7 (*Horizontalization*) The mapping h defined by Eq. (3.11) is called the *horizontalization*. The mapping $h\mathrm{d}$ is called the *horizontal derivative operator*.

We express the horizontal derivative operator in charts. Let $f : Y \to \mathbf{R}$ be a function, i.e. in charts $f = f(t, q^\sigma)$. It holds

$$\mathrm{d}f(t, q^\nu) = \frac{\partial f(t, q^\nu)}{\partial t}\,\mathrm{d}t + \frac{\partial f(t, q^\nu)}{\partial q^\sigma}\,\mathrm{d}q^\sigma\ .$$

Using relation (3.13) we obtain

$$h\,\mathrm{d}f = \left(\frac{\partial f(t, q^\nu)}{\partial t} + \frac{\partial f(t, q^\nu)}{\partial q^\sigma}\,\dot{q}^\sigma\right)\,\mathrm{d}t\ = \frac{\mathrm{d}f(t, q^\nu)}{\mathrm{d}t}\mathrm{d}t\ ,$$

where $\mathrm{d}/\mathrm{d}t$ is the well-known operator of the *total derivative*. For an arbitrary k-form ω, $k \geq 1$, it holds $h\,\mathrm{d}\omega = 0$. For a function f on J^1Y we obtain

$$h\,\mathrm{d}f = \left(\frac{\partial f(t, q^\nu, \dot{q}^\nu)}{\partial t} + \frac{\partial f(t, q^\nu, \dot{q}^\nu)}{\partial q^\sigma}\,\dot{q}^\sigma + \frac{\partial f(t, q^\nu, \dot{q}^\nu)}{\partial \dot{q}^\sigma}\,\ddot{q}^\sigma\right)\,\mathrm{d}t = \frac{\mathrm{d}f(t, q^\nu, \dot{q}^\nu)}{\mathrm{d}t}\mathrm{d}t\ .$$

3.3 Contact Ideal

A particular importance for calculus of variations have so-called *contact forms*. The contactness of forms is the property also related to the fibred structure of manifolds.

Definition 3.8 (*Contact forms*) Let $r = 0, 1, 2$ and $k \geq 1$.

(a) A k-form ω on J^rY is called *contact* if for every section γ of the projection π it holds $J^r\gamma^*\omega = 0$.
(b) A contact 1-form ω on J^rY is called *1-contact*. A k-form ω, $k > 1$, on J^rY is called *1-contact* if its contraction by an arbitrary π_r-vertical vector field on J^rY is a horizontal form. A k-form ω, $k > 1$, on J^rY is called *q-contact*, $2 \leq q \leq k$, if the $(k-1)$-form $i_\xi\omega$ is $(q-1)$-contact for every π_r-vertical vector field on J^rY. (In this context the horizontal form is considered as 0-contact.)

Example 3.2 Consider following forms and study their contactness from the point of view of Definition 3.8.

(a) Let $\omega = A_\sigma \, dq^\sigma$ be a 1-form on Y. Let γ be a section of π. Then

$$\gamma^*\omega = (A_\sigma \circ \gamma)\frac{dq^\sigma \gamma(t)}{dt}\, dt\ .$$

This expression vanishes for all γ iff $A_\sigma = 0$ for $1 \le \sigma \le m$. This means that there are no nontrivial contact 1-forms on Y.

(b) Let $\omega = A_\sigma \, dq^\sigma \wedge dt + B_{\sigma\nu}\, dq^\sigma \wedge dq^\nu$, $B_{\sigma\nu} + B_{\nu\sigma} = 0$, be a 2-form on Y, let γ be a section of π and let $\xi = \xi^\sigma \frac{\partial}{\partial q^\sigma}$ be a vertical vector field on Y. Then

$$\gamma^*\omega = (A_\sigma \circ \gamma)\frac{dq^\sigma \gamma}{dt}\, dt \wedge dt + (B_{\sigma\nu} \circ \gamma)\left(\frac{dq^\sigma \gamma}{dt}\, dt\right) \wedge \left(\frac{dq^\nu \gamma}{dt}\, dt\right) = 0\ .$$

The form ω is contact. On the other hand

$$i_\xi \omega = A_\sigma \xi^\sigma \, dt + (B_{\sigma\nu} - B_{\nu\sigma})\xi^\sigma \, dq^\nu\ ,$$

so, the form ω is neither 1-contact nor 2-contact.

(c) The 2-form $\omega = A_\sigma \, dq^\sigma \wedge dt$ on Y is 1-contact, because it is contact ($\gamma^*\omega = 0$) and $i_\xi\omega = A_\sigma \xi^\sigma \, dt$ for every π-vertical vector field on Y.

Proposition 3.5 *(a) All contact forms on J^rY form the ideal. We denote it as $\mathcal{I}_C(J^rY)$.*

(b) Let $r = 0, 1$ and $\eta \in \Lambda^k(TJ^rY)$, $k \ge 1$. Then the form

$$p\eta = \pi^*_{r+1,r}\eta - h\eta \tag{3.14}$$

is contact.

(c) A mapping

$$p : \Lambda^k(TJ^rY) \ni \eta \longrightarrow p\eta \in \Lambda^k(TJ^{r+1}Y) \tag{3.15}$$

is **R***-linear. The decomposition*

$$\pi^*_{r+1,r}\eta = h\eta + p\eta$$

is unique.

The proof of this proposition is very easy (see Problem 3.7).

Definition 3.9 (*Contactization*) The mapping p defined by relations (3.14) and (3.15) is called the *contactization*. The mapping pd is called the *contact derivative operator*.

It is clear that every k-form ω for $k > 1$ is contact, i.e. $h\omega = 0$ and thus $\pi^*_{r+1,r}\omega = p\omega$. More precisely, every k-form for $k > 1$ is *at least* $(k-1)$-contact. Differently from the horizontalization mapping the contactization mapping does not preserve the exterior product (see Problem 3.8). Now let us express the contact 1-forms in charts. Let η be a 1-form on J^2Y in general, i.e.

$$\eta = \eta_0 \,\mathrm{d}t + \eta_\sigma \,\mathrm{d}q^\sigma + \tilde{\eta}_\sigma \,\mathrm{d}\dot{q}^\sigma + \hat{\eta}_\sigma \,\mathrm{d}\ddot{q}^\sigma \ ,$$

$\eta_0, \eta_\sigma, \tilde{\eta}_\sigma, \hat{\eta}_\sigma, 1 \le \sigma \le m$, being functions on J^2Y, and let $\gamma = \big(t\gamma(t), q^\sigma\gamma(t)\big)$ be a section of the projection π. Then

$$\begin{aligned} J^2\gamma^*\eta = \Bigg(&(\eta_0 \circ J^2\gamma) + (\eta_\sigma \circ J^2\gamma)\frac{\mathrm{d}q^\sigma\gamma(t)}{\mathrm{d}t} \\ &+ (\tilde{\eta}_\sigma \circ J^2\gamma)\frac{\mathrm{d}^2q^\sigma\gamma(t)}{\mathrm{d}t^2} + (\hat{\eta}_\sigma \circ J^2\gamma)\frac{\mathrm{d}^3q^\sigma\gamma(t)}{\mathrm{d}t^3}\Bigg)\,\mathrm{d}t \ . \end{aligned}$$

The form η on J^2Y is contact if and only if the just obtained expression of $J^2\gamma^*\eta$ is zero for an arbitrary section γ, i.e.

$$\eta_0 + \eta_\sigma\dot{q}^\sigma + \tilde{\eta}_\sigma\ddot{q}^\sigma + \hat{\eta}_\sigma q^\sigma_{(3)} = 0 \ ,$$

where $q^\sigma_{(3)}$ are the highest coordinates on J^3Y. However, the components of the form η are functions on J^2Y, i.e.

$$\hat{\eta}_\sigma = 0, \quad \eta_0 = -\left(\eta_\sigma\dot{q}^\sigma + \tilde{\eta}_\sigma\ddot{q}^\sigma\right) \ .$$

The chart expression of the form η is

$$\eta = \eta_\sigma(\mathrm{d}q^\sigma - \dot{q}^\sigma\,\mathrm{d}t) + \tilde{\eta}_\sigma(\mathrm{d}\dot{q}^\sigma - \ddot{q}^\sigma\,\mathrm{d}t) \ .$$

So, every contact 1-form on J^2Y is a linear combination of contact forms

$$\omega^\sigma = \mathrm{d}q^\sigma - \dot{q}^\sigma\,\mathrm{d}t, \quad \dot{\omega}^\sigma = \mathrm{d}\dot{q}^\sigma - \ddot{q}^\sigma\,\mathrm{d}t \ . \tag{3.16}$$

We can see that for $r = 1$ every contact 1-form is a linear combination of forms ω^σ, while for $r = 0$ we have no nontrivial contact 1-form.

Example 3.3 Let η and χ be k-forms on J^1Y expressed in the base of *forms adapted to the contact structure* as follows:

$$\eta = A_{\sigma_1\ldots\sigma_{k-1}}\,\omega^{\sigma_1} \wedge \ldots \wedge \omega^{\sigma_{k-1}} \wedge \mathrm{d}t \ , \tag{3.17}$$

$$\chi = B_{\sigma_1\ldots\sigma_k}\,\omega^{\sigma_1} \wedge \ldots \wedge \omega^{\sigma_k} \ , \tag{3.18}$$

where $A_{\sigma_1\ldots\sigma_{k-1}}(t, q^\nu, \dot{q}^\nu)$ and $B_{\sigma_1\ldots\sigma_k}(t, q^\nu, \dot{q}^\nu)$ are some functions on J^1Y, anti-symmetric in all indices. The form η is $(k-1)$-contact, while the form χ is k-contact.

Conversely, every $(k-1)$-contact k-form on J^1Y can be expressed as (3.17) and every k-contact k-form on J^1Y can be expressed as (3.18). The situation is quite analogous for J^2Y, where

$$\eta = \sum_{i=0}^{k-1} A_{\sigma_1\ldots\sigma_i,\sigma_{i+1}\ldots\sigma_{k-1}}\, \omega^{\sigma_1} \wedge \ldots \wedge \omega^{\sigma_i} \wedge \dot{\omega}^{\sigma_{i+1}} \wedge \ldots \wedge \dot{\omega}^{\sigma_{k-1}} \wedge \mathrm{d}t\ ,$$

$$\chi = \sum_{i=0}^{k} B_{\sigma_1\ldots\sigma_i,\sigma_{i+1}\ldots\sigma_k}\, \omega^{\sigma_1} \wedge \ldots \wedge \omega^{\sigma_i} \wedge \dot{\omega}^{\sigma_{i+1}} \wedge \ldots \wedge \dot{\omega}^{\sigma_k}\ ,$$

where $A_{\sigma_1\ldots\sigma_{k-1}}(t, q^\nu, \dot{q}^\nu, \ddot{q}^\nu)$ and $B_{\sigma_1\ldots\sigma_k}(t, q^\nu, \dot{q}^\nu, \ddot{q}^\nu)$ are some functions on J^2Y, antisymmetric separately in first i indices and in remaining $(k-1-i)$ indices, $0 \le i \le k-1$, for functions "A", or $(k-i)$ indices, $0 \le i \le k$, for functions "B".

There is a close relation of jet prolongations of vector fields to the contact ideal:

Definition 3.10 (*Contact symmetries*) A vector field $\tilde{\xi}$ on J^1Y (or a vector field $\hat{\xi}$ on J^2Y) is called the *contact symmetry*, or the *symmetry of the contact ideal* if for every form $\omega \in \mathcal{I}_C$ on J^1Y, or J^2Y, respectively, it holds

$$\partial_{\tilde{\xi}}\omega \in \mathcal{I}_C \quad \text{or} \quad \partial_{\hat{\xi}}\omega \in \mathcal{I}_C\ .$$

Here $\mathcal{I}_C$ denotes the contact ideal on J^1Y, or alternatively on J^2Y, $\partial_{\tilde{\xi}}$ and $\partial_{\hat{\xi}}$ mean the Lie derivative with respect to the vector fields $\tilde{\xi}$ and $\hat{\xi}$, respectively.

Proposition 3.6 *A π_1-projectable vector field $\tilde{\xi}$ on J^1Y (a π_2-projectable vector field $\hat{\xi}$ on J^2Y) is a contact symmetry if and only if there exists a vector field ξ on Y such that $\tilde{\xi} = J^1\xi$ ($\hat{\xi} = J^2\xi$), respectively.*

Proof We show the proposition for $r = 1$. Let $\tilde{\xi}$ be a π_1-projectable vector field on J^1Y, i.e.

$$\tilde{\xi} = \xi^0(t)\frac{\partial}{\partial t} + \xi^\sigma(t, q^\nu, \dot{q}^\nu)\frac{\partial}{\partial q^\sigma} + \tilde{\xi}^\sigma(t, q^\nu, \dot{q}^\nu)\,\frac{\partial}{\partial \dot{q}^\sigma}.$$

Such a vector field is a symmetry of the contact ideal if the Lie derivative of every contact form ω on J^1Y with respect to this vector field is contact. It is sufficient to prove this property for 1-forms $\omega^\sigma = \mathrm{d}q^\sigma - \dot{q}^\sigma\,\mathrm{d}t$, because every contact form on J^1Y can be written as $\omega^\sigma \wedge \eta_\sigma$, where η_σ are some forms on J^1Y. On the other hand, for two arbitrary forms ω and η and every vector field $\tilde{\zeta}$ on J^1Y it holds $\partial_{\tilde{\zeta}}(\omega \wedge \eta) = \partial_{\tilde{\zeta}}\omega \wedge \eta + \omega \wedge \partial_{\tilde{\zeta}}\eta$. Then

$$\partial_{\tilde{\xi}}(\omega^\sigma \wedge \eta_\sigma) = \partial_{\tilde{\xi}}\omega^\sigma \wedge \eta_\sigma + \omega^\sigma \wedge \partial_{\tilde{\xi}}\eta_\sigma\ .$$

We express the Lie derivative of the form ω^σ in the base of 1-forms adapted to the contact structure, $(\mathrm{d}t, \omega^\nu, \mathrm{d}\dot{q}^\nu)$. For the Lie derivative it holds

$$\partial_\xi \omega^\sigma = i_\xi \,\mathrm{d}\omega^\sigma + \mathrm{d}i_\xi \omega^\sigma \,, \tag{3.19}$$

$$\partial_\xi \omega^\sigma = i_\xi(-\mathrm{d}\dot{q}^\sigma \wedge \mathrm{d}t) + \mathrm{d}i_\xi(\mathrm{d}q^\sigma - \dot{q}^\sigma\,\mathrm{d}t) = -\tilde{\xi}^\sigma\,\mathrm{d}t + \xi^0\,\mathrm{d}\dot{q}^\sigma + \mathrm{d}(\xi^\sigma - \dot{q}^\sigma\xi^0)$$
$$= \left(-\tilde{\xi}^\sigma + \frac{\mathrm{d}'\xi^\sigma}{\mathrm{d}t} - \dot{q}^\sigma\frac{\mathrm{d}\xi^0}{\mathrm{d}t}\right)\mathrm{d}t + \frac{\partial\xi^\sigma}{\partial\dot{q}^\nu}\,\mathrm{d}\dot{q}^\nu + \frac{\partial\xi^\sigma}{\partial q^\nu}\omega^\nu \,.$$

The form $\partial_\xi \omega^\sigma$ is contact if and only if it holds

$$\frac{\partial\xi^\sigma}{\partial\dot{q}^\nu} = 0, \quad \text{and} \quad -\tilde{\xi}^\sigma + \frac{\mathrm{d}'\xi^\sigma}{\mathrm{d}t} - \dot{q}^\sigma\frac{\mathrm{d}\xi^0}{\mathrm{d}t} = 0 \,.$$

The first condition means that the components ξ^σ are independent of velocities. The second condition then gives

$$\tilde{\xi}^\sigma = \frac{\mathrm{d}\xi^\sigma}{\mathrm{d}t} - \dot{q}^\sigma\frac{\mathrm{d}\xi^0}{\mathrm{d}t} \,.$$

So, the unique symmetries of the contact ideal on J^1Y are vector fields of the type $J^1\xi$, where ξ is a π-projectable vector field on Y. □

Now, we formulate one of the key theorems. It concerns the decomposition of an arbitrary differential form to its components of various degrees of contactness.

Theorem 3.1 *Let $s = 0, 1$ and $r = 1, 2$. The following holds:*

(a) For every k-form η on J^sY, there exists a unique decomposition

$$\pi^*_{s+1,s}\eta = p_{k-1}\eta + p_k\eta \,, \tag{3.20}$$

where $p_{k-1}\eta$ is the $(k-1)$-contact form and $p_k\eta$ is the k-contact form. These forms are called the $(k-1)$-contact and the k-contact component of the form η, respectively. (A k-form with zero $(k-1)$-contact component is called strongly contact.)

(b) Let $r = 1, 2$. The module $\Lambda^k_{J^{r-1}Y}(TJ^rY)$ of $\pi_{r,r-1}$-horizontal forms is the direct sum of submodules $\Lambda^{k,k-1}_{J^{r-1}Y}(TJ^rY)$ and $\Lambda^{k,k}_{J^{r-1}Y}(TJ^rY)$ of $(k-1)$-contact and k-contact forms, respectively, i.e.

$$\Lambda^k_{J^{r-1}Y}(TJ^rY) = \Lambda^{k,k-1}_{J^{r-1}Y}(TJ^rY) \oplus \Lambda^{k,k}_{J^{r-1}Y}(TJ^rY) \,.$$

(c) Let η be a $(k-1)$-contact k-form on J^rY, or, alternatively, k-contact k-form on J^rY, $r = 1, 2$. Then it is $\pi_{r,r-1}$-horizontal.

Proof The part (a) of the theorem for 1-forms is the consequence of Proposition 3.5 proved in Problem 3.7. In arbitrary fibred charts every k-form is a linear combination of products of the type $\varrho = \eta_1 \wedge \ldots \wedge \eta_k$, where $\eta_1, \ldots, \eta_k$ are 1-forms. Then for the form $\varrho = \eta_1 \wedge \ldots \wedge \eta_k$ we have

$$\begin{aligned}\pi^*_{s+1,s}\varrho &= (h\eta_1 + p\eta_1) \wedge \ldots \wedge (h\eta_k + p\eta_k)\\ &= \sum_{j=1}^{k} (-1)^{j-1} h\eta_j \wedge p\eta_1 \wedge \ldots \wedge p\eta_{j-1} \wedge p\eta_{j+1} \wedge \ldots \wedge p\eta_k\\ &\qquad + p\eta_1 \wedge \ldots \wedge p\eta_k \,.\end{aligned}$$

Denote

$$\begin{aligned}p_{k-1}\varrho &= \sum_{j=1}^{k} (-1)^{j-1} h\eta_j \wedge p\eta_1 \wedge \ldots \wedge p\eta_{j-1} \wedge p\eta_{j+1} \wedge \ldots \wedge p\eta_k \,,\\ p_k\varrho &= p\eta_1 \wedge \ldots \wedge p\eta_k \,.\end{aligned}$$

It is evident that the form $p_{k-1}\varrho$ is $(k-1)$-contact while the form $p_k\varrho$ is k-contact. Thus the existence of the decomposition of a form $\pi^*_{s+1,s}\eta$ into a sum of $(k-1)$-contact and k-contact forms is proved. Now let suppose that there exist two various decompositions of $\pi^*_{s+1,s}$,

$$\pi^*_{s+1,s} = p_{k-1}\eta + p_k\eta = \bar{p}_{k-1}\eta + \bar{p}_k\eta \;\Rightarrow\; p_{k-1}\eta - \bar{p}_{k-1}\eta = \bar{p}_k\eta - p_k\eta \,.$$

However, the unique form which is simultaneously $(k-1)$-contact and k-contact is the trivial (zero) form.

The proof of the second and third parts of the theorem is now simple. A $\pi_{r,r-1}$-horizontal k-form η on $J^rY, r = 1, 2$, is a linear combination of k-forms ϱ projectable on $J^{r-1}Y$, i.e. for every such form ϱ there exists a k-form χ on $J^{r-1}Y$ such that $\pi^*_{r,r-1}\chi = \varrho$ for $j = 0, \ldots, k$. (Coefficients of such a linear combinations are of course function on J^rY.) Then it holds

$$\varrho = \pi^*_{r,r-1}\chi = p_{k-1}\chi + p_k\chi = \varrho_{k-1} + \varrho_k \,,$$

where $\varrho_{k-1} = p_{k-1}\chi \in \Lambda^{k,k-1}_{J^{r-1}Y}(J^rY)$ and $\varrho_k = p_k\chi \in \Lambda^{k,k}_{TJ^{r-1}Y}(J^rY)$. Conversely, let η be a $(k-1)$-contact k form on J^rY. Then $p_k\eta = 0$ and $p_{k-1}\eta$ is $\pi_{r,r-1}$-horizontal. The similar consideration is relevant for k-contact k forms. This finishes the proof. □

Example 3.4 Let f be a function on J^1Y, i.e. in coordinates $f = f(t, q^\sigma, \dot{q}^\sigma)$. The exterior derivative of this function is the 1-form on J^1Y. The decomposition of $\pi^*_{2,1}\,\mathrm{d}f$ into its 0-contact (horizontal) and 1-contact components reads

$$\pi^*_{2,1}\,\mathrm{d}f = \pi^*_{2,1}\left(\frac{\partial f}{\partial t}\,\mathrm{d}t + \frac{\partial f}{\partial q^\sigma}\,\mathrm{d}q^\sigma + \frac{\partial f}{\partial \dot{q}^\sigma}\,\mathrm{d}\dot{q}^\sigma\right) .$$

Putting $\mathrm{d}q^\sigma = \omega^\sigma + \dot{q}^\sigma \, \mathrm{d}t$ and $\mathrm{d}\dot{q}^\sigma = \dot{\omega}^\sigma + \ddot{q}^\sigma \, \mathrm{d}t$ we can express this form in the base of 1-forms adapted to the contact structure:

$$\pi^*_{2,1} \, \mathrm{d}f = \frac{\mathrm{d}f}{\mathrm{d}t} \, \mathrm{d}t + \frac{\partial f}{\partial q^\sigma} \omega^\sigma + \frac{\partial f}{\partial \dot{q}^\sigma} \, \dot{\omega}^\sigma \,, \tag{3.21}$$

$$h \, \mathrm{d}f = \frac{\mathrm{d}f}{\mathrm{d}t} \, \mathrm{d}t, \quad p \, \mathrm{d}f = \frac{\partial f}{\partial q^\sigma} \omega^\sigma + \frac{\partial f}{\partial \dot{q}^\sigma} \, \dot{\omega}^\sigma \,.$$

The situation is quite analogous for a function f on J^2Y we have

$$\pi^*_{3,2} \, \mathrm{d}f = \pi^*_{3,2} \left(\frac{\partial f}{\partial t} \, \mathrm{d}t + \frac{\partial f}{\partial q^\sigma} \, \mathrm{d}q^\sigma + \frac{\partial f}{\partial \dot{q}^\sigma} \, \mathrm{d}\dot{q}^\sigma + \frac{\partial f}{\partial \ddot{q}^\sigma} \, \mathrm{d}\ddot{q}^\sigma \right) .$$

Adding to (3.16) the relation for contact 1-forms $\ddot{\omega}^\sigma = \mathrm{d}\ddot{q}^\sigma - q^\sigma_{(3)} \, \mathrm{d}t$, where $q^\sigma_{(3)}$ are highest coordinates on J^3Y (i.e. the coordinates $q^\sigma_{(3)} J^3_x \gamma$ of 3-jets of sections $\gamma : X \ni x \to \gamma(x) \in Y$), we obtain

$$\pi^*_{3,2} \, \mathrm{d}f = \frac{\mathrm{d}f}{\mathrm{d}t} \, \mathrm{d}t + \frac{\partial f}{\partial q^\sigma} \omega^\sigma + \frac{\partial f}{\partial \dot{q}^\sigma} \dot{\omega}^\sigma + \frac{\partial f}{\partial \ddot{q}^\sigma} \ddot{\omega}^\sigma \,,$$

$$h \, \mathrm{d}f = \frac{\mathrm{d}f}{\mathrm{d}t} \, \mathrm{d}t = \left(\frac{\partial f}{\partial t} + \frac{\partial f}{\partial q^\sigma} \dot{q}^\sigma + \frac{\partial f}{\partial \dot{q}^\sigma} \ddot{q}^\sigma + \frac{\partial f}{\partial \ddot{q}^\sigma} q^\sigma_{(3)} \right) \mathrm{d}t \,, \tag{3.22}$$

$$p \, \mathrm{d}f = \frac{\partial f}{\partial q^\sigma} \omega^\sigma + \frac{\partial f}{\partial \dot{q}^\sigma} \dot{\omega}^\sigma + \frac{\partial f}{\partial \ddot{q}^\sigma} \ddot{\omega}^\sigma \,. \tag{3.23}$$

The second key theorem concerning the calculus of differential forms is the *Poincaré* lemma adapted to the fibred structure. From the classical analysis we know that every *closed* k form η (i.e. such that $\mathrm{d}\eta = 0$) defined on a open convex subset of $\mathbf{R}^n$ is *exact*, i.e. there exists a $(k-1)$-form ϱ such that $\eta = \mathrm{d}\varrho$. Together with the standard version of the Poincaré lemma we can formulate also the version "adapted" to the fibred structure of underlying manifolds.

Theorem 3.2 *Poincaré lemma. Let $Y = I \times V$, where I is an open interval and $V \subset \mathbf{R}^m$ is an open ball centred in the point $O = (0, \dots, 0)$ (i.e. a convex set). Consider the fibred manifold $p : Y \to I$, where p is the projection on the first factor. Let $k > 1$.*

(a) Let η be a closed k-form on J^rY, $r = 0, 1, 2$. Then there exists a $(k-1)$-form ϱ on J^rY such that $\eta = \mathrm{d}\varrho$.
(b) Let η be a closed $(k-1)$-contact k-form on J^rY, $r = 0, 1, 2$. Then there exists a $(k-2)$-contact $(k-1)$-form ϱ on J^rY such that $\eta = \mathrm{d}\varrho$.
(c) Let η be a closed k-contact k-form on J^rY, $r = 1, 2$. Then there exists a $(k-1)$-contact $(k-1)$-form ϱ on J^rY such that $\eta = \mathrm{d}\varrho$.

Proof The assertion (a) is the standard Poincaré lemma. The assertion (c) is formulated only for $r = 1, 2$, because there are no nontrivial k-contact k-forms on Y. The proof is based on the construction of a generally defined operator $\mathcal{A} : \eta \to \mathcal{A}\eta$ assigning to a k-form η the $(k-1)$ form $A\eta$ such that (see below)

$$\eta = \mathrm{d}(\mathcal{A}\eta) + \mathcal{A}(\mathrm{d}\eta) \ . \tag{3.24}$$

Then if η is closed, i.e. $\mathrm{d}\eta = 0$, we obtain $\eta = \mathrm{d}(\mathcal{A}\eta)$. Define the mappings (for $r = 1$ and $r = 2$ denoted by the same symbol, for simplicity)

$$\begin{aligned} \chi :\ & [0,1] \times J^1Y \ni (u, t, q^\sigma, \dot{q}^\sigma) \longrightarrow \chi(u, t, q^\sigma, \dot{q}^\sigma) \\ & = (t, uq^\sigma, u\dot{q}^\sigma) \in J^1Y \ , \\ \chi :\ & [0,1] \times J^2Y \ni (u, t, q^\sigma, \dot{q}^\sigma, \ddot{q}^\sigma) \longrightarrow \chi(u, t, q^\sigma, \dot{q}^\sigma, \ddot{q}^\sigma) \\ & = (t, uq^\sigma, u\dot{q}^\sigma, u\ddot{q}^\sigma) \in J^2Y \ . \end{aligned} \tag{3.25}$$

It holds

$$\chi^*\omega^\sigma = q^\sigma \,\mathrm{d}u + u\omega^\sigma, \quad \chi^*\dot{\omega}^\sigma = \dot{q}^\sigma \,\mathrm{d}u + u\dot{\omega}^\sigma \ . \tag{3.26}$$

Consider for simplicity a form η on J^1Y which is decomposable into its $(k-1)$-contact and k-contact components on J^1Y, i.e. forms $p_{k-1}\eta$ and $p_k\eta$ defined in principle on J^2Y are projectable onto J^1Y. The chart expression of this decomposition is

$$\eta = A_{\sigma_1\dots\sigma_{k-1}}\omega^{\sigma_1} \wedge \dots \wedge \omega^{\sigma_{k-1}} \wedge \mathrm{d}t + B_{\sigma_1\dots\sigma_k}\omega^{\sigma_1} \wedge \dots \wedge \omega^{\sigma_k} \ ,$$

where $A_{\sigma_1\dots\sigma_{k-1}}$ and $B_{\sigma_1\dots\sigma_k}$ are functions on J^1Y completely antisymmetric in all indices. Now we calculate $\chi^*\eta$:

$$\begin{aligned} \chi^*\eta = {} & (A_{\sigma_1\dots\sigma_{k-1}} \circ \chi)(q^{\sigma_1}\,\mathrm{d}u + u\omega^{\sigma_1}) \wedge \dots \wedge (q^{\sigma_{k-1}}\,\mathrm{d}u + u\omega^{\sigma_{k-1}}) \wedge \mathrm{d}t \\ & + (B_{\sigma_1\dots\sigma_k} \circ \chi)(q^{\sigma_1}\,\mathrm{d}u + u\omega^{\sigma_1}) \wedge \dots \wedge (q^{\sigma_k}\,\mathrm{d}u + u\omega^{\sigma_k}) \ . \end{aligned}$$

Executing the exterior products and taking into account the antisymmetry of function coefficients in all indices we finally decompose the form $\chi^*\eta$ into two parts

$$\chi^*\eta = \mathrm{d}u \wedge \mu + \bar{\eta} \ ,$$

where

$$\begin{aligned} \mu = {} & (k-1)u^{k-2}\left[q^{\sigma_1}(A_{\sigma_1\dots\sigma_{k-1}} \circ \chi)\,\omega^{\sigma_2} \wedge \dots \wedge \omega^{\sigma_{k-1}} \wedge \mathrm{d}t\right] \\ & + ku^{k-1}\left[q^{\sigma_1}(B_{\sigma_1\dots\sigma_k} \circ \chi)\,\omega^{\sigma_2} \wedge \dots \wedge \omega^{\sigma_k}\right] \ , \\ \bar{\eta} = {} & u^{k-1}(A_{\sigma_1\dots\sigma_{k-1}} \circ \chi)\,\omega^{\sigma_1} \wedge \dots \wedge \omega^{\sigma_{k-1}} \wedge \mathrm{d}t \\ & + u^k(B_{\sigma_1\dots\sigma_k} \circ \chi)\,\omega^{\sigma_1} \wedge \dots \wedge \omega^{\sigma_k} \ . \end{aligned}$$

We anticipate that the form $\bar{\eta}$ is not meaningful for our construction, we present it only for completeness. Define the operator $\mathcal{A}$ as follows:

$$\mathcal{A}:\ \Lambda^k(TJ^1Y) \ni \eta \longrightarrow \mathcal{A}\eta = \int_0^1 \mu \,\mathrm{d}u \in \Lambda^{k-1}(TJ^1Y)\,,$$

$$\mathcal{A}\eta = \left[(k-1)q^{\sigma_1}\int_0^1 u^{k-2}(A_{\sigma_1\ldots\sigma_{k-1}}\circ\chi)\,\mathrm{d}u\right]\omega^{\sigma_2}\wedge\ldots\wedge\omega^{\sigma_{k-1}}\wedge\mathrm{d}t$$
$$+\left[kq^{\sigma_1}\int_0^1 u^{k-1}(B_{\sigma_1\ldots\sigma_k}\circ\chi)\,\mathrm{d}u\right]\omega^{\sigma_2}\wedge\ldots\wedge\omega^{\sigma_k}\,. \tag{3.27}$$

Now we have to prove relation (3.24). The proof is based on the direct calculation and it is rather technical. We present here only important steps.

First we calculate the form $\mathrm{d}\mathcal{A}\eta$. We use the antisymmetry of the exterior product of forms as well as the antisymmetry of coefficients, i.e.

$$A_{\sigma_1\ldots\sigma_i\ldots\sigma_j\ldots\sigma_{k-1}} = -A_{\sigma_1\ldots\sigma_j\ldots\sigma_i\ldots\sigma_{k-1}}\,,\quad \text{for}\ \ 1\le i,j\le k-1\,,$$
$$B_{\sigma_1\ldots\sigma_i\ldots\sigma_j\ldots\sigma_k} = -B_{\sigma_1\ldots\sigma_j\ldots\sigma_i\ldots\sigma_k}\,,\quad \text{for}\ \ 1\le i,j\le k\,,$$

and the facts that $\mathrm{d}q^\mu\wedge\mathrm{d}t = \omega^\mu\wedge\mathrm{d}t$, $\mathrm{d}\omega^\mu\wedge\mathrm{d}t = 0$.

$$\mathrm{d}\mathcal{A}\eta = (k-1)\left(\int_0^1 u^{k-2}(A_{\sigma_1\ldots\sigma_{k-1}}\circ\chi)\,\mathrm{d}u\right)\omega^{\sigma_1}\wedge\ldots\wedge\omega^{\sigma_{k-1}}\wedge\mathrm{d}t$$
$$+k\left(\int_0^1 u^{k-1}(B_{\sigma_1\ldots\sigma_k}\circ\chi)\,\mathrm{d}u\right)\omega^{\sigma_1}\wedge\ldots\wedge\omega^{\sigma_k}$$
$$+(-1)^{k-1}k\dot{q}^{\sigma_1}\left(\int_0^1 u^{k-1}(B_{\sigma_1\ldots\sigma_k}\circ\chi)\,\mathrm{d}u\right)\omega^{\sigma_2}\wedge\ldots\wedge\omega^{\sigma_k}\wedge\mathrm{d}t$$
$$+(k-1)q^{\sigma_1}\left[\int_0^1 u^{k-2}\left(\frac{\partial(A_{\sigma_1\ldots\sigma_{k-1}}\circ\chi)}{\partial q^\lambda}\mathrm{d}q^\lambda\right.\right.$$
$$\left.\left.+\frac{\partial(A_{\sigma_1\ldots\sigma_{k-1}}\circ\chi)}{\partial\dot{q}^\lambda}\mathrm{d}\dot{q}^\lambda\right)\mathrm{d}u\right]\omega^{\sigma_2}\wedge\ldots\wedge\omega^{\sigma_{k-1}}\wedge\mathrm{d}t$$

$$
\begin{aligned}
&+ kq^{\sigma_1}\left[\int_0^1 u^{k-1}\left(\frac{\partial(B_{\sigma_1\ldots\sigma_k}\circ\chi)}{\partial t}\,\mathrm{d}t + \frac{\partial(B_{\sigma_1\ldots\sigma_k}\circ\chi)}{\partial q^\lambda}\,\mathrm{d}q^\lambda\right.\right.\\
&\qquad\left.\left. + \frac{\partial(B_{\sigma_1\ldots\sigma_k}\circ\chi)}{\partial \dot{q}^\lambda}\,\mathrm{d}\dot{q}^\lambda\right)\mathrm{d}u\right]\wedge\omega^{\sigma_2}\wedge\ldots\wedge\omega^{\sigma_k}\\
&+ (-1)^k k(k-1)q^{\sigma_1}\left(\int_0^1 u^{k-1}(B_{\sigma_1\ldots\sigma_k}\circ\chi)\,\mathrm{d}u\right)\mathrm{d}\dot{q}^{\sigma_2}\wedge\omega^{\sigma_3}\wedge\ldots\wedge\omega^{\sigma_k}\wedge\mathrm{d}t\,.
\end{aligned}
$$

On the other hand, for $\mathcal{A}\,\mathrm{d}\eta$ we obtain

$$
\begin{aligned}
\mathcal{A}\,\mathrm{d}\eta =\, & q^\lambda\left[\int_0^1 u^{k-1}\left(\frac{\partial A_{\sigma_1\ldots\sigma_{k-1}}}{\partial q^\lambda}\circ\chi\right)\mathrm{d}u\right]\omega^{\sigma_1}\wedge\ldots\wedge\omega^{\sigma_{k-1}}\wedge\mathrm{d}t\\
&- (k-1)q^{\sigma_1}\left[\int_0^1 u^{k-1}\left(\frac{\partial A_{\sigma_1\ldots\sigma_{k-1}}}{\partial q^\lambda}\circ\chi\right)\mathrm{d}u\right]\mathrm{d}q^\lambda\wedge\omega^{\sigma_2}\wedge\ldots\wedge\omega^{\sigma_{k-1}}\wedge\mathrm{d}t\\
&+ \dot{q}^\lambda\left[\int_0^1 u^{k-1}\left(\frac{\partial A_{\sigma_1\ldots\sigma_{k-1}}}{\partial \dot{q}^\lambda}\circ\chi\right)\mathrm{d}u\right]\omega^{\sigma_1}\wedge\ldots\wedge\omega^{\sigma_{k-1}}\wedge\mathrm{d}t\\
&- (k-1)q^{\sigma_1}\left[\int_0^1 u^{k-1}\left(\frac{\partial A_{\sigma_1\ldots\sigma_{k-1}}}{\partial \dot{q}^\lambda}\circ\chi\right)\mathrm{d}u\right]\mathrm{d}\dot{q}^\lambda\wedge\omega^{\sigma_2}\wedge\ldots\wedge\omega^{\sigma_{k-1}}\wedge\mathrm{d}t\\
&+ (-1)^k kq^{\sigma_1}\left[\int_0^1 u^{k}\left(\frac{\partial B_{\sigma_1\ldots\sigma_k}}{\partial t}\circ\chi\right)\mathrm{d}u\right]\wedge\omega^{\sigma_2}\wedge\ldots\wedge\omega^{\sigma_k}\wedge\mathrm{d}t\\
&+ q^\lambda\left[\int_0^1 u^{k}\left(\frac{\partial B_{\sigma_1\ldots\sigma_k}}{\partial q^\lambda}\circ\chi\right)\mathrm{d}u\right]\omega^{\sigma_1}\wedge\ldots\wedge\omega^{\sigma_k}\\
&- kq^{\sigma_1}\left[\int_0^1 u^{k}\left(\frac{\partial B_{\sigma_1\ldots\sigma_k}}{\partial q^\lambda}\circ\chi\right)\mathrm{d}u\right]\mathrm{d}q^\lambda\wedge\omega^{\sigma_2}\wedge\ldots\wedge\omega^{\sigma_k}\\
&+ \dot{q}^\lambda\left[\int_0^1 u^{k}\left(\frac{\partial B_{\sigma_1\ldots\sigma_k}}{\partial \dot{q}^\lambda}\circ\chi\right)\mathrm{d}u\right]\omega^{\sigma_1}\wedge\ldots\wedge\omega^{\sigma_k}\\
&- kq^{\sigma_1}\left[\int_0^1 u^{k}\left(\frac{\partial B_{\sigma_1\ldots\sigma_k}}{\partial \dot{q}^\lambda}\circ\chi\right)\mathrm{d}u\right]\mathrm{d}\dot{q}^\lambda\wedge\omega^{\sigma_2}\wedge\ldots\wedge\omega^{\sigma_k}
\end{aligned}
$$

$$+ k(k-1)(-1)^{k-1} q^{\sigma_1} \left[\int_0^1 u^{k-1} (B_{\sigma_1 \dots \sigma_k} \circ \chi) \, \mathrm{d}u \right] \mathrm{d}\dot{q}^2 \wedge \omega^{\sigma_3} \wedge \dots \omega^{\sigma_k} \wedge \mathrm{d}t$$

$$+ k(-1)^k \dot{q}^{\sigma_1} \left[\int_0^1 u^{k-1} (B_{\sigma_1 \dots \sigma_k} \circ \chi) \, \mathrm{d}u \right] \omega^{\sigma_2} \wedge \dots \wedge \omega^{\sigma_k} \wedge \mathrm{d}t \, .$$

Calculating the sum $\mathrm{d}\mathcal{A}\eta + \mathcal{A}\,\mathrm{d}\eta$ we take into account the following: For a function $f(t, q^\lambda, \dot{q}^\lambda)$ and the mapping

$$\chi : (u, t, q^\nu, \dot{q}^\nu) \longrightarrow (t, uq^\nu, u\dot{q}^\nu)$$

it holds for all $1 \le \nu, \lambda \le m$

$$\begin{aligned}
\frac{\partial (f \circ \chi)}{\partial t} &= \frac{\partial f}{\partial t} \circ \chi \, , \\
\frac{\partial (f \circ \chi)}{\partial q^\lambda} &= \frac{f(t, uq^\nu, u\dot{q}^\nu)}{\partial q^\lambda} = \frac{f(t, uq^\nu, u\dot{q}^\nu)}{\partial u q^\lambda} \frac{\partial u q^\lambda}{\partial q^\lambda} = u \left(\frac{\partial f}{\partial q^\lambda} \circ \chi \right) , \\
\frac{\partial (f \circ \chi)}{\partial \dot{q}^\lambda} &= u \left(\frac{\partial f}{\partial \dot{q}^\lambda} \circ \chi \right) .
\end{aligned}$$

We obtain

$$\begin{aligned}
\mathrm{d}\mathcal{A}\eta + \mathcal{A}\,\mathrm{d}\eta = & \left\{ \int_0^1 \left[(k-1) u^{k-2} (A_{\sigma_1 \dots \sigma_{k-1}} \circ \chi) + q^\lambda u^{k-1} \left(\frac{\partial A_{\sigma_1 \dots \sigma_{k-1}}}{\partial q^\lambda} \circ \chi \right) \right. \right. \\
& \left. \left. + \, \dot{q}^\lambda u^{k-1} \left(\frac{\partial A_{\sigma_1 \dots \sigma_{k-1}}}{\partial \dot{q}^\lambda} \circ \chi \right) \right] \mathrm{d}u \right\} \omega^{\sigma_1} \wedge \dots \wedge \omega^{\sigma_{k-1}} \wedge \mathrm{d}t \\
& + \left\{ \int_0^1 k u^{k-1} (B_{\sigma_1 \dots \sigma_k} \circ \chi) + u^k \left[q^\lambda \left(\frac{\partial B_{\sigma_1 \dots \sigma_k}}{\partial q^\lambda} \circ \chi \right) \right. \right. \\
& \left. \left. + \, \dot{q}^\lambda \left(\frac{\partial B_{\sigma_1 \dots \sigma_k}}{\partial \dot{q}^\lambda} \circ \chi \right) \right] \mathrm{d}u \right\} \omega^{\sigma_1} \wedge \dots \wedge \omega^{\sigma_k} \, .
\end{aligned}$$

Now consider a function $f(t, q^\lambda, \dot{q}^\lambda)$ and calculate per parts the following integral:

$$\begin{aligned}
\int_0^1 s u^{s-1} (f \circ \chi) \, \mathrm{d}u &= \left[u^s (f \circ \chi) \right]_0^1 - \int_0^1 u^s \frac{\mathrm{d}(f \circ \chi)}{\mathrm{d}u} \, \mathrm{d}u \\
&= f - \int_0^1 u^s \left(q^\lambda \frac{\partial f}{\partial q^\lambda} \circ \chi + \dot{q}^\lambda \frac{\partial f}{\partial \dot{q}^\lambda} \circ \chi \right) \mathrm{d}u \, .
\end{aligned}$$

Applying this result to $s = k - 1$ and $f = A_{\sigma_1 \dots \sigma_{k-1}}(t, q^\lambda, \dot{q}^\lambda)$ and to $s = k$ and $f = B_{\sigma_1 \dots \sigma_k}(t, q^\lambda, \dot{q}^\lambda)$ we finally get relation (3.24). So, we proved that for a closed k-form η, $k > 1$, it holds $\eta = \mathrm{d}\mathcal{A}\eta$. Then there exist a form ϱ on $J^1 Y$ such that $\eta = \mathrm{d}\varrho$, $\varrho = \mathcal{A}\eta$.

Now suppose that η is a $(k-1)$-contact k-form on $J^1 Y$. (This form is $\pi_{1,0}$-horizontal.) Its chart expression is as (3.17) and $B_{\sigma_1 \dots \sigma_k} = 0$ for all $1 \leq \sigma_1, \dots, \sigma_k \leq m$. Then we obtain from (3.27)

$$\mathcal{A}\eta = \left[(k-1) q^{\sigma_1} \int_0^1 u^{k-2} (A_{\sigma_1 \dots \sigma_{k-1}} \circ \chi) \, \mathrm{d}u \right] \omega^{\sigma_2} \wedge \dots \wedge \omega^{\sigma_{k-1}} \wedge \mathrm{d}t \; .$$

This form is $(k-2)$-contact. Analogously, if η is a k-contact k-form on $J^1 Y$, it has the chart expression (3.18). Putting $A_{\sigma_1 \dots \sigma_{k-1}} = 0$ into (3.27) we get

$$\mathcal{A}\eta = \left[k q^{\sigma_1} \int_0^1 u^{k-1} (B_{\sigma_1 \dots \sigma_k} \circ \chi) \, \mathrm{d}u \right] \omega^{\sigma_2} \wedge \dots \wedge \omega^{\sigma_k} \; .$$

This form is $(k-1)$-contact. The proof for closed forms on $J^2 Y$ is quite analogous, even though technically more complicated. □

The operator $\mathcal{A}$ can be, of course, constructed in a way described in the proof of Theorem 3.2 for an arbitrary k-form η defined on $J^r Y$ (exactly speaking for $\pi^*_{r+1,r}\eta$), not only for "purely" $(k-1)$-contact or "purely" k-contact k-forms, or for k-forms decomposable into their $(k-1)$-contact and k-contact components directly on their definition space $J^r Y$.

3.4 Distributions

Distributions play an important role in solving variational problems. Usually, solutions (stationary points) of a variational problem, so-called *critical sections* or *extremals* of the functional representing such a problem, are integral sections of appropriate distributions. Let us define a distribution on a fibred manifold.

Definition 3.11 (*Distributions*) Let (Y, π, X) be a fibred manifold and $(J^1 Y, \pi_1, X)$ and $(J^2 Y, \pi_2, X)$ its first and second prolongation, respectively. Let $W_r \subset J^r Y$ be an open set. A *(local) distribution* on $J^r Y$, $r = 0, 1, 2$, is a mapping

$$\mathcal{D} : W_r \ni z \longrightarrow \mathcal{D}(z) = \mathcal{D}_z J^r Y \subset T_z J^r Y \; ,$$

assigning to a point $z \in W_r$ a vector subspace $\mathcal{D}_z J^r Y$ of the tangent space $T_z J^r Y$. (For $r = 0, 1, 2$ we denoted $z = y$, $z = \tilde{y}$, $z = \hat{y}$, respectively.) Analogously, let $U \subset X$ be an open set on X. A local distribution on X is the mapping

$$\mathcal{D} : U \ni x \longrightarrow \mathcal{D}(x) = \mathcal{D}_x X \subset T_x X$$

assigning to a point $x \in U$ a vector subspace $\mathcal{D}_x X$ of the tangent space $T_x X$. The *rank* of the distribution $\mathcal{D}$ is the mapping assigning to every point of the domain of the distribution the dimension of the corresponding vector subspace. If the rank of the distribution is constant mapping, we speak about the *constant rank distribution*. If $W_r = J^r Y$, or $U = X$, the corresponding distribution is *global*.

Example 3.5 The mapping assigning to every point $z \in J^r Y$ the corresponding tangent space $T_z J^r Y$ is the distribution of the constant rank $(r+1)m+1$. It is called the *tangent distribution*.

A distribution $\mathcal{D}$ on $J^r Y$ can be generated by a system of vector fields (ξ_ι), $\iota \in I$, where I is an index set, in such a way that at every point of $J^r Y$ the vector subspace $\mathcal{D}_z J^r Y \subset T_z J^r Y$ is given by all linear combinations of vectors $\xi_\iota(z)$, $\iota \in I$, i.e.

$$\mathcal{D}(z) = \operatorname{span}\{\xi_\iota \mid \xi_\iota \in \mathcal{D}_z J^r Y, \iota \in I\} .$$

Example 3.6 Consider the fibred manifold $(\mathbf{R} \times \mathbf{R}^3, p, \mathbf{R})$ and the π-vertical vector field on Y

$$\xi = \frac{GM}{\left((q^1)^2 + (q^2)^2 + (q^3)^2\right)^{3/2}} \left(q^1 \frac{\partial}{\partial q^1} + q^2 \frac{\partial}{\partial q^2} + q^3 \frac{\partial}{\partial q^3} \right) ,$$

where G and M are constants (gravitational constant and mass, respectively). This vector field represents the intensity of the gravitational field caused by a point mass M. It generates the central distribution with zero horizontal component. Its rank is constant and equal to 1. This distribution is not defined globally because the vector field ξ is not defined for $q^1 = q^2 = q^3 = 0$.

Example 3.7 Let (Y, π, X) be a fibred manifold of the dimension $m+1$. Let P be its submanifold defined by equations

$$f^i(t, q^\sigma) = 0, \quad 1 \le i \le k, \quad 1 \le k \le m-1, \quad \operatorname{rank}\left(\frac{\partial f^i}{\partial q^\sigma}\right) = k .$$

Then all vector fields ξ on Y obeying the condition $\mathrm{d}f^i(\xi) = 0$ generate the tangent distribution of the submanifold P. Suppose that $y \in P$ and $\xi(y) \in T_y P$. Then there exists a curve

$$C : (-\varepsilon, \varepsilon) \ni u \longrightarrow C(u) = \left(tC(u), q^1C(u), \ldots, q^mC(u)\right)$$

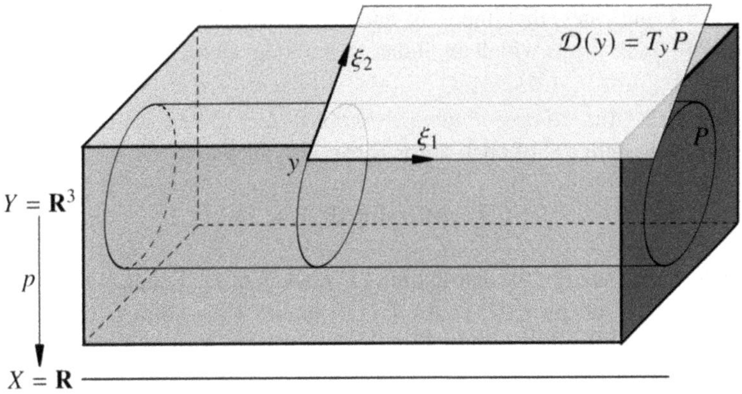

Fig. 3.7 Tangent distribution of a submanifold

lying in P and going through the point y (i.e. $C(0) = y$) such that $\xi(y)$ is its tangent vector, i.e.

$$\xi(y) = \left.\frac{\mathrm{d}tC(u)}{\mathrm{d}u}\right|_{u=0} \frac{\partial}{\partial t} + \left.\frac{\mathrm{d}q^\sigma C(u)}{\mathrm{d}u}\right|_{u=0} \frac{\partial}{\partial q^\sigma} .$$

It holds

$$\mathrm{d}f^i(\xi) = \left.\frac{\partial f^i}{\partial t}\frac{\mathrm{d}tC(u)}{\mathrm{d}u}\right|_{u=0} + \left.\frac{\partial f^i}{\partial q^\sigma}\frac{\mathrm{d}q^\sigma C(u)}{\mathrm{d}u}\right|_{u=0} = \left.\frac{\mathrm{d}(f^i \circ C)(u)}{\mathrm{d}u}\right|_{u=0} = 0 ,$$

because $C(u) \in P$ and then $(f^i \circ C)(u) = 0$. So, every vector tangent to P at the point y satisfies the condition $\mathrm{d}f^i\big(\xi(y)\big) = 0$ and vice versa, every vector satisfying this condition is tangent to P. The mapping $\mathcal{D} : Y \ni y \to \mathcal{D}(y) \subset T_yY$, where $\mathcal{D}(y)$ is a vector subspace of T_yY, is the distribution on Y. Its restriction to P is the tangent distribution of the submanifold P. Figure 3.7 shows a simple example: The fibred manifold is $(\mathbf{R} \times \mathbf{R}^2, p, \mathbf{R})$, i.e. $m = 2$. The submanifold P is defined by the equation $(k = 1)$ $f^1(t, q^1, q^2) = (q^1)^2 + (q^2)^2 - 1 = 0$, i.e. it is independent of the variable t measured along the base. It holds

$$\mathrm{d}f^1 = 2q^1\,\mathrm{d}q^1 + 2q^2\,\mathrm{d}q^2 .$$

Vector fields satisfying the condition $\mathrm{d}f^1(\xi) = 0$, i.e. $2q^1\xi^1 + 2q^2\xi^2 = 0$ are of the form

$$\xi = \xi^0\frac{\partial}{\partial t} + s\left(q^2\frac{\partial}{\partial q^1} - q^1\frac{\partial}{\partial q^2}\right), \quad s \text{ being a parameter.}$$

So, the tangent distribution to $\mathcal{D}$ is generated as

$$\mathcal{D}(y) = \operatorname{span}\left\{\frac{\partial}{\partial t}, q^2\frac{\partial}{\partial q^1} - q^1\frac{\partial}{\partial q^2}\right\} .$$

Restricted to P it defines the tangent distribution to the submanifold P.

Example 3.7 indicates that there is another possibility of how to generate a distribution—by 1-forms which annihilate generating vector fields.

Definition 3.12 (*Annihilator of the distribution*) Let $\mathcal{D}$ be a distribution on $J^r Y$, $r = 0, 1, 2$. A system $\mathcal{D}^0$ of all 1-forms η_κ on $J^r Y$ such that

$$i_\xi \eta_\kappa = 0, \quad \text{for all} \quad \xi \in \mathcal{D}$$

is called the *annihilator* of the distribution $\mathcal{D}$. We denote $\mathcal{D}^0 = \text{annih}\, \mathcal{D}$. The dimension of the vector subspace $\mathcal{D}^0 \subset \Lambda^1(T J^r Y)$ is $\dim J^r Y - \dim \mathcal{D}$ and it is called the *codimension* of the distribution $\mathcal{D}$,

$$\text{codim}\, \mathcal{D} = \dim J^r Y - \dim \mathcal{D}.$$

Recall that the annihilator of the distribution defined in Example 3.7 is $\text{annih}\, \mathcal{D} = (\mathrm{d}f^1, \dots, \mathrm{d}f^k)$. Its dimension as a vector subspace of the space $\Lambda^1(TY)$ is k, which determines the codimension of the distribution $\mathcal{D}$. The dimension of this distribution is $m - k + 1$ and its codimension is k.

Definition 3.13 (*Integral sections*) Consider a fibred manifold $(J^r Y, \pi_r, X)$, $r = 0, 1, 2$. Let $\mathcal{D}$ be a distribution on $J^r Y$. A section δ of π_r is called the *integral section* of the distribution $\mathcal{D}$ if

$$\delta^* \eta = 0 \quad \text{for every} \quad \eta \in \mathcal{D}^0 .$$

The concept of integral sections is very illustrative: Let, e.g. $\mathcal{D}$ be a distribution on $J^1 Y$ and let $\mathcal{D}^0$ be its annihilator. Let $\xi \in \mathcal{D}$ and $\eta \in \mathcal{D}^0$. Then $\eta(\xi) = 0$. Let $\delta = \big(t\delta(t), q^\sigma \delta(t), \dot{q}^\sigma \delta(t)\big)$ be a section of π_1 defined on an open set $U \subset X$. For an arbitrary vector field $\xi_0 = \xi^0 \frac{\partial}{\partial t}$ on X denote

$$\xi = T_x \delta(\xi^0) = \xi^0 \left(\frac{\partial}{\partial t} + \frac{\mathrm{d}q^\sigma \delta(t)}{\mathrm{d}t} \frac{\partial}{\partial q^\sigma} + \frac{\mathrm{d}\dot{q}^\sigma \delta(t)}{\mathrm{d}t} \frac{\partial}{\partial \dot{q}^\sigma} \right) \quad \text{and}$$
$$\delta^* \eta\big(\pi_1(\tilde{y})\big)(\xi_0) = \eta(\tilde{y})(T_{\pi_1(\tilde{y})}\delta)(\xi_0) = \eta(\tilde{y})(\xi) .$$

If δ is an integral section of the distribution $\mathcal{D}$, i.e. $\delta^* \eta = 0$, we can see that $\eta(\xi) = 0$ and $\xi \in \mathcal{D}$. On the other hand, the vector field ξ is tangent to the section δ in every point $\tilde{y} \in \delta(U)$. The situation is shown for a simple example on Fig. 3.8—a distribution of the rank 2 defined on $(\mathbf{R} \times \mathbf{R}^3, p, \mathbf{R})$. It raises a question if there exists a distribution having no integral sections. Consider a distribution $\mathcal{D}$ on $J^2 Y$. Let $\eta \in \mathcal{D}^0$. The general chart expression of η is

$$\eta = A\, \mathrm{d}t + B_\sigma\, \mathrm{d}q^\sigma + C_\sigma\, \mathrm{d}\dot{q}^\sigma + D_\sigma\, \mathrm{d}\ddot{q}^\sigma .$$

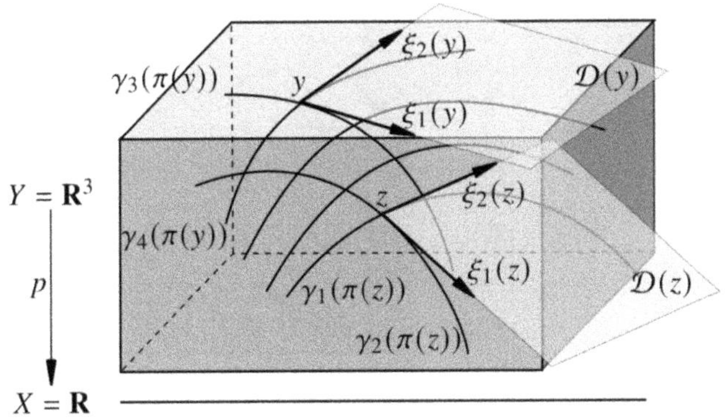

Fig. 3.8 Integral sections of a distribution

We want to find the conditions under what the equation

$$\delta^* \eta = 0 \Longrightarrow A + B_\sigma q^\sigma + C_\sigma \dot{q}^\sigma + D_\sigma \ddot{q}^\sigma = 0$$

for sections of the form

$$\delta : U \ni t \longrightarrow \delta(t) = \big(t, q^\sigma \delta(t), \dot{q}^\sigma \delta(t), \ddot{q}^\sigma \delta(t)\big) \in J^2 Y ,$$

has no solution. A special case of such a situation falls for $B_\sigma = C_\sigma = D_\sigma = 0$ for all $1 \leq \sigma \leq m$, while $A \neq 0$. This means that the annihilator of the distribution $\mathcal{D}$ do not contain the form $\mathrm{d}t$ and thus it is defined exceptionally by vertical subspaces of tangent spaces of $J^2 Y$. The following definition specifies the special type of distribution which does not have the above mentioned property.

Definition 3.14 (*Weakly horizontal distribution*) A distribution $\mathcal{D}$ on $J^r Y$, $r = 0, 1, 2$, is called *weekly horizontal* if for every $z \in \operatorname{Dom} \mathcal{D}$ it holds that $\mathcal{D}(z)$ is a complementary vector subspace to a vector subspace of $T_z J^r Y$ generated by π_r-vertical vectors.

The definition means that in $\mathcal{D}(z)$ there always exists a non-vertical vector, i.e. $\mathrm{d}t$ do not belong to $\mathcal{D}^0$. If a distribution $\mathcal{D}$ *is not* weekly horizontal, then $\mathcal{D}(z)$ contains only π_r-vertical vectors, i.e. $\mathrm{d}t \in \mathcal{D}^0$. Of course, this distribution has no integral sections, because $\delta^* \mathrm{d}t = \mathrm{d}t \neq 0$. Note that the distribution in Example 3.6 is not weekly horizontal, because it is generated by the π-vertical vector field, and thus its annihilator contains the form $\mathrm{d}t$. Figure 3.9 shows an example of a distribution which is not weekly horizontal.

Now we define some special types of distributions meaningful in the calculus of variations.

Definition 3.15 (*Cartan distribution*) The distribution on $J^1 Y$ defined by the annihilator $\mathcal{D}^0 = \operatorname{span}\{\omega^\sigma\}$, $1 \leq \sigma \leq m$, is called the *Cartan distribution*.

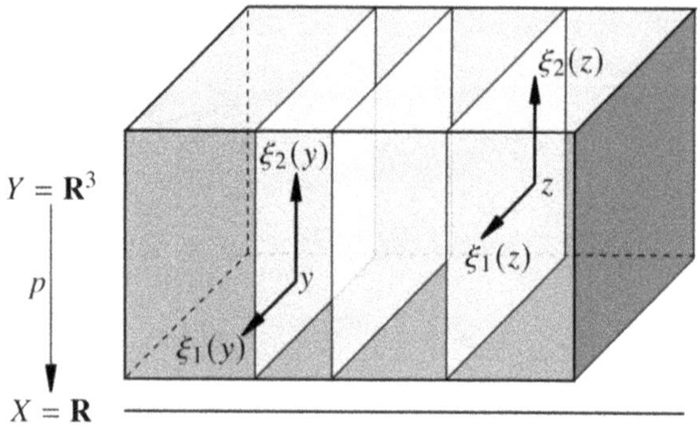

Fig. 3.9 Distribution is not weekly horizontal

Let us characterize vector fields generating the Cartan distribution on J^1Y. They are given by conditions $i_{\tilde{\xi}}(\omega^\sigma) = 0$, $1 \le \sigma \le m$.

$$\tilde{\xi} = \xi^0 \frac{\partial}{\partial t} + \xi^\nu \frac{\partial}{\partial q^\nu} + \tilde{\xi}^\nu \frac{\partial}{\partial \dot{q}^\nu},$$

$$\omega^\sigma(\tilde{\xi}) = 0 \implies \xi^\sigma - \dot{q}^\sigma \xi^0 = 0 \implies \xi^\sigma = \dot{q}^\sigma \xi^0,$$

$$\mathcal{D} = \operatorname{span}\left\{\frac{\partial}{\partial t} + \dot{q}^\sigma \frac{\partial}{\partial q^\sigma}, \frac{\partial}{\partial \dot{q}^\sigma}\right\}.$$

Note that the integral sections of the Cartan distribution are just all holonomic section of the projection π_1, i.e. $\delta = J^1\gamma$, where γ is a section of π.

Now we will study some properties of dynamical forms on J^2Y affine in variables $\ddot{q}^\nu$. These forms are especially meaningful for physics. Denote such a form as E,

$$E = E_\sigma\, \omega^\sigma \wedge \mathrm{d}t, \quad E_\sigma = A_\sigma(t, q^\lambda, \dot{q}^\lambda) + B_{\sigma\nu}(t, q^\lambda, \dot{q}^\lambda)\ddot{q}^\nu. \tag{3.28}$$

It can be easily shown that E is the 1-contact component of the form (defined on J^1Y)

$$\alpha = A_\sigma\, \omega^\sigma \wedge \mathrm{d}t + B_{\sigma\nu}\, \omega^\sigma \wedge \mathrm{d}\dot{q}^\nu + F_{\sigma\nu}\, \omega^\sigma \wedge \omega^\nu, \tag{3.29}$$

where $F_{\sigma\nu}$ is an arbitrary function on J^1Y antisymmetric in indices σ and ν. So all forms α on J^1Y differing only by a 2-contact form $F_{\sigma\nu}\,\omega^\sigma \wedge \omega^\nu$ are considered as equivalent in the sense that their 1-contact components are the same and equal to the dynamical form E. Denote the equivalence class as $[\alpha]$. Define the distribution Δ_α on J^1Y by its annihilator

$$\Delta_\alpha^0 = \operatorname{span}\left\{i_{\tilde{\xi}}\alpha \mid \tilde{\xi} \in \text{the set of } \pi_1\text{-vertical vector fields on } J^1Y\right\}. \tag{3.30}$$

Let

$$\tilde{\xi} = \xi^\nu \frac{\partial}{\partial q^\nu} + \tilde{\xi}^\nu \frac{\partial}{\partial \dot{q}^\nu},$$

then

$$i_{\tilde{\xi}}\alpha = A_\sigma \xi^\sigma \,\mathrm{d}t - \left(B_{\sigma\nu}\tilde{\xi}^\nu + 2F_{\sigma\nu}\xi^\nu\right)\omega^\sigma + B_{\sigma\nu}\xi^\sigma \,\mathrm{d}\dot{q}^\nu .$$

The annihilator of the distribution is then (recall that $F_{\sigma\nu} = -F_{\nu\sigma}$)

$$\Delta^0_\alpha = \left(A_\sigma \,\mathrm{d}t + 2F_{\sigma\nu}\,\omega^\nu + B_{\sigma\nu}\,\mathrm{d}\dot{q}^\nu,\ B_{\sigma\nu}\omega^\nu\right) . \tag{3.31}$$

In addition, define the distribution D_α by its annihilator

$$D^0_\alpha = \mathrm{span}\left\{i_{\tilde{\xi}}\alpha \mid \tilde{\xi} \in \text{the set of all vector fields on } J^1Y\right\} . \tag{3.32}$$

Definition 3.16 (*Dynamical system, dynamical distribution*) Let E be a dynamical form on J^2Y affine in variables $\ddot{q}^\sigma$, $1 \le \sigma \le m$. The class $[\alpha]$ of forms associated with E is called the *dynamical system*, alternatively *mechanical system*. Every representative of this class is called the *Lepage 2-form associated with* E, or alternatively the *Lepage equivalent of* E. The distribution annihilated by (3.31) is called the *dynamical distribution* of the representative $\alpha \in [\alpha]$. The distribution annihilated by (3.32) is called the *characteristic distribution* of the representative α. Denote as $[\Delta_\alpha]$ the class of all dynamical distributions for $\alpha \in [\alpha]$. The mechanical system is called *regular* if there exists a representative $\Delta_\alpha \in [\Delta_\alpha]$ of the constant rank rank $\Delta_\alpha = 1$ on J^1Y.

Proposition 3.7 *Let E be a dynamical form on J^2Y affine in variables $\ddot{q}^\sigma$, $1 \le \sigma \le m$. Then it holds the following:*

(a) The sets of all holonomic integral sections of all dynamical distributions $\Delta_\alpha \in [\Delta_\alpha]$ coincide.
(b) A section δ of π_1 is a holonomic section of distributions $\Delta_\alpha \in [\Delta_\alpha]$ if and only if it holds $E \circ J^2\gamma = 0$, where $\delta = J^1\gamma$.

The proof of this proposition is very simple (see Problem 3.15).

Proposition 3.8 *Let E be a dynamical form on J^2Y affine in variables $\ddot{q}^\sigma$, $1 \le \sigma \le m$. Then the following conditions are equivalent:*

(a) The associated mechanical system $[\alpha]$ is regular.
(b) The matrix $(B_{\sigma\nu})$, $1 \le \sigma, \nu \le m$, is regular on the whole domain of definition of the form E.
(c) All dynamical distributions $\Delta_\alpha \in [\Delta_\alpha]$ are of the rank 1.
(d) All dynamical distributions $\Delta_\alpha \in [\Delta_\alpha]$ coincide and it holds

$$\Delta_\alpha = \mathrm{span}\left\{\frac{\partial}{\partial t} + \dot{q}^\sigma \frac{\partial}{\partial q^\sigma} - (\mathscr{B}^{\sigma\nu}A_\nu)\frac{\partial}{\partial \dot{q}^\sigma}\right\}, \quad (\mathscr{B}^{\sigma\nu}) = (B_{\sigma\nu})^{-1} ,$$
$$\Delta^0_\alpha = \mathrm{span}\left\{A_\sigma \,\mathrm{d}t + B_{\sigma\nu}\,\mathrm{d}\dot{q}^\sigma,\ B_{\sigma\nu}\omega^\nu\right\} .$$

Proof (a) ⇒ (b): Suppose that the mechanical system $[\alpha]$ is regular, i.e. the class $[\Delta_\alpha]$ contains a representative Δ_{α_0} of the rank 1 (see Definition 3.16). This means that for this representative the annihilator $\Delta^0_{\alpha_0}$ (which is at every fixed point $\tilde{y} \in J^1Y$ the vector subspace of $\Lambda^1_{\tilde{y}}(TJ^1Y)$) is of the dimension $(2m + 1) - \operatorname{rank} \Delta^0_{\alpha_0} = 2m$. So, the forms (3.31) generating the annihilator $\Delta^0_{\alpha_0}$ are linearly independent in every point $\tilde{y}$, i.e. the following equation has only the trivial solution $\beta^\sigma = 0$, $\gamma^\sigma = 0$, $1 \leq \sigma \leq m$:

$$\beta^\sigma \left(A_\sigma \, dt + 2F_{\sigma\nu}\omega^\nu + B_{\sigma\nu} \, d\dot{q}^\nu \right) + \gamma^\sigma B_{\sigma\nu} \, \omega^\nu = 0$$
$$\Longrightarrow \quad \beta^\sigma A_\sigma = 0, \quad 2\beta^\sigma F_{\sigma\nu} + \gamma^\sigma B_{\sigma\nu} = 0, \quad \beta^\sigma B_{\sigma\nu} = 0\,,$$

for $1 \leq \sigma, \nu \leq m$. We thus have the homogeneous system of $2m + 1$ linear equations for $2m$ unknown coefficients β^σ and γ^σ for which we require the unique solution—the trivial one. The third set of equations $\gamma^\sigma B_{\sigma\nu}$ has the unique (trivial) solution $\gamma^\sigma = 0$ for all $1 \leq \sigma \leq m$ iff the matrix $(B_{\sigma\nu})$ is regular. Taking into account this fact we can conclude that $\beta^\sigma = 0$ for all σ as well.
(b) ⇒ (c): The conclusion just obtained is independent of functions $F_{\sigma\nu}$ and thus it holds for all distributions belonging to the class $[\Delta_\alpha]$. So, rank of all these distributions is constant and equals to 1 on their domain of definition.
(c) ⇒ (d): For vector field ξ on J^1Y generating the distribution Δ_α of a form $\alpha \in [\alpha]$ we obtain from (3.31)

$$A_\sigma \xi^0 + 2F_{\sigma\nu}(\xi^\nu - \dot{q}^\nu \xi^0) + B_{\sigma\nu}\tilde{\xi}^\nu = 0, \quad B_{\sigma\nu}(\xi^\nu - \dot{q}^\nu \xi^0) = 0\,.$$

We have $2m$ equations for $(2m + 1)$ unknown components of the vector field ξ. Their solution generates in every point $\tilde{y} \in \operatorname{Dom} \Delta_\alpha$ the one-dimensional vector subspace of $T_{\tilde{y}}J^1Y$. So the rank of the matrix of the system is $2m$ which implies that the matrix $(B_{\sigma\nu})$ is regular and thus the second set of the above mentioned equations $B_{\sigma\nu}(\xi^\nu - \dot{q}^\nu \xi^0) = 0$ has only the trivial solution. Then $\xi^\nu = \dot{q}^\nu \xi^0$. From the first set of equations we obtain the components $\tilde{\xi}^\sigma = -\mathcal{B}^{\sigma\nu} A_\nu \xi^0$. Finally

$$\Delta_\alpha = \operatorname{span} \left\{ \frac{\partial}{\partial t} + \dot{q}^\sigma \frac{\partial}{\partial q^\sigma} - \mathcal{B}^{\sigma\nu} A_\nu \frac{\partial}{\partial \dot{q}^\sigma} \right\}. \tag{3.33}$$

We can see that the generator of the distribution Δ_α is independent of functions $F_{\sigma\nu}$, i.e. it is the same for all representatives of the mechanical system $[\alpha]$. The annihilator is then independent of $F_{\sigma\nu}$ too. We can choose $F_{\sigma\nu} = 0$ and obtain

$$\Delta^0_\alpha = \operatorname{span} \left\{ A_\sigma \, dt + B_{\sigma\nu} \, d\dot{q}^\nu, \, B_{\sigma\nu} \, \omega^\nu \right\}. \tag{3.34}$$

(d) ⇒ (a): Suppose that all dynamical distributions of the class $[\Delta_\alpha]$ are identical, i.e. independent of $F_{\sigma\nu}$ and relations (3.33) and (3.34) hold. The existence of the inverse matrix to $(B_{\sigma\nu})$ itself implies its regularity. So, it is evident that $\operatorname{rank} \Delta_\alpha = 1$.

□

Note that holonomic sections $\delta = J^1\gamma$ of the (uniquely defined) dynamical distribution of a regular mechanical system are solutions of the equation

$$\ddot{q}^\sigma \circ J^2\gamma = -(\mathcal{B}^{\sigma\nu} A_\nu) \circ J^2\gamma \,. \tag{3.35}$$

Indeed, we proved for a regular mechanical system $[\alpha]$ that for every holonomic section $J^1\gamma$ of the dynamical distribution Δ_α it holds $(A_\sigma + B_{\sigma\nu}\ddot{q}^\nu) \circ J^2\gamma = 0$, independently of $F_{\sigma\nu}$. Nevertheless, from Proposition 3.8 it follows that the matrix $(B_{\sigma\nu})$ is regular and then Eq. (3.35) holds.

At the end of the section let us briefly comment on the case of dynamical forms on J^1Y affine in velocities (first derivatives of coordinates). Their physical meaning will be explained later. The chart expression of such a form is

$$E = E_\sigma\, \omega^\sigma \wedge \mathrm{d}t, \quad E_\sigma = A_\sigma(t, q^\lambda) + B_{\sigma\nu}(t, q^\lambda)\dot{q}^\nu \,.$$

Also for such a form there can exist the *dynamical system* $[\alpha]$ in the sense presented above for dynamical forms of the type (3.28), generated by $\pi_{1,0}$-projectable representatives α such that $E = p_1\alpha$. If we suppose such a representative to be of the form $\alpha = a_\sigma(t, q^\lambda)\, \mathrm{d}q^\sigma \wedge \mathrm{d}t + b_{\sigma\nu}(t, q^\lambda)\, \mathrm{d}q^\sigma \wedge \mathrm{d}q^\nu$, where $b_{\sigma\nu} + b_{\nu\sigma} = 0$, we obtain the conditions for a_σ and $b_{\sigma\nu}$: $a_\sigma = A_\sigma$, $2b_{\sigma\nu} = B_{\sigma\nu}$. The second condition means that the dynamical system exists if and only if the coefficients $B_{\sigma\nu}$ of the form E are antisymmetrical. In such a case we obtain

$$\pi_{1,0}^*\alpha = \left(A_\sigma + B_{\sigma\nu}\dot{q}^\nu\right) \omega^\sigma \wedge \mathrm{d}t + \frac{1}{2} B_{\sigma\nu}\, \omega^\sigma \wedge \omega^\nu \,, \tag{3.36}$$

$$\alpha = A_\sigma\, \mathrm{d}q^\sigma \wedge \mathrm{d}t + \frac{1}{2} B_{\sigma\nu}\, \mathrm{d}q^\sigma \wedge \mathrm{d}q^\nu \,. \tag{3.37}$$

Thus, the dynamical system is given by the sole representative. The definition of the dynamical distribution is completely analogous to that presented by relation (3.30), i.e.

$$\Delta_\alpha^0 = \mathrm{span}\left\{i_\xi\alpha \;\middle|\; \xi \in \text{ the set of } \pi\text{-vertical vector fields on } Y\right\} ,$$

$$\Delta_\alpha^0 = \mathrm{span}\left\{A_\sigma\, \mathrm{d}t + B_{\sigma\nu}\, \mathrm{d}q^\nu\right\} . \tag{3.38}$$

Analogously, a dynamical form $E = (A_\sigma + B_{\sigma\nu}\dot{q}^\nu)\, \mathrm{d}q^\sigma \wedge \mathrm{d}t$ on J^1Y is called *regular*, if the distribution Δ_α has the constant rank $\Delta_\alpha = 1$. This leads to condition $\det(B_{\sigma\nu}) \neq 0$.

3.5 Problems

Problem 3.1 Let ξ be a vector field on an open set $W \subset Y$ and $\mathcal{G}_W = \{\alpha_u\}$ its local one-parameter group. By the direct calculation in fibred coordinates (V, ψ), $\psi = (t, q^\nu)$, (U, φ), $U = \pi(V)$, $\varphi = (t)$, prove the relation $T\alpha_u \circ \xi = \xi \circ \alpha_u$, $u \in (-\varepsilon, \varepsilon)$.

Problem 3.2 Since Proposition 3.2 has only local character, it is sufficient to prove it by the direct calculation in coordinates for the fibred manifold $(\mathbf{R} \times \mathbf{R}^m, p, \mathbf{R})$, where p is the projection on the first factor. Present this proof.

Problem 3.3 Let $\mathcal{G}_W = \{\alpha_u\}$, $u \in (-\varepsilon, \varepsilon)$ be a local one-parameter group of isomorphisms of a fibred manifold (Y, π, X), and denote α_{0u} the π-projection of the isomorphism α_u, $u \in (-\varepsilon, \varepsilon)$. Prove that the families of isomorphisms $\{(J^1\alpha_u, \alpha_{0u})\}$ and $\{(J^2\alpha_u, \alpha_{0u})\}$ of the fibred manifolds (J^1Y, π_1, X) and (J^2Y, π_2, X), respectively, are local one-parameter groups of transformations of these fibred manifolds.

Problem 3.4 Prove all parts of Proposition 3.3.

Problem 3.5 Prove the following assertion: Let $\omega \in \Lambda^k_X(TJ^rY)$, $r = 0, 1, 2$, (i.e. ω is a π_r-horizontal form) and let $J^r\gamma^*\omega = 0$. Then ω is trivial (zero).

Problem 3.6 Prove Proposition 3.4 for $r = 1$, i.e. for k-forms on J^1Y.

Problem 3.7 Prove the following assertions:

(a) Let $\omega \in \Lambda^1(TJ^rY)$, $r = 0, 1$, be an arbitrary 1-form. Prove that the form $p\omega = \pi^*_{r+1,r}\omega - h\omega$ is contact. Prove that the mapping $p : \Lambda^1TJ^rY \ni \omega \to p\omega \in \Lambda^1(TJ^{r+1}Y)$ is $\mathbf{R}$-linear.
(b) A set of all contact forms on J^rY is the (differential) ideal.
(c) Prove that every k-form ω on J^rY, $r = 0, 1, 2$, for $k > 1$, is contact.
(d) Every contact k-form η, $k \geq 1$, on J^1Y can be written in the form $\eta = \omega^\sigma \wedge \eta_\sigma$, where η_σ are some $(k-1)$-forms. Formulate the corresponding assertion for a contact k-form on J^2Y.
(e) Regardless of the fact that all k-forms on Y for $k > 1$ are contact, there are no nontrivial (i.e. nonzero) k-contact k-forms, on Y. Find the general expression for $(k-1)$-contact k-forms on Y for $k > 1$.

Problem 3.8 Prove that the contactization mapping does not preserve the exterior product of forms in general. Characterize the situations when this difference is zero.

Problem 3.9 Prove the assertions presented in Example 3.2.

Problem 3.10 Let η be a k-form on Y. Express $\pi_{1,0}^{*}\eta$ in the basis of forms adapted to the contact structure $(dt, \omega^{\sigma}, d\dot{q}^{\sigma})$ and decompose this form into its $(k-1)$-contact and k-contact components. Show that coefficients in these expressions are functions on J^1Y affine in variables $\dot{q}^{\sigma}$. Perform the same for a k-form η on J^1Y.

Problem 3.11 Prove Proposition 3.6 for $r = 2$, i.e. that prolongations $J^2\xi$ of π-projectable vector fields are the (unique) symmetries of the contact ideal on J^2Y.

Problem 3.12 Determine whether the Lie derivative preserves the π_r-horizontality and contactness of forms, i.e. whether the Lie derivative of a π_r-horizontal, or $\pi_{r,s}$-horizontal form on J^rY, $r = 0, 1, 2$, $s < r$, with respect a vector field ξ on J^rY is horizontal and whether the Lie derivative of a contact form on J^rY with respect to a vector field ξ on J^rY is contact. If the answer is negative in general, specify vector fields for which it is positive.

Problem 3.13 Let (α, α_0) be a local isomorphism of a fibred manifold (Y, π, X). Prove that $J^r\alpha^*$, $r = 1, 2$, preserves the decomposition of a form $\eta \in \Lambda^k(TJ^rY)$ into its $(k-1)$ and k-contact component, i.e.

$$J^r\alpha^*(p_{k-1}\eta + p_k\eta) = p_{k-1}J^{r-1}\alpha^*\eta + p_k J^{r-1}\alpha^*\eta\,.$$

Problem 3.14 Formulate and prove the analogue of the Poincaré lemma (Theorem 3.2) for closed 1-forms.

Problem 3.15 Differential 2-forms are extraordinarily important in calculus of variations. For example 1-contact 2-forms on J^2Y affine in variables $\ddot{q}^{\sigma}$ are *dynamical forms* describing the dynamics of mechanical systems—they are related directly to equations of motion of these systems. The meaning of the Poincaré lemma for these forms is also evident.

(a) Characterize all closed 1-contact and 2-contact 2-forms on J^1Y and J^2Y. Characterize closed 2-forms on J^1Y and J^2Y in general (without the requirement of the "pure" 1-contactness or the "pure" 2-contactness).
(b) Prove the Poincaré lemma for 2-forms on J^1Y and J^2Y performing all steps in detail.
(c) Let η be a 2-form on J^1Y, i.e.

$$\eta = A_{\sigma}\omega^{\sigma} \wedge dt + B_{\sigma\nu}\omega^{\sigma} \wedge d\dot{q}^{\nu} + F_{\sigma\nu}\omega^{\sigma} \wedge \omega^{\nu} + P_{\sigma}\dot{\omega}^{\sigma} \wedge dt + Q_{\sigma\nu}\, d\dot{q}^{\sigma} \wedge d\dot{q}^{\nu}\,,$$

where A_{σ}, P_{σ}, $B_{\sigma\nu}$, $F_{\sigma\nu}$, $Q_{\sigma\nu}$ are functions on J^1Y, $F_{\sigma\nu} + F_{\nu\sigma} = 0$, $Q_{\sigma\nu} + Q_{\nu\sigma} = 0$. (Note: $d\dot{q}^{\sigma} \wedge dt = \dot{\omega}^{\sigma} \wedge dt$.) Construct the form $\mathcal{A}\eta$ following the Poincaré lemma.

Problem 3.16 Considering the properties of dynamical distributions of mechanical systems present the proof of Proposition 3.7.

Problem 3.17 Following the idea of the proof of Proposition 3.1 prove the parts (a) and (c).

Problem 3.18 Prove the following assertions.

(a) Let the vector field $\tilde{\xi}$ on $J^1 Y$ be $\pi_{1,0}$-projectable and let ξ be its $\pi_{1,0}$-projection. Then

$$\xi = \xi^0(t, q^\nu) \frac{\partial}{\partial t} + \xi^\sigma(t, q^\nu) \frac{\partial}{\partial q^\sigma},$$
$$\tilde{\xi} = \xi^0(t, q^\nu) \frac{\partial}{\partial t} + \xi^\sigma(t, q^\nu) \frac{\partial}{\partial q^\sigma} + \tilde{\xi}^\sigma(t, q^\nu, \dot{q}^\nu) \frac{\partial}{\partial \dot{q}^\sigma}.$$

(b) Let the vector field $\hat{\xi}$ on $J^2 Y$ be $\pi_{2,0}$-projectable and let ξ be its $\pi_{2,0}$-projection. Then

$$\xi = \xi^0(t, q^\nu) \frac{\partial}{\partial t} + \xi^\sigma(t, q^\nu) \frac{\partial}{\partial q^\sigma},$$
$$\hat{\xi} = \xi^0(t, q^\nu) \frac{\partial}{\partial t} + \xi^\sigma(t, q^\nu) \frac{\partial}{\partial q^\sigma} + \tilde{\xi}^\sigma(t, q^\nu, \dot{q}^\nu, \ddot{q}^\nu) \frac{\partial}{\partial \dot{q}^\sigma}$$
$$+ \hat{\xi}^\sigma(t, q^\nu, \dot{q}^\nu, \ddot{q}^\nu) \frac{\partial}{\partial \ddot{q}^\sigma}.$$

(c) Let the vector field $\hat{\xi}$ on $J^2 Y$ be $\pi_{2,1}$-projectable and let $\tilde{\xi}$ be its $\pi_{2,1}$-projection. Then

$$\tilde{\xi} = \xi^0(t, q^\nu, \dot{q}^\nu) \frac{\partial}{\partial t} + \xi^\sigma(t, q^\nu, \dot{q}^\nu) \frac{\partial}{\partial q^\sigma} + \tilde{\xi}^\sigma(t, q^\nu, \dot{q}^\nu) \frac{\partial}{\partial \dot{q}^\sigma},$$
$$\hat{\xi} = \xi^0(t, q^\nu, \dot{q}^\nu) \frac{\partial}{\partial t} + \xi^\sigma(t, q^\nu, \dot{q}^\nu) \frac{\partial}{\partial q^\sigma}$$
$$+ \tilde{\xi}^\sigma(t, q^\nu, \dot{q}^\nu) \frac{\partial}{\partial \dot{q}^\sigma} + \hat{\xi}^\sigma(t, q^\nu, \dot{q}^\nu, \ddot{q}^\nu) \frac{\partial}{\partial \ddot{q}^\sigma}.$$

Problem 3.19 Let ξ be a π_r-vertical vector field defined on an open set $W_r \subset J^r Y$, $r = 0, 1, 2$. Let (V, ψ), $\psi = (t, q^\sigma)$, be fibred chart on Y, (U, φ), the associated chart on X and (V_r, ψ_r) the associated fibred chart on $J^r Y$ for $r = 1, 2$. Let $W_r \cap V_r \neq \emptyset$. Express the chart representation of the vector field ξ on $W_r \cap V_r$. Answer the same question for a $\pi_{r,s}$-projectable vector field ξ on W_r, $r = 1, 2$, $s = 0, 1$.

Problem 3.20 Consider the fibred manifold $(\mathbf{R} \times \mathbf{R}^2, p, \mathbf{R})$. Let

$$\alpha_u : \mathbf{R} \times \mathbf{R}^2 \ni (t, q^1, q^2) \longrightarrow \alpha_u(t, q^1, q^2)$$
$$= (t + u, q^1 \cos u + q^2 \sin u, -q^1 \sin u + q^2 \cos u) \in \mathbf{R} \times \mathbf{R}^2.$$

Verify whether the mappings α_u are the isomorphisms of the given fibred manifold. If the answer is positive, find their projections and the components of the corresponding π-projectable vector field ξ and its projection ξ_0. Calculate the components of jet prolongations $J^1\xi$ and $J^2\xi$ of the π-projectable vector field ξ.

Problem 3.21 Calculate the first and second prolongations of the π-projectable vector field ξ defined on an $(m+1)$-dimensional fibred manifold (Y, π, X), with the chart expression

$$\xi = t^2 \frac{\partial}{\partial t} + (t^2 q^\nu \beta^\sigma_\nu) \frac{\partial}{\partial q^\sigma} ,$$

where $\beta^\sigma_\nu(t, q^\lambda)$ are some functions on Y, $1 \le \sigma, \nu \le m$. Rewrite the results for a special case when β^σ_ν are constants.

Problem 3.22 Write down the general expression of a k-form ω on X,Y,J^1Y and J^2Y. Determine the dimensions of spaces $\Lambda^k(T_x X)$, $\Lambda^k(T_y Y)$, $\Lambda^k(T_{\tilde{y}} J^1 Y)$ and $\Lambda^k(T_{\hat{y}} J^2 Y)$. For X,Y,J^1Y and J^2Y determine the maximum order k of nontrivial (nonzero) forms.

Problem 3.23 Let (V, ψ), $\psi = (t, q^\sigma)$, and $(\bar{V}, \bar{\psi})$, $\bar{\psi} = (\bar{t}, \bar{q}^\sigma)$, be fibred charts on Y, (U, φ), $\varphi = (t)$, and $(\bar{U}, \bar{\varphi})$, $\bar{\varphi} = (\bar{t})$, the associated charts on X. Suppose that $V \cap \bar{V} \neq \emptyset$. Let (V_1, ψ_1), $\psi_1 = (t, q^\sigma, \dot{q}^\sigma)$, $(\bar{V}_1, \bar{\psi}_1)$, $\bar{\psi}_1 = (\bar{t}, \bar{q}^\sigma, \dot{\bar{q}}^\sigma)$, and (V_2, ψ_2), $\psi_2 = (t, q^\sigma, \dot{q}^\sigma, \ddot{q}^\sigma)$, $(\bar{V}_2, \bar{\psi}_2)$, $\bar{\psi}_2 = (\bar{t}, \bar{q}^\sigma, \dot{\bar{q}}^\sigma, \ddot{\bar{q}}^\sigma)$ be the associated fibred chats on J^1Y and J^2Y, respectively. Derive the transformation relations for components of vector fields and differential k-forms on the intersection $V \cap \bar{V}$.

Problem 3.24 Find the chart expressions of 2-forms of all possible degrees of contactness on J^1Y and J^2Y,

(a) in standard bases $(dt, dq^\sigma, d\dot{q}^\sigma)$ on J^1Y and $(dt, dq^\sigma, d\dot{q}^\sigma, \ddot{q}^\sigma)$ on J^2Y, respectively, and
(b) in bases adapted to the contact structure, i.e. $(dt, d\omega^\sigma, d\dot{q}^\sigma)$ on J^1Y and $(dt, d\omega^\sigma, d\dot{\omega}^\sigma, d\ddot{q}^\sigma)$ on J^2Y, respectively.

Problem 3.25 Let η be a $(k-1)$-contact k-form on Y with the chart expression

$$\eta = A_{\sigma_1 \dots \sigma_{k-1}} \, dq^{\sigma_2} \wedge \dots \wedge dq^{\sigma_{k-1}} \wedge dt ,$$

where functions $A_{\sigma_1 \dots \sigma_{k-1}} = A_{\sigma_1 \dots \sigma_{k-1}}(t, q^\nu)$ are antisymmetric in all indices. Prove that

$$i_{\xi_1} \dots i_{\xi_{k-1}} \eta = k! A_{\sigma_1 \dots \sigma_{k-1}} \xi_1^{\sigma_1} \cdots \xi_{k-1}^{\sigma_{k-1}} \, dt ,$$

where $\xi_1, \dots, \xi_{k-1}$ are π-vertical vector fields on Y. What can we conclude as for the degree of contactness of this form?

Problem 3.26 Find the decomposition of a k-form defined on $J^r Y, r = 0, 1$, into its $(k-1)$-contact and k-contact components explicitly in charts, separately for $r = 0$ and $r = 1$, i.e. for

$$\eta = A_{\sigma_1 \ldots \sigma_{k-1}} \, dq^{\sigma_1} \wedge \ldots \wedge dq^{\sigma_{k-1}} \wedge dt + B_{\sigma_1 \ldots \sigma_k} \, dq^{\sigma_1} \wedge \ldots \wedge dq^{\sigma_k} ,$$

where $A_{\sigma_1 \ldots \sigma_{k-1}}$ and $B_{\sigma_1 \ldots \sigma_k}$ are functions on Y antisymmetric in all indices, and

$$\begin{aligned} \eta = & A_{\sigma_1 \ldots \sigma_j, \sigma_{j+1}, \ldots, \sigma_{k-1}} \, dq^{\sigma_1} \wedge \ldots \wedge dq^{\sigma_j} \wedge d\dot{q}^{\sigma_{j+1}} \wedge \ldots \wedge d\dot{q}^{\sigma_{k-1}} \wedge dt \\ & + B_{\sigma_1 \ldots \sigma_\ell, \sigma_{\ell+1} \ldots \sigma_k} \, dq^{\sigma_1} \wedge \ldots \wedge dq^{\sigma_\ell} \wedge d\dot{q}^{\sigma_{\ell+1}} \wedge \ldots \wedge d\dot{q}^{\sigma_k} \wedge dt , \end{aligned}$$

where $A_{\sigma_1 \ldots \sigma_j, \sigma_{j+1} \ldots \sigma_{k-1}}$ and $B_{\sigma_1 \ldots \sigma_\ell, \sigma_{\ell+1} \ldots \sigma_k}$ are functions on $J^1 Y$. Functions "A" are antisymmetric in indices $\sigma_1, \ldots, \sigma_j$ and indices $\sigma_{j+1}, \ldots, \sigma_{k-1}$, $0 \le j \le k-1$, functions "B" are antisymmetric in indices $\sigma_1, \ldots, \sigma_\ell$ and indices $\sigma_{\ell+1}, \ldots, \sigma_k$, $0 \le \ell \le k$.

As a special case express the 1-contact and 2-contact components of a 2-form on Y, resp. on $J^1 Y$, i.e. calculate the decomposition $\pi^*_{1,0} \eta = p_1 \eta + p_2 \eta$ for $\eta \in \Lambda^2(TY)$, resp. $\pi^*_{2,1} \eta = p_1 \eta + p_2 \eta$ for $\eta \in \Lambda^2(T J^1 Y)$.

Problem 3.27 Express the horizontal and 1-contact component of the exterior derivative of a function f defined on Y.

Problem 3.28 Let $r = 0, 1$.

(a) Prove that the horizontal component $h\eta$ of a 1-form η on $J^r Y$ is a Lagrangian of the order $r + 1$ in general. Formulate the conditions under which the form $h\eta$ is projectable on $J^r Y$, i.e. the order of this Lagrangian is effectively lower than in a general case.
(b) Let η be a 2-form on $J^r Y$. Express its 1-contact component $p_1 \eta$ and formulate the conditions under what it is a dynamical form (on $J^{r+1} Y$ in general). Under what conditions this dynamical form is projectable on $J^r Y$?

Problem 3.29 Let $(\mathbf{R} \times \mathbf{R}, p, \mathbf{R})$ be the fibred manifold and let $\eta = A_1 \, \omega^1 \wedge dt + B_{11} \, \omega^1 \wedge d\dot{q}^1$ be a form on $J^1 Y$. Find the conditions for the closeness of the form η and construct the form ϱ such that $\eta = d\varrho$ (via the operator $\mathcal{A}$ – Poincaré lemma). Verify whether the forms $kq^1 \, \omega^1 \wedge dt + m \, \omega^1 \wedge d\dot{q}^1$ (linear harmonic oscillator) and $\frac{g}{l} \sin q^1 \, \omega^1 \wedge dt + \omega^1 \wedge d\dot{q}^1$ (simple pendulum) are closed and construct the corresponding forms ϱ. (k, m, g and l are positive constants.)

Problem 3.30 Write down the matrix of the system of linear equations for components $(\xi^0, \xi^\nu, \tilde{\xi}^\nu)$, $1 \le \nu \le m$, of a vector field ξ belonging to the dynamical distribution Δ_α of a regular dynamical system $[\alpha]$ (see the proof of Proposition 3.8, c) $\Rightarrow$ d)).

Problem 3.31 Derive expressions (3.36) and (3.37) for the dynamical system $[\alpha]$ and for the corresponding dynamical distribution (3.38). Prove the regularity condition for the matrix $(B_{\sigma\nu})$.

Problem 3.32 Consider dynamical forms on J^1Y in their general form $E_\sigma \omega^\sigma \wedge dt$, where $E_\sigma = E_\sigma(t, q^\nu, \dot{q}^\nu)$ are general functions on J^1Y (not necessarily affine in variables $\dot{q}^\nu$). Examine the types of 2-forms α such that $E = p_1\alpha$.

References

1. O. Krupková *The Geometry of Ordinary Variational Equations, Lecture Notes in Mathematics*, vol. 1678 (Springer, Berlin, 1997)
2. J. Musilová, P. Musilová, *Mathematics for Understanding and Praxis III* (VUTIUM, Brno, 2017). ((In Czech.))
3. D. Krupka: *Introduction to Global Variational Geometry* (Atlantis Press, 2015)
4. M. Spivak, *Calculus on Manifolds. A Modern Approach to Classical Theorems of Advancer Calculus.* 27 edn. (Perseus Books Publishing, L. L. C., Massachusetts, 1998)

4 Calculus of Variations

There are many geometrical and physical problems which can be effectively solved by methods of calculus of variations. Most popular of them is the *brachistochrone problem* formulated in 1696 by Johann Bernoulli as a task for world mathematicians: *A small body (a mass particle) slides without friction along a "chute" placed in the homogeneous gravitational field of the Earth from the point A to the point B. What should be the shape of the chute for assuring the minimum time of the motion of the body between points A and B.* It is well known that the brachistochrone is a cycloid. What is the mathematical formulation of this problem? Imagine the chute as a curve in the xy-plane, with the axis y directed along the gravitational intensity $\mathbf{g}$. Such a curve represents the graph of a function $y = f(x)$. For solving this problem it is necessary to minimize the time of the motion from A to B not as a function of a certain variable, but as a "function" of $f(x)$. Such a mapping assigning to functions $f(x)$ from a certain set (i.e. having prescribed properties) a real number is a *functional*. The image of the function $f(x)$ by this mapping is mostly expressed by certain integral. The task then is to find functions leading to minima, maxima or saddle points of various types of the functional. Such functions are called *stationary points*, or *extremals* of the functional. Typical examples: find the shortest line connecting two given points in the Euclidean space, find the shortest line connecting two given points on a sphere, find the shape of a chain fixed on its ends in the homogeneous gravitational field of the Earth, find the curve $y = f(x)$ which forms the surface with the minimal area during the rotation around the x-axis, etc. There are also important physical principles based on studying stationary points of appropriate functionals: the Fermat principle for light propagation resulting in the laws of reflection and refraction, the Maupertuis principle or the Hamilton principle in mechanics. Moreover, many key physical theories are based on the problem of stationary points of functionals with various domains of definition given not only by curves, but also by sets of curves, surfaces or other objects. So, we

J. Musilová et al., *Calculus of Variations on Fibred Manifolds and Variational Physics*, Lecture Notes in Physics 1033, https://doi.org/10.1007/978-3-031-77408-9_4

have variational formulation of classical mechanics, electrodynamics and quantum mechanics. What means "variational"? A stationary point of a functional has the characteristic property: Varying slightly the object (a curve, set of curves, surface, etc.) which represents a stationary point of the functional, the corresponding change (variation) of the value of the functional is zero in linear approximation. The modern calculus of variations uses the integral calculus based on differential forms. In theories based on fibred manifolds as underlying geometric structures it works, of course, with differential forms closely connected with the fibred structure of these manifolds—as it was developed in Chap. 3. In the actual Chap. 4 we present the calculus of variation in its version appropriate for theories developed on fibred manifolds with one-dimensional bases, called *mechanics*. (Note that in classical mechanics time is measured along the base of the underlying fibred manifold.) The domain of definition of functionals in mechanics is formed by sections of fibred manifolds representing admissible trajectories of corresponding mechanical systems parametrized by time, the real trajectories being stationary points (extremals). An important part of the calculus of variations is devoted to the study of symmetries of mechanical systems and resulting conservation laws.

Prerequisites Transformation theorem for a Riemann integral, integral of a differential form, general Stokes theorem, Lie derivative and its properties, preceding Chaps. 2 and 3. □

4.1 Lagrange Structure and Variational Integrals

In this section we consider a $(m+1)$-dimensional fibred manifold (Y, π, X) with the one-dimensional base. Let Ω be a compact submanifold of X, most frequently $\Omega = [a, b]$ (a closed interval). Denote as $\Gamma_\Omega(\pi)$ a set of all sections γ of the projection π such that $\Omega \subset \operatorname{Dom} \gamma$. Now we introduce basic concepts of the variational problem. We concentrate on the *first-order* problem.

Definition 4.1 (*Variational integral*) Let λ be a first-order Lagrangian. The pair (π, λ) is called the *first-order Lagrange structure*. Let $\Omega = [a, b]$. The mapping

$$S : \Gamma_\Omega(\pi) \ni \gamma \longrightarrow S[\gamma] = \int_\Omega J^1\gamma^*\lambda \in \mathbf{R} \tag{4.1}$$

is called the *variational integral*, or *action function* of the Lagrange structure (π, λ) on Ω.

Note: For general r-th prolongation (J^rY, π_r, X) of (Y, π, X) (in our situations more frequently $r = 2$) Lagrangians are horizontal 1-forms on J^rY, $\lambda \in \Lambda^1_X(TJ^rY)$, $\lambda = L\,dt$, where $L = L(t, q^\sigma, \dots, q^\sigma_r)$, (π, λ) is then *r-th-order Lagrange structure*, the formal expression of the variational integral is as in (4.1) with $J^r\gamma$ instead of $J^1\gamma$.

Definition 4.1 introduces the standard concept of the first-order variational integral. This concept can be generalized (see Problem 3.28): If ϱ is an arbitrary 1-form on J^1Y, then $\pi_{2,1}^*\varrho = h\varrho + p\varrho$ (Theorem 3.1). The horizontal component $h\varrho$ of the form ϱ is the horizontal form on J^2Y affine in variables $\ddot{q}^\sigma$, $1 \leq \sigma \leq m$, i.e. the second-order Lagrangian in general. Then we can write

$$\int_\Omega J^1\gamma^*\varrho = \int_\Omega J^2\gamma^*\pi_{2,1}^*\varrho = \int_\Omega J^2\gamma^* h\varrho \,,$$

because $J^2\gamma^* p\varrho = 0$. For a general form ϱ we obtain the second-order variational integral for the Lagrangian affine in variables $\ddot{q}^\sigma$. (The conditions under what the form $h\varrho$ is projectable on J^1Y was derived in the Problem 3.28.) Consider a first-order Lagrangian $\lambda \in \Lambda_X^1(TJ^1Y)$. Adding to it an arbitrary contact 1-form η on J^1Y we obtain the same variational integral, i.e. for $\varrho = \lambda + \eta$ we have

$$\int_\Omega J^1\gamma^*\varrho = \int_\Omega J^1\gamma^*(\lambda + \eta) = \int_\Omega J^1\gamma^*\lambda \,.$$

This means that every 1-form on J^1Y differing from the Lagrangian λ by a contact 1-form on J^1Y leads to the same variational integral as the Lagrangian λ itself. Now we introduce the concept of variations—variation of a section and the corresponding variation of an action function. Variations of sections are interpreted with the help of projectable vector fields.

Consider a π-projectable vector field ξ on Y and $\{\alpha_u\}$ its local one-parameter group with the projection $\{\alpha_{0u}\}$. For a given value of the parameter u the image $\alpha_{0u}(\Omega)$ is a compact set. Consider a section $\gamma \in \Gamma_\Omega(\pi)$. This means that the section γ is defined on a "wider" domain U, containing Ω. Then $\alpha_{0u}(\Omega) \subset \alpha_{0u}(U)$. So, the section $\gamma_u = \alpha_u \circ \gamma \circ \alpha_{0u}^{-1}$ is defined on $\alpha_{0u}(U)$ and then $\gamma_u \in \Gamma_{\alpha_{0u}(\Omega)}(\pi)$. The above mentioned considerations hold for certain interval $(\varepsilon, \varepsilon)$ of parameters u. We have a one-parameter family of sections $\{\gamma_u\}$, $u \in (-\varepsilon, \varepsilon)$. The situation is schematically shown in Fig. 4.1.

Definition 4.2 (*Variations of sections*) Let ξ be a π-projectable vector field on J^1Y and $\{\alpha_u\}$ its local one-parameter group with the projection $\{\alpha_{0u}\}$. Let $\{\gamma_u\}$, $u \in (-\varepsilon, \varepsilon)$, be a one-parameter family of sections such that $\gamma_u = \alpha_u \circ \gamma \circ \alpha_{0u}^{-1} \in \Gamma_{\alpha_{0u}(\Omega)}(\pi)$. This family is called the *variation* or *deformation* of the section γ *induced by the vector field* ξ. The vector field ξ is called the *variation*.

Consider a mapping

$$(-\varepsilon, \varepsilon) \ni u \longrightarrow S[\gamma_u] = \int_{\alpha_{0u}(\Omega)} J^1\gamma_u^*\lambda \in \mathbf{R} \,. \tag{4.2}$$

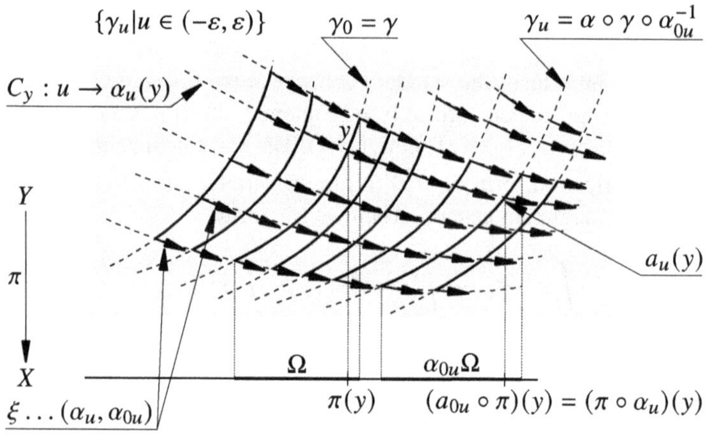

Fig. 4.1 Projectable vector fields as deformations of sections

Calculate the derivation of this mapping with respect to the parameter u.

$$\left.\frac{\mathrm{d}S[\gamma_u]}{\mathrm{d}u}\right|_{u=0} = \left[\frac{\mathrm{d}}{\mathrm{d}u}\int_{\alpha_{0u}(\Omega)} J^1\gamma_u^*\lambda\right]_{u=0} = \left[\frac{\mathrm{d}}{\mathrm{d}u}\int_{\alpha_{0u}(\Omega)} J^1(\alpha_u\circ\gamma\circ\alpha_{0u}^{-1})^*\lambda\right]_{u=0}$$

$$= \left[\frac{\mathrm{d}}{\mathrm{d}u}\int_{\alpha_{0u}(\Omega)} \left(J^1\alpha_u\circ J^1\gamma\circ\alpha_{0u}^{-1}\right)^*\lambda\right]_{u=0} = \left[\frac{\mathrm{d}}{\mathrm{d}u}\int_{\alpha_{0u}(\Omega)} \alpha_{0u}^{-1*}\circ J^1\gamma^*\circ J^1\alpha_u^*\,\lambda\right]_{u=0}$$

$$= \int_\Omega J^1\gamma^*\left[\frac{\mathrm{d}}{\mathrm{d}u}\left(J^1\alpha_u^*\lambda\right)\right]_{u=0} = \int_\Omega J^1\gamma^*\partial_{J^1\xi}\lambda\,.$$

In the fourth step of the above calculation we used the integral transformation theorem, and in the last step the definition of the Lie derivation was applied.

Definition 4.3 (*Variational derivative*) Let ξ be a variation. The mapping

$$\Gamma_\Omega(\pi)\ni\gamma \;\longrightarrow\; \left[\frac{\mathrm{d}}{\mathrm{d}u}\int_{\alpha_{0u}(\Omega)} J^1\gamma_u^*\lambda\right]_{u=0} = \int_\Omega J^1\gamma^*\partial_{J^1\xi}\lambda$$

is called the *variational derivative* or *first variation* of the action function (of the variational integral).

Note that the $\partial_{J^1\xi}\lambda$ is the horizontal 1-form on J^1Y, i.e. the first-order Lagrangian (see also Problem 3.12). Thus the variational derivative of the action function of the Lagrangian λ is the action function of the Lagrangian $\partial_{J^1\xi}$. In an analogous way, we can define variational derivatives of higher orders: Let (π, λ) be a first-order Lagrange structure and let ξ and ζ be the variations. Then integrals

$$\int_\Omega J^1\gamma^* \partial_{J^1\zeta} \partial_{J^1\xi} \lambda \quad \text{and} \quad \int_\Omega J^1\gamma^* \partial_{J^1\xi} \partial_{J^1\zeta} \lambda$$

are the second-order variational derivatives of the action function (4.2).

4.2 First Variational Formula, Lepagean Forms

In this section we will study the variational derivative of an action function in detail. We present the integral *first variational formula* which represents the invariant decomposition of the variational derivative into the sum of two parts: (1) the integral on the domain Ω dependent only on the variation ξ (not on components of its prolongation on J^1Y) and (2) the integral on $\partial\Omega$ called the *boundary term*. The sense of such a decomposition becomes clear in the next sections concerning extremals of action functions. An important role in this decomposition plays the *Lepage equivalent* of the Lagrangian. For the decompositions we use relation (3.19) in its general version (formally the same for an arbitrary form η). We can write the variational derivative of the action function as follows:

$$\int_\Omega J^1\gamma^* \partial_{J^1\xi} \lambda = \int_\Omega J^1\gamma^* i_{J^1\xi}\, \mathrm{d}\lambda + \int_\Omega J^1\gamma^* \mathrm{d} i_{J^1\xi} \lambda \, .$$

Using in the second summand the commutativity of pullback and exterior derivative, and applying the general Stokes theorem we obtain

$$\int_\Omega J^1\gamma^* \partial_{J^1\xi} \lambda = \int_\Omega J^1\gamma^* i_{J^1\xi}\, \mathrm{d}\lambda + \int_{\partial\Omega} J^1\gamma^* i_{J^1\xi} \lambda \, .$$

However, this decomposition of the variational derivative of action function does not fulfil the condition of independence of the first integral on prolonged components of the variation ξ. Using the fact that the addition to the Lagrangian a completely free contact 1-form $\eta = \eta_\sigma \omega^\sigma$ on J^1Y leads to the unchanged action function, and the fact that Lie derivative of a contact form with respect to a vector field $J^1\xi$ is again contact (see Problem 3.12), we can work with the decomposition of the variational derivative corresponding to the form $\lambda + \eta$.

$$\begin{aligned} \int_\Omega J^1\gamma^* \partial_{J^1\xi} \lambda &= \int_\Omega J^1\gamma^* \partial_{J^1\xi} (\lambda + \eta) \\ &= \int_\Omega J^1\gamma^* i_{J^1\xi}\, \mathrm{d}(\lambda + \eta) + \int_{\partial\Omega} J^1\gamma^* i_{J^1\xi} (\lambda + \eta) \, . \end{aligned} \tag{4.3}$$

We try to find such a form η for which the prolonged components of the variation ξ are eliminated from the first summand in relation (4.3). This requires some technical calculations in coordinates. We proceed on J^2Y and use relation (3.21). For the chart expressions of the variation and its prolongations we use notation (3.1). It holds

$$\int_\Omega J^1\gamma^* i_{J^1\xi}\,\mathrm{d}(\lambda+\eta) = \int_\Omega J^2\gamma^*\pi^*_{2,1} i_{J^1\xi}\,\mathrm{d}(\lambda+\eta) = \int_\Omega J^2\gamma^* i_{J^2\xi}\,\mathrm{d}\pi^*_{2,1}(\lambda+\eta)\,, \tag{4.4}$$

$$\begin{aligned} \mathrm{d}\pi^*_{2,1}(\lambda+\eta) &= \pi^*_{2,1}\mathrm{d}\left(L\,\mathrm{d}t + \eta_\sigma\omega^\sigma\right) \\ &= \left(\frac{\partial L}{\partial q^\sigma} - \frac{\mathrm{d}\eta_\sigma}{\mathrm{d}t}\right)\omega^\sigma\wedge\mathrm{d}t + \left(\frac{\partial L}{\partial \dot q^\sigma} - \eta_\sigma\right)\dot\omega^\sigma\wedge\mathrm{d}t \\ &+ \left\{\frac{\partial \eta_\sigma}{\partial q^\nu}\right\}_{\mathrm{alt}\,(\sigma,\nu)}\omega^\nu\wedge\omega^\sigma + \frac{\partial\eta_\sigma}{\partial\dot q^\nu}\dot\omega^\nu\wedge\omega^\sigma\,. \end{aligned} \tag{4.5}$$

We obtained the decomposition of the form $\mathrm{d}(\lambda+\eta)$ into its 1-contact and 2-contact component,

$$\begin{aligned} p_1\,\mathrm{d}(\lambda+\eta) &= \left(\frac{\partial L}{\partial q^\sigma} - \frac{\mathrm{d}\eta_\sigma}{\mathrm{d}t}\right)\omega^\sigma\wedge\mathrm{d}t + \left(\frac{\partial L}{\partial \dot q^\sigma} - \eta_\sigma\right)\dot\omega^\sigma\wedge\mathrm{d}t\,, \\ p_2\,\mathrm{d}(\lambda+\eta) &= \left\{\frac{\partial \eta_\sigma}{\partial q^\nu}\right\}_{\mathrm{alt}\,(\sigma,\nu)}\omega^\nu\wedge\omega^\sigma + \frac{\partial\eta_\sigma}{\partial\dot q^\nu}\dot\omega^\nu\wedge\omega^\sigma\,. \end{aligned}$$

Taking into account that the contraction of a 2-contact 2-form by a vector field is a 1-contact 1-form and that the pullback of a contact form on J^rY by a section $J^r\gamma$ vanishes, we can write

$$J^2\gamma^* i_{J^2\xi}\pi^*_{2,1}\,\mathrm{d}(\lambda+\eta) = \left[\left(\frac{\partial L}{\partial q^\sigma} - \frac{\mathrm{d}\eta_\sigma}{\mathrm{d}t}\right)(\xi^\sigma - \dot q^\sigma\xi^0) + \left(\frac{\partial L}{\partial\dot q^\sigma} - \eta_\sigma\right)(\tilde\xi^\sigma - \ddot q^\sigma\xi^0)\right]\mathrm{d}t.$$

Requiring the independence of this form on components $\tilde\xi^\sigma$ we obtain

$$\eta_\sigma = \frac{\partial L}{\partial\dot q^\sigma} \implies \eta = \frac{\partial L}{\partial\dot q^\sigma}\,\omega^\sigma\,. \tag{4.6}$$

On the basis of the above calculations, we can formulate the following proposition:

Proposition 4.1 *Let (π,λ) be a first-order Lagrange structure, $\lambda = L(t,q^\sigma,\dot q^\sigma)\,\mathrm{d}t$, $1\le\sigma\le m$. Then there exists the unique 1-form θ_λ on J^1Y such that*

$$\begin{aligned} h\theta_\lambda &= \pi^*_{2,1}\lambda\,, \\ p_1\,\mathrm{d}\theta_\lambda &= \left(\frac{\partial L}{\partial q^\sigma} - \frac{\mathrm{d}}{\mathrm{d}t}\frac{\partial L}{\partial\dot q^\sigma}\right)\omega^\sigma\wedge\mathrm{d}t\,. \end{aligned} \tag{4.7}$$

The chart expression of this form is

$$\theta_\lambda = L\,\mathrm{d}t + \frac{\partial L}{\partial \dot{q}^\sigma}\,\omega^\sigma \,.$$

Definition 4.4 (*Lepage equivalent, Lepage mapping, Euler-Lagrange (EL) form, EL mapping, regularity*) The form θ_λ given by Proposition 4.1 is called the *Lepage equivalent* of the Lagrangian λ. The mapping assigning to a Lagrangian its Lepage equivalent is called the *Lepage mapping* (of the first type). The dynamical form $E_\lambda = p_1\,\mathrm{d}\theta_\lambda$ on J^2Y,

$$E_\lambda = E_\sigma\,\omega^\sigma \wedge \mathrm{d}t = \left(\frac{\partial L}{\partial q^\sigma} - \frac{\mathrm{d}}{\mathrm{d}t}\frac{\partial L}{\partial \dot{q}^\sigma}\right)\omega^\sigma \wedge \mathrm{d}t$$

is called the *Euler-Lagrange form* of the Lagrangian λ. The mapping assigning to a Lagrangian its Euler-Lagrange form is called *Euler-Lagrange mapping*. A Lagrange structure is called *regular*, if the dynamical system $[\alpha]$ associated with its Euler-Lagrange form is regular.

Note that the conclusion concerning the uniqueness of the Lepage equivalent of the Lagrangian is not valid for field theories (developed on fibred manifolds with multidimensional bases—see Chap. 8). Another note concerns regularity of a (first order) Lagrange structure. In general it holds

$$E_\sigma = \frac{\partial L}{\partial q^\sigma} - \frac{\mathrm{d}}{\mathrm{d}t}\frac{\partial L}{\partial \dot{q}^\sigma} = \left(\frac{\partial L}{\partial q^\sigma} - \frac{\mathrm{d}'}{\mathrm{d}t}\frac{\partial L}{\partial \dot{q}^\sigma}\right) + \left(-\frac{\partial^2 L}{\partial \dot{q}^\sigma \partial \dot{q}^\nu}\right)\ddot{q}^\nu \,,$$

where we introduced the operator $\frac{\mathrm{d}'}{\mathrm{d}t}$ as

$$\frac{\mathrm{d}'}{\mathrm{d}t} f(t, q^\nu, \dot{q}^\nu) = \frac{\partial f}{\partial t} + \frac{\partial f}{\partial q^\sigma}\dot{q}^\sigma = \frac{\mathrm{d}f}{\mathrm{d}t} - \frac{\partial f}{\partial \dot{q}^\sigma}\ddot{q}^\sigma \,.$$

So, the criterion of regularity of such a Lagrange structure coincides with the regularity of the matrix $(B_{\sigma\nu}) = \left(-\frac{\partial^2 L}{\partial \dot{q}^\sigma \partial \dot{q}^\nu}\right)$, $1 \le \sigma, \nu \le m$, (see Proposition 3.8). This is the commonly used concept of regularity. Nevertheless, it is not satisfactorily general from the geometrical point of view. Namely, there are first-order Lagrangians not regular in the standard sense, but regular from the point of view of regularity of the corresponding dynamical system α. Consider a first-order Lagrangian affine in velocities $\lambda = (L_0 + L_\nu \dot{q}^\nu)\mathrm{d}t$, where L_0 and L_ν, $1 \le \nu \le m$, are functions on Y. Its Euler-Lagrange form is also of the first order, concretely

$$E_\lambda = E_\sigma\,\mathrm{d}q^\sigma \wedge \mathrm{d}t, \quad E_\sigma(t, q^\mu, \dot{q}^\mu) = \left(\frac{\partial L_0}{\partial q^\sigma} - \frac{\partial L_\nu}{\partial t}\right) + \left(\frac{\partial L_\nu}{\partial q^\sigma} - \frac{\partial L_\sigma}{\partial q^\nu}\right)\dot{q}^\nu \,.$$

In agreement with the note at the end of Sect. 3.4 the corresponding dynamical system $[\alpha]$ is regular if and only if the matrix $(B_{\sigma\nu})$, $B_{\sigma\nu} = \left(\frac{\partial L_\nu}{\partial q^\sigma} - \frac{\partial L_\sigma}{\partial q^\nu}\right)$, $1 \leq \sigma, \nu \leq m$, is regular. (In detail see Chap. 6.)

The following theorem states the *first variational formula* in the integral and differential form.

Theorem 4.1 *First variational formula. Let (π, λ) be a first-order Lagrange structure and let θ_λ and E_λ be corresponding Lepage equivalent and Euler-Lagrange form of the Lagrangian, respectively. Then for every section $\gamma \in \Gamma_\Omega(\pi)$ the following equivalent relations hold:*

$$\int_\Omega J^1\gamma^* \partial_{J^1\xi}\lambda = \int_\Omega J^1\gamma^* i_{J^1\xi}\, \mathrm{d}\theta_\lambda + \int_{\partial\Omega} J^1\gamma^* i_{J^1\xi}\theta_\lambda \,, \tag{4.8}$$

$$\int_\Omega J^1\gamma^* \partial_{J^1\xi}\lambda = \int_\Omega J^2\gamma^* i_{J^2\xi} E_\lambda + \int_{\partial\Omega} J^1\gamma^* i_{J^1\xi}\theta_\lambda \,, \tag{4.9}$$

$$\partial_{J^1\xi}\lambda = h i_{J^1\xi}\, \mathrm{d}\theta_\lambda + h\, \mathrm{d} i_{J^1\xi}\theta_\lambda \,. \tag{4.10}$$

Proof Relations (4.8) and (4.9) follow from the calculations above. We will derive relation (4.10) using the fact that the Lie derivative of a horizontal form on J^1Y by the first prolongation $J^1\xi$ of a variation ξ is again a horizontal form, and the Lie derivative of a contact form on J^1Y by $J^1\xi$ is a contact form.

$$\partial_{J^1\xi}\lambda = h\partial_{J^1\xi}(\lambda + \eta) \,,$$

(up to a projection, because $h\partial_{J^1\xi}(\lambda + \eta)$ is projectable on J^1Y), where η is given by relation (4.6) and $\theta_\lambda = \lambda + \eta$. Then

$$\partial_{J^1\xi}\lambda = h\partial_{J^1\xi}\theta_\lambda = h i_{J^1\xi}\, \mathrm{d}\theta_\lambda + h\, \mathrm{d} i_{J^1\xi}\theta_\lambda \,.$$

This is just relation (4.10). □

Relation (4.10) is the first variational formula in the *infinitesimal form*, while relations (4.8) and (4.9) represent the first variational formula in the *integral form*.

Let $[\alpha]$ be the mechanical system associated with the Euler-Lagrange form. Its representatives have the form given by relation (3.29) and the explicit expressions for A_σ and $B_{\sigma\nu}$ result from the following relation:

$$\left(\frac{\partial L}{\partial q^\sigma} - \frac{\mathrm{d}'}{\mathrm{d}t}\frac{\partial L}{\partial \dot{q}^\sigma}\right)\omega^\sigma \wedge \mathrm{d}t - \frac{\partial^2 L}{\partial \dot{q}^\sigma \partial \dot{q}^\nu}\omega^\sigma \wedge \mathrm{d}\dot{q}^\sigma + F_{\sigma\nu}\,\omega^\sigma \wedge \omega^\nu \,,$$

with arbitrary functions $F_{\sigma\nu}$ on J^1Y.

4.3 Extremals, Euler-Lagrange Equations

In this section we formulate the variational problem itself. We maintain all notations from the previous section and present the concept of stationary points (called extremals or critical sections) of a variational integral.

Definition 4.5 (*Variation with fixed endpoints*) Let $\Omega = [a, b]$. A π-vertical variation ξ on J^1Y defined on an open set W such that $\Omega \subset \pi(W)$ is called the *variation with fixed endpoints on* Ω if the restriction of the vector field ξ on the set $\pi^{-1}(\partial\Omega)$ vanishes, i.e. $\xi|_{\pi^{-1}(\partial\Omega)} = 0$. (See schematic Fig. 4.2.)

Definition 4.6 (*Extremals*) Let $\Omega = [a, b]$. A section $\gamma \in \Gamma_\Omega(\pi)$ is called the *extremal*, or alternatively the *critical section* of the Lagrange structure (π, λ) *on the set* Ω if for every variation ξ with fixed endpoints the variational derivative of the action function vanishes, i.e.

$$\int_\Omega J^1\gamma^* \partial_{J^1\xi}\lambda = 0\,. \tag{4.11}$$

A section $\gamma \in \Gamma(\pi)$ is called the *extremal*, or the *critical section* of the Lagrange structure (π, λ) if it is the extremal on every closed interval $\Omega \subset X$ such that $\Omega \subset \operatorname{Dom}\gamma$.

The definition of the extremal means that the integral on the left-hand side of relation (4.11) is zero for all integration limits lying inside the domain of the section γ. The following proposition formulates equivalent conditions for extremals.

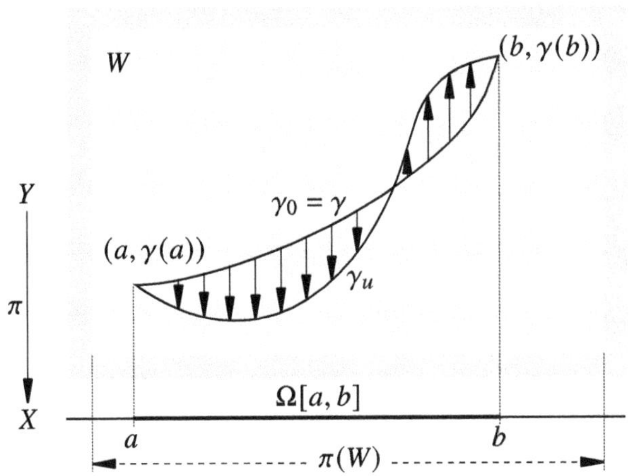

Fig. 4.2 Variation with fixed endpoints

Theorem 4.2 *Let (π, λ) be a first-order Lagrange structure, θ_λ and E_λ the corresponding Lepage equivalent and Euler-Lagrange form, respectively. Let γ be a section of the projection π. Then the following conditions are equivalent:*

(a) The section γ is an extremal of the given Lagrange structure.
(b) For every π-vertical vector field ξ on J^1Y it holds

$$J^1\gamma^* i_{J^1\xi}\, \mathrm{d}\theta_\lambda = 0\,. \tag{4.12}$$

(c) For every π-projectable vector field ξ relation (4.12) is satisfied.
(d) The Euler-Lagrange form vanishes along $J^2\gamma$, i.e.

$$E_\lambda \circ J^2\gamma = 0\,. \tag{4.13}$$

(e) The section γ is a solution of the system of m second-order ordinary Euler-Lagrange differential equations

$$\left(\frac{\partial L}{\partial q^\sigma} - \frac{\mathrm{d}}{\mathrm{d}t}\frac{\partial L}{\partial \dot{q}^\sigma}\right) \circ J^2\gamma = 0, \quad 1 \le \sigma \le m\,. \tag{4.14}$$

Proof Suppose that γ is an extremal of the given Lagrange structure. Then for every variation with fixed endpoints ζ and every closed interval $\Omega \subset \operatorname{Dom}\gamma$ it holds relation (4.11), and thus, using the integral first variation formula (4.8) we obtain

$$\int_\Omega J^1\gamma^* i_{J^1\zeta}\, \mathrm{d}\theta_\lambda + \int_{\partial\Omega} J^1\gamma^* i_{J^1\zeta}\theta_\lambda = 0\,. \tag{4.15}$$

The vector field ζ is a variation with fixed endpoints, i.e. it vanishes on $\partial\Omega$, and then

$$\int_\Omega J^1\gamma^* i_{J^1\zeta}\, \mathrm{d}\theta_\lambda = 0\,.$$

Expressing the previous relation in coordinates we obtain step by step

$$\int_\Omega J^1\gamma^* i_{J^1\zeta}\, \mathrm{d}\theta_\lambda = \int_\Omega J^2\gamma^* \pi_{2,1}^* \left(i_{J^1\zeta}\, \mathrm{d}\theta_\lambda\right) = \int_\Omega J^2\gamma^* \left(i_{J^2\zeta} \pi_{2,1}^*\, \mathrm{d}\theta_\lambda\right)$$

(components $\hat{\zeta}^\sigma$ do not affect the expression $i_{J^2\zeta}\pi_{2,1}^*\, \mathrm{d}\theta_\lambda$),

$$\begin{aligned}
\pi_{2,1}^*\, \mathrm{d}\theta_\lambda &= \pi_{2,1}^*\, \mathrm{d}\left(L\, \mathrm{d}t + \frac{\partial L}{\partial \dot{q}^\sigma}\omega^\sigma\right) \\
&= \frac{\partial L}{\partial q^\sigma}\omega^\sigma \wedge \mathrm{d}t + \frac{\partial L}{\partial \dot{q}^\sigma}\dot{\omega}^\sigma \wedge \mathrm{d}t - \frac{\mathrm{d}}{\mathrm{d}t}\left(\frac{\partial L}{\partial \dot{q}^\sigma}\right)\omega^\sigma \wedge \mathrm{d}t - \frac{\partial L}{\partial \dot{q}^\sigma}\dot{\omega}^\sigma \wedge \mathrm{d}t \\
&\quad + \left.\frac{\partial^2 L}{\partial q^\nu \partial \dot{q}^\sigma}\right|_{\mathrm{alt}\,(\sigma,\nu)} \omega^\nu \wedge \omega^\sigma + \frac{\partial^2 L}{\partial \dot{q}^\nu \partial \dot{q}^\sigma}\dot{\omega}^\nu \wedge \omega^\sigma\,,
\end{aligned}$$

$$J^1\gamma^* i_{J^1\zeta}\, \mathrm{d}\theta_\lambda = J^2\gamma^* i_{J^2\zeta} \pi^*_{2,1}\, \mathrm{d}\theta_\lambda = \left[\frac{\partial L}{\partial q^\sigma} - \frac{\mathrm{d}}{\mathrm{d}t}\left(\frac{\partial L}{\partial \dot{q}^\sigma}\right)\right](\zeta^\sigma - \dot{q}^\sigma \zeta^0)\, \mathrm{d}t$$

$$\Longrightarrow \int_\Omega \left[E_\sigma(\zeta^\sigma - \dot{q}^\sigma \zeta^0)\right] \circ J^2\gamma\, \mathrm{d}t = 0\,.$$

Recall that this is valid for *every* $\Omega \subset \mathrm{Dom}\,\gamma$ and *every* variation with fixed endpoints ζ. Even though there is the limiting condition for the vector field ζ (it vanishes on $\partial\Omega$) the expressions $u^\sigma(t) = (\zeta^\sigma - \dot{q}^\sigma \zeta^0) \circ J^1\gamma$ are arbitrary functions of the variable t vanishing on $\partial\Omega$ (i.e. at points a and b). So, following the fundamental lemma of the calculus of variations (see Problem 4.2) we can conclude that the expressions E_σ are zero along the extremal γ, i.e. the condition (e) is a consequence of (a). It could seem that this result is limited to special types of vector fields on Y (π-vertical and vanishing on $\partial\Omega$). Nonetheless, it is general: We obtained the conclusion $E_\sigma \circ J^2\gamma = 0$ by considerations based on the assumption of arbitrary functions $u^\sigma(t)$ vanishing on $\partial\Omega$, i.e. for a subset of all (of course, integrable) functions on Ω. On the other hand, our result concerning the properties of expressions E_σ along extremals does not depend on these functions, i.e. it is valid universally (which immediately implies the condition (d)). So, we can conclude that the condition $J^1\gamma^* i_{J^1\xi}\, \mathrm{d}\theta_\lambda = 0$ is valid not only for a (vertical) variation with fixed endpoints, but also for an arbitrary π-vertical variation ξ—the condition (b), as well as for an quite arbitrary variation ξ (i.e. a π-projectable vector field on Y)—the condition (c). Moreover this condition can be generalized for an arbitrary vector field $\tilde{\xi}$ on J^1Y, i.e. if γ is an extremal, then $J^1\gamma^* i_{\tilde{\xi}}\, \mathrm{d}\theta_\lambda = 0$ for an arbitrary vector field on J^1Y—in detail see, e.g. [1]. (Note that such a vector field is not the variation from the point of view of Definition 4.2.) The "reverse" part of the proof is quite simple—every of the conditions (b), (c), (d) and (e) leads to the conclusion that γ is an extremal in the sense of Definition 4.6. □

Now, discuss the condition for extremals more generally. Suppose that ξ is an arbitrary variation. If we generalize the definition of an extremal (a critical section) to such a case by the formally identical condition as (4.11) we obtain relation (4.15). Now, the boundary integral is not zero, in general. Thus for extremals we have the additional condition

$$\int_{\partial\Omega} J^1\gamma^* i_{J^1\xi}\theta_\lambda = 0 \Longrightarrow \left[L\xi^0 + \frac{\partial L}{\partial \dot{q}^\sigma}(\xi^\sigma - \dot{q}^\sigma \xi^0)\right]_a^b = 0$$

(in coordinates given for arbitrary π-projectable vector field ξ on Y). For an arbitrary π-vertical vector field ξ on Y this leads to the condition

$$\left[\frac{\partial L}{\partial \dot{q}^\sigma}\xi^\sigma\right]_a^b = 0\,. \tag{4.16}$$

Relation (4.16) represents so-called *natural boundary conditions*.

Proposition 4.2 *Let (π, λ) be a first-order Lagrange structure, θ_λ and E_λ the corresponding Lepage equivalent and the Euler-Lagrange form, respectively. Let $[\alpha]$ be the mechanical system associated with E_λ. The section $\gamma \in \Gamma(\pi)$ is the extremal of the given Lagrange structure iff*

$$J^1\gamma^* i_{J^1\xi}\alpha = 0 \tag{4.17}$$

for every variation ξ and an arbitrary representative $\alpha \in [\alpha]$.

The proof of this proposition is based on Theorem 4.2 and it is very easy (see Problem 4.3).

Proposition 4.3 *Let (π, λ) be a Lagrange structure and α the mechanical system associated with its Euler-Lagrange form E_λ. A section $\gamma \in \Gamma(\pi)$ is an extremal of the Lagrange structure (π, λ) if and only if $J^1\gamma$ is an integral section of the corresponding dynamical distribution Δ_α.*

Proof Let γ be an extremal of the given Lagrange structure. By calculating the pullbacks of 1-forms annihilating the dynamical distribution given by relation (3.31) we will prove that it is also the integral section of the dynamical distribution.

$$\begin{aligned}
J^1\gamma^* & \left(A_\sigma \, dt + 2F_{\sigma\nu}\omega^\nu + B_{\sigma\nu} \, d\dot{q}^\nu\right) \\
&= J^2\gamma^*\pi_{2,1}^*\left(A_\sigma \, dt + 2F_{\sigma\nu}\omega^\nu + B_{\sigma\nu} \left(\dot{\omega}^\sigma + \ddot{q}^\sigma \, dt\right)\right) \\
&= J^2\gamma^*\pi_{2,1}^*\left(A_\sigma \, dt + B_{\sigma\nu}\ddot{q}^\nu \, dt\right) = E_\sigma \circ J^2\gamma \, dt = 0 \, .
\end{aligned}$$

Now suppose that $J^1\gamma$ is an integral section of the dynamical distribution Δ_α, i.e.

$$J^1\gamma^*(A_\sigma \, dt + F_{\sigma\nu}\omega^\nu + B_{\sigma\nu} \, d\dot{q}^\nu) = 0.$$

Calculating the expression on the left-hand side of this condition we obtain $E_\sigma \circ J^2\gamma = 0$ (The pullback $J^1\gamma^* F_{\sigma\nu}\omega^\nu$ is zero for every holonomic section of the projection π_1, i.e. for every section γ of the projection π.) □

In the following we present some examples of the first-order variational problem.

Example 4.1 Shortest curve connecting two points in $\mathbf{R}^m$. Let $A = (q_A^1, \ldots, q_A^m)$ and $B = (q_B^1, \ldots, q_B^m)$ be two points in the Euclidean space $\mathbf{R}^m$. Determine the shortest curve connecting these points. We suppose that curve should be obtained in the parametric form $[a, b] \ni t \to \left(q^1(t), \ldots, q^m(t)\right)$, i.e. as an extremal

$$\gamma : \mathbf{R} \supset [a, b] \longrightarrow \left(t, q^1(t), \ldots, q^m(t)\right) \in \mathbf{R}^{m+1}$$

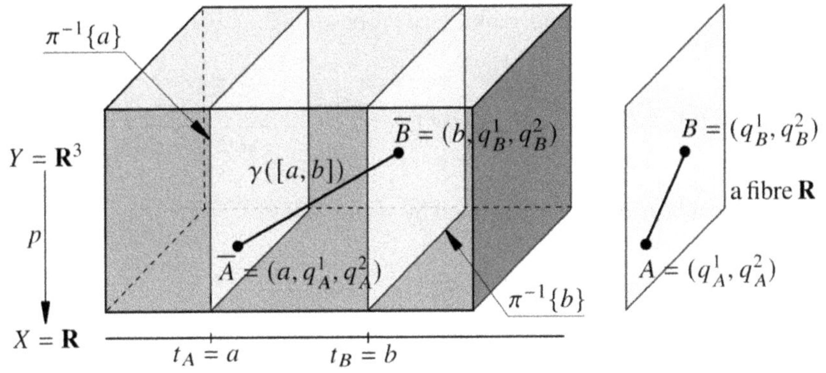

Fig. 4.3 The shortest line connecting two points in $\mathbf{R}^2$

of the Lagrange structure defined on the fibred manifold $(Y, \pi, X) = (\mathrm{R}^{m+1}, p, \mathrm{R})$. The action function is given by the length of the curve, i.e.

$$S[\gamma] = \int_a^b J^1\gamma^* \sqrt{(\dot{q}^1)^2 + \cdots + (\dot{q}^m)^2}\, \mathrm{d}t \ .$$

The Euler-Lagrange form of the above given Lagrangian is

$$E_\lambda = -\frac{\mathrm{d}}{\mathrm{d}t} \frac{\dot{q}^\sigma}{\sqrt{(\dot{q}^1)^2 + \cdots + (\dot{q}^m)^2}}\, \omega^\sigma \wedge \mathrm{d}t \ .$$

The solution of Euler-Lagrange equations is easy, we obtain

$$\frac{\mathrm{d}}{\mathrm{d}t} \frac{\dot{q}^\sigma}{\sqrt{(\dot{q}^1)^2 + \cdots + (\dot{q}^m)^2}} = 0 \Longrightarrow \frac{\dot{q}^\sigma}{\sqrt{(\dot{q}^1)^2 + \cdots + (\dot{q}^m)^2}} = \text{const.} \Longrightarrow \dot{q}^\sigma = k^\sigma = \text{const.}$$

and then

$$q^\sigma \gamma(t) = k^\sigma t + c^\sigma, \ c^\sigma = \text{const.} \quad \text{for} \quad 1 \leq \sigma \leq m \ .$$

The shortest curve connecting the points A and B is the right line going through A and B (the boundary conditions for obtaining the constants k^σ and c^σ). The situation is shown in Fig. 4.3 for $m = 2$. Finally we obtain

$$q^\sigma \gamma(t) = \frac{q_B^\sigma - q_A^\sigma}{b - a}(t - a) + q_A^\sigma, \quad 1 \leq \sigma \leq m \ .$$

The dynamical distribution Δ_α is annihilated by 1-forms (3.31). Because the Lagrange structure is regular (verify this fact) we can anticipate that all dynamical distributions are of the constant rank 1 (see Proposition 3.8). For our case is $A_\sigma = 0$, then

$$\Delta_\alpha^0 = \text{span}\,\{F_{\sigma\nu}\omega^\nu + B_{\sigma\nu}\,\mathrm{d}\dot{q}^\nu,\, B_{\sigma\nu}\omega^\nu\},\quad B_{\sigma\nu} = -\frac{\dot{q}^\sigma\dot{q}^\nu}{\left[(\dot{q}^1)^2 + \cdots + (\dot{q}^m)^2\right]^{3/2}}\,.$$

The dynamical distribution is generated by vector fields

$$\tilde{\xi} = \xi^0\frac{\partial}{\partial t} + \xi^\sigma\frac{\partial}{\partial q^\sigma} + \tilde{\xi}^\sigma\frac{\partial}{\partial \dot{q}^\sigma}$$

on J^1Y such that

$$i_{\tilde{\xi}}(F_{\sigma\nu}\omega^\nu + B_{\sigma\nu}\,\mathrm{d}\dot{q}^\nu) = 0,\quad i_{\tilde{\xi}}(B_{\sigma\nu}\omega^\nu) = 0$$

$$\Longrightarrow\quad F_{\sigma\nu}(\xi^\nu - \dot{q}^\nu\xi^0) + B_{\sigma\nu}\tilde{\xi}^\nu = 0,\quad B_{\sigma\nu}(\xi^\nu - \dot{q}^\nu\xi^0) = 0,\quad 1 \le \sigma \le m\,.$$

The second set of m linear equations for m unknown variables $\xi^\sigma - \dot{q}^\sigma\xi^0$ has only the trivial solution, because the matrix $(B_{\sigma\nu})$ is regular, i.e. $\xi^\sigma = \dot{q}^\sigma\xi^0$. The first set of equations then gives $\tilde{\xi}^\sigma = 0$. The dynamical distribution is then generated by the vector field (putting $\xi^0 = 1$)

$$\tilde{\xi} = \frac{\partial}{\partial t} + \dot{q}^\sigma\frac{\partial}{\partial q^\sigma}$$

and the sections $J^1\gamma = (t, k^\sigma t + c^\sigma, k^\sigma)$ are its integral sections,

$$\tilde{\xi}\circ J^1\gamma = T_t J^1\gamma\left(\frac{\partial}{\partial t}\right),$$

$$J^1\gamma^*(F_{\sigma\nu}\omega^\sigma + B_{\sigma\nu}\,\mathrm{d}\dot{q}^\nu) = B_{\sigma\nu}\circ J^1\gamma\,\frac{\mathrm{d}\dot{q}^\nu J^1\gamma}{\mathrm{d}t}\,\mathrm{d}t,\quad J^1\gamma^*(B_{\sigma\nu}\omega^\nu) = 0\,,$$

$$B_{\sigma\nu}\circ J^1\gamma = -\frac{\dot{q}^\sigma\dot{q}^\nu}{\left[(\dot{q}^1)^2 + \cdots + (\dot{q}^m)^2\right]^{3/2}}\circ J^1\gamma = 0$$

for $J^1\gamma(t) = (t, k^\sigma t + c^\sigma, c^\sigma)$. It is verified that all sections $\gamma(t) = (t, k^\sigma t + c^\sigma)$ obtained as extremals of the given Lagrangian are integral sections of corresponding dynamical distribution.

Example 4.2 Shortest curve connecting two points on a sphere. Let A and B be two given points on the unit sphere $x^2 + y^2 + z^2 = 1$ in $\mathbf{R}^3$. The problem is to find among all the lines connecting these points the shortest one. This is a typical variational problem. The question now is: what is the underlying fibred manifold and what is the Lagrangian. It seems that it is a constrained problem: Lines connecting points A and B lie on the sphere, i.e. coordinates x, y and z of their points are not independent. We can transform this problem into an unconstrained one, taking into account the fact that a point on the sphere is given by two independent coordinates—the spherical angle ϑ (geographic width) and the azimuthal angle φ (geographic length). A line on the sphere can be described, e.g. by the azimuthal angle φ as a function of the spherical angle ϑ. So, the underlying fibred manifold is $(Y, \pi, X) = (\mathbf{R} \times \mathbf{R}, p, \mathbf{R})$, i.e. $m = 1$. We find solutions of the problem as sections $\varphi = \varphi\gamma(\vartheta)$, $\gamma \in \Gamma_\Omega$, $\Omega = [-\pi/2, \pi/2]$. The Lagrangian is

$$\lambda = \sqrt{1 + \dot{\varphi}^2 \cos^2 \vartheta}\, \mathrm{d}\vartheta \ .$$

Geometrically, this expression represents the elementary arc length in spherical coordinates. The (single) Euler-Lagrange equation reads

$$\frac{\mathrm{d}}{\mathrm{d}\vartheta} \frac{\dot{\varphi} \cos^2 \vartheta}{\sqrt{1 + \dot{\varphi}^2 \cos^2 \vartheta}} = 0 \ .$$

This leads to the first-order differential equation with separated variables:

$$\frac{\mathrm{d}\varphi}{\mathrm{d}\vartheta} = \frac{K}{\cos \vartheta \sqrt{\cos^2 \vartheta - K^2}} \ ,$$

where K is an integration constant. Integrating this equation (the substitution $u = \frac{K}{\sqrt{1-K^2}} \operatorname{tg} \vartheta$, $|K| < 1$) we obtain

$$\sin \varphi \cos \vartheta \cos C + \cos \varphi \cos \vartheta \sin C = \frac{K}{\sqrt{1 - K^2}} \sin \vartheta$$

where C is again an integration constant. Finally we have

$$x \sin C + y \cos C - \frac{Kz}{\sqrt{1 - K^2}} = 0, \quad x = \cos \varphi \cos \vartheta, \ y = \sin \varphi \cos \vartheta, \ z = \sin \vartheta \ .$$

This equation represents a plane in $\mathbf{R}^3$ going through the origin of coordinates. This plane intersects the sphere in a principal circle (see Fig. 4.4). Constants K and C are obtained from the given coordinates of points A and B (boundary conditions).

Example 4.3 (*Brachistochrone problem*) A particle of mass m can move in the xy-plane from point $A = (0, 0)$ to point $B = (d, h)$ along a curve $y = \gamma(x)$ in the homogeneous Earth gravitational field with the free-fall acceleration $\mathbf{g} = (0, g)$ (see Fig. 4.5). The problem is to find the function $y = \gamma(x)$ such that the time of the motion is minimized. Suppose that the initial velocity of the particle is zero. The underlying fibred manifold is $(Y, \pi, X) = (\mathbf{R} \times \mathbf{R}, p, \mathbf{R})$, the fibred chart on Y, the associated chart on X and the associated fibred chart on J^1Y are global and they are

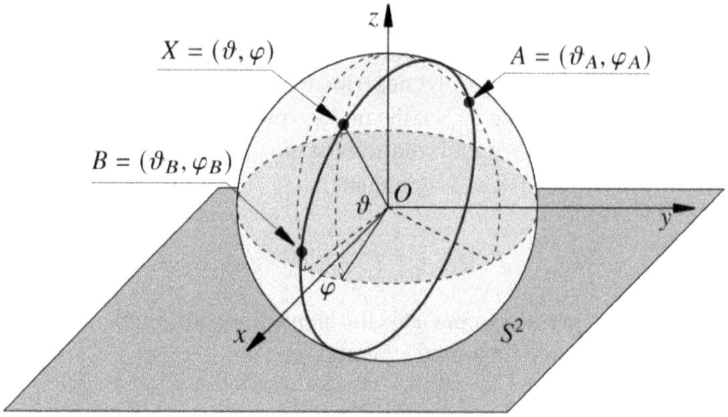

Fig. 4.4 The shortest line connecting two points on S^2

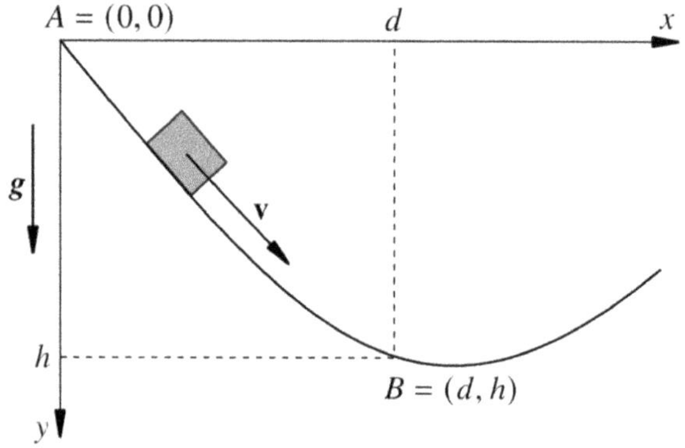

Fig. 4.5 The brachistochrone problem

successively $(\mathbf{R}^2, \psi)$, $\psi = (x, y)$, $(\mathbf{R}, \varphi)$, $\varphi = (x)$, $(\mathbf{R}^3, \psi_1)$, $\psi_1 = (x, y, \dot{y})$. (By $\dot{y}\gamma(x)$ for a section

$$\gamma : \Omega = [0, d] \ni x \to \gamma(x) = \big(x, y\gamma(x)\big) \in \mathbf{R}^2$$

we mean the derivative $\frac{\mathrm{d}y\gamma(x)}{\mathrm{d}x}$.) The speed of the particle at the point $\big(x, y\gamma(x)\big)$ can be determined using the energy conservation law

$$\frac{1}{2}mv^2 = mgy \implies v = \sqrt{2gy}\,.$$

The time τ of the motion of the particle from point A to point B is given by the integral

$$\tau = \int_C \frac{\mathrm{d}\ell}{v(y)}, \quad C = \big\{(x, y) \in \mathbf{R}^2 \mid x \in \Omega, y = y\gamma(x)\big\}$$

where $\mathrm{d}\ell$ is the length element of the curve. This means that τ depends on the section γ as follows:

$$S[\gamma] = \int_{\Omega} J^1\gamma^*\lambda = \int_0^d \frac{\sqrt{1+\dot{y}^2}}{\sqrt{2gy}}\,\mathrm{d}t, \quad \text{i.e.} \quad \lambda = \frac{\sqrt{1+\dot{y}^2}}{\sqrt{2gy}}\,\mathrm{d}t\,.$$

The corresponding Euler-Lagrange equation reads, after some technical calculations

$$1 + \dot{y}^2 + 2y\ddot{y} = 0\,. \tag{4.18}$$

Multiplying this equation by $\dot{y}$ we obtain

$$\frac{\mathrm{d}}{\mathrm{d}x}\left\{y\left[1+\left(\frac{\mathrm{d}y}{\mathrm{d}x}\right)^2\right]\right\} = 0 \implies y\left[1+\left(\frac{\mathrm{d}y}{\mathrm{d}x}\right)^2\right] = \text{const.}$$

The general solution of this equation in the parametric form represents the family of cycloids,

$$x(t) = a(t - \sin t), \quad y(t) = a(1 - \cos t), \quad a = \text{const.}$$

The required particular solution is limited by two boundary conditions $y\gamma(0) = 0$, $y\gamma(d) = h$. (Note that the "false solution" $y(x) = \text{const.}$ arising from the multiplying equation (4.18) by $\dot{y}$ does not fulfil the boundary conditions if $d \neq 0$—free fall.)

Example 4.4 (*Geodesics and Newton equations of motion*) Consider a Lagrange structure (π, λ) with m degrees of freedom in general. Suppose that there is a symmetric metric on fibres $\pi^{-1}(x)$, $x \in X$, $g = g_{\sigma\nu}(q^\lambda)\,\mathrm{d}q^\sigma \otimes \mathrm{d}q^\nu$. This means that every fibre is a metric space. Let the Lagrangian be of the form

$$\lambda = L\,\mathrm{d}t, \quad L = \frac{1}{2}g_{\sigma\nu}(q^\mu)\,\dot{q}^\sigma\dot{q}^\nu - V(t, q^\mu)\,.$$

Then the Euler-Lagrange equations are

$$g_{\mu\nu}\ddot{q}^\nu + \Gamma_{\mu\sigma\nu}\dot{q}^\sigma\dot{q}^\nu + \frac{\partial V}{\partial q^\mu} = 0, \quad \text{where} \quad \Gamma_{\mu\sigma\nu} = \frac{1}{2}\left(\frac{\partial g_{\mu\sigma}}{\partial q^\nu} + \frac{\partial g_{\nu\mu}}{\partial q^\sigma} - \frac{\partial g_{\sigma\nu}}{\partial q^\mu}\right)$$

are *Christoffel symbols*, $\lambda = 1, \ldots, m$. There are two special cases with an extraordinary geometrical or physical meaning, respectively: For $V(t, q^\varrho) = 0$ these equations read

$$g_{\mu\nu}\ddot{q}^\nu + \Gamma_{\mu\sigma\nu}\dot{q}^\sigma\dot{q}^\nu = 0\,.$$

Their solutions represent the shortest lines connecting two given points in the given metric space, so-called *geodesics*. As a very special example we can consider the fibred manifold $(\mathbf{R} \times \mathbf{R}^3, p, \mathbf{R})$, $m = 3$, where the fibres $\mathbf{R}^3$ are Euclidean spaces.

Denoting $q^1 = x$, $q^2 = y$, $q^3 = z$ (Cartesian coordinates) we can write the metric as

$$g = \delta_{\sigma\nu}\, \mathrm{d}q^\sigma \otimes \mathrm{d}q^\nu = \mathrm{d}x \otimes \mathrm{d}x + \mathrm{d}y \otimes \mathrm{d}y + \mathrm{d}z \otimes \mathrm{d}z \, .$$

The equations of geodesics are

$$\ddot{x} = 0, \quad \ddot{y} = 0, \quad \ddot{z} = 0 \, ,$$

which is in agreement with the fact derived in Example 4.1: The shortest line connecting two given points in the Euclidean space is the right line. Considering again the fibred manifold $(\mathbf{R} \times \mathbf{R}^3, p, \mathbf{R})$, $m = 3$, with Euclidean spaces as fibres, but the generally nonzero potential $V(t, x, y, z)$ and $g = m\ (\mathrm{d}x \otimes \mathrm{d}y + \mathrm{d}y \otimes \mathrm{d}x + \mathrm{d}z \otimes \mathrm{d}z)$ we obtain the well-known *Newton equations of motion* for a mass particle,

$$m\ddot{x} = -\frac{\partial V}{\partial x}, \quad m\ddot{y} = -\frac{\partial V}{\partial y}, \quad m\ddot{z} = -\frac{\partial V}{\partial z} \implies m\mathbf{a} = -\mathrm{grad}\, V(t, \mathbf{r}) \, ,$$

where m is the particle mass, $\mathbf{r} = (x, y, z)$ and $\mathbf{a} = (\ddot{x}, \ddot{y}, \ddot{z})$ its position vector and acceleration, respectively, and $\mathbf{F} = -\mathrm{grad}\, V$ is the force acting on the particle. Note that this is not the most general expression of the Newton equations. There are two significant differences between the Newton original formulation of his second law from the above presented equation: The second Newton law states that the time derivative of the particle momentum $\mathbf{p} = m\mathbf{v}$ equals to the resultant of all forces exerted on the particle by its surrounding, i.e.

$$\frac{\mathrm{d}\mathbf{p}}{\mathrm{d}t} = \sum_k \mathbf{F}_k \, .$$

This formulation takes into account the possibility of non-constant mass (the motion of a particle or a body with variable mass, leading to the Meščerskij equation). Moreover, the force laws, i.e. the expressions for forces $\mathbf{F}$, can be general functions of time, position and velocity of the particle, i.e. they need not be of the gradient (variational) type. Such types of forces are, e.g. frictional ones. This means that not all equations of motion can be derived from the variational principle. We will characterize the general expression for *variational forces* which allows to construct a Lagrangian (and thus to derive the motion equations from the variational principle) in Chap. 5.

Finally we will discuss two special variational problems: the *trivial* variational problem and the *singular* variational problem.

Definition 4.7 (*Trivial Lagrange structure*) A Lagrange structure (π, λ) as well as the Lagrangian are called *trivial* if $E_\lambda = 0$.

Proposition 4.4 *A (first order) Lagrange structure* (π, λ) *is trivial iff there exists a function* f *on* Y *such that*

$$\lambda = \frac{\mathrm{d}f}{\mathrm{d}t}\,\mathrm{d}t\,.$$

This proposition can be easily proved by the direct calculations. On the other hand, it has a deeper meaning from the point of view of the variational sequence (see Chap. 7).

Definition 4.8 (*Singular Lagrange structure*) A Lagrange structure (π, λ) as well as the Lagrangian are called *singular* if $\det\left(\frac{\partial^2 L}{\partial \dot{q}^\sigma \partial \dot{q}^\nu}\right) = 0$, $1 \le \sigma, \nu \le m$.

(Note that there exists a more general definition of a regular Lagrange structure than that based on the nonzero determinant formed by second derivatives of the Lagrangian with respect to velocities—see Chap. 6 or [1] in detail. This general definition determines singular Lagrange structure as complementary to those defined as regular.)

4.4 Symmetries and Conservation Laws, Noether Theorem

Conservation laws are a very effective tool for solving problems in various physical theories. What is the origin of conservation laws in variational physics, or, in our case, in mechanics? It lies in symmetries of the variational problem itself—in the concrete Lagrange structure. As an introductory remark, let us consider the Lagrangian and the Euler-Lagrange equations (4.14) for a mass particle in a potential field with potential energy $V(t, \mathbf{r})$

$$L(t, \mathbf{r}) = \frac{1}{2}m\mathbf{v}^2 - V(t, \mathbf{r}), \quad \text{where} \quad \mathbf{v} = \dot{r}, \quad m\dot{\mathbf{v}} = -\mathrm{grad}\, V(t, \mathbf{r})\,.$$

One of the types of symmetries occurs when the Lagrangian does not depend on certain coordinate, so-called *cyclic coordinate*. Let it be the coordinate x, for certainty. The corresponding Euler-Lagrange equation has the form $m\ddot{x} = 0$, i.e. $p_x = mv_x = \text{const}$. So, the corresponding momentum p_x is conserved along extremals. If the potential function V is time independent, we have

$$m\dot{\mathbf{v}} + \mathrm{grad}\, V(\mathbf{r}) = 0 \Longrightarrow m\mathbf{v}\dot{\mathbf{v}} + \mathbf{v}\,\mathrm{grad}\, V(\mathbf{r}) = 0 \Longrightarrow \frac{\mathrm{d}}{\mathrm{d}t}\left(\frac{1}{2}mv^2 + V(\mathbf{r})\right) = 0\,.$$

This means that the quantity $E = \frac{1}{2}mv^2 + V(\mathbf{r})$ called the *mechanical energy* of the particle is conserved along extremals.

Now let us describe symmetries and resulting conservation laws of a Lagrange structure in a quite general way. We know that differential forms on a fibred manifold vary along vector fields. The "measure" of every such variation is the Lie derivative with respect to the corresponding vector field.

Definition 4.9 (*Invariance transformation, symmetry, Noether equation*) Consider a fibred manifold (Y, π, X) and a Lagrange structure (π, λ). Let (α, α_0) be a local isomorphism of the given fibred manifold. It is called the *invariance transformation* of the given Lagrange structure if it holds

$$J^1\alpha^*\lambda = \lambda \,.$$

Let ξ be a π-projectable vector field on Y and (α_u, α_{0u}) its local one-parameter group. The vector field ξ is called the *point symmetry* or simply the *symmetry* of the given Lagrange structure, if the elements of its local one-parameter group are invariance transformations of the Lagrange structure, i.e.

$$J^1\alpha_u^*\lambda - \lambda = 0 \implies \partial_{J^1\xi}\lambda = 0 \,.$$

This condition is called the *Noether equation*.

We can realize its dual meaning: on the one hand it enables us to determine the symmetries of a given Lagrange structure and, on the other hand, to determine the Lagrange structures for which the given vector field is their symmetry. Let us express the Noether equation in coordinates. Let (V, ψ), $\psi = (t, q^\sigma)$, $1 \leq \sigma \leq m$, be a fibred chart on Y, $(\pi(V), \varphi)$, $\varphi = (t)$, the associated chart on X and $(\pi^{-1}(V), \psi_1)$, $\psi_1 = (t, q^\sigma, \dot{q}^\sigma)$, the associated fibred chart on J^1Y. The coordinate expressions of the Lagrangian and the vector field ξ and its first jet prolongation are (see Sect. 3.1)

$$\lambda = L(t, q^\sigma, \dot{q}^\sigma)\,\mathrm{d}t, \quad \xi = \xi^0(t)\frac{\partial}{\partial t} + \xi^\sigma(t, q^\nu)\frac{\partial}{\partial q^\sigma} \,,$$

$$J^1\xi = \xi^0(t)\frac{\partial}{\partial t} + \xi^\sigma(t, q^\nu)\frac{\partial}{\partial q^\sigma} + \tilde{\xi}^\sigma(t, q^\nu, \dot{q}^\nu)\frac{\partial}{\partial \dot{q}^\sigma} \,,$$

$$\tilde{\xi}^\sigma(t, q^\nu, \dot{q}^\nu) = \frac{\mathrm{d}\xi^\sigma(t, q^\nu)}{\mathrm{d}t} - \dot{q}^\sigma\frac{\mathrm{d}\xi^0(t)}{\mathrm{d}t}.$$

Using formula (3.19) in its general version we obtain after some calculations

$$\partial_{J^1\xi}\lambda = i_{J^1\xi}\,\mathrm{d}\lambda + \mathrm{d}i_{J^1\xi}\lambda = 0$$
$$\implies \left(\frac{\partial L}{\partial t}\xi^0 + \frac{\partial L}{\partial q^\sigma}\xi^\sigma + \frac{\partial L}{\partial \dot{q}^\sigma}\tilde{\xi}^\sigma + L\frac{\mathrm{d}\xi^0}{\mathrm{d}t}\right)\mathrm{d}t = 0 \,,$$

or, substituting $\tilde{\xi}^\sigma$ by its expression through ξ^0 and ξ^σ,

$$\left(\frac{\partial L}{\partial t}\xi^0 + \frac{\partial L}{\partial q^\sigma}\xi^\sigma + \frac{\partial L}{\partial \dot{q}^\sigma}\left(\frac{\mathrm{d}\xi^\sigma}{\mathrm{d}t} - \dot{q}^\sigma\frac{\mathrm{d}\xi^0}{\mathrm{d}t}\right) + L\frac{\mathrm{d}\xi^0}{\mathrm{d}t}\right)\mathrm{d}t = 0 \,. \tag{4.19}$$

The expected fact is that the Lie derivative of the Lagrangian is horizontal.

Now we explain the meaning of symmetries from the point of view of the first variational formula (4.8)–(4.10). It is contained in the following theorem, which presents the formulation of the classical *Emmy Noether theorem* on the base of fibred manifolds. Note that the Emmy Noether theorem is one of deep results of the calculus of variations.

Theorem 4.3 *Theorem Emmy Noether. Let* (π, λ) *be a first-order Lagrange structure and* θ_λ *the Lepage equivalent of the Lagrangian. Suppose that the* π*-projectable vector field* ξ *is a symmetry of the Lagrange structure. Then* $(i_{J^1\xi}\theta_\lambda) \circ J^1\gamma = \mathrm{const.}$ *for every extremal* γ *of the mentioned Lagrange structure.*

Proof Let ξ be a symmetry of the Lagrangian λ. Then the left-hand side of the first variational formula (4.8) is zero and we have

$$0 = \int_\Omega J^1\gamma^* i_{J^1\xi}\, \mathrm{d}\theta_\lambda + \int_{\partial\Omega} J^1\gamma^* i_{J^1\xi}\theta_\lambda \, .$$

The first integral on the right-hand side vanishes along extremals and thus the same holds for the second integral, i.e.

$$\int_{\partial\Omega} J^1\gamma^* i_{J^1\xi}\theta_\lambda = 0 \, .$$

Using the Stokes theorem and the commutativity of the pullback mapping and the exterior derivative we obtain

$$\int_\Omega \mathrm{d}\left(i_{J^1\xi}\theta_\lambda \circ J^1\gamma\right) = 0$$

for arbitrary Ω. This means that $\mathrm{d}J^1\gamma^* i_{J^1\xi}\theta_\lambda = \mathrm{d}\left(i_{J^1\xi}\theta_\lambda \circ J^1\gamma\right) = 0$ (see Problem 4.2). Then we obtain

$$(i_{J^1\xi}\theta_\lambda) \circ J^1\gamma = \mathrm{const.}\, ,$$

which is the *conservation law* corresponding to the symmetry ξ. □

The quantity $\Phi(\xi) = i_{J^1\xi}\theta_\lambda$ conserved along prolongations of extremals of the Lagrange structure is called the *flow* or *Noether flow* corresponding to the symmetry ξ. Let us express the flow explicitly in coordinates:

$$\Phi(\xi) = i_{J^1\xi}\theta_\lambda = i_{J^1\xi}\left(L\,\mathrm{d}t + \frac{\partial L}{\partial \dot{q}^\sigma}\omega^\sigma\right) = L\xi^0 + \frac{\partial L}{\partial \dot{q}^\sigma}(\xi^\sigma - \dot{q}^\sigma \xi^0)\, ,$$

$$\Phi(\xi) = \left(L - \dot{q}^\sigma \frac{\partial L}{\partial \dot{q}^\sigma}\right)\xi^0 + \frac{\partial L}{\partial \dot{q}^\sigma}\xi^\sigma \, . \tag{4.20}$$

Important examples of conservation laws in mechanics concern the momenta and the mechanical energy, as it is well known from classical Newtonian mechanics and as it was mentioned in the introduction of this section. These conservation laws are closely related to fundamental symmetries of the time-space: the homogeneity of time and space and the isotropy of space. These universal properties of time-space mean that the laws of motion of a *free* mass particle are invariant with respect to time translations, space translations and space rotations. These operations are represented by the following transformations $(t, x^i) \to (\bar{t}, \bar{x}^i)$, $i = 1, 2, 3$, of the underlying fibred manifold $(\mathbf{R} \times \mathbf{R}^3, p, \mathbf{R})$ endowed by charts with Cartesian coordinates (t, x^1, x^2, x^3):

$$t \longrightarrow t + \tau, \quad \tau = \text{const.}, \quad \text{i.e.} \quad \bar{t} = t + \tau\ , \tag{4.21}$$

$$x^i \longrightarrow x^i + \beta^i, \quad \beta^i = \text{const.}, \quad \text{i.e.} \quad \bar{x}^i = x^i + \beta^i\ , \tag{4.22}$$

$$x^i \longrightarrow x^k \gamma_k^i \quad \gamma_k^i = \text{const.}, \quad \text{i.e.} \quad \bar{x}^i = x^k \gamma_k^i\ , \tag{4.23}$$

where $G = (\gamma_k^i)$ is an orthogonal matrix (only three elements of this matrix are independent due to orthogonality relations). The general transformation under which the motion laws of a free particle are invariant is an arbitrary composition of operations (4.21), (4.22) and (4.23). These operations form the 7-parameter group (parameters τ, β^i and γ^i, $i = 1, 2, 3$) called the *Euclidean group of transformations*. The following example shows the context with the Noether equation and Emmy Noether theorem.

Example 4.5 (*Generators of time-space symmetries*) Consider the fibred manifold $(\mathbf{R} \times \mathbf{R}^3, p, \mathbf{R})$ representing the Euclidean time-space. Let $(\mathbf{R} \times \mathbf{R}^3, \psi)$, $\psi = (t, x^i)$, x^i, $i = 1, 2, 3$, being Cartesian coordinates, be the global fibred chart on $\mathbf{R} \times \mathbf{R}^3$, $(\mathbf{R}, \varphi)$, $\varphi = (t)$ the associated chart on $\mathbf{R}$ and $(J^1(\mathbf{R} \times \mathbf{R}^3), \psi_1)$, $\psi_1 = (t, x^i, \dot{x}^i)$, the associated fibred chart on $J^1(\mathbf{R} \times \mathbf{R}^3)$. It is well known from physics that the time homogeneity means that the motion of a free mass particle is invariant with respect to time translations generated by the vector field $\frac{\partial}{\partial t}$. Similarly, the space homogeneity represents the invariance with respect to space translations $\frac{\partial}{\partial x^i}$, and the space isotropy means the invariance of the free-particle motion with respect to space rotations $\epsilon_{ijk} x^k \frac{\partial}{\partial x^j}$. Let

$$\lambda = L(t, x^i, \dot{x}^i)\, dt, \quad L(t, x^i, \dot{x}^i) = \frac{1}{2} m \left[(\dot{x}^1)^2 + (\dot{x}^2)^2 + (\dot{x}^3)^2 \right] - V(t, x^i)$$

be the Lagrangian of a mass particle in a potential field. We will derive the additional conditions for this Lagrangian, under which the mentioned time-space generators are its point symmetries.

(a) Time translations. Putting the components of the vector field

$$\xi = \frac{\partial}{\partial t}, \quad \xi^0 = 1, \ \ \xi^i = 0, \ \ \tilde{\xi}^i = 0$$

into the Noether equation (4.19) we obtain

$$\frac{\partial L}{\partial t} = 0 \Longrightarrow \frac{\partial V}{\partial t} = 0 ,$$

i.e. the potential function does not depend on time explicitly. The corresponding flow is following relation (4.20),

$$\begin{aligned}\Phi = L - \dot{x}^i \frac{\partial L}{\partial \dot{x}^i} &= -\frac{1}{2} m \left((\dot{x}^1)^2 + (\dot{x}^2)^2 + (\dot{x}^3)^2\right) - V(x^1, x^2, x^3) \\ &= -\left[\frac{1}{2} m \left((\dot{x}^1)^2 + (\dot{x}^2)^2 + (\dot{x}^3)^2\right) + V(x^1, x^2, x^3)\right] .\end{aligned}$$

Except for the sign "minus" this is the mechanical energy of the particle.

(b) Space translations. The generator of a space translation along the x^i-axis is

$$\xi = \frac{\partial}{\partial x^i}, \quad \xi^0 = 0, \ \xi^i = 1, \ \xi^j = 0 \text{ for } j \neq i, \ \tilde{\xi}^i = 0 .$$

The Noether equation gives

$$\frac{\partial L}{\partial x^i} = 0 \Longrightarrow \frac{\partial V}{\partial x^i} = 0 .$$

This means that x^i is the cyclic coordinate. The corresponding flow is

$$\Phi = \frac{\partial L}{\partial \dot{x}^i} = m\dot{x}^i .$$

(c) Space rotations. Generators of space rotations are

$$\xi_i = \epsilon_{ijk} x^k \frac{\partial}{\partial x^j}, \quad i = 1, 2, 3,$$

where $\epsilon = (\epsilon_{ijk})$, $1 \leq i, j, k$ is the Levi-Civita symbol.

For illustration let us derive, e.g. ξ_1, i.e. the generator of the rotation around the x^1-axis (see also Example 2.8). The rotation around the x^1-axis by the angle (parameter) u is given by the orthogonal matrix T as follows:

$$(\bar{x}^1, \bar{x}^2, \bar{x}^3) = (x^1, x^2, x^3) \begin{pmatrix} 1 & 0 & 0 \\ 0 & \cos u & -\sin u \\ 0 & \sin u & \cos u \end{pmatrix} ,$$

i.e.

$$\bar{x}^1 = x^1, \quad \bar{x}^2 = x^2 \cos u + x^3 \sin u, \quad \bar{x}^3 = -x^2 \sin u + x^3 \cos u .$$

These equations define the one-parameter group of the vector field

$$\xi_1 = (\xi_1^0, \xi_1^1, \xi_1^2, \xi_1^3), \quad \xi_1^0 = 0, \quad \xi_1^i = \left.\frac{\mathrm{d}\bar{x}^i}{\mathrm{d}u}\right|_{u=0},$$

$$\xi_1^1 = 0, \quad \xi_1^2 = -x^2 \sin u + x^3 \cos u|_{u=0} = x^3, \quad \xi_1^3 = -x^2 \cos u - x^3 \sin u|_{u=0} = -x^2,$$

$$\xi_1 = x^3 \frac{\partial}{\partial x^2} - x^2 \frac{\partial}{\partial x^3}.$$

The first jet prolongations of vector fields ξ_1, ξ_2 and ξ_3 are

$$J^1\xi_1 = \left(x^3 \frac{\partial}{\partial x^2} - x^2 \frac{\partial}{\partial x^3}\right) + \left(\dot{x}^3 \frac{\partial}{\partial \dot{x}^2} - \dot{x}^2 \frac{\partial}{\partial \dot{x}^3}\right),$$

$$J^1\xi_2 = \left(x^1 \frac{\partial}{\partial x^3} - x^3 \frac{\partial}{\partial x^1}\right) + \left(\dot{x}^1 \frac{\partial}{\partial \dot{x}^3} - \dot{x}^3 \frac{\partial}{\partial \dot{x}^1}\right),$$

$$J^1\xi_3 = \left(x^2 \frac{\partial}{\partial x^1} - x^1 \frac{\partial}{\partial x^2}\right) + \left(\dot{x}^2 \frac{\partial}{\partial \dot{x}^1} - \dot{x}^1 \frac{\partial}{\partial \dot{x}^2}\right).$$

For example for ξ_1 (rotation around the x^1-axis, see Fig. 4.6) the Noether equation gives

$$\left(x^3 \frac{\partial L}{\partial x^2} - x^2 \frac{\partial L}{\partial x^3}\right) + \left(\dot{x}^3 \frac{\partial L}{\partial \dot{x}^2} - \dot{x}^2 \frac{\partial L}{\partial \dot{x}^3}\right) = 0, \quad \text{i.e.}$$

$$\left(x^2 \frac{\partial V}{\partial x^3} - x^3 \frac{\partial V}{\partial x^2}\right) = 0.$$

Because $\mathbf{F} = -\text{grad}\, V$ is the force acting on the particle in the potential field, and $\mathbf{M} = \mathbf{r} \times \mathbf{F}$ is the torque of this force with respect to the origin of the coordinate system, we can see that the last condition means that the third component of the torque is zero. The corresponding flow is

$$\Phi(\xi_1) = x^3 \frac{\partial L}{\partial \dot{x}^2} - x^2 \frac{\partial L}{\partial \dot{x}^3} = m(x^3\dot{x}^2 - x^2\dot{x}^3) = (\mathbf{r} \times m\mathbf{v})_1 = (\mathbf{r} \times \mathbf{p})_1.$$

The conserved quantity (flow) is the first component of the impulse momentum.

The case when all mentioned generators are the symmetries of the Lagrangian is that of a free particle. All our conclusions are in agreement with the well-known results of the classical Newtonian mechanics.

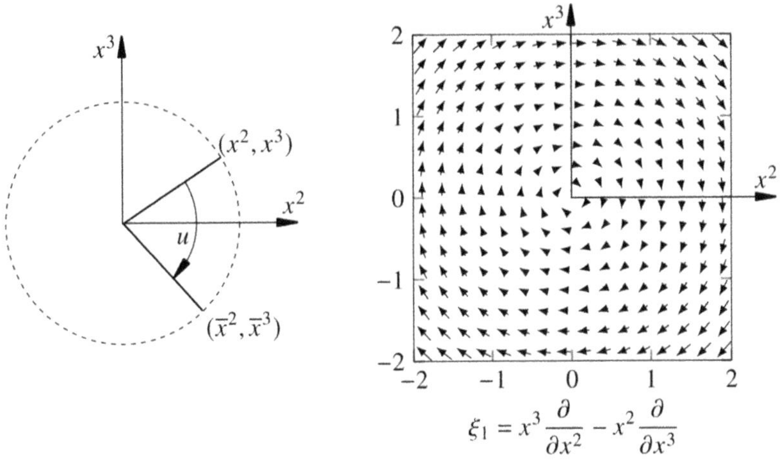

Fig. 4.6 Rotations around the x^1-axis

Conversely, if we do not suppose a concrete form of the Lagrange function $L(t, \mathbf{r}, \dot{\mathbf{r}})$ and we require for all seven above mentioned generators to be its symmetries, we obtain the conditions

$$\frac{\partial L}{\partial t} = 0, \quad \frac{\partial L}{\partial x^i} = 0, \quad \epsilon_{ijk}\left(x^k \frac{\partial L}{\partial x^j} + \dot{x}^k \frac{\partial L}{\partial \dot{x}^j}\right) = 0 \, .$$

These conditions are satisfied simultaneously if and only if $L = L(\dot{\mathbf{r}}^2)$. This completely agrees with the time-space symmetries, i.e. time and space homogeneity and space isotropy.

Example 4.6 Let us solve the reversed problem: We wish to find all symmetries of the Lagrange structure corresponding to a free mass particle. We of course anticipate that there will be combinations of time translations, space translations and space rotations. As in Example 4.5 the fibred manifold under consideration is $(\mathbf{R} \times \mathbf{R}^3, p, \mathbf{R})$, with the global fibred chart using Cartesian coordinates. The free-particle Lagrangian is

$$\lambda = \frac{1}{2} m \left((\dot{x})^1 + (\dot{x})^2 + (\dot{x})^3\right) \mathrm{d}t = \frac{1}{2} m \delta_{ij} \dot{x}^i \dot{x}^j \, \mathrm{d}t \, .$$

Putting this Lagrangian into the Noether equation (4.19) we obtain after some calculations the polynomial in variables $\dot{x}^i, i = 1, 2, 3$. The coefficients of this polynomial are functions on Y, i.e. functions of variables t and x^i. For symmetries of the given Lagrange structure this polynomial vanishes:

$$\left(\frac{\partial \xi^i}{\partial x^j} - \frac{1}{2} \delta_{ij} \frac{\mathrm{d}\xi^0}{\mathrm{d}t}\right) \dot{x}^i \dot{x}^j + \frac{\partial \xi^i}{\partial t} \dot{x}^i = 0, \quad 1 \leq i, j \leq 3 \, .$$

This means that all coefficients are zero, i.e.

$$\frac{\partial \xi^i}{\partial t} = 0, \quad \left[\frac{\partial \xi^i}{\partial x^j}\right]_{\text{sym}\,(i,j)} - \frac{1}{2}\delta_{ij}\frac{\mathrm{d}\xi^0}{\mathrm{d}t} = 0, \quad i = 1, 2, 3\,.$$

First set of these conditions means that the components ξ^i are independent of time, i.e. $\xi^i = \xi^i(x^j)$. The second set gives

$$\frac{\partial \xi^i}{\partial x^j} + \frac{\partial \xi^j}{\partial x^i} = 0 \text{ for } i \neq j, \quad \text{and} \quad \frac{\partial \xi^i}{\partial x^i} = \frac{1}{2}\frac{\mathrm{d}\xi^0}{\mathrm{d}t} \quad \text{(without summation).}$$

As the component ξ^0 depends only on t (the vector field ξ is π-projectable), both sides of the second condition must be a constant, say k. So, $\xi^0 = 2kt + q$ and ξ^i have the form

$$\xi^1 = kx^1 + \zeta^1(x^2, x^3), \quad \xi^2 = kx^2 + \zeta^2(x^1, x^3), \quad \xi^3 = kx^3 + \zeta^3(x^1, x^2)\,,$$

$$\frac{\partial \zeta^i}{\partial x^j} + \frac{\partial \zeta^j}{\partial x^i} = 0 \quad \text{for} \quad i \neq j\,.$$

Let us discuss the antisymmetry relations for derivatives of components ζ^i with respect to variables x^j. For example, the left-hand side of the relation $\frac{\partial \zeta^1}{\partial x^2} = -\frac{\partial \zeta^2}{\partial x^1}$ does not depend on x^1 and the right-hand side does not depend on x^2. Both sides are equal to a function $A^1_2(x^3)$ of the variable x^3. Integrating the obtained equations we obtain

$$\zeta^1 = A^1_2(x^3)x^2 + B^1(x^3), \quad \zeta^2 = -A^1_2(x^3)x^1 + B^2(x^3)\,.$$

Using all these antisymmetry relations (Problem 4.16) we finally obtain the general expressions for components of symmetries:

$$\begin{aligned}
\xi^1 &= (kx^1 + a^1) + A^3x^2 - A^2x^3\,,\\
\xi^2 &= (kx^2 + a^2) - A^3x^1 + A^1x^3\,,\\
\xi^3 &= (kx^3 + a^3) + A^2x^1 - A^1x^2\,,\\
\xi &= (2kt + q)\frac{\partial}{\partial t} + (kx^i + a^i)\frac{\partial}{\partial x^i} + \epsilon^i_{jk}A^jx^k\frac{\partial}{\partial x^i}\,,
\end{aligned}$$

where k, q, a^i, A^i, 1, 2, 3 are constants. All generators of symmetries of the free-particle Lagrange structure are contained in this general expression: For $k = 0, q = 1$, $a^i = 0$, $A^i = 0, i = 1, 2, 3$, we obtain the generator of the time translation. For $k = 0$, $q = 0$, $a^i = 1$, $a^j = 0$ for $j \neq i$, $A^i = 0$ for $i = 1, 2, 3$, we obtain the generator of the space translation. Finally, for $k = 0$, $q = 0$, $a^i = 0$ for $i = 1, 2, 3$, $A^1 = 0$, $A^2 = A^3 = 1$, we have the generator of the space rotation around the x^1-axis. By the cyclic change we obtain the remaining generators of space rotations.

Example 4.7 Let us generalize the standard quadratic Lagrangian by considering metrics $f_X = f_{tt}(t)\,\mathrm{d}t \otimes \mathrm{d}t$ on the base X of the underlying fibred manifold (Y, π, X), i.e.

$$f_Y = f_{tt}(t)\,\mathrm{d}t \otimes \mathrm{d}t + g_{\sigma\nu}(q^\mu)\,\mathrm{d}q^\sigma \otimes \mathrm{d}q^\nu \,.$$

Denoting $F = \det(f_{tt})$ we can write (see [3])

$$\lambda = L\,\mathrm{d}t + \frac{\mathrm{d}\chi}{\mathrm{d}t}\,\mathrm{d}t = -\frac{1}{2}\sqrt{F}\left(f^{tt} g_{\sigma\nu}\,\dot{q}^\sigma \dot{q}^\nu + m^2c^2\right)\mathrm{d}t + \frac{\mathrm{d}\chi}{\mathrm{d}t}\,\mathrm{d}t \,,$$

where by the expression $\frac{\mathrm{d}\chi}{\mathrm{d}t}\,\mathrm{d}t$, $\chi = \chi(t, q^\mu)$, we take into account the possible differences of λ by a trivial Lagrangian. The Lepage equivalent is then

$$\theta_\lambda = \frac{1}{2}\sqrt{F}\left\{\left(f^{tt}\, g_{\sigma\nu}\,\dot{q}^\sigma \dot{q}^\nu - m^2c^2\right)\mathrm{d}t - 2f^{tt}\, g_{\sigma\nu}\,\dot{q}^\nu\,\mathrm{d}q^\sigma\right\} + \mathrm{d}\chi \,.$$

Putting L into (4.20) we obtain the flows

$$\Phi(\xi) = \frac{1}{2}\sqrt{F}\left\{\left(f^{tt}\, g_{\sigma\nu}\,\dot{q}^\sigma \dot{q}^\nu - m^2c^2\right)\xi^0 - 2f^{tt}\, g_{\sigma\nu}\,\dot{q}^\sigma \xi^\nu\right\} + \Phi_0 \,,$$

where

$$\Phi_0 = \frac{\partial\chi}{\partial t}\xi^0 + \frac{\partial\chi}{\partial q^\sigma}\xi^\sigma \tag{4.24}$$

and $\xi = \xi^0 \frac{\partial}{\partial t} + \xi^\sigma \frac{\partial}{\partial q^\sigma}$ are again invariance transformations given by the Noether equation. Using the condition

$$f_{tt} \cdot f^{tt} = 1 \implies \dot{f}_{tt}\, f^{tt} + \dot{f}^{tt}\, f_{tt} = 0$$

we obtain after some calculations the following condition for invariance transformations:

$$\begin{aligned}\partial_{J^1\xi}\lambda = -\frac{1}{2}\sqrt{F}\Bigg\{\dot{q}^\sigma \dot{q}^\nu \Bigg[\Bigg(f^{tt}\frac{\partial g_{\sigma\nu}}{\partial q^\mu}\,\xi^\mu + f^{tt}\, g_{\mu\sigma}\frac{\partial \xi^\mu}{\partial q^\nu} + f^{tt}\, g_{\mu\nu}\frac{\partial \xi^\mu}{\partial q^\sigma}\Bigg)\\ - f^{tt}\, g_{\sigma\nu}\left(\frac{\mathrm{d}\xi^0}{\mathrm{d}t} - \frac{1}{2}\xi^0\frac{\dot{f}^{tt}}{f^{tt}}\right)\Bigg] + \dot{q}^\sigma\left(2f^{tt}\, g_{\mu\sigma}\frac{\partial\xi^\mu}{\partial t} - \frac{2}{\sqrt{F}}\frac{\partial\Phi_0}{\partial q^\sigma}\right)\\ + m^2c^2\left(\frac{\mathrm{d}\xi^0}{\mathrm{d}t} - \frac{1}{2}\xi^0\frac{\dot{f}^{tt}}{f^{tt}} - \frac{2}{m^2c^2\sqrt{F}}\frac{\partial\Phi_0}{\partial t}\right)\Bigg\} = 0 \,.\end{aligned}$$

All coefficients of this polynomial in velocities must vanish, i.e.

$$\frac{\mathrm{d}\xi^0}{\mathrm{d}t} - \frac{1}{2}\xi^0\frac{\dot{f}^{tt}}{f^{tt}} - \frac{2}{m^2c^2\sqrt{F}}\frac{\partial\Phi_0}{\partial t} = 0, \quad 2f^{tt}\, g_{\mu\sigma}\frac{\partial\xi^\mu}{\partial t} - \frac{2}{\sqrt{F}}\frac{\partial\Phi_0}{\partial q^\sigma} = 0 \,,$$

$$\left(\frac{\partial g_{\sigma\nu}}{\partial q^\mu}\xi^\mu + g_{\mu\sigma}\frac{\partial \xi^\mu}{\partial q^\nu} + g_{\mu\nu}\frac{\partial \xi^\mu}{\partial q^\sigma}\right) - g_{\sigma\nu}\left(\frac{\mathrm{d}\xi^0}{\mathrm{d}t} - \frac{1}{2}\xi^0\frac{\dot{f}^{tt}}{f^{tt}}\right) = 0\,.$$

Recall that Φ_0 is given by (4.24), i.e. it contains unknown components of invariance transformations. For the minimal Lagrangian we put $\chi = 0$ and obtain ($1 \le \sigma, \nu \le m$)

$$\frac{\mathrm{d}\xi^0}{\mathrm{d}t} - \frac{1}{2}\xi^0\frac{\dot{f}^{tt}}{f^{tt}} = 0, \quad g_{\sigma\nu}\frac{\partial \xi^\sigma}{\partial t} = 0, \quad \frac{\partial g_{\sigma\nu}}{\partial q^\mu}\xi^\mu + g_{\mu\nu}\frac{\partial \xi^\mu}{\partial q^\sigma} + g_{\sigma\mu}\frac{\partial \xi^\mu}{\partial q^\nu} = 0\,. \tag{4.25}$$

We can see that the third condition is the equation for Killing vector field, the solution of the first two equations is

$$\xi^0(t) = K\sqrt{f^{tt}} = \frac{K}{\sqrt{F}} \quad \text{where } K \text{ is a constant,} \qquad \frac{\partial \xi^\sigma}{\partial t} = 0\,.$$

Thus, we can choose metrics f_{tt} arbitrarily. The choice of a general Lagrangian (i.e. with an arbitrary function $\chi \neq 0$) changes conditions (4.25) only slightly. The main difference occurs in the equation for ξ^σ, now the equation for the component of a homothetic Killing field.

4.5 Generalized Theorem Emmy Noether

In the preceding section we have studied Lagrangians invariant under some transformations of the underlying fibred manifold and the corresponding conserving flows. Nevertheless, for the motion of a mechanical system the equations of motion are decisive. It is evident that if a Lagrangian is invariant under a transformation group (the one-parameter group of the point symmetry of the Lagrangian), then the corresponding Euler-Lagrange equations are invariant as well. From the point of view of physics the (more general) reversed problem is relevant: what are the symmetries leaving unchanged the Euler-Lagrange equations. The following definition specifies such situations.

Definition 4.10 (*Generalized symmetry, Noether-Bessel-Hagen equation*) Let (Y, π, X) be a fibred manifold, (π, λ) a (first order) Lagrange structure and E_λ the Euler-Lagrange form of the Lagrangian. Let (α, α_0) be a a local isomorphism of the given fibred manifold. It is called the *generalized invariance transformation* of the given Lagrange structure if it holds

$$J^2\alpha^* E_\lambda = E_\lambda\,.$$

Let ξ be a π-projectable vector field on Y and (α_u, α_{0u}) its local one-parameter group. The vector field is called the *generalized symmetry* of the given Lagrange structure, if the elements of its local one-parameter group are generalized invariance transformations of the Lagrange structure, i.e.

$$J^2\alpha_u^* E_\lambda - E_\lambda = 0 \implies \partial_{J^2\xi} E_\lambda = 0\,. \tag{4.26}$$

This condition is called the *Noether-Bessel-Hagen equation—NBH*.

Using relation (3.19) (in its general version) and expressing the Lie derivative $\partial_{J^2\xi} E_\lambda$ in coordinates we obtain from (4.26) after some calculations

$$\begin{aligned}\partial_{J^2\xi} E_\lambda &= i_{J^2\xi}\,\mathrm{d}E_\lambda + \mathrm{d}i_{J^2\xi} E_\lambda = i_{J^2\xi}(\mathrm{d}E_\sigma \wedge \mathrm{d}q^\sigma \wedge \mathrm{d}t) + \mathrm{d}(E_\sigma \xi^\sigma\,\mathrm{d}t - E_\sigma \xi^0\,\mathrm{d}q^\sigma) \\ &= \left(\frac{\partial E_\sigma}{\partial t}\xi^0 + \frac{\partial E_\sigma}{\partial q^\nu}\xi^\nu + \frac{\partial E_\sigma}{\partial \dot q^\nu}\tilde\xi^\nu + \frac{\partial E_\sigma}{\partial \ddot q^\nu}\hat\xi^\nu + E_\nu\frac{\partial \xi^\nu}{\partial q^\sigma} + E_\sigma\frac{\mathrm{d}\xi^0}{\mathrm{d}t}\right)\mathrm{d}q^\sigma \wedge \mathrm{d}t\,.\end{aligned}$$

Analogously as in the preceding section it is not surprising that the Lie derivative of the (π_1-horizontal) Euler-Lagrange form is again π_1-horizontal. Thus, the Noether-Bessel-Hagen equation expressed in coordinates gives the system m partial differential equations for generalized symmetries ξ:

$$\frac{\partial E_\sigma}{\partial t}\xi^0 + \frac{\partial E_\sigma}{\partial q^\nu}\xi^\nu + \frac{\partial E_\sigma}{\partial \dot q^\nu}\tilde\xi^\nu + \frac{\partial E_\sigma}{\partial \ddot q^\nu}\hat\xi^\nu + E_\nu\frac{\partial \xi^\nu}{\partial q^\sigma} + E_\sigma\frac{\mathrm{d}\xi^0}{\mathrm{d}t} = 0\,, \tag{4.27}$$

where

$$\tilde\xi^\sigma = \frac{\mathrm{d}\xi^\sigma}{\mathrm{d}t} - \dot q^\sigma\frac{\mathrm{d}\xi^0}{\mathrm{d}t}, \quad \hat\xi^\sigma = \frac{\mathrm{d}\tilde\xi^\sigma}{\mathrm{d}t} - \ddot q^\sigma\frac{\mathrm{d}\xi^0}{\mathrm{d}t}\,. \tag{4.28}$$

We have stated that each symmetry of a Lagrangian is obviously the symmetry of its Euler-Lagrange form, i.e. it is the generalized symmetry of Lagrange structure. The reversed question is answered in the following proposition.

Proposition 4.5 *Let (Y, π, X) be a fibred manifold and (π, λ) a Lagrange structure.*

(a) *If ξ is a symmetry of the Lagrange structure (π, λ), then it is its generalized symmetry.*
(b) *If ξ is a generalized symmetry of the Lagrange structure (π, λ), then $\partial_{J^1\xi}\lambda$ is the trivial (null) Lagrangian.*

Proof As the proof of the part (a) of the proposition is quite trivial, we will keep our mind on the part (b). Suppose that ξ is a generalized symmetry of the given Lagrange structure, i.e. $\partial_{J^2\xi} E_\lambda = 0$. First, we will prove the identity

$$J^2\alpha^* E_\lambda = E_{J^1\alpha^*\lambda} \tag{4.29}$$

for an arbitrary local isomorphism α of the fibred manifold (Y, π, X). We can use the results of Problem 3.13 where we proved that the pullback of a form on $J^r Y, r = 1, 2$

by $J^r\alpha$ preserves the decomposition of a form into its components of various degrees of contactness. Let θ_λ be the Lepage equivalent of the Lagrangian λ. Then

$$J^2\alpha^* E_\lambda = J^2\alpha^*(p_1\, \mathrm{d}\theta_\lambda) = p_1(J^1\alpha^*\, \mathrm{d}\theta_\lambda) = p_1\, \mathrm{d}(J^1\alpha^*\theta_\lambda)\,. \tag{4.30}$$

On the other hand, for $\theta_\lambda = \lambda + \eta$, where η is uniquely determined contact 1-form, we have

$$J^1\alpha^*\theta_\lambda = J^1\alpha^*\lambda + J^1\alpha^*\eta, \quad \theta_{J^1\alpha^*\lambda} = J^1\alpha^*\lambda + \bar{\eta}\,,$$

where $\bar{\eta}$ is the contact 1-form uniquely defined by the requirement that $\theta_{J^1\alpha^*\lambda}$ is the Lepage equivalent of the Lagrangian $J^1\alpha^*\lambda$. Thus, the Lepage forms $J^1\alpha^*\theta_\lambda$ and $\theta_{J^1\alpha^*\lambda}$ are Lepage equivalents of the same Lagrangian $J^1\alpha^*\lambda$. However, the Lepage equivalent of a Lagrangian is unique, which gives

$$J^1\alpha^*\theta_\lambda = \theta_{J^1\alpha^*\lambda}, \quad \text{and then} \quad p_1\, \mathrm{d}(J^1\alpha^*\theta_\lambda) = p_1\, \mathrm{d}\theta_{J^1\alpha^*\lambda} = E_{J^1\alpha^*\lambda}\,.$$

Finally, using (4.30) we obtain the identity (4.29), and because of linearity of the mapping $\lambda \to E_\lambda$, we conclude

$$\partial_{J^2\xi} E_\lambda = \lim_{u\to 0} \frac{J^2\alpha_u^* E_\lambda - E_\lambda}{u} = \lim_{u\to 0} \frac{E_{J^1\alpha_u^*\lambda} - E_\lambda}{u} = \lim_{u\to 0} \frac{E_{J^1\alpha_u^*\lambda - \lambda}}{u} = E_{\partial_{J^1\xi}\lambda}\,.$$

The Noether-Bessel-Hagen equation then gives

$$E_{\partial_{J^1\xi}\lambda} = 0\,.$$

Thus $\partial_{J^1\xi}\lambda$ is a trivial Lagrangian. This means that a symmetry of the Euler-Lagrange form is not in general the symmetry of the corresponding Lagrangian, i.e. $\partial_{J^1\xi}\lambda \neq 0$ in general. In coordinates

$$\left[\frac{\partial L}{\partial t}\xi^0 + \frac{\partial L}{\partial q^\sigma}\xi^\sigma + \frac{\partial L}{\partial \dot{q}^\sigma}\left(\frac{\mathrm{d}\xi^\sigma}{\mathrm{d}t} - \dot{q}^\sigma\frac{\mathrm{d}\xi^0}{\mathrm{d}t}\right) + L\frac{\mathrm{d}\xi^0}{\mathrm{d}t}\right]\mathrm{d}t = \frac{\mathrm{d}\chi(t, q^\sigma)}{\mathrm{d}t}\,\mathrm{d}t\,, \tag{4.31}$$

where $\chi(t, q^\sigma)$ is a function on Y. □

The question now is, what generalized conservation laws follow from the generalized symmetries.

Theorem 4.4 *Generalized theorem Emmy Noether. Let* (π, λ) *be a first-order Lagrange structure and* θ_λ *the Lepage equivalent of the Lagrangian. Suppose that the* π*-projectable vector field* ξ *is a generalized symmetry of the Lagrange structure. Then* $(i_{J^1\xi}\theta_\lambda - \chi) \circ J^1\gamma = \text{const.}$ *for every extremal* γ *of the Lagrange structure and the function* χ *on* Y *corresponds to the symmetry* ξ *via the Noether-Bessel-Hagen equation (4.31).*

Proof Let ξ be a generalized symmetry of the Lagrange structure (π, λ). Then $\partial_{J^1\xi}\lambda = \frac{d\chi}{dt}\, dt$, where χ is a function on Y and we obtain from the first variational formula (4.8)

$$\int_{\Omega} J^1\gamma^* \, d\chi = \int_{\Omega} J^1\gamma^* i_{J^1\xi} \, d\theta_\lambda + \int_{\partial\Omega} J^1\gamma^* i_{J^1\xi}\theta_\lambda \, .$$

The first integral on the right-hand side vanishes along extremals and thus

$$\int_{\partial\Omega} J^1\gamma^* i_{J^1\xi}\theta_\lambda - \int_{\Omega} J^1\gamma^* \, d\chi = 0 \, .$$

Using the Stokes theorem and the commutativity of the pullback mapping and the exterior derivative we obtain

$$\int_{\Omega} d\left(i_{J^1\xi}\theta_\lambda - \chi\right) \circ J^1\gamma = 0$$

for arbitrary Ω. This means that $d\left(i_{J^1\xi}\theta_\lambda - \chi\right) \circ J^1\gamma = 0$ (see Problem 4.2). Then we obtain

$$\left(i_{J^1\xi}\theta_\lambda - \chi\right) \circ J^1\gamma = \text{const.} \, , \tag{4.32}$$

in components

$$\Phi(\xi) = \left(L - \dot{q}^\sigma \frac{\partial L}{\partial \dot{q}^\sigma}\right)\xi^0 + \frac{\partial L}{\partial \dot{q}^\sigma}\xi^\sigma - \chi = \text{const.}$$

This is the *generalized conservation law* corresponding to the generalized symmetry ξ. □

In the following example we solve the problem of generalized symmetries of the Lagrange structure of a free mass particle.

Example 4.8 Consider again the fibred manifold $(\mathbf{R} \times \mathbf{R}^3, p, \mathbf{R})$ with the global Cartesian chart and the Lagrange structure (p, λ),

$$\lambda = \frac{1}{2} m \delta_{ij} \dot{x}^i \dot{x}^j , \quad E_\lambda = E_i \, dx^i \wedge dt , \quad E_i = -m\delta_{ij}\ddot{x}^j \, .$$

Putting these expressions into the Noether-Bessel-Hagen equation (4.27) we obtain

$$\frac{\partial^2 \xi^i}{\partial x^k \partial x^l}\dot{x}^k \dot{x}^l + \left(\frac{\partial \xi^i}{\partial x^k} + \frac{\partial \xi^k}{\partial x^i} - \delta_{ik}\frac{d\xi^0}{dt}\right)\ddot{x}^k + \left(2\frac{\partial^2 \xi^i}{\partial t \partial x^k} - \delta_{ik}\frac{d^2\xi^0}{dt^2}\right)\dot{x}^k + \frac{\partial^2 \xi^i}{\partial t^2} = 0 \, .$$

This is the polynomial in variables $\dot{x}^i$ and $\ddot{x}^i$, $i = 1, 2, 3$, with coefficients defined on Y. It vanishes if and only if all coefficients are zero, i.e.

$$0 = \frac{\partial^2 \xi^i}{\partial x^k \partial x^l} , \tag{4.33}$$

$$0 = \frac{\partial \xi^i}{\partial x^k} + \frac{\partial \xi^k}{\partial x^i} - \delta_{ik} \frac{\mathrm{d}\xi^0}{\mathrm{d}t} , \tag{4.34}$$

$$0 = 2 \frac{\partial^2 \xi^i}{\partial t \partial x^k} - \delta_{ik} \frac{\mathrm{d}^2 \xi^0}{\mathrm{d}t^2} , \tag{4.35}$$

$$0 = \frac{\partial^2 \xi^i}{\partial t^2} . \tag{4.36}$$

Condition (4.33) implies that the components $\xi^i(t, x^j)$ are linear functions of coordinates, i.e.

$$\xi^i = A^i_j(t) x^j + B^i(t), \quad i = 1, 2, 3 .$$

Condition (4.36) gives

$$\ddot{A}^i_j(t) x^j + \ddot{B}^i(t) = 0 \implies A^i_j(t) = a^i_j t + k^i_j, \quad B^i(t) = b^i t + a^i ,$$

where a^i_j, k^i_j, b^i, a^i, $i, j = 1, 2, 3$, are integration constants. From condition (4.35) for $i \neq j$ and $i = j$, respectively, we obtain

$$\dot{A}^i_j(t) = 0 \Rightarrow a^i_j = 0, \quad 2\dot{A}^i_i = \frac{\mathrm{d}^2 \xi^0}{\mathrm{d}t^2} \Rightarrow a^i_i = a \ \text{ for } i = 1, 2, 3 .$$

Finally, condition (4.34) for $i \neq j$ and $i = j$, respectively, leads to relations

$$A^i_j(t) + A^j_i(t) = 0 \implies k^i_j + k^j_i = 0 ,$$

$$2(at + k^i_i) = \frac{\mathrm{d}\xi^0}{\mathrm{d}t} \implies k^i_i = k, \ \text{ for } i = 1, 2, 3, \text{ and } \xi^0 = at^2 + 2kt + q ,$$

where q is an integration constant. Summarizing all obtained relations for components of generalized symmetries we obtain

$$\begin{aligned}
\xi^0 &= at^2 + 2kt + q , \\
\xi^1 &= (at + k)x^1 + k^1_2 x^2 - k^3_1 x^3 + b^1 t + a^1 , \\
\xi^2 &= -k^1_2 x^1 + (at + k)x^2 + k^2_3 x^3 + b^2 t + a^2 , \\
\xi^3 &= k^3_1 x^1 - k^2_3 x^2 + (at + k)x^3 + b^3 t + a^3 .
\end{aligned}$$

Denoting $k^2_3 = A^1$, $k^3_1 = A^2$, $k^1_2 = A^3$, we can write

$$\xi = (at^2 + 2kt + q)\frac{\partial}{\partial t} + (atx^i + kx^i + b^i t + a^i)\frac{\partial}{\partial x^i} + \epsilon^i_{jk} A^j x^k \frac{\partial}{\partial x^i} . \tag{4.37}$$

All symmetries of the Lagrange structure are contained in (4.37). We obtain them by putting $a = 0$, $b^i = 0$, $i = 1, 2, 3$.

Now we will calculate the function $\chi(t, x^i)$ corresponding to just derived generalized symmetries putting them into (4.31). After some calculations we obtain

$$\chi(t, x^1, x^2, x^3) = \frac{1}{2}ma\left((x^1)^2 + (x^2)^2 + (x^2)^2\right) + m\left(b^1x^1 + b^2x^2 + b^3x^3\right).$$

For $a = 0$, $b^i = 0$ for $i = 1, 2, 3$ the function $\chi(t, x^1, x^2, x^3)$ vanishes, which is consistent with the fact that for this choice of constants a and b^i we obtain the symmetries of the Lagrange structure (see Example 4.6). The last task of this example is to find the generalized flows. We obtain them from relation (4.32):

$$\begin{aligned}\Phi(\xi) &= -\frac{1}{2}m\delta_{ij}\dot{x}^i\dot{x}^j\xi^0 + m\delta_{ij}\dot{x}^j\xi^i - \chi \\ &= -\frac{1}{2}m\delta_{ij}\Big[(at^2 + 2kt + q)\dot{x}^i\dot{x}^j \\ &\quad - 2(atx^i + kx^i + b^it + a^i + \epsilon^i_{jk}A^jx^k)\dot{x}^j + (ax^ix^j + 2b^ix^j)\Big].\end{aligned} \tag{4.38}$$

For special cases of generators of time translations, space translations and space rotations, corresponding to symmetries (solutions of the Noether equation (4.19)) we obtain from (4.38) the well-known conservation laws: conservation of the kinetic energy $\frac{1}{2}m\delta_{ij}\dot{x}^i\dot{x}^j$, conservation of the momenta $m\dot{x}^i$ and conservation of the angular momenta $m\epsilon_{ijk}x^i\dot{x}^j$, $i = 1, 2, 3$.

4.6 Problems

Problem 4.1 Let $\varrho = \lambda + \eta$, where λ is a first-order Lagrangian and η is a contact 1-form on J^1Y. Let ξ be a π-projectable vector field on J^1Y. Prove relation (4.4).

Problem 4.2 Prove the following proposition called the *fundamental lemma of the calculus of variations*: Let $f(t)$ be a function continuous on a closed interval $\Omega = [a, b]$ (at points a and b it is right-continuous and left-continuous, respectively). Let for every continuous function $u(t)$ defined on Ω and satisfying the conditions $u(a) = u(b) = 0$ the integral

$$\int_\Omega f(t)u(t)\,\mathrm{d}t$$

vanishes. Then the function $f(t)$ is identically zero on Ω.

Problem 4.3 Prove Proposition 4.2.

Problem 4.4 Prove Proposition 4.4.

Problem 4.5 Problem of the minimal rotational surface. Find the smooth function $q(t)$ defined on the closed interval $[a, b]$ such that the surface formed by rotation of its graph around the t-axis is minimal.

Problem 4.6 Solve the problem of a shortest line connecting two given points on the sphere S^2 (Example 4.2) using geodesics (Example 4.4).

Problem 4.7 On the fibred manifold $(\mathbf{R} \times \mathbf{R}^2, p, \mathbf{R})$ with global Cartesian fibred chart $(\mathbf{R} \times \mathbf{R}^2, \psi)$, $\psi = (t, x^1, x^2)$, two Lagrange structures are given (p, λ_1) and (p, λ_2), with Lagrangians $\lambda_1 = L_1\, \mathrm{d}t$, $\lambda_2 = L_2\, \mathrm{d}t$,

$$L_1 = \frac{m}{2}\left((\dot{x}^1)^2 + (\dot{x}^2)^2\right) - \frac{1}{2}m\omega^2\left((x^1)^2 + (x^2)^2\right), \quad L_2 = m\dot{x}^1\dot{x}^2 - m\omega^2 x^1 x^2 .$$

Prove that both these Lagrange structures lead to the same equations of motion (equations of extremals). On the other hand, the difference $\lambda_1 - \lambda_2$ is *not* the trivial Lagrangian. Explain this result.

Problem 4.8 Let $(\mathbf{R} \times \mathbf{R}^3, p, \mathbf{R})$ be a fibred manifold endowed with the global Cartesian chart $(\mathbf{R} \times \mathbf{R}^3, \psi)$, $\psi = (t, x^1, x^2, x^3)$. Consider the Lagrangian structure (π, λ), where λ is the *Cawley* Lagrangian [1,4]

$$\lambda = L(t, x^i, \dot{x}^i)\, \mathrm{d}t = \left(\dot{x}^1\dot{x}^3 + \frac{1}{2}(x^2)^2 x^3\right) \mathrm{d}t .$$

Find all symmetries and corresponding flows of this Lagrange structure.

Problem 4.9 Execute all calculations concerning symmetries in Example 4.6, using the antisymmetry relations for ζ.

Problem 4.10 By the direct calculation in coordinates prove that the first integral in the decomposition

$$\int_{\Omega} J^1\gamma^* \partial_{J^1\xi}\lambda = \int_{\Omega} J^1\gamma^* i_{J^1\xi}\, \mathrm{d}\lambda + \int_{\partial\Omega} J^1\gamma^* i_{J^1\xi}\lambda$$

depends on the prolonged components $\tilde{\xi}^\sigma$, $1 \le \sigma \le m$, of the variation ξ.

Problem 4.11 Perform the detailed calculation leading to relation (4.5).

Problem 4.12 Let (p, λ) be a Lagrange structure on $(\mathbf{R} \times \mathbf{R}, p, \mathbf{R})$. Let $\lambda = L(q, \dot{q})\, \mathrm{d}t$, i.e. the Lagrange function does not depend on t. Consider the action function on a closed interval $\Omega = [a, b]$. Reformulate the variational problem for a

Lagrange structure $(\tilde{p}, \tilde{\lambda})$ for $(\mathbf{R} \times \mathbf{R}, \tilde{p}, \mathbf{R})$, where $\tilde{p}$ is the projection on the second factor, i.e.

$$\int_a^b L(q, \dot{q})\,\mathrm{d}t = \int_{q(a)}^{q(b)} \tilde{L}(q, t, \acute{t})\,\mathrm{d}q \ .$$

Characterize a special form of the Lagrangian $\tilde{L}$ appropriate for finding the extremals.

Problem 4.13 The Lagrangian λ in the brachistochrone problem (Example 4.3) does not depend on the variable x. Reformulate the problem following the results of Problem 4.12, i.e. consider the Lagrange structure $(\tilde{p}, \tilde{\lambda})$ instead of (p, λ), $\tilde{\lambda} = \tilde{L}(y, \acute{x}) = \acute{x} L(y, \dot{y})$.

Problem 4.14 Consider a homogeneous perfectly flexible wire of the linear density μ and length l. The wire hangs in the homogeneous gravitational field of the Earth and it is fixed in two points $A = (0, h)$ and $B = (d, h)$ at the same height h over the ground, the distance AB being $d < l$. (The free-fall acceleration is denoted as g). The shape of the wire (a graph $y(x)$ on the interval $[0, d]$) is such that its gravitational potential energy is minimal. Find the function $y(x)$. Note: The variable along the base of the underlying manifold is here denoted as x, instead of t.

Problem 4.15 Consider a first-order singular Lagrange structure on (Y, π, X) for $m = 1$ (a system with only one degree of freedom). Discuss the possible form of the Lagrangian. Write the corresponding Lepage equivalent θ_λ of this Lagrangian, its Euler-Lagrange form E_λ and the associated mechanical system $[\alpha]$. Discuss the solutions of corresponding variational problem (including the possible boundary conditions).

Problem 4.16 Execute in detail all calculations in Example 4.8.

References

1. O. Krupková, *The Geometry of Ordinary Variational Equations*. Lectures in Mathematical Notes, vol. 1678 (Springer, Berlin, 1997)
2. M. Spivak, *Calculus on Manifolds. A Modern Approach to Classical Theorems of Advancer Calculus*. 27 edn. (Perseus Books Publishing, L. L. C., Massachusetts, 1998)
3. M. Lenc, J. Musilová, L. Czudková, Lepage forms theory applied. Arch. Math. (Brno) **45**, 279–287 (2009)
4. R. Cawley, Determination of the Hamiltonian in the presence of constraints. Phys. Rev. Lett. **42**, 413–416 (1979)

5 Dynamical Forms and the Inverse Problem

In the previous chapter, we studied *variational* mechanical systems within the first-order calculus of variations on fibred manifolds and their prolongations. The motion of such systems follows from the variational principle known as the stationary (more frequently extremal) action principle. The action is given by the integral (4.1) connected with the corresponding Lagrange structure. The stationary points (extremals) of the Lagrange structure are solutions of equivalent sets of equations of motion (4.11, 4.12, 4.13, 4.14, 4.15, 4.17). The information about the motion of a variational mechanical system is carried i.a. by the Euler-Lagrange form, which is the special type *dynamical form* on the second prolongation of the underlying fibred manifold in general: (1) It is affine in accelerations (second derivatives of coordinates) and (2) it is closely related to the (first order) Lagrange structure on the given fibred manifold. Main properties of dynamical forms were described in Sect. 3.4, including corresponding dynamical distributions. So, for describing the motion of variational mechanical systems the following scheme is used:

→ Lagrange structure (π, λ)
→ Lepage equivalent θ_λ
→ Euler-Lagrange form E_λ and/or the mechanical system $[\alpha]$
→ equations of motion $E_\lambda \circ J^2\gamma = 0$.

In general, the equations of motion of classical mechanical systems with m degrees of freedom have the form $A_\sigma(t, q^\nu, \dot{q}^\nu) + B_\sigma(t, q^\nu, \dot{q}^\nu)\ddot{q}^\nu = 0, 1 \leq \sigma, \nu \leq m$, i.e. they form the system of second-order ordinary differential equations. We can consider the left-hand sides of these equations as components E_σ of the dynamical form $E = E_\sigma(t, q^\nu, \dot{q}^\nu, \ddot{q}^\nu)\omega^\sigma \wedge dt$. Trajectories of the corresponding mechanical system considered as sections of the underlying fibred manifold can be described by the relation $E \circ J^2\gamma = 0$. Nevertheless, the form E need not necessarily come from

J. Musilová et al., *Calculus of Variations on Fibred Manifolds and Variational Physics*, Lecture Notes in Physics 1033, https://doi.org/10.1007/978-3-031-77408-9_5

some Lagrangian. This means that for the form E there may be no Lagrangian. So, the problem is to find the criteria under what the dynamical form E affine in accelerations is *variational*, i.e. it comes from a Lagrangian. If the answer is positive, there is a second task—to find all Lagrangians corresponding to this dynamical form. Obviously, such Lagrangians differ from one another by a trivial Lagrangian. As the "basic" the Lagrangian of the minimum possible order is considered, called the *minimal Lagrangian*. The just formulated problem is called the *inverse problem of calculus of variations*. In this chapter, we present the complete solution of the local version of this problem for second-order dynamical forms affine in accelerations. The variationality of first-order dynamical forms will be discussed as well.

Prerequisites vector fields, distributions, differential forms, closed and exact differential forms, Chap. 4. □

5.1 A Note to Higher Order Lagrange Structures

We formulated the problem of variationality of a dynamical form not too precisely as the requirement to decide whether a dynamical form (defined on J^2Y and affine in accelerations) arises from a (standardly first order) Lagrangian as its Euler-Lagrange form. In Chap. 4 we have seen that a first-order Lagrangian regular in the sense of the "historical" concept of regularity, i.e. the regularity of the matrix formed by its second derivatives with respect to variables $\dot{q}^\sigma$, leads to the Euler Lagrange form on J^2Y affine in variables $\ddot{q}^\sigma$, $1 \leq \sigma \leq m$. On the other hand, it was to some extent surprising, that the Euler-Lagrange form of a first-order Lagrangian (affine in variables $\dot{q}^\sigma$) can be of the first order too. Such a Lagrange structure can be again regular, in the sense of regularity of the dynamical system. In general, for a first-order Lagrange structure the Euler-Lagrange form must be either of the second order and affine in accelerations, or of the first order. However, on the basis of these facts some questions arise:

- Are there some special *second order* (or higher order, in general) Lagrangians leading to Euler-Lagrange forms on J^2Y affine in acceleration?
- Are there simultaneously first-order Lagrangians leading to this Euler-Lagrange form?

If the answers are positive, then there obviously exist higher order *trivial* Lagrangians—the differences between higher order and first-order Lagrangians. Since we wish to solve the inverse problem of calculus of variations lying in finding the criteria for the second-order dynamical form affine in accelerations to be the Euler-Lagrange form of a first-order Lagrangian, we must answer the above questions. This requires to generalize the concept of the Lagrange structure to higher jet prolongations of fibred manifolds—for our purposes, this is specifically the second order. We perform the generalization rather formally, only for the purpose to answer the first of both questions and to formulate the precise definition of variationality of a dynamical form. (Note that all possible questions concerning the variationality problem are answered in detail in the monograph [1].)

In full analogy with the approach in Sect. 4.2 we introduce r-th Lagrange structure, the Lepage equivalent of the Lagrangian and the Euler-Lagrange form by definition, calculations in charts are made for $r = 2$. Generalization to arbitrary order r is straightforward, but the calculations are formally somewhat more complicated in specific situations.

Definition 5.1 (*Lagrange structure, Lagrangian, Euler-Lagrange (EL) form of the r-th order*) Let (Y, π, X) be a fibred manifold. The *r-th order Lagrange structure* is the pair (π, λ), where λ is a horizontal 1-form on J^rY, $\lambda \in \Lambda^1_X(TJ^rY)$. The *Lepage equivalent* of the Lagrangian λ is the 1-form θ_λ on J^rY, such that $p_1\,\mathrm{d}\theta_\lambda$ is $\pi_{(r+1,0)}$-horizontal form and $h\theta_\lambda = \pi^*_{(r+1,r)}\lambda$. The form $E_\lambda = p_1\,\mathrm{d}\theta_\lambda$ is called the *Euler-Lagrange form* of the Lagrangian λ.

We prove the existence and uniqueness of the Lepage equivalent of a second-order Lagrangian by the quite analogous calculation as in Sect. 4.2 (relations (4.4))–(4.6)). Let $(\mathrm{d}t, \omega^\sigma, \tilde{\omega}^\sigma, \hat{\omega}^\sigma)$ be the basis of 1-forms on J^3Y adapted to the contact structure. We search a contact 1-form η on J^2Y such that the form $i_{J^3\xi}\,p_1\,\mathrm{d}(\lambda + \eta)$ depend only on components ξ^0 and ξ^ν, $1 \le \nu \le m$, of a π-projectable vector field on Y. Denote $\eta = \eta_\sigma\omega^\sigma + \tilde{\eta}_\sigma\tilde{\omega}^\sigma$, the components η_σ, $\tilde{\eta}_\sigma$ being functions on J^2Y. After technical calculations we obtain

$$\begin{aligned} p_1\,\mathrm{d}(\lambda + \eta) = {} & \left(\frac{\partial L}{\partial q^\sigma} - \frac{\mathrm{d}\eta_\sigma}{\mathrm{d}t}\right)\omega^\sigma \wedge \mathrm{d}t + \left(\frac{\partial L}{\partial \dot{q}^\sigma} - \eta_\sigma - \frac{\mathrm{d}\tilde{\eta}_\sigma}{\mathrm{d}t}\right)\tilde{\omega}^\sigma \wedge \mathrm{d}t \\ & + \left(\frac{\partial L}{\partial \ddot{q}^\sigma} - \tilde{\eta}_\sigma\right)\hat{\omega}^\sigma \wedge \mathrm{d}t\,. \end{aligned}$$

The requirement of the dependence of the form $i_{J^3\xi}\,p_1\,\mathrm{d}(\lambda + \eta)$ only on components of the vector field ξ itself leads to the equations

$$\frac{\partial L}{\partial \dot{q}^\sigma} - \eta_\sigma - \frac{\mathrm{d}\tilde{\eta}_\sigma}{\mathrm{d}t} = 0, \quad \frac{\partial L}{\partial \ddot{q}^\sigma} - \tilde{\eta}_\sigma = 0, \quad \sigma = 1, \dots, m\,.$$

The form $\theta_\lambda = \lambda + \eta$ is

$$\theta_\lambda = L\,\mathrm{d}t + \left(\frac{\partial L}{\partial \dot{q}^\sigma} - \frac{\mathrm{d}}{\mathrm{d}t}\frac{\partial L}{\partial \ddot{q}^\sigma}\right)\omega^\sigma + \frac{\partial L}{\partial \ddot{q}^\sigma}\tilde{\omega}^\sigma\,.$$

We obtained the Lepage equivalent of the Lagrangian λ. The corresponding Euler-Lagrange form is

$$E_\lambda = p_1\,\mathrm{d}\theta_\lambda = \left(\frac{\partial L}{\partial q^\sigma} - \frac{\mathrm{d}}{\mathrm{d}t}\frac{\partial L}{\partial \dot{q}^\sigma} + \frac{\mathrm{d}^2}{\mathrm{d}t^2}\frac{\partial L}{\partial \ddot{q}^\sigma}\right)\mathrm{d}q^\sigma \wedge \mathrm{d}t\,.$$

The double application of the total derivative operator means that the form E_λ is of fourth order, in general. We are interested in the conditions under which the form E_λ is only second order, and moreover affine in coordinates. We will calculate the components of the Euler-Lagrange form as polynomials in third and fourth derivatives of coordinates, the coefficients being functions on J^2Y. We use the total derivative operator for functions f on J^2Y and g on J^3Y, respectively, as follows:

$$\frac{\mathrm{d}f}{\mathrm{d}t} = \left(\frac{\partial f}{\partial t} + \frac{\partial f}{\mathrm{d}q^\sigma}\dot{q}^\sigma + \frac{\partial f}{\partial \dot{q}^\sigma}\ddot{q}^\sigma\right) + \frac{\partial f}{\partial \ddot{q}^\sigma} q^\sigma_{(3)} = \frac{\mathrm{d}' f}{\mathrm{d}t} + \frac{\partial f}{\partial \ddot{q}^\sigma} q^\sigma_{(3)} ,$$
$$\frac{\mathrm{d}g}{\mathrm{d}t} = \left(\frac{\partial g}{\partial t} + \frac{\partial g}{\mathrm{d}q^\sigma}\dot{q}^\sigma + \frac{\partial g}{\partial \dot{q}^\sigma}\ddot{q}^\sigma + \frac{\partial g}{\partial \ddot{q}^\sigma} q^\sigma_{(3)}\right) + \frac{\partial g}{\partial q^\sigma_{(3)}} q^\sigma_{(4)} = \frac{\mathrm{d}' g}{\mathrm{d}t} + \frac{\partial g}{\partial q^\sigma_{(3)}} q^\sigma_{(4)} ,$$

where $q^\sigma_{(3)}$ and $q^\sigma_{(4)}$ denote the third and fourth derivative. For E_σ we obtain after some uncomfortable calculations (perform them in detail).

$$\begin{aligned}
E_\sigma = {} & \frac{\partial L}{\partial q^\sigma} - \frac{\mathrm{d}}{\mathrm{d}t}\frac{\partial L}{\partial \dot{q}^\sigma} + \frac{\mathrm{d}^2}{\mathrm{d}t^2}\frac{\partial L}{\partial \ddot{q}^\sigma} = \left(\frac{\partial L}{\partial q^\sigma} - \frac{\mathrm{d}'}{\mathrm{d}t}\frac{\partial L}{\partial \dot{q}^\sigma} + \frac{\mathrm{d}'^2}{\mathrm{d}t^2}\frac{\partial L}{\partial \ddot{q}^\sigma}\right) \\
& + \left(-\frac{\partial^2 L}{\partial \dot{q}^\sigma \partial \ddot{q}^\nu} + \frac{\partial}{\partial \ddot{q}^\nu}\left(\frac{\mathrm{d}'}{\mathrm{d}t}\frac{\partial L}{\partial \ddot{q}^\sigma}\right) + \frac{\mathrm{d}'}{\mathrm{d}t}\frac{\partial^2 L}{\partial \ddot{q}^\nu \partial \ddot{q}^\sigma}\right) q^\nu_{(3)} \\
& + \frac{\partial^3 L}{\partial \ddot{q}^\mu \partial \ddot{q}^\nu \partial \ddot{q}^\sigma} q^\mu_{(3)} q^\nu_{(3)} + \frac{\partial^2 L}{\partial \ddot{q}^\nu \ddot{q}^\sigma} q^\nu_{(4)} .
\end{aligned}$$

The conditions for vanishing the terms of the fourth and third order are

$$0 = \frac{\partial^2 L}{\partial \ddot{q}^\nu \partial \ddot{q}^\sigma} \implies L = a(t, q^\nu, \dot{q}^\nu) + b_\sigma(t, q^\nu, \dot{q}^\nu)\ddot{q}^\sigma , \tag{5.1}$$
$$0 = \frac{\partial^3 L}{\partial \ddot{q}^\mu \partial \ddot{q}^\nu \partial \ddot{q}^\sigma} , \tag{5.2}$$
$$0 = -\frac{\partial^2 L}{\partial \dot{q}^\sigma \partial \ddot{q}^\nu} + \frac{\partial}{\partial \ddot{q}^\nu}\left(\frac{\mathrm{d}'}{\mathrm{d}t}\frac{\partial L}{\partial \ddot{q}^\sigma}\right) + \frac{\mathrm{d}'}{\mathrm{d}t}\frac{\partial^2 L}{\partial \ddot{q}^\nu \partial \ddot{q}^\sigma} , \tag{5.3}$$

for all $1 \le \sigma, \nu, \mu \le m$. If conditions (5.1) are satisfied, then as a consequence conditions (5.2) are fulfilled and conditions (5.3) reduce to

$$\frac{\partial^2 L}{\partial \dot{q}^\sigma \partial \ddot{q}^\nu} - \frac{\partial^2 L}{\partial \dot{q}^\nu \partial \ddot{q}^\sigma} = 0 .$$

The Lagrangian then reads

$$\lambda = (a + b_\sigma \ddot{q}^\sigma)\,\mathrm{d}t, \quad a = a(t, q^\nu, \dot{q}^\nu), \quad b = b(t, q^\nu, \dot{q}^\nu), \quad \frac{\partial b_\sigma}{\partial \dot{q}^\nu} - \frac{\partial b_\nu}{\partial \dot{q}^\sigma} = 0 .$$

The Euler-Lagrange form is then of the second order. After calculating its components we even find that they are affine in accelerations. Thus we can conclude that a second-order Euler-Lagrange forms affine in accelerations arise not only from first-order Lagrangians, but also from special second-order Lagrangians. However, every such second-order Lagrangian is a sum of a first-order Lagrangian and a trivial Lagrangian, the second of them being the total derivative of a function $f = f(t, q^{\nu}, \dot{q}^{\nu})$. Concretely, f can be an arbitrary function satisfying the condition $b_{\sigma} = \frac{\partial f}{\partial \dot{q}^{\sigma}}$ for $1 \leq \sigma \leq m$.

Proposition 5.1 *Let λ be a Lagrangian. Suppose that its Euler-Lagrange form E_{λ} is of the second order, i.e. $E_{\lambda} \in \Lambda^2_Y(TJ^2Y)$, and that it is affine in variables $\ddot{q}^{\sigma}$, $1 \leq \sigma \leq m$. Then the class $[\lambda]$ of all Lagrangians producing this form contains a first-order representative.*

Proof The calculations and conclusions preceding the proposition represent the proof. □

The consideration in this section confirms that *all* second-order Euler-Lagrange forms affine in accelerations arise from first-order Lagrange structures. Solving the inverse problem for dynamical forms of this type we will completely answer the questions concerning the variationality of equations of motion relevant for classical mechanics.

5.2 Variational Dynamical Forms

This section introduces the geometrical concept of the variationality of a dynamical form E on J^2Y, based on the properties of the dynamical system $[\alpha]$ assigned to this dynamical form. With respect to result obtained in the previous section we consider exclusively dynamical forms on J^2Y affine in variables $\ddot{q}^{\sigma}$, $1 \leq \sigma \leq m$, i.e. of the type (3.28):

$$E = E_{\sigma}\, dq^{\sigma} \wedge dt = \left(A_{\sigma} + B_{\sigma\nu}\ddot{q}^{\nu}\right) dq^{\sigma} \wedge dt \ ,$$

where $A_{\sigma} = A_{\sigma}(t, q^{\nu}, \dot{q}^{\nu})$ and $B_{\sigma\nu} = B_{\sigma\nu}(t, q^{\nu}, \dot{q}^{\nu})$ are functions.

Definition 5.2 (*Variationality*) A dynamical form $E \in \Lambda^2_Y(TJ^2Y)$, $E = E_{\sigma}\, dq^{\sigma} \wedge dt$ is called *globally variational*, if there exists an integer $r \geq 1$ and a *global* Lagrangian λ on J^rY such that $E = E_{\lambda}$, up to a possible projection. This dynamical form E is called *locally variational*, if for every point $\hat{y} \in J^2Y$ there exists its open neighborhood W_2 such that the restriction E_{W_2} is variational.

Taking into account the analysis in the preceding Sect. 5.1 we know that for every Euler-Lagrange form the class of Lagrangians leading to this form necessarily contains first-order Lagrangians. However, this class contains also higher order Lagrangians having the same Euler-Lagrange form. The uncertainty of the equality $E = E_{\lambda}$ up to a projection means the following: If the Lagrangian is of the r-th order then its Lepage equivalent is formally of the order $(2r-1)$ and the Euler-Lagrange form $E_\lambda = p_1\,\mathrm{d}\theta_\lambda$ is of the order $2r$. So, for $r=2$, $E_\lambda \in \Lambda^2_Y(TJ^4Y)$ for the variational dynamical form E. In this chapter, we will study only local variationality of dynamical forms. The global variationality will only be mentioned in Chap. 7. (General discussion of global existence of local structures on manifolds is one of the rather difficult problems connected with topological questions solved, e.g. in [2–4] and others.)

Theorem 5.1 *Let (Y, π, X) be a fibred manifold and $E \in \Lambda^2_Y(TJ^2Y)$ a dynamical form affine in variables $\ddot{q}^\sigma$, $1 \le \sigma \le m$. The form E is locally variational if and only if its dynamical system contains a closed representative. Such a representative (denoted as α_E) is then unique.*

Proof First suppose that the form E is locally variational, i.e. every point $\hat{y} \in J^2Y$ has an open neighborhood W_2 such that the restriction of the form E on this neighborhood is variational. This means that there exists a Lagrangian such that its Euler-Lagrange form equals this restriction up to a possible projection. However, by Proposition 5.1 we know that between all such Lagrangians there exists a *first order* one. Denote it as λ and as θ_λ its Lepage equivalent. We can of course choose such a neighborhood W_2 for which λ is defined on $\pi_{2,1}(W_2)$. Then we can write $E_{W_2} = E_\lambda$ (strictly). Denote $[\alpha]$ the dynamical system of the form E_{W_2}. The proof of the existence of its close representative is very simple, because $\mathrm{d}\theta_\lambda \in [\alpha]$. Indeed, it holds

$$\begin{aligned}\mathrm{d}\theta_\lambda = &\left(\frac{\partial L}{\partial q^\sigma} - \frac{\mathrm{d}'}{\mathrm{d}t}\frac{\partial L}{\partial \dot{q}^\sigma}\right)\omega^\sigma \wedge \mathrm{d}t + \left(-\frac{\partial^2 L}{\partial \dot{q}^\sigma \partial \dot{q}^\nu}\right)\omega^\sigma \wedge \mathrm{d}\dot{q}^\nu \\ &+ \left\{\frac{\partial^2 L}{\partial q^\sigma \partial \dot{q}^\nu}\right\}_{\mathrm{alt}(\sigma,\nu)} \omega^\sigma \wedge \omega^\nu \,.\end{aligned}$$

Let us study the uniqueness. Suppose that α is another closed representative of the dynamical system. We prove that $\alpha = \mathrm{d}\theta_\lambda$. As it holds $p_1\alpha = p_1\,\mathrm{d}\theta_\lambda$, then

$$\alpha - \mathrm{d}\theta_\lambda = G_{\sigma\nu}\,\omega^\sigma \wedge \omega^\nu \,,$$

where $G_{\sigma\nu}$ are some functions on J^1Y, and $G_{\sigma\nu} + G_{\nu\sigma} = 0$. As α is supposed to be a closed representative as well, we obtain $\mathrm{d}\,(G_{\sigma\nu}\,\omega^\sigma \wedge \omega^\nu) = 0$, i.e.

$$\begin{aligned}&\frac{\mathrm{d}' G_{\sigma\nu}}{\mathrm{d}t}\omega^\sigma \wedge \omega^\nu \wedge \mathrm{d}t + \left\{\frac{\partial G_{\sigma\nu}}{\partial q^\mu}\right\}_{\mathrm{alt}\,(\nu,\mu)} \omega^\sigma \wedge \omega^\nu \wedge \omega^\mu \\ &+ \frac{\partial G_{\sigma\nu}}{\partial \dot{q}^\mu}\omega^\sigma \wedge \omega^\nu \wedge \mathrm{d}\dot{q}^\mu + (G_{\sigma\nu} - G_{\nu\sigma})\,\omega^\sigma \wedge \mathrm{d}\dot{q}^\nu \wedge \mathrm{d}t = 0 \,.\end{aligned}$$

All components of this form vanish. The equation $G_{\sigma\nu} - G_{\nu\sigma} = 0$ together with the antisymmetry relation $G_{\sigma\nu} + G_{\nu\sigma} = 0$ gives $G_{\sigma\nu} = 0$ for $1 \le \sigma, \nu \le m$. Thus $\alpha_E = \mathrm{d}\theta_\lambda$ which proves the uniqueness of the closed representative of the dynamical system of an arbitrary Lagrange structure.

Conversely, suppose that the dynamical system $[\alpha]$ of the form E_{W_2} contains a closed representative α_E, i.e. $\mathrm{d}\alpha_E = 0$, supposedly on $\pi_{2,1}(W_2)$. Then by the Poincaré lemma (Theorem 3.2a)) there exists a 1-form ϱ such that $\alpha_E = \mathrm{d}\varrho$. For obtaining this form we use the results of Problem 3.15 putting $P_\sigma = 0$, $Q_{\sigma\nu} = 0$, $1 \leq \sigma, \nu \leq m$, into relations (9.16) or (9.17). We obtain

$$\begin{aligned}
\pi^*_{2,1}\varrho = & \left[q^\sigma \int_0^1 (E_\sigma \circ \chi)\, \mathrm{d}u \right] \mathrm{d}t \\
& + \left[2q^\nu \int_0^1 u(F_{\nu\sigma} \circ \chi)\, \mathrm{d}u - \dot{q}^\nu \int_0^1 u(B_{\sigma\nu} \circ \chi)\, \mathrm{d}u \right] \omega^\sigma \\
& + \left[q^\nu \int_0^1 u(B_{\nu\sigma} \circ \chi)\, \mathrm{d}u \right] \dot{\omega}^\sigma \,, \\
\varrho = & \left[q^\sigma \int_0^1 (A_\sigma \circ \chi)\, \mathrm{d}u \right] \mathrm{d}t \\
& + \left[2q^\nu \int_0^1 u(F_{\nu\sigma} \circ \chi)\, \mathrm{d}u - \dot{q}^\nu \int_0^1 u(B_{\sigma\nu} \circ \chi)\, \mathrm{d}u \right] \omega^\sigma \\
& + \left[q^\nu \int_0^1 u(B_{\nu\sigma} \circ \chi)\, \mathrm{d}u \right] \mathrm{d}\dot{q}^\sigma \,.
\end{aligned}$$

Evidently, it holds $E_{W_2} = p_1\alpha_E = p_1\, \mathrm{d}\varrho$. Denote

$$\hat{\lambda} = h\varrho = h\mathcal{A}\alpha = \left[q^\sigma \int_0^1 (E_\sigma \circ \chi)\, \mathrm{d}u \right] \mathrm{d}t \,.$$

This is the Lagrangian defined on W_2, i.e. second-order Lagrangian, called the *Vainberg-Tonti Lagrangian* and it will be discussed in Sect. 5.4. It holds $E_{W_2} = E_{\hat{\lambda}}$. This proves that the form E is locally variational. The uniqueness of the closed representative α_E was shown in the first part of the proof. □

Definition 5.3 (*Lepage equivalent of Euler-Lagrange form, Lepage mapping*) Let E be a (locally) variational dynamical form. The closed representative α_E of its dynamical system is called the *Lepage equivalent* of the form E. The mapping assigning to a locally variational form its Lepage equivalent is called the *Lepage mapping* (of the second type).

Lepage forms and Lepage mappings are closely connected with the concept of variational sequences. Some details concerning these forms and mappings will be given in Chap. 7.

Let us consider a generalization of the second-order inverse problem. Suppose that $E = E_\sigma \omega^\sigma \wedge \mathrm{d}t$ be a general dynamical form on J^2Y, i.e. $E_\sigma = E_\sigma(t, q^\nu, \dot{q}^\nu, \ddot{q}^\nu)$ are functions on J^2Y (without other specification of their form). The following theorem gives the criterion of variationality of such a form.

Theorem 5.2 *A dynamical form*

$$E = E_\sigma \omega^\sigma \wedge \mathrm{d}t \in \Lambda^2_Y(TJ^2Y), \; E_\sigma = E_\sigma(t, q^\nu, \dot{q}^\nu, \ddot{q}^\nu),$$

is locally variational if and only if to every point $\hat{y} \in J^2Y$ there exists an open neighborhood W_2 and a 2-contact 2-form F such that $\mathrm{d}(E + F) = 0$ *on* W_2.

Proof Suppose that E is locally variational. Then by definition for an arbitrarily chosen point $\hat{y}$ there exists an open neighborhood W_2 and a Lagrangian λ such that $E_{W_2} = E_\lambda$ up to a possible projection. Then $E_{W_2} = p_1 \, \mathrm{d}\theta_\lambda$, where θ_λ is the Lepage equivalent (Cartan form) of the Lagrangian λ, again up to a projection. Denoting $F = p_2 \, \mathrm{d}\theta_\lambda$ we obtain

$$\mathrm{d}\theta_\lambda = p_1 \mathrm{d}\theta_\lambda + p_2 \mathrm{d}\theta_\lambda = E_{W_2} + F \; ,$$

up to a projection. The 2-form F is 2-contact and the form $E_{W_2} + F$ is closed, $\mathrm{d}(E_{W_2} + F) = \mathrm{d}^2\theta_\lambda = 0$.

Conversely, suppose that to every point $\hat{y} \in J^2Y$ there exists an open neighborhood W_2 and a 2-contact 2-form F such that $\mathrm{d}(E_{W_2} + F) = 0$. Denote $\alpha = E_{W_2} + F$. This form is $\pi_{2,1}$-projectable and $p_1\alpha = E_{W_2}$. Due to its closeness and by the Poincaré lemma (Theorem 3.2) there exists a form ϱ such that $\mathrm{d}\varrho = \alpha$. It holds $\varrho = \mathcal{A}\alpha$, where $\mathcal{A}$ is the operator specified by the Poincaré lemma. The form $\lambda = h\varrho = h\mathcal{A}\alpha$ is the Lagrangian and E_{W_2} is its Euler-Lagrange form. So, the form E is locally variational. □

5.3 Helmholtz Conditions of Variationality

In the previous section, we have characterized locally variational dynamical forms E on J^2Y (for E on J^1Y see Problem 5.1). Theorem 5.2 gives the geometrical condition leading to requirements for coefficients of a dynamical form ensuring its variationality. We now derive these requirements. Let $E = E_\sigma \omega^\sigma \wedge \mathrm{d}t$, where $E_\sigma = A_\sigma + B_{\sigma\nu}\ddot{q}^\nu$ and $A_\sigma = A_\sigma(t, q^\mu, \dot{q}^\mu)$ and $B_{\sigma\nu} = B_{\sigma\nu}(t, q^\mu, \dot{q}^\mu)$, $1 \le \sigma, \nu, \mu \le m$, are functions on J^1Y. The general representative of the corresponding dynamical system is of the form (3.29), i.e.

$$\alpha = A_\sigma \omega^\sigma \wedge \mathrm{d}t + B_{\sigma\nu}\omega^\sigma \wedge \mathrm{d}\dot{q}^\nu + F_{\sigma\nu}\omega^\sigma \wedge \omega^\nu \; ,$$

where $F_{\sigma\nu} = F_{\sigma\nu}(t, q^\mu, \dot{q}^\mu)$ are free functions on $J^1 Y$ antisymmetric in indices σ and ν. Putting $\mathrm{d}\alpha = 0$ we obtain the conditions for functions A_σ and $B_{\sigma\nu}$ under which the dynamical form E is locally variational. Moreover, we determine the functions $F_{\sigma\nu}$. We proceed on $J^1 Y$ and use the relations of the type

$$\begin{aligned} \mathrm{d}f &= \frac{\mathrm{d}' f}{\mathrm{d}t}\,\mathrm{d}t + \frac{\partial f}{\partial q^\sigma}\,\omega^\sigma + \frac{\partial f}{\partial \dot{q}^\sigma}\,\mathrm{d}\dot{q}^\sigma\,, \\ \mathrm{d}\alpha &= \mathrm{d}A_\sigma \wedge \omega^\sigma \wedge \mathrm{d}t + \mathrm{d}B_{\sigma\nu} \wedge \omega^\sigma \wedge \mathrm{d}\dot{q}^\nu + B_{\sigma\nu}\,\mathrm{d}\omega^\sigma \wedge \mathrm{d}\dot{q}^\nu \\ &\quad + \mathrm{d}F_{\sigma\nu} \wedge \omega^\sigma \wedge \omega^\nu + F_{\sigma\nu}\,\mathrm{d}\omega^\sigma \wedge \omega^\nu - F_{\sigma\nu}\omega^\sigma \wedge \mathrm{d}\omega^\nu\,. \end{aligned}$$

After calculations and rearrangements of expressions we obtain

$$\begin{aligned} \mathrm{d}\alpha \; = & \left(\left\{ \frac{\partial A_\nu}{\partial q^\sigma} \right\}_{\mathrm{alt}\,(\sigma,\nu)} + \frac{\mathrm{d}' F_{\sigma\nu}}{\mathrm{d}t} \right) \omega^\sigma \wedge \omega^\nu \wedge \mathrm{d}t \\ & + \left(\frac{\mathrm{d}' B_{\sigma\nu}}{\mathrm{d}t} - \frac{\partial A_\sigma}{\partial \dot{q}^\nu} + 2F_{\sigma\nu} \right) \omega^\sigma \wedge \mathrm{d}\dot{q}^\nu \wedge \mathrm{d}t \\ & + \{B_{\sigma\nu}\}_{\mathrm{alt}(\sigma,\nu)}\, \mathrm{d}\dot{q}^\sigma \wedge \mathrm{d}\dot{q}^\nu \wedge \mathrm{d}t + \left\{ \frac{\partial F_{\sigma\nu}}{\partial q^\mu} \right\}_{\mathrm{alt}\,(\sigma,\nu)} \omega^\sigma \wedge \omega^\nu \wedge \omega^\mu \\ & + \left(\left\{ \frac{\partial B_{\nu\mu}}{\partial q^\sigma} \right\}_{\mathrm{alt}\,(\sigma,\nu)} + \frac{\partial F_{\sigma\nu}}{\partial \dot{q}^\mu} \right) \omega^\sigma \wedge \omega^\nu \wedge \mathrm{d}\dot{q}^\mu + \left\{ \frac{\partial B_{\sigma\nu}}{\partial \dot{q}^\mu} \right\}_{\mathrm{alt}\,(\mu,\nu)} \omega^\sigma \wedge \mathrm{d}\dot{q}^\nu \wedge \mathrm{d}\dot{q}^\mu\,. \end{aligned}$$

The condition $\mathrm{d}\alpha = 0$ leads to zero components, i.e.

$$0 = \left\{ \frac{\partial A_\nu}{\partial q^\sigma} \right\}_{\mathrm{alt}\,(\sigma,\nu)} + \frac{\mathrm{d}' F_{\sigma\nu}}{\mathrm{d}t}\,, \tag{5.4}$$

$$0 = \frac{\mathrm{d}' B_{\sigma\nu}}{\mathrm{d}t} - \frac{\partial A_\sigma}{\partial \dot{q}^\nu} + 2F_{\sigma\nu} \implies \left[\frac{\partial A_\sigma}{\partial \dot{q}^\nu} - \frac{\mathrm{d}' B_{\sigma\nu}}{\mathrm{d}t} \right]_{\mathrm{sym}\,(\sigma,\nu)} = 0\,, \tag{5.5}$$

$$0 = \{B_{\sigma\nu}\}_{\mathrm{alt}\,(\sigma,\nu)} \implies B_{\sigma\nu} = B_{\nu\sigma}\,, \tag{5.6}$$

$$0 = \left\{ \frac{\partial F_{\sigma\nu}}{\partial q^\mu} \right\}_{\mathrm{alt}\,(\sigma,\nu)}\,, \tag{5.7}$$

$$0 = \left\{ \frac{\partial B_{\nu\mu}}{\partial q^\sigma} \right\}_{\mathrm{alt}\,(\sigma,\nu)} + \frac{\partial F_{\sigma\nu}}{\partial \dot{q}^\mu}\,, \tag{5.8}$$

$$0 = \left\{ \frac{\partial B_{\sigma\nu}}{\partial \dot{q}^\mu} \right\}_{\mathrm{alt}\,(\mu,\nu)} \implies \frac{\partial B_{\sigma\nu}}{\partial \dot{q}^\mu} = \frac{\partial B_{\sigma\mu}}{\partial \dot{q}^\nu}\,.$$

These conditions are not independent. For obtaining independent conditions we put (5.6) into (5.5) and express $F_{\sigma\nu}$ from (5.5). Then put $F_{\sigma\nu}$ into (5.4). Conditions (5.7) and (5.8) depend on remaining ones. Finally we obtain the independent set of *Helmholtz conditions of variationality*.

Proposition 5.2 *Let E be a dynamical form on J^2Y, $E = E_\sigma \omega^\sigma \wedge \mathrm{d}t$, $E_\sigma = A_\sigma + B_{\sigma\nu}\ddot{q}^\nu$, A_σ and $B_{\sigma\nu}$ being functions on J^1Y. The form E is locally variational if and only if it holds*

$$B_{\sigma\nu} = B_{\nu\sigma}\,, \tag{5.9}$$

$$\frac{\partial B_{\sigma\nu}}{\partial \dot{q}^\mu} = \frac{\partial B_{\sigma\mu}}{\partial \dot{q}^\nu}\,, \tag{5.10}$$

$$\frac{\partial A_\sigma}{\partial \dot{q}^\nu} + \frac{\partial A_\nu}{\partial \dot{q}^\sigma} = 2\frac{\mathrm{d}' B_{\sigma\nu}}{\mathrm{d}t}\,, \tag{5.11}$$

$$\frac{\partial A_\sigma}{\partial q^\nu} - \frac{\partial A_\nu}{\partial q^\sigma} = \frac{1}{2}\frac{\mathrm{d}'}{\mathrm{d}t}\left(\frac{\partial A_\sigma}{\partial \dot{q}^\nu} - \frac{\partial A_\nu}{\partial \dot{q}^\sigma}\right)\,. \tag{5.12}$$

The function $F_{\sigma\nu}$ are then given uniquely and

$$F_{\sigma\nu} = \frac{1}{4}\left(\frac{\partial A_\sigma}{\partial \dot{q}^\nu} - \frac{\partial A_\nu}{\partial \dot{q}^\sigma}\right)\,. \tag{5.13}$$

The (unique) closed representative of a locally variational dynamical form $(A_\sigma + B_{\sigma\nu}\ddot{q}^\nu)\omega^\sigma \wedge \mathrm{d}t$, $A_\sigma, B_{\sigma\nu} \in \Lambda^0(TJ^1Y)$ (functions on J^1Y), then reads

$$\alpha_E = A_\sigma \omega^\sigma \wedge \mathrm{d}t + B_{\sigma\nu}\omega^\sigma \wedge \mathrm{d}\dot{q}^\nu + \frac{1}{4}\left(\frac{\partial A_\sigma}{\partial \dot{q}^\nu} - \frac{\partial A_\nu}{\partial \dot{q}^\sigma}\right)\omega^\sigma \wedge \omega^\nu\,.$$

Another version of Helmholtz conditions of variationality is expressed directly via components E_σ of the dynamical form E under investigation:

Proposition 5.3 *Let E be a dynamical form on J^2Y, $E = E_\sigma \omega^\sigma \wedge \mathrm{d}t$, $E_\sigma = A_\sigma + B_{\sigma\nu}\ddot{q}^\nu$, A_σ and $B_{\sigma\nu}$ being functions on J^1Y. The form E is locally variational if and only if it holds*

$$\frac{\partial E_\sigma}{\partial \ddot{q}^\nu} - \frac{\partial E_\nu}{\partial \ddot{q}^\sigma} = 0\,, \tag{5.14}$$

$$\frac{\partial E_\sigma}{\partial \dot{q}^\nu} + \frac{\partial E_\nu}{\partial \dot{q}^\sigma} - \frac{\mathrm{d}}{\mathrm{d}t}\left(\frac{\partial E_\sigma}{\partial \ddot{q}^\nu} + \frac{\partial E_\nu}{\partial \ddot{q}^\sigma}\right) = 0\,, \tag{5.15}$$

$$\frac{\partial E_\sigma}{\partial q^\nu} - \frac{\partial E_\nu}{\partial q^\sigma} - \frac{1}{2}\frac{\mathrm{d}}{\mathrm{d}t}\left(\frac{\partial E_\sigma}{\partial \dot{q}^\nu} - \frac{\partial E_\nu}{\partial \dot{q}^\sigma}\right) = 0\,. \tag{5.16}$$

Proof This proposition can be easily proved by the direct calculation. We express the derivatives of expressions E_σ as follows:

$$\frac{\partial E_\sigma}{\partial \ddot{q}^\nu} = B_{\sigma\nu}, \quad \frac{\partial E_\sigma}{\partial \dot{q}^\nu} = \frac{\partial A_\sigma}{\partial \dot{q}^\nu} + \frac{\partial B_{\sigma\mu}}{\partial \dot{q}^\nu}\ddot{q}^\mu, \quad \frac{\partial E_\sigma}{\partial q^\nu} = \frac{\partial A_\sigma}{\partial q^\nu} + \frac{\partial B_{\sigma\mu}}{\partial q^\nu}\ddot{q}^\mu\,.$$

Then we express the left-hand sides of conditions (5.14), (5.15) and (5.16) and take into account Helmholtz conditions from Proposition 5.2 to make certain that they vanish. □

We will show that Proposition 5.3 can be generalized for an *arbitrary* dynamical form $E = E_\sigma \omega^\sigma \wedge dt$ on J^2Y, i.e. without anticipating that the expressions E_σ are affine in variables $\ddot{q}^\nu$. On the other hand, we proved in Sect. 5.1 that every second-order Euler-Lagrange form (i.e. the variational dynamical form from the point of view of the definition of variationality) is affine in $\ddot{q}^\nu$. We will see that this fact will follow from the generalized Proposition 5.3 as a consequence.

Theorem 5.3 *Let $E = E_\sigma \omega^\sigma \wedge dt \in \Lambda^2_Y(TJ^2Y)$ be a variational dynamical form, E_σ being (arbitrary) functions on J^2Y. Then*

(a) functions E_σ satisfy Helmholtz conditions (5.14), (5.15) and (5.16),
(b) there exists a unique closed 2-form α projectable on J^1Y such that $E = p^1\alpha$,

$$\alpha = \left(E_\sigma - \ddot{q}^\nu \left[\frac{\partial E_\sigma}{\partial \ddot{q}^\nu} \right]_{\mathrm{sym}\,(\sigma,\nu)} \right) \omega^\sigma \wedge dt \tag{5.17}$$
$$+ \left[\frac{\partial E_\sigma}{\partial \ddot{q}^\nu} \right]_{\mathrm{sym}\,(\sigma,\nu)} \omega^\sigma \wedge d\dot{q}^\nu + \frac{1}{2} \left\{ \frac{\partial E_\sigma}{\partial \dot{q}^\nu} \right\}_{\mathrm{alt}\,(\sigma,\nu)} \omega^\sigma \wedge \omega^\nu ,$$

(c) the form E is affine in coordinates $\ddot{q}^\nu$, i.e. $E_\sigma = A_\sigma + B_{\sigma\nu}\ddot{q}^\nu$, where A_σ and $B_{\sigma\nu}$ are functions on J^1Y.

Proof Let F be a 2-contact 2-form on J^2Y such that $d(E + F) = 0$. Denote $\alpha = E + F$, i.e. $\pi^*_{3,2}E = p_1\alpha$, $\pi^*_{3,2}F = p_2\alpha$. The decomposition of the form α into its 1-contact and 2-contact components is projectable on J^2Y. Thus, the most general chart expression of α reads (see Problem 5.2)

$$\pi^*_{3,2}\alpha = E_\sigma \omega^\sigma \wedge dt + B_{\sigma\nu}\omega^\sigma \wedge \dot{\omega}^\nu + F_{\sigma\nu}\omega^\sigma \wedge \omega^\nu + Q_{\sigma\nu}\dot{\omega}^\sigma \wedge \dot{\omega}^\nu ,$$

where E_σ, $B_{\sigma\nu}$, $F_{\sigma\nu}$ and $Q_{\sigma\nu}$, $1 \leq \sigma, \nu \leq m$, are functions on J^2Y, $F_{\sigma\nu}$ and $Q_{\sigma\nu}$ being antisymmetric in indices σ, ν. (Pullback $\pi^*_{3,2}\alpha$ is here formal, for the purpose of continued calculations.) Calculating $\pi^*_{3,2}d\alpha$ and taking into account that it is zero we obtain the conditions for these functions. We use relations (3.22) and (3.23) for decomposition of the exterior derivative operator d.

$$\pi^*_{3,2}d\alpha = \left(\left\{ \frac{\partial E_\sigma}{\partial q^\nu} \right\}_{\mathrm{alt}\,(\sigma,\nu)} + \frac{dF_{\sigma\nu}}{dt} \right) \omega^\sigma \wedge \omega^\nu \wedge dt$$
$$+ \left(-\frac{\partial E_\sigma}{\partial \dot{q}^\nu} + \frac{dB_{\sigma\nu}}{dt} + 2F_{\sigma\nu} \right) \omega^\sigma \wedge \dot{\omega}^\nu \wedge dt + \left(-\frac{\partial E_\sigma}{\partial \ddot{q}^\nu} + B_{\sigma\nu} \right) \omega^\sigma \wedge \ddot{\omega}^\nu \wedge dt$$
$$+ \left(\{B_{\sigma\nu}\}_{\mathrm{alt}\,(\sigma,\nu)} + \frac{dQ_{\sigma\nu}}{dt} \right) \dot{\omega}^\sigma \wedge \dot{\omega}^\nu \wedge dt + 2Q_{\sigma\nu}\dot{\omega}^\sigma \wedge \ddot{\omega}^\nu \wedge dt$$
$$+ \left\{ \frac{\partial F_{\sigma\nu}}{\partial q^\mu} \right\}_{\mathrm{alt}\,(\sigma,\mu)} \omega^\sigma \wedge \omega^\nu \wedge \omega^\mu + \left(\left\{ \frac{\partial B_{\nu\mu}}{\partial q^\sigma} \right\}_{\mathrm{alt}\,(\sigma,\nu)} + \frac{\partial F_{\sigma\nu}}{\partial \dot{q}^\mu} \right) \omega^\sigma \wedge \omega^\nu \wedge \dot{\omega}^\mu$$

$$+\frac{\partial F_{\sigma\nu}}{\partial \ddot{q}^{\mu}}\omega^{\sigma}\wedge\omega^{\nu}\wedge\ddot{\omega}^{\mu}+\left(\left\{\frac{\partial B_{\sigma\nu}}{\partial \dot{q}^{\mu}}\right\}_{\mathrm{alt}\,(\mu,\nu)}+\frac{\partial Q_{\nu\mu}}{\partial q^{\sigma}}\right)\omega^{\sigma}\wedge\dot{\omega}^{\nu}\wedge\dot{\omega}^{\mu}$$
$$+\frac{\partial B_{\sigma\nu}}{\partial \ddot{q}^{\mu}}\omega^{\sigma}\wedge\dot{\omega}^{\nu}\wedge\ddot{\omega}^{\mu}+\left\{\frac{\partial Q_{\sigma\nu}}{\partial \dot{q}^{\mu}}\right\}_{\mathrm{alt}\,(\sigma,\mu)}\dot{\omega}^{\sigma}\wedge\dot{\omega}^{\nu}\wedge\dot{\omega}^{\nu}+\frac{\partial Q_{\sigma\nu}}{\partial \ddot{q}^{\mu}}\dot{\omega}^{\sigma}\wedge\dot{\omega}^{\nu}\wedge\ddot{\omega}^{\mu}\,.$$

As α is closed, all components of the above form must vanish. We immediately see that $Q_{\sigma\nu}=0$ which simplifies other conditions:

$$0=\left\{\frac{\partial E_{\sigma}}{\partial q^{\nu}}\right\}_{\mathrm{alt}\,(\sigma,\nu)}+\frac{\mathrm{d}F_{\sigma\nu}}{\mathrm{d}t}\,, \tag{5.18}$$
$$0=-\frac{\partial E_{\sigma}}{\partial \dot{q}^{\nu}}+\frac{\mathrm{d}B_{\sigma\nu}}{\mathrm{d}t}+2F_{\sigma\nu}\,, \tag{5.19}$$
$$0=-\frac{\partial E_{\sigma}}{\partial \ddot{q}^{\nu}}+B_{\sigma\nu}\,, \tag{5.20}$$
$$0=\{B_{\sigma\nu}\}_{\mathrm{alt}\,(\sigma,\nu)}\,, \tag{5.21}$$
$$0=\left\{\frac{\partial F_{\sigma\nu}}{\partial q^{\mu}}\right\}_{\mathrm{alt}\,(\sigma,\mu)}\,,$$
$$0=\left\{\frac{\partial B_{\nu\mu}}{\partial q^{\sigma}}\right\}_{\mathrm{alt}\,(\sigma,\nu)}+\frac{\partial F_{\sigma\nu}}{\partial \dot{q}^{\mu}}\,,$$
$$0=\frac{\partial F_{\sigma\nu}}{\partial \ddot{q}^{\mu}}\,, \tag{5.22}$$
$$0=\left\{\frac{\partial B_{\sigma\nu}}{\partial \dot{q}^{\mu}}\right\}_{\mathrm{alt}\,(\mu,\nu)}\,,$$
$$0=\frac{\partial B_{\sigma\nu}}{\partial \ddot{q}^{\mu}}\,. \tag{5.23}$$

Conditions (5.22) and (5.23) mean that functions $B_{\sigma\nu}$ and $F_{\sigma\nu}$ are projectable on J^1Y. Condition (5.21) says that functions $B_{\sigma\nu}$ are symmetric and from condition (5.20) we obtain the explicit expression for them, indicating that functions E_{σ} are affine in accelerations and their partial derivatives with respect the variables $\ddot{q}^{\nu}$ are symmetric. So, we have obtained Helmholtz condition (5.14). It holds

$$\frac{\partial E_{\sigma}}{\partial \ddot{q}^{\nu}}-\frac{\partial E_{\nu}}{\partial \ddot{q}^{\sigma}}=0,\quad B_{\sigma\nu}=\frac{1}{2}\left(\frac{\partial E_{\sigma}}{\partial \ddot{q}^{\nu}}+\frac{\partial E_{\nu}}{\partial \ddot{q}^{\sigma}}\right),\quad E_{\sigma}=A_{\sigma}+B_{\sigma\nu}\ddot{q}^{\nu}\,,$$

where A_{σ} are functions on J^1Y. Furthermore, putting $B_{\sigma\nu}$ into symmetrized condition (5.19) we obtain Helmholtz condition (5.15)

$$\frac{\partial E_{\sigma}}{\partial \dot{q}^{\nu}}+\frac{\partial E_{\nu}}{\partial \dot{q}^{\sigma}}-\frac{\mathrm{d}}{\mathrm{d}t}\left(\frac{\partial E_{\sigma}}{\partial \ddot{q}^{\nu}}+\frac{\partial E_{\nu}}{\partial \ddot{q}^{\sigma}}\right)=0\,.$$

Finally, alternation of condition (5.19) together with the symmetry of $B_{\sigma\nu}$ leads to the expression for $F_{\sigma\nu}$ which put into (5.18) leads to Helmholtz condition (5.16),

$$\left(\frac{\partial E_\sigma}{\partial q^\nu} - \frac{\partial E_\nu}{\partial q^\sigma}\right) - \frac{1}{2}\frac{\mathrm{d}}{\mathrm{d}t}\left(\frac{\partial E_\sigma}{\partial \dot{q}^\nu} - \frac{\partial E_\nu}{\partial \dot{q}^\sigma}\right) = 0, \quad F_{\sigma\nu} = \frac{1}{4}\left(\frac{\partial E_\sigma}{\partial \dot{q}^\nu} - \frac{\partial E_\nu}{\partial \dot{q}^\sigma}\right).$$

For the form α we obtain expression (5.17) taking into account that $A_\sigma = E_\sigma - B_{\sigma\nu}\ddot{q}^\nu$. Even though the components of α are through function E_σ defined effectively on J^2Y, they are projectable on J^1Y. So, all assertions of the theorem are proved. □

The conclusions of this section and of the previous one confirm the anticipation presented in Sect. 5.1: The only variational dynamical forms on J^2Y are those affine in accelerations ("highest" variables $\ddot{q}^\sigma$). No other types of second-order variational dynamical forms exist. This is also consistent with classical mechanics, where the Lagrangian of a system of mass particles is the difference of kinetic and potential energy. As the kinetic energy is standardly a quadratic form in velocities (coordinates $\dot{q}^\sigma$) and the potential energy is a function of time and coordinates, the left-hand sides of variational equations of motion are affine in accelerations. (Note that the previous conclusion can be generalized for an arbitrary order of the prolongation of the underlying fibred manifold, as proved in [1]).

Example 5.1 Consider two very simple mechanical systems executing the one-dimensional motion (denote $q^1 = x$):

- a linear harmonic oscillator with the equation of motion $m\ddot{x} + kx = 0$,
- a damped linear oscillator with the equation of motion $m\ddot{x} + b\dot{x} + kx = 0$.

Physicists know that in the case a) the corresponding dynamical form is variational, the first-order Lagrangian being $L = \frac{1}{2}m\dot{x}^2 - \frac{1}{2}kx^2$, which is not the case b). Let us test the variationality in both cases. The underlying fibred manifold is $(\mathbf{R} \times \mathbf{R}, p, \mathbf{R})$, the dynamical form is of the second order, $E = (-kx - m\ddot{x})\,\omega^1 \wedge dt$ in case a) and $E = (-kx - b\dot{x} - m\ddot{x})\,\omega^1 \wedge dt$ in case b). In case a) we have $A_1(x, \dot{x}) = -kx$ and $B_{11} = -m$. All Helmholtz conditions (5.9), (5.10), (5.11), (5.12) are fulfilled. In case b) it holds $A_1 = A_1(x, \dot{x}) = -kx - b\dot{x}$ and $B_{11} = -m =$ const. Helmholtz condition (5.11) is not satisfied. The dynamical form corresponding to a linear damped oscillator is not variational.

Example 5.2 Consider a system of two second-order ordinary differential equations (coordinates are denoted as x and y, respectively, for simplicity)

$$\ddot{x} + \beta\dot{x} + \gamma^2 x = 0, \quad \ddot{y} - \beta\dot{y} + \gamma^2 y = 0,$$

where β and $\gamma^2 > 0$ are constants. The question is whether this system of equations is variational or not. We will see that the formulation of the question is not precise. Suppose that left-hand sides of equations are components of the dynamical form

$$E = -(\ddot{x} + \beta\dot{x} + \gamma^2 x)\omega^1 \wedge \mathrm{d}t - (\ddot{y} - \beta\dot{y} + \gamma^2 y)\omega^2 \wedge \mathrm{d}t\,,$$

$$A_1 = -(\beta\dot{x} + \gamma^2 x), \quad A_2 = -(-\beta\dot{y} + \gamma^2 y), \quad B_{11} = B_{22} = -1, \quad B_{12} = B_{21} = 0\,.$$

Helmholtz condition (5.11) is not satisfied for $\sigma = \nu = 1$ as well as for $\sigma = \nu = 2$. So it seems that the system of equations is not variational, because the dynamical form E is not variational. On the other hand, estimate that the Lagrangian

$$\lambda = L\,\mathrm{d}t, \quad L = \dot{x}\dot{y} - \frac{1}{2}\beta(y\dot{x} - x\dot{y}) - \gamma^2 xy$$

leads exactly to the given system of equations. How is it possible? Let us construct the Euler-Lagrange form of the Lagrangian λ. We obtain

$$E_\lambda = -\left[(\ddot{y} - \beta\dot{y} + \gamma^2 y)\omega^1 \wedge \mathrm{d}t + (\ddot{x} + \beta\dot{x} + \gamma^2 x)\omega^2 \wedge \mathrm{d}t\right]\,.$$

This dynamical form is, of course, variational. It fulfills all Helmholtz conditions. It differs from the form E by the order (enumeration) of components. We can see that the shape of the dynamical form as a whole is decisive for its variability. (The requirement to decide on the variability of a dynamical form as a whole represents the so-called *weak inverse problem* of calculus of variations. The *strong inverse problem* solves the question of whether there exists a variational dynamic form for a given system of ODEs, or a equivalent system. This problem has not yet been resolved in full generality.)

5.4 Lagrangians of Variational Dynamical Forms

In this short section we study the construction of Lagrangians of (second order) variational dynamical forms. We have proved in the previous part of this chapter that

(a) if a dynamical form E on J^2Y is variational, then it is obligatorily affine in variables $\ddot{q}^\sigma$, $1 \leq \sigma \leq m$,
(b) to every variational second-order dynamical form there exist a first-order Lagrangian λ (called the *minimal order Lagrangian*) such that $E = E_\lambda$.

All Lagrangians $\tilde{\lambda}$ for which $E_{\tilde{\lambda}} = E_\lambda$ differ form the minimal Lagrangian λ by trivial (null) Lagrangian. We describe the construction of a Lagrangian directly from the expressions E_σ of the given variational dynamical form, as well as the construction of minimal order Lagrangian.

Proposition 5.4 *Let $E = E_\sigma \omega \wedge \mathrm{d}t$ be a locally variational dynamical form on J^2Y. Then*

$$\hat{\lambda} = \left(q^\sigma \int_0^1 (E_\sigma \circ \chi) \, \mathrm{d}u \right) \mathrm{d}t \,, \tag{5.24}$$

where

$$\chi : [0,1] \times J^2(I \times V)(u, t, q^\sigma, \dot{q}^\sigma, \ddot{q}^\sigma) \longrightarrow (t, uq^\sigma, u\dot{q}^\sigma, u\ddot{q}^\sigma) \in J^2(I \times V) \,,$$

(for $I = [0,1]$ and $V \subset \mathbf{R}^m$ an open ball centered in the origin of coordinates) is the mapping defined by relation (3.25).

Proof Because a second-order variational dynamical form must be affine in accelerations we refer to the proof of Theorem 5.1. There the Vainberg-Tonti Lagrangian was constructed. □

Now we will construct a minimal (first order) Lagrangian for a second-order variational dynamical form E affine in accelerations. The result of the construction is formulated in the following proposition.

Proposition 5.5 *Let $E = E_\sigma \omega \wedge \mathrm{d}t$, $E_\sigma = (A_\sigma + B_{\sigma\nu}\ddot{q}^\nu)\omega^\sigma \wedge \mathrm{d}t$ be a locally variational dynamical form on J^2Y. Then there is a minimal (first order) Lagrangian λ*

$$\lambda = \left\{ q^\sigma \int_0^1 (E_\sigma \circ \chi) \, \mathrm{d}u - \frac{\mathrm{d}}{\mathrm{d}t} \left[q^\sigma \dot{q}^\nu \int_0^1 \left(\int_0^1 (B_{\sigma\nu} \circ \chi) u \, \mathrm{d}u \right) \circ \bar{\chi} \, \mathrm{d}s \right] + \frac{\mathrm{d}\bar{f}}{\mathrm{d}t} \right\} \mathrm{d}t \,, \tag{5.25}$$

where

$$\bar{\chi} : [0,1] \times J^1(I \times V) \ni (s, t, q^\sigma, \dot{q}^\sigma) \longrightarrow \in (t, q^\sigma, s\dot{q}^\sigma) \in J^1(I \times V) \,. \tag{5.26}$$

Proof Vainberg-Tonti Lagrangian is of the second order. Nevertheless, it is affine in variables $\ddot{q}^\sigma$, as we proved in Sect. 5.1. We proved that the second-order Lagrangian has the Euler-Lagrange form of the second order if and only if it is of the form $\hat{\lambda} = a + b_\sigma \ddot{q}^\sigma$, where $a = a(t, q^\nu, \dot{q}^\nu)$ and $b_\sigma = b_\sigma(t, q^\nu, \dot{q}^\nu)$ are functions on J^1Y, and $\frac{\partial b_\sigma}{\partial \dot{q}^\nu} - \frac{\partial b_\nu}{\partial \dot{q}^\sigma} = 0$. We wish to find the function $f = f(t, q^\nu, \dot{q}^\nu)$ such that $\frac{\partial f}{\partial \dot{q}^\sigma} = b_\sigma$. Subtracting the form $h\,\mathrm{d}f = \frac{\mathrm{d}f}{\mathrm{d}t}\,\mathrm{d}t$ from the Vainberg-Tonti Lagrangian we obtain a first-order Lagrangian. The function f will be not, of course, unique, thus we will obtain a class of first-order Lagrangians differing by first-order trivial Lagrangians.

Putting $E_\sigma = A_\sigma + B_{\sigma\nu}\ddot{q}^\nu$ into relation (5.24) for the Vainberg-Tonti Lagrangian we obtain

$$\hat{\lambda} = \left(q^\sigma \int_0^1 (A_\sigma \circ \chi)\,\mathrm{d}u + q^\sigma \int_0^1 \left((B_{\sigma\nu}\ddot{q}^\nu) \circ \chi \right) \mathrm{d}u \right) \mathrm{d}t$$
$$= \left(q^\sigma \int_0^1 (A_\sigma \circ \chi)\,\mathrm{d}u + q^\sigma \ddot{q}^\nu \int_0^1 (B_{\sigma\nu} \circ \chi) u\,\mathrm{d}u \right) \mathrm{d}t\,,$$

i.e.

$$a = q^\sigma \int_0^1 (A_\sigma \circ \chi)\,\mathrm{d}u, \quad b_\sigma = q^\nu \int_0^1 (B_{\nu\sigma} \circ \chi) u\,\mathrm{d}u\,.$$

(We can easily verify that due to Helmholtz condition (5.10) the symmetry relations $\frac{\partial b_\sigma}{\partial \dot{q}^\nu} - \frac{\partial b_\nu}{\partial \dot{q}^\sigma} = 0$ are satisfied.) For a function f which is to be found we have

$$\mathrm{d}f = \frac{\mathrm{d}' f}{\mathrm{d}t}\,\mathrm{d}t + \frac{\partial f}{\partial q^\sigma}\omega^\sigma + b_\sigma\,\mathrm{d}\dot{q}^\sigma, \quad \text{because} \quad b_\sigma = \frac{\partial f}{\partial \dot{q}^\sigma}\,.$$

This 1-form is closed, i.e. we can use the operator $\mathcal{A}$ from the Poincaré lemma to find f. For the construction we use mapping (5.26). Then

$$\bar{\chi}^*\,\mathrm{d}f = \bar{\chi}^* \left(\frac{\mathrm{d}' f}{\mathrm{d}t}\,\mathrm{d}t + \frac{\partial f}{\partial q^\sigma}\omega^\sigma \right) + (b_\sigma \circ \bar{\chi})(\dot{q}^\sigma \mathrm{d}s + s\,\mathrm{d}\dot{q}^\sigma) = \mathrm{d}s \wedge (b_\sigma \circ \bar{\chi})\dot{q}^\sigma + \bar{\eta}\,,$$

where the form $\bar{\eta}$ does not contain $\mathrm{d}s$ and thus it is not relevant for the construction of the operator $\mathcal{A}$. Following the Poincaré lemma we obtain

$$f = \mathcal{A}\,\mathrm{d}f = \dot{q}^\sigma \int_0^1 (b_\sigma \circ \bar{\chi})\,\mathrm{d}s = \dot{q}^\sigma q^\nu \int_0^1 \left(\int_0^1 (B_{\sigma\nu} \circ \chi) u\,\mathrm{d}u \right) \circ \bar{\chi}\,\mathrm{d}s\,.$$

Finally we obtain

$$\lambda = \hat{\lambda} - \left(\frac{\mathrm{d}f}{\mathrm{d}t} - \frac{\mathrm{d}\bar{f}}{\mathrm{d}t} \right) \mathrm{d}t\,,$$

where $\bar{f}$ is a function on Y representing an uncertainty of the minimal order Lagrangian given by first-order trivial Lagrangians. Putting $\hat{\lambda}$ and f into the expression for λ we obtain relation (5.25). □

Note that the construction of a minimal Lagrangian from the Vainberg-Tonti one is an application of the solution of the more general problem called the *order reduction*. This problem for mechanics is completely solved in [1], the general solution for field theory was outlined in [5].

Example 5.3 We find Vainberg-Tonti Lagrangians and minimal Lagrangians for variational systems in Examples 5.1 and 5.2

(a) In Example 5.1a) we have the variational dynamical form $E = -(kx + m\ddot{x})\omega^1 \wedge dt$, i.e. $A_1 = -kx$, $B_{11} = -m$. The Vainberg-Tonti Lagrangian is

$$\hat{\lambda} = \hat{L}\,dt, \quad \hat{L} = x\int_0^1 ((-kx - m\ddot{x})) \circ \chi\, du = -(kx^2 + mx\ddot{x})\int_0^1 u\,du = -\frac{1}{2}(kx^2 + mx\ddot{x}) .$$

This Lagrangian is of the second order, as expected. Let us construct the reduction to a first-order Lagrangian, using (5.25). We obtain

$$\begin{aligned} L &= \hat{L} - \frac{\mathrm{d}}{\mathrm{d}t}\left(x\dot{x}\int_0^1\left(\int_0^1 -mu\,\mathrm{d}u\right)\mathrm{d}s\right) + \frac{\mathrm{d}\bar{f}}{\mathrm{d}t} \\ &= -\frac{1}{2}(kx^2 + mx\ddot{x}) - \frac{\mathrm{d}}{\mathrm{d}t}\left(-\frac{1}{2}mx\dot{x}\right) + \frac{\mathrm{d}\bar{f}}{\mathrm{d}t} \\ &= \frac{1}{2}m\dot{x}^2 - \frac{1}{2}kx^2 + \frac{\mathrm{d}\bar{f}}{\mathrm{d}t} . \end{aligned}$$

(b) In Example 5.2 the variational dynamical form reads

$$E = -(\ddot{y} - \beta\dot{y} + \gamma^2 y)\omega^1 \wedge \mathrm{d}t - (\ddot{x} + \beta\dot{x} + \gamma^2 x)\omega^2 \wedge \mathrm{d}t ,$$

i.e. $A_1 = -(-\beta\dot{y} + \gamma^2 y)$, $A_2 = -\beta\dot{x} - \gamma^2 x$, $B_{11} = B_{22} = 0$, $B_{12} = B_{21} = -1$. The Vainberg-Tonti Lagrangian is $\hat{\lambda} = \hat{L}\,dt$, where

$$\begin{aligned} \hat{L} &= x\int_0^1 \big((-\ddot{y} + \beta\dot{y} - \gamma^2 y) \circ \chi\big)\,du + y\int \big((-\ddot{x} - \beta\dot{x} - \gamma^2 x) \circ \chi\big)\,du \\ &= -\frac{1}{2}(y\ddot{x} + x\ddot{y}) - \frac{1}{2}\beta(y\dot{x} - x\dot{y}) - \gamma^2 xy . \end{aligned}$$

The reduction construction gives

$$\begin{aligned} L &= \hat{L} - \frac{\mathrm{d}}{\mathrm{d}t}\left[\left(x\dot{y}\int_0^1 (-u\,\mathrm{d}u)\,\mathrm{d}s\right) + \left(y\dot{x}\int_0^1 (-u\,\mathrm{d}u)\,\mathrm{d}s\right)\right] + \frac{\mathrm{d}\bar{f}}{\mathrm{d}t} \\ &= \dot{x}\dot{y} - \frac{1}{2}\beta(y\dot{x} - x\dot{y}) - \gamma^2 xy + \frac{\mathrm{d}\bar{f}}{\mathrm{d}t} . \end{aligned}$$

5.5 Variational Forces in Mechanics

In classical (Newtonian) mechanics of particles and particle systems the equations of motion read

$$m_0 g_{\sigma\nu} \ddot{q}^\nu = F_\sigma, \quad 1 \leq \sigma, \nu \leq m,$$

where m_0 is a constant (e.g. the mass of a particle), m is the number of degrees of freedom of the studied system, $(q^1, \ldots, q^m)$, $(\dot{q}^1, \ldots, \dot{q}^m)$ and $(\ddot{q}^1, \ldots, \ddot{q}^m)$ are the *generalized coordinates, velocities and accelerations*, respectively, $F_\sigma = F_\sigma(t, q^\nu, \dot{q}^\nu)$ are the *generalized forces*. The matrix $G = (g_{\sigma\nu})$, $g_{\sigma\nu} = g_{\sigma\nu}(q^\mu)$ (i.e. it depends only on coordinates), is considered as symmetric and positively definite. It represents the metric tensor. We wish to derive the conditions for generalized forces under which the equations of motion are variational, i.e. they arise from a variational principle. Such forces are called the *variational forces*. The criterion for such a property is the variationality of the dynamical form

$$E = \left(F_\sigma - m_0 g_{\sigma\nu} \ddot{q}^\nu \right) \omega^\sigma \wedge dt .$$

For ensure the variationality of the form E, the Helmholtz conditions (5.9)–(5.12), or equivalently (5.14)–(5.16), are to be satisfied. Testing Helmholtz conditions (5.9)–(5.12) we immediately see that (5.9) and (5.10) are fulfilled due to properties of the matrix G (conversely, condition (5.9) itself poses the requirement of symmetry of the matrix G). So, it remains to treat conditions (5.11) and (5.12), i.e.

$$\frac{\partial F_\sigma}{\partial \dot{q}^\nu} + \frac{\partial F_\nu}{\partial \dot{q}^\sigma} = -2m_0 \frac{\partial g_{\sigma\nu}}{\partial q^\mu} \dot{q}^\mu, \quad \frac{\partial F_\sigma}{\partial q^\nu} - \frac{\partial F_\nu}{\partial q^\sigma} = \frac{1}{2} \frac{d'}{dt} \left(\frac{\partial F_\sigma}{\partial \dot{q}^\nu} - \frac{\partial F_\nu}{\partial \dot{q}^\sigma} \right) . \tag{5.27}$$

The first of both conditions means that functions F_σ have the form of a second degree polynomial in variables $\dot{q}^\sigma$, i.e.

$$F_\sigma = \Gamma_{\sigma\nu\mu} \dot{q}^\nu \dot{q}^\mu + \phi_{\sigma\nu} \dot{q}^\nu + \psi_\sigma , \tag{5.28}$$

where $\Gamma_{\sigma\nu\mu}$ are functions on Y symmetric in indices ν and μ, $\phi_{\sigma\nu}$ and ψ are functions on Y as well. Putting this expression into conditions (5.27) we obtain after rather technical calculations (in detail see Problem 5.3)

$$\Gamma_{\sigma\nu\mu} = -\frac{m_0}{2} \left(\frac{\partial g_{\sigma\mu}}{\partial q^\nu} + \frac{\partial g_{\sigma\nu}}{\partial q^\mu} - \frac{\partial g_{\nu\mu}}{\partial q^\sigma} \right) , \tag{5.29}$$

$$0 = \phi_{\sigma\nu} + \phi_{\nu\sigma} , \tag{5.30}$$

$$0 = \frac{\partial \phi_{\sigma\mu}}{\partial q^\nu} + \frac{\partial \phi_{\mu\nu}}{\partial q^\sigma} + \frac{\partial \phi_{\nu\sigma}}{\partial q^\mu} , \tag{5.31}$$

$$0 = \frac{\partial \psi_\sigma}{\partial q^\nu} - \frac{\partial \psi_\nu}{\partial q^\sigma} - \frac{\partial \phi_{\sigma\nu}}{\partial t} . \tag{5.32}$$

Now let us study a special example important from the point of view of physics—the equations of motion of a mass particle under a force field **F** in the Euclidean space.

This means that the underlying fibred manifold is $(\mathbf{R} \times \mathbf{R}^3, p, \mathbf{R})$ with the global fibred chart $(\mathbf{R} \times \mathbf{R}^3, \psi)$ with Cartesian coordinates $\psi = (t, x^1, x^2, x^3)$. Enumerate them by indices $1 \le i, j, \ldots \le 3$, instead of $\sigma, \nu, \ldots$ The Euclidean metrics is $G = (\delta_{ij})$, so $\Gamma_{ijk} = 0$ for $1 \le i, j, k \le 3$ and $F_i = \phi_{ij}\dot{x}^j + \psi_i$. Conditions (5.30), (5.31) and (5.32) remain unchanged except for the change of indices $(\sigma, \nu, \mu) \to (i, j, k)$. Condition (5.30) enables us to consider three nonzero quantities as components of a vector $\boldsymbol{\phi}$, denoting $\phi_{12} = -\phi_{21} = \phi^3$, $\phi_{31} = -\phi_{13} = \phi^2$, $\phi_{23} = -\phi_{32} = \phi^1$. Condition (5.31) then gives $\operatorname{div} \boldsymbol{\phi} = 0$ which means that there is a vector function $\mathbf{A}$ on Y such that

$$\phi^i = \epsilon^{ijk} \frac{\partial A_k}{\partial x^j} \quad \text{i.e.} \quad \boldsymbol{\phi} = \operatorname{rot} \mathbf{A} .$$

Condition (5.32) then gives

$$\frac{\partial \boldsymbol{\phi}}{\partial t} = -\operatorname{rot} \boldsymbol{\psi} \implies \frac{\partial \operatorname{rot} \mathbf{A}}{\partial t} = -\operatorname{rot} \boldsymbol{\psi} \implies \operatorname{rot} \left(\boldsymbol{\psi} + \frac{\partial \mathbf{A}}{\partial t} \right) = 0 .$$

The last relation means that there exists a function V on Y such that

$$\boldsymbol{\psi} + \frac{\partial \boldsymbol{\phi}}{\partial t} = -\operatorname{grad} V .$$

Denoting $\mathbf{v} = (\dot{x}^1, \dot{x}^2, \dot{x}^3)$ (the particle velocity) we can write the general form of a variational force $\mathbf{F}$:

$$\mathbf{F} = \mathbf{v} \times \operatorname{rot} \mathbf{A} - \frac{\partial \mathbf{A}}{\partial t} - \operatorname{grad} V . \tag{5.33}$$

Moreover, if we denote $\mathbf{B} = \operatorname{rot} \mathbf{A}$ and $\mathbf{E} = -\operatorname{grad} V$, we can write

$$\mathbf{F} = \mathbf{v} \times \mathbf{B} + \mathbf{E}, \quad \operatorname{div} \mathbf{B} = 0, \quad \operatorname{rot} \mathbf{E} = -\frac{\partial \mathbf{B}}{\partial t} . \tag{5.34}$$

The last set of equations can be interpreted, e.g. as the Lorentz electromagnetic force and second and third Maxwell's equations if we consider the vectors $\mathbf{E}$ and $\mathbf{B}$ as electric intensity and magnetic induction, respectively. Then $\mathbf{A}$ and V represent the vector and scalar potential, respectively. On the other hand the general expression (5.33) includes also other types of forces. As an example we can mention fictive forces manifested in non-inertial reference frames in mechanics (see Problem 5.4).

5.6 Problems

Problem 5.1 Characterize and prove the variationality of first-order dynamical forms affine in velocities. For a dynamical form $E = E_\sigma \omega^\sigma \wedge dt$, $E_\sigma = A_\sigma + B_{\sigma\nu}\dot{q}^\nu$, where A_σ and $B_{\sigma\nu}$ are functions on Y, derive the Helmholtz conditions of variationality.

Problem 5.2 Let $E = E_\sigma \omega^\sigma \wedge \mathrm{d}t$ be a dynamical form on J^2Y. Suppose that there exists a 2-contact 2-form F on J^2Y such that $\mathrm{d}(E + F) = 0$, i.e. $\alpha = E + F$ is the closed form on J^2Y and it holds $p_1\alpha = \pi^*_{3,2}E$. (The form α is given on J^2Y and thus $p_1\alpha$ is formally on J^3Y. Nevertheless, by definition of α, its decomposition is projectable onto J^2Y.) Prove that the most general chart expression of the form α is such that

$$\pi^*_{3,2}\alpha = E_\sigma \omega^\sigma \wedge \mathrm{d}t + B_{\sigma\nu}\omega^\sigma \wedge \dot{\omega}^\nu + F_{\sigma\nu}\omega^\sigma \wedge \omega^\nu + Q_{\sigma\nu}\dot{\omega}^\sigma \wedge \dot{\omega}^\nu \,,$$

where E_σ, $B_{\sigma\nu}$, $F_{\sigma\nu}$ and $Q_{\sigma\nu}$, $1 \le \sigma, \nu \le m$, are functions on J^2Y, $F_{\sigma\nu}$ and $Q_{\sigma\nu}$ being antisymmetric in indices σ, ν.

Problem 5.3 Execute in detail all calculations leading to the expressions of functions $\Gamma_{\sigma\nu\mu}$ via partial derivatives of the metric tensor (relation (5.29)).

Problem 5.4 Prove that the general expressions (5.33) or equivalently (5.34) for variational forces acting on a mass particle moving in the Euclidean time-space include all types of fictive forces (these forces take place in non-inertial reference frames).

Problem 5.5 Let (Y, π, X) be a fibred manifold. Generalize the concepts of r-equivalent sections of this manifold, r-jets of a section at a given point $x \in X$ and r-jet prolongation of the fibred manifold via r-jets, for $r > 2$, especially for $r = 3$ and $r = 4$. Follow formally the corresponding procedures in Sect. 2.4.

Problem 5.6 Let (Y, π, X) be a fibred manifold. Generalizing the procedure in Sect. 3.1 write down the recurrent relations for components of prolongations of a π-projectable vector field on Y (the analogues of relations (3.8)).

Problem 5.7 Write down the basis of 1-forms adapted to the contact structure of a fibred manifold (Y, π, X) for $r > 2$ (i.e. use the contact 1-forms ω^σ, $\tilde{\omega}^\sigma$, $\hat{\omega}^\sigma$, ...)

Problem 5.8 Perform all calculations leading to the Lepage equivalent and the Euler-Lagrange form of the second-order Lagrangian in detail.

Problem 5.9 Perform in detail all calculations leading to conclusions concerning the order of Lagrangians producing the second-order Euler-Lagrange forms affine in accelerations, i.e. derive conditions (5.1), (5.2) and (5.3).

Problem 5.10 Execute in detail all calculations proving Theorem 5.3.

Problem 5.11 Decide whether the system of equations

$$qB\dot{y} - m\ddot{x} - 0, \quad -qB\dot{x} - m\ddot{y} = 0$$

is variational. These are the equations of motion of a charged mass particle with mass m and charge q in the homogeneous magnetic field of induction **B** directed along the positive z-axis. If the system is variational, construct a first-order Lagrangian.

Problem 5.12 Construct the Vainberg-Tonti Lagrangian for the (variational) dynamical form $E = -m\ddot{x}\,\mathrm{d}x \wedge \mathrm{d}t - m\ddot{y}\,\mathrm{d}y \wedge \mathrm{d}t$ representing a free mass particle. Construct the minimal Lagrangian using the reduction procedure.

Problem 5.13 Consider the fibred manifold $(\mathbf{R} \times \mathbf{R}^3, p, \mathbf{R})$ with the global fibred chart $\mathbf{R} \times \mathbf{R}^3, \psi, \psi = (t, x, y, z)$, and the dynamical form

$$E = (-\ddot{z})\,\mathrm{d}x \wedge \mathrm{d}t + (yz)\,\mathrm{d}y \wedge \mathrm{d}t + \left(\frac{1}{2}y^2 - \ddot{x}\right)\mathrm{d}z \wedge \mathrm{d}t\ .$$

Decide whether this dynamical form is variational. If the answer is positive, construct the Vainberg-Tonti Lagrangian and a minimal Lagrangian.

Problem 5.14 For each of variational forms presented in Examples 5.1 and 5.2 and Problems 5.11, 5.12 and 5.13 construct its dynamical system and dynamical distribution.

References

1. O. Krupková, The Geometry of Ordinary Variational Equations. Lectures in Mathematical Notes, vol. 1678 (Springer, Berlin, 1997)
2. M.A. Malakhaltsev, De Rham cohomology, in *Handbook of Global Analysis* (Elsevier Sci. B. V., Amsterdam, 2008), pp. 953–981
3. R. Vitolo, Variational sequences, in *Handbook of Global Analysis* (Elsevier Sci. B. V., Amsterdam, 2008), pp. 1115–1163
4. D. Krupka, *Introduction to Global Variational Geometry* (Atlantis Press and the Author, 2015)
5. O. Rossi, The Lagrangian order reduction in field theories. Preprint 2017, 1–22

Hamiltonian Systems and the Hamilton–Jacobi Theory

6

We have learned about two pillars of the classical calculus of variations: the Lagrange theory and Noether theory of symmetries and conservation laws. However, there is a third solid pillar, due to Hamilton and Jacobi.

The idea about Hamilton equations is smart and elegant, has a rich geometrical background and a strong potential with respect to applications. It is mainly this part of the calculus of variations which had a substantial influence on rush developments in theoretical and mathematical physics. Moreover, as a part of mathematics it was developed independently and gave birth to symplectic and Poisson geometry and modern theory of dynamical systems, highly influenced developments in differential equations, spectral theory, control theory, or algebraic topology and geometry, to name at least a few.

Hamilton equations in the classical calculus of variations represent a system of equations which are equivalent with Euler-Lagrange equations, hence represent another tool for finding extremals. However, the geometry behind makes them extremely interesting and useful. As for practical calculations, the difference of Lagrangian and Hamiltonian approach is as follows: While extremals of the variational functional are solutions of m (ordinary) differential equations of the second order in the Lagrangian approach in first-order mechanics, the (equivalent) Hamiltonian approach represents the solution of $2m$ equations of the first order.

Although the interpretation of variational physics (mechanics and field theory) presented in this book is based on the Lagrangian approach, for completeness we also present the basic elements of Hamiltonian theory, but only for mechanics. The interpretation is again based on a geometric basis using fibred manifolds. We focus on the regularity of the variational problem. Compared to standard approaches in Hamilton theory, our approach is enriched with a generalized definition of regularity (first introduced by O. Rossi-Krupková in the eighties and elaborated particularly in the key paper [1]) and its consequences, including examples. The standard definition

J. Musilová et al., *Calculus of Variations on Fibred Manifolds and Variational Physics*, Lecture Notes in Physics 1033, https://doi.org/10.1007/978-3-031-77408-9_6

of regularity based on the properties of the Lagrangian is generalized by introducing the so-called Hamilton extremals and mechanical system (A generalized approach to regularity has already been outlined in Chap. 3).

Prerequisites geometrical background (fibred manifolds and their jet prolongations, sections, vector fields, differential forms (Chap. 3)), Lagrangian mechanics on fibred manifolds and their prolongations (Chaps. 4 and 5). □

6.1 Regular Variational Problem in the First-Order Mechanics and Hamilton Equations

First, we present the concept of the standardly defined regularity of the first-order variational problem in mechanics.

Definition 6.1 (*Regularity*) Let (Y, π, X), $\dim X = 1$, $\dim Y = m$, $m \geq 1$. Let (V, ψ), $\psi = (t, q^\sigma)$, $1 \leq \sigma \leq m$, be a fibred chart on Y, $\big(\pi(V), \varphi\big)$, $\varphi = (t)$, the associated chart on X and (V_1, ψ_1), $V_1 = \pi^{-1}(V)$, $\psi_1 = (t, q^\sigma, \dot{q}^\sigma)$, the associated fibred chart on J^1Y. Let (π, λ) be a 1st order Lagrange structure with Lagrangian $\lambda \in \Lambda^1_X(TJ^1Y)$ defined on an open set $W_1 \subset J^1Y$ such that $W_1 \subset V_1$. Lagrangian $\lambda = L\,dt$ (or, equivalently, Lagrange structure (π, λ)) is called *regular* if the matrix

$$\mathcal{M} = \left(\frac{\partial^2 L}{\partial \dot{q}^\sigma \partial \dot{q}^\nu}\right), \quad 1 \leq \sigma, \nu \leq m, \tag{6.1}$$

is regular, i.e. $\det \mathcal{M} \neq 0$.

Proposition 6.1 *(Generalized momenta, Legendre transformation) Let (Y, π, X), $\dim X = 1$, $\dim Y = m + 1$, be a fibred manifold. Let (π, λ) be a first-order Lagrange structure with Lagrangian $\lambda \in \Lambda^1_X(TJ^1Y)$ regular in the sense of Definition 6.1 defined on an open set $W_1 \subset J^1Y$. Let (V, ψ), $\psi = (t, q^\sigma)$, $(\pi(V), \varphi)$, $\varphi = (t)$, $(\pi^{-1}_{1,0}(V), \psi_1)$, $\psi_1 = (t, q^\sigma, \dot{q}^\sigma)$, $1 \leq \sigma \leq m$, be a fibred chart on Y, associated chart on X and associated fibred chart on J^1Y, respectively, such that $\pi^{-1}_{1,0}(V) \supset W_1$. For $\lambda = L(t, q^\sigma, \dot{q}^\sigma)dt$ define*

$$p_\sigma = \frac{\partial L}{\partial \dot{q}^\sigma}, \quad 1 \leq \sigma \leq m, \tag{6.2}$$

so called generalized momenta. Denote $V_1 = \pi^{-1}_{1,0}(V)$. Let $(V_1, \tilde{\psi}_1)$, with

$$\tilde{\psi}_1 = (t, q^1, \ldots, q^m, p_1, \ldots, p_m) \tag{6.3}$$

be the associated fibred chart on J^1Y such that coordinates p_σ, $1 \le \sigma \le m$, are given by (6.2). Then the mapping

$$\mathcal{L} : V_1 \ni (t, q^1, \ldots, q^m, \dot{q}^1, \ldots, \dot{q}^m) \longrightarrow (t, q^1, \ldots, q^m, p_1, \ldots, p_m) \in \tilde{\psi}_1(V_1), \tag{6.4}$$

(supposing that $\tilde{\psi}_1(V_1) \subset W_1$) is a transformation of coordinates. This mapping is called the Legendre transformation.

Proof The Jacobi matrix of the mapping $\mathcal{L} \circ \psi_1^{-1}$ is

$$D\mathcal{L} = \begin{pmatrix} 1 & 0 & \ldots & 0 & \frac{\partial p_1}{\partial t} & \ldots & \frac{\partial p_m}{\partial t} \\ 0 & 1 & \ldots & 0 & \frac{\partial p_1}{\partial q^1} & \ldots & \frac{\partial p_m}{\partial q^1} \\ \ldots & \ldots & \ldots & \ldots & \ldots & \ldots & \ldots \\ 0 & 0 & \ldots & 1 & \frac{\partial p_1}{\partial q^m} & \ldots & \frac{\partial p_m}{\partial q^m} \\ 0 & 0 & \ldots & 0 & \frac{\partial p_1}{\partial \dot{q}^1} & \ldots & \frac{\partial p_m}{\partial \dot{q}^1} \\ \ldots & \ldots & \ldots & \ldots & \ldots & \ldots & \ldots \\ 0 & 0 & \ldots & 0 & \frac{\partial p_1}{\partial \dot{q}^m} & \ldots & \frac{\partial p_m}{\partial \dot{q}^m} \end{pmatrix} = \underbrace{\begin{pmatrix} 1 & 0 & \frac{\partial p_\sigma}{\partial t} \\ 0 & E_m & \frac{\partial p_\sigma}{\partial q^\nu} \\ 0 & 0 & \frac{\partial p_\sigma}{\partial \dot{q}^\nu} \end{pmatrix}}_{\text{the matrix of order } 2m+1} .$$

It is evident that $\det D\mathcal{L} = \det \mathcal{M} \neq 0$. So, the mapping $\mathcal{L}$ is a transformation of coordinates. □

The (regular) mapping $\mathcal{L}$ introduced in Proposition 6.2 is called the *Legendre transformation.*

Definition 6.2 (*Hamilton function*) Consider the conditions presented in Definition 6.1 and assumptions of Proposition 6.1. The function

$$H = \left(-L + p_\sigma \dot{q}^\sigma\right) \circ \mathcal{L}, \quad \text{i.e.} \tag{6.5}$$

$$H(t, q^\sigma, p_\sigma) = -L\big(t, q^\sigma, \dot{q}^\sigma(t, q^\nu, p_\nu)\big) + p_\sigma \dot{q}^\sigma(t, q^\nu, p_\nu) \tag{6.6}$$

is so called *Hamilton function.*

Proposition 6.2 *Let (π, λ) be a regular 1st order Lagrange structure on a fibred manifold (Y, π, X), $\dim X = 1$, with the Lagrangian $\lambda \in \Lambda^1_X(TJ^1Y)$ defined on an open set $W_1 \subset J^1Y$. Let (V, ψ), $W_1 \subset \pi^{-1}_{1,0}(V)$, $\psi = (t, q^\sigma)$, $\big(\pi(V), \varphi\big)$, $\varphi = (t)$, and (V_1, ψ_1), $V_1 = \pi^{-1}_{1,0}(V)$, $\psi_1 = (t, q^\sigma, \dot{q}^\sigma)$, (V_2, ψ_2), $V_2 = \pi^{-1}_{2,0}(V)$, $\psi_2 = (t, q^\sigma, \dot{q}^\sigma, \ddot{q}^\sigma)$, $1 \le \sigma \le m$, a fibred chart on Y, the associated chart on X, and the associated fibred charts on J^1Y and J^2Y, respectively. The section*

$$\delta : \pi(V) \ni t \longrightarrow \delta(t) = \big(t, q^\sigma \circ \delta(t), p_\sigma \circ \delta(t)\big) \in J^1Y$$

of the projection π_1 is a solution of (so called) Hamilton equations

$$\frac{\mathrm{d}(q^\sigma \circ \delta)}{\mathrm{d}t} = \frac{\partial H}{\partial p_\sigma} \circ \delta, \quad \frac{\mathrm{d}(p_\sigma \circ \delta)}{\mathrm{d}t} = -\frac{\partial H}{\partial q^\sigma} \circ \delta, \quad 1 \le \sigma \le m, \tag{6.7}$$

if and only if there exists an extremal γ of λ such that $\delta = J^1\gamma$.

Proof The proof is based on simple coordinate calculations. (If the assumption $W_1 \subset \pi_{1,0}^{-1}(V)$ is not fulfilled, we work in coordinates on the set $W_1 \cap \pi_{1,0}^{-1}(V) \neq \emptyset$.) The assumption of regularity ensures that the Legendre mapping (6.4) is a coordinate transformation. First, suppose that the section γ is an extremal of λ. Then

$$\left[\frac{\partial L}{\partial q^\sigma} - \frac{\mathrm{d}}{\mathrm{d}t}\left(\frac{\partial L}{\partial \dot{q}^\sigma}\right)\right] \circ J^2\gamma = 0 \implies \frac{\partial L}{\partial q^\sigma} \circ J^1\gamma = \left[\frac{\mathrm{d}}{\mathrm{d}t}\left(\frac{\partial L}{\partial \dot{q}^\sigma}\right)\right] \circ J^2\gamma.$$

Taking into account the Legendre transformation and the definition of generalized momenta and Hamilton function we can rewrite the left-hand and right-hand sides of the last equation into the form

$$\left(\frac{\partial L}{\partial q^\sigma} \circ \mathcal{L}\right) \circ J^1\gamma = -\frac{\partial H}{\partial q^\sigma} \circ J^1\gamma, \quad \frac{\mathrm{d}p_\sigma}{\mathrm{d}t} \circ J^2\gamma = \frac{\mathrm{d}}{\mathrm{d}t}\left(p_\sigma \circ J^1\gamma\right). \tag{6.8}$$

Thus we obtain

$$\frac{\mathrm{d}(p_\sigma \circ J^1\gamma)}{\mathrm{d}t} = -\frac{\partial H}{\partial q^\sigma} \circ J^1\gamma.$$

On the other hand it holds

$$\dot{q}^\sigma \circ \mathcal{L} = \frac{\partial H}{\partial p_\sigma} \implies \frac{\mathrm{d}(q^\sigma \circ J^1\gamma)}{\mathrm{d}t} = \frac{\partial H}{\partial p_\sigma} \circ J^1\gamma. \tag{6.9}$$

So, the section $J^1\gamma$ is a solution of Hamilton equations.

Now suppose that a section δ of the projection π_1 is a solution of Eq. (6.7). Then from their first set we can see that

$$\frac{\partial H}{\partial p_\sigma} \circ \delta = \frac{\mathrm{d}q^\sigma \circ \delta}{\mathrm{d}t} = \dot{q}^\sigma \circ \delta.$$

This means that the section δ is holonomic, i.e. there exists a section γ of π such that $\delta = J^1\gamma$. The second set of Hamiltonian equations gives

$$\frac{\mathrm{d}(p_\sigma \circ \delta)}{\mathrm{d}t} = -\frac{\partial H}{\partial q^\sigma} \circ \delta \implies \frac{\mathrm{d}}{\mathrm{d}t}\left(\frac{\partial L}{\partial \dot{q}^\sigma} \circ J^1\gamma\right) = \frac{\partial L}{\partial q^\sigma} \circ J^1\gamma.$$

Thus, the section γ is an extremal of the Lagrangian λ. □

Example 6.1 Proposition 6.2 shows the relationship between extremals of a given regular Lagrangian (solutions of Euler-Lagrange equations—i.e. m ordinary 2nd order differential equations) and solutions of Hamilton equations ($2m$ ordinary 1st order differential equations). The geometrical interpretation is as follows: Consider the fibred manifold $(Y, \pi, X) = (\mathbf{R} \times \mathbf{R}^m, p, \mathbf{R})$ and the 1st order Lagrange structure (π, λ), $\lambda = L(t, q^\sigma, \dot{q}^\sigma)\,dt$. Extremals of this Lagrangian are sections

$$\gamma : t \longrightarrow \left(t, q^1\gamma(t), \ldots, q^m\gamma(t)\right)$$

of the projection π such that they are solutions of Euler-Lagrange equations of λ (m second order differential equations). These sections represent the trajectories of the system in the configuration space. On the other hand, consider the fibred manifold $(\tilde{Y}, \tilde{\pi}, \tilde{X}) = (\mathbf{R} \times \mathbf{R}^{2m}, p, \mathbf{R})$ and the 1st order Lagrange structure $(\tilde{\pi}, \tilde{\lambda})$, where

$$\tilde{\lambda} = \tilde{L}\,\mathrm{d}t, \quad \tilde{L} = p_\sigma \dot{q}^\sigma - H.$$

Solutions of Hamilton equations are sections of the projection $\tilde{\pi}$,

$$\tilde{\gamma} : t \longrightarrow \left(t, q^1\tilde{\gamma}(t), \ldots, q^m\tilde{\gamma}(t), p_1\tilde{\gamma}(t), \ldots, p_m\tilde{\gamma}(t)\right).$$

We are going to see that the Hamilton equations are Euler-Lagrange equations of the Lagrangian $\tilde{\lambda} \in \Lambda^1_{\tilde{X}}(TJ^1\tilde{Y})$, noting that the coordinates of a point $z \in J^1\tilde{Y}$ are

$$(t, q^\nu, \dot{q}^\nu, p_\nu, \dot{p}_\nu), \quad 1 \le \nu \le m,$$

$$\frac{\partial \tilde{L}}{\partial q^\sigma} - \frac{\mathrm{d}}{\mathrm{d}t}\left(\frac{\partial \tilde{L}}{\partial \dot{q}^\sigma}\right) = -\frac{\partial H}{\partial q^\sigma} - \frac{\mathrm{d}p_\sigma}{\mathrm{d}t},$$

$$\frac{\partial \tilde{L}}{\partial p_\sigma} - \frac{\mathrm{d}}{\mathrm{d}t}\left(\frac{\partial \tilde{L}}{\partial \dot{p}_\sigma}\right) = \dot{q}^\sigma - \frac{\partial H}{\partial p_\sigma}.$$

So, the Euler-Lagrange equations of $\tilde{\lambda}$ (equations for a section $\tilde{\gamma}$) read

$$\left(-\frac{\mathrm{d}p_\sigma}{\mathrm{d}t} - \frac{\partial H}{\partial q^\sigma}\right) \circ J^1\tilde{\gamma} = 0, \quad \left(\dot{q}^\sigma - \frac{\partial H}{\partial p_\sigma}\right) \circ J^1\tilde{\gamma} = 0.$$

The Euler-Lagrange form of $\tilde{\lambda}$ is

$$E_{\tilde{\lambda}} = \left(-\frac{\mathrm{d}p_\sigma}{\mathrm{d}t} - \frac{\partial H}{\partial q^\sigma}\right) \mathrm{d}q^\sigma \wedge \mathrm{d}t + \left(\dot{q}^\sigma - \frac{\partial H}{\partial p_\sigma}\right) \mathrm{d}p_\sigma \wedge \mathrm{d}t.$$

Example 6.2 Consider two functions defined on an open set $W_1 \subset J^1Y$, $f = f(t, q^\sigma, p_\sigma)$ and $g = g(t, q^\sigma, p_\sigma)$, $1 \le \sigma \le m$, and their *Poisson bracket*

$$\{f, g\} = \frac{\partial f}{\partial q^\sigma}\frac{\partial g}{\partial p_\sigma} - \frac{\partial f}{\partial p_\sigma}\frac{\partial g}{\partial q^\sigma}.$$

Calculating e.g. the total derivative of $f(t, q^\sigma, p_\sigma)$ we obtain

$$\frac{\mathrm{d}f}{\mathrm{d}t} = \frac{\partial f}{\partial t} + \frac{\partial f}{\partial q^\sigma}\dot{q}^\sigma + \frac{\partial f}{\partial p_\sigma}\dot{p}_\sigma.$$

Let $\delta = J^1\gamma$ be a holonomic section along which the Hamilton equations (6.7) are fulfilled. Then

$$\dot{q}^\sigma \circ J^1\gamma = \frac{\partial H}{\partial p_\sigma} \circ J^1\gamma, \quad \dot{p}_\sigma \circ J^\gamma = -\frac{\partial H}{\partial q^\sigma} \circ J^1\gamma$$

$$\Longrightarrow \quad \frac{\mathrm{d}f}{\mathrm{d}t} \circ J^1\gamma = \left(\frac{\partial f}{\partial t} + \{f, H\}\right) \circ J^1\gamma.$$

Consider a special situation when f does not depend on t explicitly and the Poisson bracket $\{f, H\}$ vanishes. In such a situation it holds

$$\frac{\mathrm{d}f}{\mathrm{d}t} \circ J^1\gamma = 0.$$

This means that the function $f = f(q^\sigma, p_\sigma)$ is constant along extremals (holonomic sections being solutions of Hamilton equations and thus being extremals of the Lagrangian, see Proposition 6.2). Such quantities are called *first integrals of Hamilton equations*. Especially, if the Hamilton function does not depend on t (the variable along the base of the corresponding fibred manifold), it itself is a first integral of Hamilton equations.

6.2 Hamilton Extremals

The regularity of a Lagrangian (6.1) represents necessary and sufficient condition for the mapping (6.4) to be a transformation of coordinates. Even though we are interested primarily in 1st order mechanics, we now notice a generalization of extremal which is needed especially for higher order Hamilton mechanics. Proposition 6.2 states that solutions of Hamilton equations in the 1st order mechanics are just 1st prolongations of extremals of the Lagrangian, i.e. holonomic sections $\delta = J^1\gamma$ of the projection π_1, where γ is an extremal of λ. Generalization of regularity lies in the idea based on a more general concept of extremals. Extremals γ of a Lagrangian λ are characterized, roughly speaking, by the equation

$$J^1\gamma^* i_{J^1\xi}\, \mathrm{d}\theta_\lambda = 0,$$

where θ_λ is the Lepage equivalent of λ and ξ is an arbitrary π-vertical (more generally π-projectable) vector field on Y. We generalize this condition by taking to account a section δ of π_1, not necessarily holonomic, and a vector field ξ on J^1Y, not necessarily the prolongation of a vector field on Y.

Definition 6.3 Hamilton extremals. Let (Y, π, X) be a fibred manifold, $\dim X = 1$, $\dim Y = m$, and (π, λ) the 1st order Lagrange structure with the Lagrangian $\lambda \in \Lambda^1_X(TJ^1Y)$ defined on an open set $W_1 \subset J^1Y$. A section $\delta \in \Gamma_\Omega(\pi_1)$, $\Omega \subset \pi_1(W_1)$ being a compact set (e.g. a closed interval) is called the *Hamilton extremal* of the given Lagrange structure, if

$$\delta^* i_\xi \, d\theta_\lambda = 0, \tag{6.10}$$

where ξ is an arbitrary π_1-vertical vector field defined on W_1 and θ_λ is the Lepage equivalent (Cartan form) of the Lagrangian λ.

We will notice the connection between newly defined Hamilton extremals and extremals of the corresponding Lagrange structure. By step-by-step calculations we get for a general vector field ζ on J^1Y (meanwhile without the assumption of verticality)

$$\zeta = \zeta^0 \frac{\partial}{\partial t} + \zeta^\sigma \frac{\partial}{\partial q^\sigma} + \zeta^\sigma_1 \frac{\partial}{\partial \dot{q}^\sigma};$$

$$\lambda = L(t, q^\nu, \dot{q}^\nu) \, dt, \quad \theta_\lambda = L \, dt + \frac{\partial L}{\partial \dot{q}^\sigma}(dq^\sigma - \dot{q}^\sigma \, dt),$$

$$d\theta_\lambda = dL \wedge dt + d\left(\frac{\partial L}{\partial \dot{q}^\sigma}\right) \wedge (dq^\sigma - \dot{q}^\sigma \, dt) - \frac{\partial L}{\partial \dot{q}^\sigma} d\dot{q}^\sigma \wedge dt$$

$$= \left(\frac{\partial L}{\partial t} dt + \frac{\partial L}{\partial q^\sigma} dq^\sigma + \frac{\partial L}{\partial \dot{q}^\sigma} d\dot{q}^\sigma\right) \wedge dt$$

$$+ \left(\frac{\partial^2 L}{\partial t \partial \dot{q}^\sigma} dt + \frac{\partial^2 L}{\partial q^\nu \partial \dot{q}^\sigma} dq^\nu + \frac{\partial^2 L}{\partial \dot{q}^\nu \partial \dot{q}^\sigma} d\dot{q}^\nu\right) \wedge (dq^\sigma - \dot{q}^\sigma \, dt)$$

$$- \frac{\partial L}{\partial \dot{q}^\sigma} d\dot{q}^\sigma \wedge dt$$

$$= \left(\frac{\partial L}{\partial q^\sigma} - \frac{\partial^2 L}{\partial t \partial \dot{q}^\sigma} - \frac{\partial^2 L}{\partial q^\sigma \partial \dot{q}^\nu} \dot{q}^\nu\right) dq^\sigma \wedge dt$$

$$- \frac{\partial^2 L}{\partial \dot{q}^\sigma \partial \dot{q}^\nu} \dot{q}^\nu d\dot{q}^\sigma \wedge dt - \frac{\partial^2 L}{\partial \dot{q}^\sigma \partial q^\nu} dq^\sigma \wedge dq^\nu - \frac{\partial^2 L}{\partial \dot{q}^\sigma \partial \dot{q}^\nu} dq^\sigma \wedge d\dot{q}^\nu,$$

$$
\begin{aligned}
i_\zeta\, \mathrm{d}\theta_\lambda = & \left(\frac{\partial L}{\partial q^\sigma} - \frac{\partial^2 L}{\partial t \partial \dot q^\sigma} - \frac{\partial^2 L}{\partial q^\sigma \partial \dot q^\nu}\dot q^\nu\right)(\zeta^\sigma\, \mathrm{d}t - \zeta^0\, \mathrm{d}q^\sigma) \\
& - \frac{\partial^2 L}{\partial \dot q^\sigma \partial \dot q^\nu}\dot q^\nu(\zeta_1^\sigma\, \mathrm{d}t - \zeta^0\, \mathrm{d}\dot q^\sigma) - \frac{\partial^2 L}{\partial q^\nu \partial \dot q^\sigma}(\zeta^\sigma\, \mathrm{d}q^\nu - \zeta^\nu\, \mathrm{d}q^\sigma) \\
& - \frac{\partial^2 L}{\partial \dot q^\sigma \partial \dot q^\nu}(\zeta^\sigma\, \mathrm{d}\dot q^\nu - \zeta_1^\nu\, \mathrm{d}q^\sigma).
\end{aligned}
$$

Taking into account that for a section δ of the projection π_1 it holds

$$
\delta^*\, \mathrm{d}t = \mathrm{d}t, \quad \delta^*\, \mathrm{d}q^\sigma = \frac{\mathrm{d}q^\sigma \delta(t)}{\mathrm{d}t}\, \mathrm{d}t, \quad \delta^*\, \mathrm{d}\dot q^\sigma = \frac{\mathrm{d}\dot q^\sigma \delta(t)}{\mathrm{d}t}\, \mathrm{d}t
$$

we obtain after some technical calculations including the needed changes of indices

$$
\delta^* i_\zeta\, \mathrm{d}\theta_\lambda = \big(Z_0(t)(\zeta^0 \circ \delta) + Z_\sigma(t)(\zeta^\sigma \circ \delta) + Z_\sigma^1(t)(\zeta_1^\sigma \circ \delta)\big)\, \mathrm{d}t,
$$

where (noting that in general $\dot q^\sigma \delta(t) \neq \frac{\mathrm{d}q^\sigma \delta(t)}{\mathrm{d}t}$)

$$
Z_\sigma^1 = -\left(\frac{\partial^2 L}{\partial \dot q^\sigma \partial \dot q^\nu} \circ \delta\right)\left[\dot q^\nu \delta(t) - \frac{\mathrm{d}q^\nu \delta(t)}{\mathrm{d}t}\right], \tag{6.11}
$$

$$
\begin{aligned}
Z_\sigma = & \left(\frac{\partial L}{\partial q^\sigma} - \frac{\partial^2 L}{\partial t \partial \dot q^\sigma} - \frac{\partial^2 L}{\partial q^\sigma \partial \dot q^\nu}\dot q^\nu\right) \circ \delta + \left[\left(\frac{\partial^2 L}{\partial q^\sigma \partial \dot q^\nu} - \frac{\partial^2 L}{\partial q^\nu \partial \dot q^\sigma}\right) \circ \delta\right] \frac{\mathrm{d}q^\nu \delta(t)}{\mathrm{d}t} \\
& - \left(\frac{\partial^2 L}{\partial \dot q^\sigma \partial \dot q^\nu} \circ \delta\right) \frac{\mathrm{d}\dot q^\nu \delta(t)}{\mathrm{d}t},
\end{aligned} \tag{6.12}
$$

$$
Z_0 = -\left[\left(\frac{\partial L}{\partial q^\sigma} - \frac{\partial^2 L}{\partial t \partial \dot q^\sigma} - \frac{\partial^2 L}{\partial q^\sigma \partial \dot q^\nu}\dot q^\nu\right) \circ \delta\right] \frac{\mathrm{d}q^\sigma \delta(t)}{\mathrm{d}t} \tag{6.13}
$$

$$
+ \left[\left(\frac{\partial^2 L}{\partial \dot q^\sigma \partial \dot q^\nu}\dot q^\nu\right) \circ \delta\right] \frac{\mathrm{d}\dot q^\sigma \delta(t)}{\mathrm{d}t}. \tag{6.14}
$$

The definition of a Hamilton extremals (6.10) requires $Z_\sigma^1(t) = 0$, $Z_\sigma(t) = 0$ for $1 \le \sigma \le m$. The condition $Z_\sigma^1(t) = 0$, see (6.11), can be interpreted as a set of m homogeneous linear equations for m unknown functions $\dot q^\nu \delta(t) - \frac{\mathrm{d}q^\nu \delta(t)}{\mathrm{d}t}$ with the matrix $\left(\frac{\partial^2 L}{\partial \dot q^\sigma \partial \dot q^\nu}\right)$, $1 \le \sigma, \nu \le m$. If this matrix is regular (which means the regularity of λ in the standard sense) the set of equations (6.11) has only the trivial solution, i.e.

$$
\dot q^\nu \circ \delta = \frac{\mathrm{d}q^\nu \delta(t)}{\mathrm{d}t}, \quad 1 \le \sigma, \nu \le m.
$$

This means that the section δ of the projection π_1 is holonomic, i.e. it is the prolongation of a certain section γ of π, $\delta = J^1\gamma$. In such a situation we have

$$
\delta = J^1\gamma \implies \dot q^\sigma J^1\gamma(t) = \frac{\mathrm{d}q^\sigma J^1\gamma(t)}{\mathrm{d}t}, \quad \ddot q^\sigma J^2\gamma = \frac{\mathrm{d}\dot q^\sigma J^1\gamma}{\mathrm{d}t} - \frac{\mathrm{d}^2 q^\sigma \gamma(t)}{\mathrm{d}t},
$$

and the sets of equations $Z_\sigma(t) = 0$ and the equation $Z_0(t) = 0$ with $Z_\sigma(t)$ and $Z_0(t)$ given by (6.12) and (6.14), respectively, obtain (after some technical calculations—see Problem 6.4) the form

$$Z_\sigma(t) = \left[\frac{\partial L}{\partial q^\sigma} - \frac{\mathrm{d}}{\mathrm{d}t}\left(\frac{\partial L}{\partial \dot{q}^\sigma}\right)\right] \circ J^2\gamma = 0 \quad \text{for } 1 \leq \sigma \leq m,$$

$$Z_0(t) = -\left[\frac{\partial L}{\partial q^\sigma} - \frac{\mathrm{d}}{\mathrm{d}t}\left(\frac{\partial L}{\partial \dot{q}^\sigma}\right)\right] \circ J^2\gamma \times (\dot{q}^\sigma J^1\gamma) = 0.$$

On the basis of these results we can formulate the following proposition characterizing the relationship between the extremals of the regular Lagrange structure and the newly defined Hamilton extremals.

Proposition 6.3 *Every Hamilton extremal δ of a regular Lagrange structure (regular Lagrangian in the sense of Definition 6.1) is the prolongation of an extremal γ of the Lagrangian, i.e. $\delta = J^1\gamma$.*

(Note that the assumption of verticality of the vector field ζ did not need to be used.) We will show the influence of regularity on the following example.

Example 6.3 Let $(Y, \pi, X) = (\mathbf{R} \times \mathbf{R}^3, p, \mathbf{R})$ and let

$$\lambda = \left(\frac{1}{2}(\dot{x} + \dot{y})^2 + \frac{1}{2}\dot{z}^2\right) \mathrm{d}t$$

be the Lagrangian (i.e. $q^1 = x, q^2 = y, q^3 = z$). The matrix $\mathcal{M}$:

$$\mathcal{M} = \begin{pmatrix} \frac{\partial^2 L}{\partial \dot{x}^2} & \frac{\partial^2 L}{\partial \dot{x}\partial \dot{y}} & \frac{\partial^2 L}{\partial \dot{x}\partial \dot{z}} \\ \frac{\partial^2 L}{\partial \dot{x}\partial \dot{y}} & \frac{\partial^2 L}{\partial \dot{y}^2} & \frac{\partial^2 L}{\partial \dot{y}\partial \dot{z}} \\ \frac{\partial^2 L}{\partial \dot{x}\partial \dot{z}} & \frac{\partial^2 L}{\partial \dot{y}\partial \dot{z}} & \frac{\partial^2 L}{\partial \dot{z}^2} \end{pmatrix} = \begin{pmatrix} 1 & 1 & 0 \\ 1 & 1 & 0 \\ 0 & 0 & 1 \end{pmatrix}.$$

This Lagrangian is not regular in the sense of the standard Definition 6.1. Its Euler-Lagrange form is

$$E_\lambda = -(\ddot{x} + \ddot{y})\,(\mathrm{d}x + \mathrm{d}y) \wedge \mathrm{d}t - \ddot{z}\,\mathrm{d}z \wedge \mathrm{d}t,$$

and the Euler-Lagrange equations (2nd order differential equations for extremals γ) read

$$\begin{aligned} &(\ddot{x} + \ddot{y}) \circ J^2\gamma = 0, \quad \ddot{z} \circ J^2\gamma = 0 \\ &\quad \Longrightarrow \; \gamma : t \longrightarrow (t, x\gamma(t), At + B - x\gamma(t), Ct + D), \end{aligned}$$

where $x\gamma(t)$ is an arbitrary function (at least of the class C^2) and A, B, C and D are arbitrary constants (Let us remember that $\dot{x}J^1\gamma = \frac{\mathrm{d}x\gamma(t)}{\mathrm{d}t}$, $\ddot{x}J^2\gamma = \frac{\mathrm{d}^2x\gamma(t)}{\mathrm{d}t^2}$).

Now let us solve the problem of Hamilton extremals (perform all calculations in detail). Let ζ be a vector field on $J^1Y = \mathbf{R} \times \mathbf{R}^3 \times \mathbf{R}^3$ and δ a section of the projection p_1. Then step-by-step calculations give

$$\zeta = \zeta^0 \frac{\partial}{\partial t} + \zeta^x \frac{\partial}{\partial x} + \zeta^y \frac{\partial}{\partial y} + \zeta^z \frac{\partial}{\partial z} + \zeta_1^x \frac{\partial}{\partial \dot{x}} + \zeta_1^y \frac{\partial}{\partial \dot{y}} + \zeta_1^z \frac{\partial}{\partial \dot{z}},$$

$$\begin{aligned}
\theta_\lambda &= -\frac{1}{2}(\dot{x}^2 + \dot{y}^2 + \dot{z}^2)\,\mathrm{d}t + (\dot{x} + \dot{y})(\mathrm{d}x + \mathrm{d}y) + \dot{z}\,\mathrm{d}z, \\
\mathrm{d}\theta_\lambda &= -(\dot{x}\,\mathrm{d}\dot{x} + \dot{y}\,\mathrm{d}\dot{y} + \dot{z}\,\mathrm{d}\dot{z}) \wedge \mathrm{d}t + (\mathrm{d}\dot{x} + \mathrm{d}\dot{y}) \wedge (\mathrm{d}x + \mathrm{d}y) + \mathrm{d}\dot{z} \wedge \mathrm{d}z, \\
i_\zeta\,\mathrm{d}\theta_\lambda &= \zeta^0(\dot{x}\,\mathrm{d}\dot{x} + \dot{y}\,\mathrm{d}\dot{y} + \dot{z}\,\mathrm{d}\dot{z}) - (\zeta^x + \zeta^y)(\mathrm{d}\dot{x} + \mathrm{d}\dot{y}) - \zeta^z\,\mathrm{d}\dot{z} \\
&\quad + \zeta_1^x(-\dot{x}\,\mathrm{d}t + \mathrm{d}x + \mathrm{d}y) + \zeta_1^y(-\dot{y}\,\mathrm{d}t + \mathrm{d}x + \mathrm{d}y) + \zeta_1^z(-\dot{z}\,\mathrm{d}t + \mathrm{d}z), \\
\delta^* i_\zeta\,\mathrm{d}\theta_\lambda &= \Big[Z_0(t)(\zeta^0 \circ \delta) + Z_x(t)(\zeta^x \circ \delta) + Z_y(t)(\zeta^y \circ \delta) + Z_z(t)(\zeta^z \circ \delta) \\
&\quad + Z_x^1(t)(\zeta_1^x \circ \delta) + Z_y^1(t)(\zeta_1^y \circ \delta) + Z_z^1(t)(\zeta_1^z \circ \delta)\Big]\,\mathrm{d}t \\
Z_x^1(t) &= -\dot{x} \circ \delta + \left(\frac{\mathrm{d}x\delta(t)}{\mathrm{d}t} + \frac{\mathrm{d}y\delta(t)}{\mathrm{d}t}\right), \\
Z_y^1(t) &= -\dot{y} \circ \delta + \left(\frac{\mathrm{d}x\delta(t)}{\mathrm{d}t} + \frac{\mathrm{d}y\delta(t)}{\mathrm{d}t}\right), \\
Z_z^1(t) &= -\dot{z} \circ \delta + \frac{\mathrm{d}z\delta(t)}{\mathrm{d}t}, \\
Z_x(t) = Z_y(t) &= -\frac{\mathrm{d}\dot{x}\delta(t)}{\mathrm{d}t} - \frac{\mathrm{d}\dot{y}\delta(t)}{\mathrm{d}t}, \quad Z_z(t) = -\frac{\mathrm{d}\dot{z}\delta(t)}{\mathrm{d}t}, \\
Z_0(t) &= \left[(\dot{x} \circ \delta)\frac{\mathrm{d}\dot{x}\delta(t)}{\mathrm{d}t} + (\dot{y} \circ \delta)\frac{\mathrm{d}\dot{y}\delta(t)}{\mathrm{d}t} + (\dot{z} \circ \delta)\frac{\mathrm{d}\dot{z}\delta(t)}{\mathrm{d}t}\right].
\end{aligned}$$

The solution of seven conditions $Z_x(t) = Z_y(t) = Z_z(t) = Z_x^1(t) = Z_y^1(t) = Z_z^1(t) = 0$ gives the result (coefficient $Z_0(t)$ is not relevant because the definition of Hamilton extremals assumes that the vector field ζ is π_1-vertical, i.e. $\zeta^0 = 0$):

$$\delta : t \longrightarrow \big(t, x\delta(t), At + B - x\delta(t), Ct + D, \dot{x}\delta(t), \dot{x}\delta(t), C\big),$$

where $x\delta(t)$ is an arbitrary function (at least differentiable) and A, B, C and D are arbitrary constants. Note that (unlike extremals of the Lagrangian) in general

$$\dot{x}\delta(t) \neq \frac{\mathrm{d}x\delta(t)}{\mathrm{d}t}, \quad \text{and} \quad \dot{y}\delta(t) \neq \frac{\mathrm{d}y\delta(t)}{\mathrm{d}t}.$$

Especially, $\dot{x}\delta(t) = \dot{y}\delta(t)$, but $\frac{\mathrm{d}y\delta(t)}{\mathrm{d}t} = A - \frac{\mathrm{d}x\delta(t)}{\mathrm{d}t}$. So, there are Hamilton extremals that are not extremals of the Lagrangian.

6.3 The Concept of Regularity Revisited

Although we remain in the field of mechanics with the geometric basis of fibred manifolds up to the second order, we will notice the idea of a generalization of the concept of regularity, which is especially important for higher-order mechanics (We have already generalized the concept of regularity in Chaps. 3 and 4 (see Definitions 3.16 and 4.4) with respect to mechanical systems corresponding to Euler-Lagrange forms. Now we will take into account the point of view of extremals).

Consider a 2nd order Lagrange structure (π, λ) on a fibred manifold (Y, π, X), $\dim X = 1$, $\dim Y = m + 1$, with the Lagrangian $\lambda \in \Lambda^1_X(TJ^2Y)$ defined on an open set $W_2 \subset J^2Y$,

$$\lambda = L(t, q^\sigma, \dot{q}^\sigma, \ddot{q}^\sigma)\, dt.$$

The standard and generally accepted definition of regularity of the mentioned Lagrange structure (or equivalently regularity of the Lagrangian) is based on the analogy with the first order situation—see Definition 6.1. This means that a 2nd order Lagrange structure is considered to be regular, if the matrix

$$\mathcal{M} = \left(\frac{\partial^2 L}{\partial \ddot{q}^\sigma \partial \ddot{q}^\nu}\right), \quad 1 \le \sigma, \nu \le m, \tag{6.15}$$

is regular. The same is established in higher order (rth order) mechanics, i.e. the regularity of the matrix formed by 2nd order partial derivatives of the Lagrange function $L = L(t, q^\sigma, \dot{q}^\sigma, \dots, q^\sigma_r)$ with respect to the highest order variables. Such a definition of regularity is tied to the concrete Lagrangian. Its suitability for the 1st order mechanics is clear from Proposition 6.2. This proposition states that for regular 1st order Lagrange structure (in the sense of the standard Definition 6.1) holds: All Hamilton extremals of a regular Lagrange structure are simultaneously extremals of the Lagrange structure and vice versa. For higher order mechanics arises a question: Consider two Lagrangians, λ_1 and λ_2, where λ_1 is 1st order Lagrangian regular in the sense of Definition 6.1 and λ_2 is a 2nd order Lagrangian

$$\lambda_2 = \lambda_1 + \frac{df(t, q^\sigma, \dot{q}^\sigma)}{dt}\, dt.$$

As

$$\frac{df(t, q^\sigma, \dot{q}^\sigma)}{dt}\, dt = \left(\frac{\partial f}{\partial t} + \frac{\partial f}{\partial q^\nu}\dot{q}^\nu + \frac{\partial f}{\partial \dot{q}^\nu}\ddot{q}^\nu\right) dt$$

is the trivial 2nd order Lagrangian (affine in 2nd order variables $(\ddot{q}^1, \dots, \ddot{q}^m)$), the Lagrangian λ_2 is not regular in the sense of relation (6.15), while λ_1 is regular in the sense of Definition 6.1. At the same time, both mentioned Lagrangians have identical Euler-Lagrange form and identical extremals. Although we could define the regularity of a Lagrangian with help of any of the minimal Lagrangians (Lagrangians of the minimum order with identical Euler-Lagrange form) which are either all regular or not in the sense of Definition 6.1, it seems more natural to formulate the concept of regularity rather using Euler-Lagrange forms (variational dynamical forms). The basic idea was already outlined in Sect. 3.4, Definition 3.16 and Proposition 3.8.

Let us look at an alternative approach to geometry related to the concept of Hamilton extremals. Let (Y, π, X), $\dim X = 1$, $\dim Y = m + 1$. Consider a fibred manifold $(\tilde{Y}, \tilde{\pi}, X)$ with the same base X and the total space $\tilde{Y} = J^1 Y$ and the projection $\tilde{\pi} = \pi_1$. Its dimension is $\dim \tilde{Y} = 2m + 1$. Next, consider the first prolongation of $(\tilde{Y}, \tilde{\pi}, X)$, i.e.

$$(\tilde{J}^1 \tilde{Y}, \tilde{\pi}_1, X) = \big(\tilde{J}^1 (J^1 Y), \pi_{1,1}, X\big), \quad \dim J^1 \tilde{Y} = 4m + 1.$$

(For better distinction of this "additional" prolongation from the standard one we denote it by tilde.) Let (V, ψ), $\psi = (t, q^\sigma)$, $\big(\pi(V), \varphi\big)$, $\varphi = (t)$, $\big(\pi_{1,0}^{-1}(V), \psi_1\big)$, $\psi_1 = (t, q^\sigma, \dot{q}^\sigma)$ be a fibred chart on Y, the associated chart on X and associated fibred chart on $J^1 Y$, respectively. For further considerations denote

$$q^\sigma = q_0^\sigma, \quad \dot{q}^\sigma = q_1^\sigma, \quad \text{etc. for } 1 \le \sigma \le m,$$

and for the basis adapted to the contact structure

$$\omega^\sigma = \omega_0^\sigma = \mathrm{d}q_0^\sigma - q_1^\sigma \,\mathrm{d}t, \quad \omega_1^\sigma = \mathrm{d}q_1^\sigma - q_2^\sigma \,\mathrm{d}t, \quad \text{etc. for } 1 \le \sigma \le m.$$

The corresponding fibred chart on $\tilde{Y}$ is thus $(\tilde{V}, \tilde{\psi})$, $\tilde{V} = \pi_{1,0}^{-1}(V)$, $\tilde{\psi} = (t, q_0^\sigma, q_1^\sigma)$, the associated fibred chart on $\tilde{J}^1 \tilde{Y} = \tilde{J}^1 (J^1 Y)$ is

$$\big(\tilde{\pi}_{1,0}^{-1}(\tilde{V}), \tilde{\psi}_1\big), \quad \tilde{\psi}_1 = (t, q_0^\sigma, q_1^\sigma, q_{0,1}^\sigma, q_{1,1}^\sigma).$$

Proposition 6.4 *Let (π, λ) be a 1st order Lagrange structure on (Y, π, X) with the Lagrangian λ defined on an open set $W_1 \subset J^1 Y$ and let δ be a section of the projection π_1 defined on $\pi_1(W_1)$. Then the condition*

$$\tilde{J}^1 \delta^* \tilde{h} i_\zeta \,\mathrm{d}\theta_\lambda = 0$$

is equivalent to condition (6.10). Here ζ is an arbitrary π_1-vertical vector field on $\tilde{Y} = J^1 Y$.

Proof We prove the generally valid relation: For every form ω it holds

$$\delta^* \omega = \tilde{J}^1 \delta^* \tilde{h} \omega. \tag{6.16}$$

Let us express both sides of Eq. (6.16) for forms $\mathrm{d}q_0^\sigma$ and $\mathrm{d}q_1^\sigma$ (for $\omega = \mathrm{d}t$ this relation is evident). We obtain

$$\delta^* \,\mathrm{d}q_0^\sigma = \frac{\mathrm{d}q_0 \delta(t)}{\mathrm{d}t} \,\mathrm{d}t, \quad \delta^* \,\mathrm{d}q_1^\sigma = \frac{\mathrm{d}q_1^\sigma \delta(t)}{\mathrm{d}t} \,\mathrm{d}t,$$

$$\tilde{h}\,\mathrm{d}q_0^\sigma = q_{0,1}^\sigma\,\mathrm{d}t,\quad \tilde{h}\,\mathrm{d}q_1^\sigma = q_{1,1}^\sigma\,\mathrm{d}t$$

$$\Longrightarrow \tilde{J}^1\delta^*\tilde{h}\,\mathrm{d}q_0^\sigma = \tilde{J}^1\delta^*(q_{0,1}^\sigma\,\mathrm{d}t) = \frac{\mathrm{d}q_0^\sigma\delta(t)}{\mathrm{d}t}\,\mathrm{d}t,$$
$$\tilde{J}^1\delta^*\tilde{h}\,\mathrm{d}q_1^\sigma = \tilde{J}^1\delta^*(q_{1,1}^\sigma\,\mathrm{d}t) = \frac{\mathrm{d}q_1^\sigma\delta(t)}{\mathrm{d}t}\,\mathrm{d}t.$$

Then relation (6.16) for arbitrary ω is evident. This finishes the proof. □

Example 6.4 Let (Y,π,X) be a fibred manifold, (J^1Y,π_1,X) its 1st prolongation and $\omega_0^\sigma = \mathrm{d}q_0^\sigma - q_1^\sigma\,\mathrm{d}t$ a contact 1-form belonging to a fibred chart adapted to the contact structure of (Y,π,X). It is evident that $h\omega_0^\sigma = 0$. Let $\big((\tilde{J}^1(J^1Y),\tilde{\pi}_1,X)\big)$ be the 1st prolongation of (J^1Y,π_1,X), $J^1Y=\tilde{Y}$, $\pi_1=\tilde{\pi}$, and $\tilde{h}$ and $\tilde{p}$ the corresponding horizontalization and contactization mapping, respectively. Denote

$$\tilde{\omega}_0 = \mathrm{d}q_0^\sigma - q_{0,1}^\sigma\,\mathrm{d}t,\quad \tilde{\omega}_1 = \mathrm{d}q_1^\sigma - q_{1,1}^\sigma\,\mathrm{d}t.$$

Let δ be a section of the projection $\tilde{\pi}=\pi_1$. Then

$$\delta^*\omega_0^\sigma = \delta^*(\mathrm{d}q_0^\sigma - q_1^\sigma\,\mathrm{d}t) = \left(\frac{\mathrm{d}q_0^\sigma\delta(t)}{\mathrm{d}t} - q_1^\sigma\delta(t)\right)\mathrm{d}t.$$

This expression is in general nonzero, because δ is considered as a general section of π_1, not necessarily holonomic, i.e. $q_1^\sigma\delta(t)\neq\frac{\mathrm{d}q_0^\sigma\delta(t)}{\mathrm{d}t}$ in general. So, ω_0^σ are not contact forms on $(J^1Y,\pi_1,X)=(\tilde{Y},\tilde{\pi},X)$. On the other hand, it holds

$$\tilde{J}^1\delta^*\tilde{\omega}_0^\sigma = \tilde{J}^1\delta^*(\mathrm{d}q_0^\sigma - q_{0,1}^\sigma\,\mathrm{d}t) = \left(\frac{\mathrm{d}q_0^\sigma\delta(t)}{\mathrm{d}t} - q_{0,1}^\sigma\tilde{J}^1\delta(t)\right)\mathrm{d}t = 0,$$

because $q_{0,1}^\sigma\tilde{J}^1\delta(t) = \frac{\mathrm{d}q_0^\sigma\delta(t)}{\mathrm{d}t}$. And similarly

$$\tilde{J}^1\delta^*\tilde{\omega}_1^\sigma = \tilde{J}^1\delta^*(\mathrm{d}q_1^\sigma - q_{1,1}^\sigma\,\mathrm{d}t) = \left(\frac{\mathrm{d}q_1^\sigma\delta(t)}{\mathrm{d}t} - q_{1,1}^\sigma\tilde{J}^1\delta(t)\right)\mathrm{d}t = 0.$$

Forms $\tilde{\omega}_0^\sigma$ and $\tilde{\omega}_1^\sigma$, $1\le\sigma\le m$, defined on $(\tilde{Y},\tilde{\pi},X)$ are contact. This means that $\tilde{h}\tilde{\omega}_0^\sigma = 0$ and $\tilde{h}\tilde{\omega}_1^\sigma = 0$, while

$$\tilde{h}\omega_0^\sigma = \tilde{h}(\mathrm{d}q_0^\sigma - q_1^\sigma\,\mathrm{d}t) = \tilde{h}\left(\tilde{\omega}_0^\sigma + q_{0,1}^\sigma\,\mathrm{d}t - q_1^\sigma\,\mathrm{d}t\right) = (q_{0,1}^\sigma - q_1^\sigma)\,\mathrm{d}t.$$

The situation is schematically depicted on Fig. 6.1.

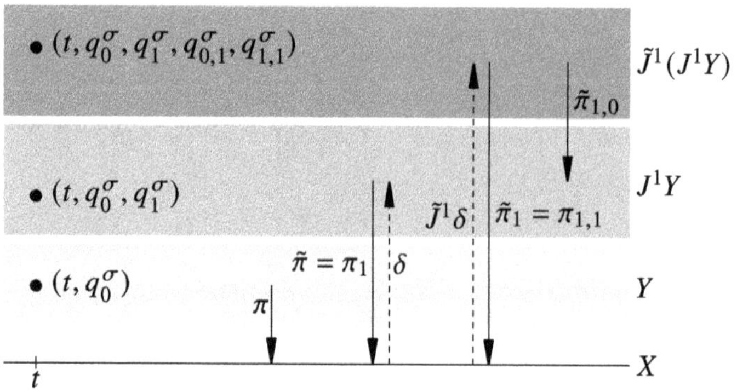

Fig. 6.1 Jet prolongation $(\tilde{J}^1\tilde{Y}, \tilde{\pi}_1, X)$ of the fibred manifold $(\tilde{Y}, \tilde{\pi}, X) = (J^1Y, \pi_1, X)$

We now express $\tilde{J}^1\delta^*\tilde{h}i_\zeta\, \mathrm{d}\theta_\lambda$: First we express $\tilde{h}i_\zeta\, \mathrm{d}\theta_\lambda$ for a π_1-vertical vector field ζ on J^1Y, i.e. $\zeta^0 = 0$, and then with the above mentioned notation $q^\sigma = q_0^\sigma, \dot{q}^\sigma = q_1^\sigma$, $1 \le \sigma \le m$, we gradually obtain

$$\begin{aligned}
\tilde{h}i_\zeta\, \mathrm{d}\theta_\lambda &= \left[\left(\frac{\partial L}{\partial q_0^\sigma} - \frac{\partial^2 L}{\partial t \partial q_1^\sigma} - \frac{\partial^2 L}{\partial q_0^\sigma \partial q_1^\nu} q_1^\nu\right)\zeta^\sigma - \frac{\partial^2 L}{\partial q_1^\sigma \partial q_1^\nu} q_1^\nu \zeta_1^\sigma\right.\\
&\quad \left. - \frac{\partial^2 L}{\partial q_0^\nu \partial q_1^\sigma}(\zeta^\sigma q_{0,1}^\nu - \zeta^\nu q_{0,1}^\sigma) - \frac{\partial^2 L}{\partial q_1^\sigma \partial q_1^\nu}(\zeta^\sigma q_{1,1}^\nu - \zeta_1^\nu q_{0,1}^\sigma)\right] \mathrm{d}t,\\
\tilde{h}i_\zeta\, \mathrm{d}\theta_\lambda &= \left(H_\sigma^0 \zeta^\sigma + H_\sigma^1 \zeta_1^\sigma\right) \mathrm{d}t,\\
H_\sigma^0 &= \left(\frac{\partial L}{\partial q_0^\sigma} - \frac{\partial^2 L}{\partial t \partial q_1^\sigma} - \frac{\partial^2 L}{\partial q_0^\sigma \partial q_1^\nu} q_1^\nu\right) - \frac{\partial^2 L}{\partial q_1^\sigma \partial q_1^\nu} q_{1,1}^\nu \qquad (6.17)\\
&\quad + q_{0,1}^\nu \left(\frac{\partial^2 L}{\partial q_0^\sigma \partial q_1^\nu} - \frac{\partial^2 L}{\partial q_0^\nu \partial q_1^\sigma}\right)\\
&= \left(\frac{\partial L}{\partial q_0^\sigma} - \frac{\partial^2 L}{\partial t \partial q_1^\sigma} - \frac{\partial^2 L}{\partial q_1^\sigma \partial q_0^\nu} q_1^\nu - \frac{\partial^2 L}{\partial q_1^\sigma \partial q_1^\nu} q_2^\nu\right) + \left(\frac{\partial^2 L}{\partial q_1^\sigma \partial q_0^\nu} - \frac{\partial^2 L}{\partial q_0^\sigma \partial q_1^\nu}\right) q_1^\nu\\
&\quad + \frac{\partial^2 L}{\partial q_1^\sigma \partial q_1^\nu}(q_2^\nu - q_{1,1}^\nu) + \left(\frac{\partial^2 L}{\partial q_0^\sigma \partial q_1^\nu} - \frac{\partial^2 L}{\partial q_0^\nu \partial q_1^\sigma}\right) q_{0,1}^\nu,\\
H_\sigma^0 &= E_\sigma(L) + \left(\frac{\partial^2 L}{\partial q_0^\sigma \partial q_1^\nu} - \frac{\partial^2 L}{\partial q_0^\nu \partial q_1^\sigma}\right)(q_{0,1}^\nu - q_1^\nu) - \frac{\partial^2 L}{\partial q_1^\sigma \partial q_1^\nu}(q_{1,1}^\nu - q_2^\nu), \qquad (6.18)\\
H_\sigma^1 &= \frac{\partial^2 L}{\partial q_1^\sigma \partial q_1^\nu}(q_{0,1}^\nu - q_1^\nu). \qquad (6.19)
\end{aligned}$$

We obtain the set of $2m$ linear equations for Hamilton extremals δ (sections of π_1), with $2m$ unknown quantities $(q_{0,1}^\sigma - q_1^\sigma) \circ \tilde{J}^1\delta$ and $(q_{1,1}^\sigma - q_2^\sigma) \circ \tilde{J}^1\delta$

$$H_\sigma^0 \circ \tilde{J}^1\delta = 0, \quad H_\sigma^1 \circ \tilde{J}^1\delta = 0, \quad 1 \le \sigma \le m. \qquad (6.20)$$

Definition 6.4 (*Hamilton form*) Let (π, λ) be a 1st order Lagrange structure on the fibred manifold (Y, π, X), $\dim X = 1$, with the Lagrangian $\lambda \in \Lambda^1_X(TJ^1Y)$ defined on na open set $W_1 \subset J^1Y$. The differential form

$$H_\lambda = H^0_\sigma \, dq^\sigma_0 \wedge dt + H^1_\sigma \, dq^\sigma_1 \wedge dt$$

defined on $\tilde{J}^1(J^1Y)$, where functions H^0_σ and H^1_σ, $1 \le \sigma \le m$, are given by relations (6.18) and (6.19), is called the *Hamilton form of the Lagrangian* λ.

On the base of above presented considerations and calculations we can formulate the following proposition.

Proposition 6.5 *Suppose that the assumptions of Definition 6.4 are fulfilled. The Hamilton form H_λ is unique 2-form on $\tilde{J}^1(J^1Y)$ for which the relation $i_{\tilde{J}^1\zeta} H_\lambda = \tilde{h}(i_\zeta \, d\theta_\lambda)$ hold, where ζ is an arbitrary π_1-vertical vector field on J^1Y. Moreover, a section $\delta \in \Gamma_{\pi_1(W_1)}(\pi_1)$ is the Hamilton extremal if and only if $H_\lambda \circ \tilde{J}^1\delta = 0$.*

We will now proceed to reformulate the concept of regularity.

Definition 6.5 (*Regular Hamilton extremals*) Let the assumptions of Definition 6.4 be again valid. A Hamilton extremal $\delta \in \Gamma_{\pi_1(W_1)}(\pi_1)$ of the Lagrangian λ is called *regular*, if there exists an extremal $\gamma \in \Gamma_{\pi_1(W_1)}(\pi)$ of the Lagrangian such that it holds $\delta = J^1\gamma$.

The relation of regularity of Hamilton extremals to the standard regularity of the corresponding Lagrangian follows from Eq. (6.20) with functions H^0_σ and H^1_σ given by (6.17)–(6.19). The set of equations $H^1_\sigma \circ \tilde{J}^1\delta = 0$, $1 \le \sigma \le m$, can be interpreted as the set of m homogeneous equations for m unknown variables $(q^\nu_{0,1} - q^\nu_1) \circ \tilde{J}^1\delta$, $1 \le \nu \le m$. Trivial solution of this set is $q^\nu_{0,1} \circ \tilde{J}^1\delta = q^\nu_1 \circ \delta$, i.e.

$$q^\nu_1 \circ \delta(t) = \frac{dq^\nu_0\delta(t)}{dt}.$$

This means that the section δ of π_1 is holonomic, i.e. there exists a section γ of π such that $\delta = J^1\gamma$. The necessary and sufficient condition for such a case is the regularity of the matrix

$$\mathcal{M} = \left(\frac{\partial^2 L}{\partial q^\sigma_1 \partial q^\nu_1} \right), \quad 1 \le \sigma, \nu \le m.$$

This is, of course, condition (6.1) of "standard" regularity of the Lagrangian. Let us take a look at the set of equations $H^0_\sigma \circ \tilde{J}^1\delta = 0$, $1 \le \sigma \le m$. If the matrix $\mathcal{M}$ is regular, then for a corresponding Hamilton extremal δ it holds $\delta = J^1\gamma$ for certain section γ, and then

$$q^{\nu}_{1,1}\tilde{J}(J^1\gamma) = q^{\nu}_2 \circ J^2\gamma, \quad 1 \leq \nu \leq m.$$

Thus the equations $H^0_\sigma \circ \tilde{J}^1\delta = 0$ are fulfilled for $E_\sigma(L) \circ J^2\gamma = 0$, $1 \leq \sigma \leq m$, which means that γ is the extremal of the Lagrangian λ. So, the sufficient condition for regularity of a Hamilton extremal reads:

Proposition 6.6 *Let (π, λ) be a 1st order Lagrange structure on (Y, π, X) with the Lagrangian λ regular on an open set $W_1 \subset J^1Y$ in the sense of definition (6.1). Let $\delta \in \Gamma_{\pi_1(W_1)}(\pi_1)$ be a Hamilton extremal of the Lagrangian λ. Then δ is regular Hamilton extremal.*

(Proposition 6.6 presents the regularity (6.1) as the sufficient, not necessary condition, because there could be another holonomic Hamilton extremals than those resulting from trivial solution of the second set of Eq. (6.20)). This means that a Hamilton extremal can be regular even if condition (6.1) is not fulfilled (Such an exceptional situation can take place if all second order partial derivatives $\frac{\partial^2 L}{\partial q^\sigma_1 \partial q^\nu_1}$ vanish as it will be shown below in Example 6.5).

For the time being, let us assume the Lagrange structure to be regular according to (6.1). We have introduced the Legendre transformation $\mathcal{L}$ (6.4) on J^1Y for the Lagrangian regular in the sense of (6.1). (In the general case of the rth order mechanics, $r \geq 1$, this transformation is carried out on the definition manifold of the Lepage equivalent, i.e. $J^{2r-1}Y$, which for $r = 1$ is coincidentally the same as the definition manifold of the Lagrangian.) Express a π_1-vertical vector field ζ and the Hamilton function in Legendre coordinates as

$$\zeta = \zeta^\sigma_0 \frac{\partial}{\partial q^\sigma_0} + \tilde{\zeta}^0_\sigma \frac{\partial}{\partial p^0_\sigma},$$

$$p^0_\sigma = \frac{\partial L}{\partial q^\sigma_1}, \quad H(t, q^\nu_0, p^0_\nu) = -L \circ \mathcal{L} + p^0_\nu (q^\nu_1 \circ \mathcal{L}).$$

For Hamilton extremals in Legendre coordinates we obtain by step by step calculations:

$$\tilde{h}\,(i_\zeta\, d\theta_\lambda) = \left[\left(-\frac{\partial H}{\partial q^\sigma_0} - p^{0,1}_\sigma\right)\zeta^\sigma_0 + \left(-\frac{\partial H}{\partial p^0_\sigma} + q^\sigma_{0,1}\right)\tilde{\zeta}^0_\sigma\right] dt. \tag{6.21}$$

A section $\delta \in \Gamma_{\pi_1(W_1)}(\pi_1)$ is a Hamilton extremal if and only if $\tilde{J}^1\delta^*\tilde{h}i_\zeta\, d\theta_\lambda = 0$ for an arbitrary π_1-vertical vector field ζ on J^1Y (Proposition 6.5). So, we obtain equations of Hamilton extremals (*Hamilton equations*) on $\tilde{J}^1(J^1Y)$:

$$\left(-\frac{\partial H}{\partial q^\sigma_0} - p^{0,1}_\sigma\right) \circ \tilde{J}^1\delta = 0, \quad \left(-\frac{\partial H}{\partial p^0_\sigma} + q^\sigma_{0,1}\right) \circ \tilde{J}^1\delta = 0. \tag{6.22}$$

Definition 6.6 (*Hamilton vector field, Euler-Lagrange distribution*) Consider a 1st order Lagrange structure on (Y, π, X) with the Lagrangian $\lambda \in \Lambda^1_X(TJ^1Y)$ defined on an open set $W_1 \subset J^1Y$ and regular in the sense of (6.1). The vector field (expressed in charts)

$$\xi = \frac{\partial}{\partial t} + \left(\frac{\partial H}{\partial p^0_\sigma} \frac{\partial}{\partial q^\sigma_0} - \frac{\partial H}{\partial q^\sigma_0} \frac{\partial}{\partial p^0_\sigma} \right) \tag{6.23}$$

is called the *Hamilton vector field*. The (one-dimensional) distribution Δ_λ generated by this vector field is called the *Euler-Lagrange distribution*.

The relation of the Hamilton vector field to Hamilton equations shows the following proposition.

Proposition 6.7 *Let (π, λ) be a 1st order Lagrange structure as specified above, with the Lagrangian regular in the sense of (6.1). The Hamilton vector field is defined on $W_1 \subset J^1Y$ as well. The following conditions are equivalent:*

(a) A section $\delta \in \Gamma_{\pi_1(W_1)}(\pi_1)$ is a Hamilton extremal if and only if it obeys the Hamilton equations (6.22)
(b) A section $\delta \in \Gamma_{\pi_1(W_1)}(\pi_1)$ is a Hamilton extremal if and only if it is an integral section of the Euler-Lagrange distribution Δ_λ.

Proof The validity of condition (a) is evident from the above presented calculations made on the basis of the definition of Hamilton extremals and leading to resulting Hamilton equations (6.22) (Recall that in the case of a Lagrange structure regular in the sense of (6.1) every Hamilton extremal is regular—see Definition 6.5). To prove (b) calculate the annihilator Δ^0_λ of the Euler-Lagrange distribution (Definition 3.12). Let $\eta \in \Lambda^1(TJ^1Y)$ be a 1-form annihilating the Hamilton vector field ξ, i.e. $i_\xi \eta = 0$. Suppose η in the chart expression as follows:

$$\eta = \eta_0 \, dt + \eta_\sigma \, dq^\sigma_0 + \tilde{\eta}^\sigma \, dp^0_\sigma,$$

where $\eta_0(t, q^\nu_0, p^0_\nu)$, $\eta_\sigma(t, q^\nu_0, p^0_\nu)$, $\tilde{\eta}^\sigma(t, q^\nu_0, p^0_\nu)$, $1 \leq \sigma, \nu \leq m$, are functions on J^1Y. The condition $i_\xi \eta = 0$ leads to the following relation between these functions

$$\eta_0 + \eta_\sigma \frac{\partial H}{\partial p^0_\sigma} - \tilde{\eta}^\sigma \frac{\partial H}{\partial q^\sigma_0} = 0 \implies \eta_0 = -\eta_\sigma \frac{\partial H}{\partial p^0_\sigma} + \tilde{\eta}^\sigma \frac{\partial H}{\partial q^\sigma_0}$$

$$\implies \eta = \eta_\sigma \left(-\frac{\partial H}{\partial p^0_\sigma} dt + dq^\sigma_0 \right) + \tilde{\eta}^\sigma \left(\frac{\partial H}{\partial q^\sigma_0} dt + dp^0_\sigma \right).$$

Let $\delta \in \Gamma_{\pi_1(W_1)}(\pi_1)$ be a section. Then

$$\tilde{J}^1\delta^*\eta = \left\{ \left[\eta_\sigma \left(-\frac{\partial H}{\partial p^0_\sigma} + q^\sigma_{0,1} \right) + \tilde{\eta}^\sigma \left(\frac{\partial H}{\partial q^\sigma_0} + p^{0,1}_\sigma \right) \right] \circ \tilde{J}^1\delta \right\} dt.$$

If the Hamilton equations are fulfilled, then $\tilde{J}^1\delta^*\eta = 0$, and vice versa. □

In addition, the Hamilton vector field has the close relation to the corresponding Lagrangian, especially it belongs to the Euler-Lagrange distribution Δ_λ, i.e. $i_\xi \, d\theta_\lambda \in \Delta_\lambda^0$, where θ_λ is the Lepage equivalent of the Lagrangian (see Problem 6.11).

Example 6.5 Suppose that

$$\frac{\partial^2 L}{\partial q_1^\sigma \partial q_1^\nu} = \frac{\partial^2 L}{\partial \dot{q}^\sigma \partial \dot{q}^\nu} = 0, \quad 1 \leq \sigma, \nu \leq m.$$

This means that the 1st order Lagrange function is linear in velocities (variables q_1^σ, $1 \leq \sigma \leq m$),

$$L(t, q_0^\nu, q_1^\nu) = a(t, q_0^\nu) + b_\sigma(t, q_0^\nu) q_1^\sigma$$

and the corresponding Euler-Lagrange form is of the 1st order only,

$$E = E_\sigma \, dq_0^\sigma \wedge dt, \quad E_\sigma = \left(\frac{\partial a}{\partial q_0^\sigma} - \frac{\partial b_\sigma}{\partial t} \right) + \left(\frac{\partial b_\nu}{\partial q_0^\sigma} - \frac{\partial b_\sigma}{\partial q_0^\nu} \right) q_1^\nu.$$

Then all functions H_σ^1 given by (6.19) are zero and functions H_σ^0 take the form

$$\begin{aligned} H_\sigma^0 &= E_\sigma(L) + \left(\frac{\partial^2 L}{\partial q_0^\sigma \partial q_1^\nu} - \frac{\partial^2 L}{\partial q_0^\nu \partial q_1^\sigma} \right) (q_{0,1}^\nu - q_1^\nu) \\ &= \left(\frac{\partial a}{\partial q_0^\sigma} - \frac{\partial b_\sigma}{\partial t} \right) + \left(\frac{\partial b_\nu}{\partial q_0^\sigma} - \frac{\partial b_\sigma}{\partial q_0^\nu} \right) q_{0,1}^\nu . \end{aligned}$$

The condition $H_\sigma^0 \circ \tilde{J}^1 \delta = 0$ for $1 \leq \sigma \leq m$ can be thus interpreted as he set of m inhomogeneous linear equations for unknowns variables $q_{0,1}^\nu \circ \tilde{J}^1 \delta$, $1 \leq \nu \leq m$. The matrix and the extended matrix of this set of equations are

$$(\beta \mid \alpha) = (\beta_{\nu\sigma} \mid \alpha_{\sigma 1}), \quad (\beta_{\nu\sigma}) = \left(\frac{\partial^2 L}{\partial q_0^\sigma \partial q_1^\nu} - \frac{\partial^2 L}{\partial q_0^\nu \partial q_1^\sigma} \right), \quad \alpha_{\sigma 1} = \frac{\partial a}{\partial q_0^\sigma} - \frac{\partial b_\sigma}{\partial t}.$$

The necessary and sufficient condition for this set of equations to have a solution is rank (β) = rank $(\beta|\alpha)$. Let us examine the possibility that the Hamilton extremal δ is holonomic, i.e. $\delta = J^1\gamma$ for certain section γ of π. In general situation (when functions $a(t, q_0^\nu)$ and $b_\sigma(t, q_0^\nu)$ are not related to each other) this means

$$\det(\beta) = \det \left(\frac{\partial^2 L}{\partial q_0^\sigma \partial q_1^\nu} - \frac{\partial^2 L}{\partial q_0^\nu \partial q_1^\sigma} \right) \neq 0.$$

In such a case the set of equations $H_\sigma^0 \circ \tilde{J}^1\delta = 0$ has the unique solution

$$q_{0,1}^\sigma \circ \tilde{J}^1\delta = \frac{dq_0^\sigma \tilde{J}^1\delta(t)}{dt}, \quad \text{and} \quad E_\sigma(L) \circ J^1\gamma = 0,$$

i.e. $\delta = J^1\gamma$ where γ is an extremal of the Lagrangian λ.

Example 6.5 shows the situation when the considered Lagrange structure is not regular in the standard sense (6.1) but there exist regular Hamilton extremals (Note that a similar situation occurs in higher-order Hamiltonian mechanics when the order of the Euler-Lagrange form is odd. In the case of even order of Euler-Lagrange form, Hamilton extremals are regular if and only if in the case of standard regularity. A detailed advanced analysis can be found in works of O. Krupková—see in References below). We can summarize results obtained above, including Example 6.5.

Proposition 6.8 *Let (π, λ) be a 1st order Lagrange structure on a fibred manifold (Y, π, X) with the Lagrangian $\lambda \in \Lambda_X^1(TJ^1Y)$ defined on an open set $W_1 \subset J^1Y$, $\lambda = L(t, q^\sigma, q_1^\sigma)\,dt$. A Hamilton extremal $\delta \in \Gamma_{\pi_1(W_1)}(\pi_1)$ is regular if and only if just one of the following conditions is fulfilled:*

(a)

$$\det\left(\frac{\partial^2 L}{\partial q_1^\sigma \partial q_1^\nu}\right) \neq 0, \quad 1 \leq \sigma, \nu \leq m,$$

(b)

$$\det\left(\frac{\partial^2 L}{\partial q_1^\sigma \partial q_1^\nu}\right) = 0, \quad \det\left(\frac{\partial^2 L}{\partial q_1^\sigma \partial q_0^\nu} - \frac{\partial^2 L}{\partial q_0^\sigma \partial q_1^\nu}\right) \neq 0, \quad 1 \leq \sigma, \nu \leq m.$$

This proposition states that there exist variational problems which are not regular in the sense of the standard regularity (6.1) but they become regular in the sense of the regularity of Hamilton extremals. We will go even further in generalizing the concept of regularity and relate it to the Euler-Lagrange form. In Sect. 3.4 we have introduced the concept of a dynamical system $[\alpha]$ associated with the dynamical form E affine in accelerations (the highest coordinates on J^2Y). Let E_λ be the Euler-Lagrange form of a 1st order Lagrangian λ, i.e. variational dynamical form of 2nd order, in general (with the exception shown in Example 6.5). Denote $[\alpha_E]$ the dynamical system of E_λ. First consider a general dynamical form affine in accelerations,

$$E_\lambda = (A_\sigma + B_{\sigma\nu}\, q_2^\nu)\omega_0^\sigma \wedge dt,$$

where $A_\sigma(t, q_0^\mu, q_1^\mu)$ and $B_{\sigma\nu}(t, q_0^\mu, q_1^\mu)$, $1 \leq \sigma, \nu, \mu \leq m$, are functions on J^1Y. As we know (see Sect. 3.4) the mechanical system associated to the dynamical form E is the class $[\alpha]$ of 2-forms on J^1Y for which it holds $p_1\alpha = E$, i.e.

$$\alpha_E = A_\sigma \omega_0^\sigma \wedge dt + B_{\sigma\nu}\omega_0^\sigma \wedge dq_1^\nu + F_{\sigma\nu}\omega_0^\sigma \wedge \omega_0^\nu, \quad F_{\sigma\nu} + F_{\nu\sigma} = 0,$$

$$F_{\sigma\nu} = F_{\sigma\nu}(t, q_0^\mu, q_1^\mu).$$

Various representatives of the class $[\alpha_E]$ differ by 2-contact 2-forms $F = F_{\sigma\nu}\omega_0^\sigma \wedge \omega_0^\nu$. The dynamical distribution Δ_α associated with the concrete representative α_E of the mechanical system $[\alpha_E]$ has the annihilator

$$\Delta_\alpha^0 = \left(A_\sigma \, \mathrm{d}t + 2F_{\sigma\nu}\omega_0^\nu + B_{\sigma\nu} \, \mathrm{d}q_1^\nu, \, B_{\sigma\nu}\omega_0^\nu\right), \quad 1 \le \sigma, \nu \le m.$$

(All these concepts connected with a dynamical form E affine in accelerations and their properties were explained in detail in Sect. 3.4—see Definition 3.16 and Propositions 3.7 and 3.8).

Let (Y, π, X) be a fibred manifold, (V, ψ), $\psi = (t, q_0^\nu)$ a fibred chart on Y, $(\pi(V), \varphi)$, $\varphi = (t)$, $(\pi_{1,0}^{-1}(V), \psi_1)$, $\psi_1 = (t, q_0^\nu, q_1^\nu)$ and $(\pi_{2,0}^{-1}(V), \psi_2)$, $\psi_2 = (t, q_0^\nu, q_1^\nu, q_2^\nu)$ the associated chart on X, associated fibred chart on J^1Y and associated fibred chart on J^2Y, respectively. Let $\lambda \in \Lambda_X^1(TJ^1Y)$ be a Lagrangian defined on an open set $W_1 \subset J^1Y$, supposing that $W_1 \subset \pi_{1,0}^{-1}(V)$ or at least $W_1 \cap \pi_{1,0}^{-1}(V) \neq \emptyset$ (in such a case the coordinate calculations hold on $W_1 \cap \pi_{1,0}^{-1}(V)$), and $E_\lambda = E_\sigma \omega_0^\sigma \wedge dt$ its Euler-Lagrange form. Consider two situations:

(a) Suppose that λ is regular in the sense of (6.1). Then E_λ is the dynamical form affine in accelerations and it obeys Helmholtz conditions (5.9)–(5.12),

$$E_\lambda = (A_\sigma + B_{\sigma\nu}q_2^\nu)\,\omega_0^\sigma \wedge \mathrm{d}t, \quad B_{\sigma\nu} + B_{\nu\sigma} = 0, \quad 1 \le \sigma, \nu \le m.$$

The mechanical system $[\alpha]$ contains the unique closed representative α_E,

$$\alpha_E = A_\sigma \omega_0^\sigma \wedge \mathrm{d}t + B_{\sigma\nu}\omega_0^\sigma \wedge \mathrm{d}q_1^\nu + \frac{1}{4}\left(\frac{\partial A^\sigma}{\partial q_1^\nu} - \frac{\partial A^\nu}{\partial q_1^\sigma}\right)\omega_0^\sigma \wedge \omega_0^\nu.$$

(b) Suppose that λ is affine in velocities,

$$L = a(t, q_0^\nu) + b_\sigma(t, q^\nu)q_1^\sigma,$$

i.e. it is not regular in the sense of (6.1). Moreover, suppose that the condition (b) of Proposition 6.8 is fulfilled. Then the corresponding Euler-Lagrange form is of 1st order, i.e.

$$E_\lambda = E_\sigma(t, q_0^\nu, q_1^\nu)\,\mathrm{d}q_0^\sigma \wedge \mathrm{d}t = E_\sigma(t, q_0^\nu, q_1^\nu)\,\omega_0^\sigma \wedge \mathrm{d}t,$$

$$E_\sigma = \left(\frac{\partial a}{\partial q_0^\sigma} - \frac{\partial b_\sigma}{\partial t}\right) + \left(\frac{\partial b_\nu}{\partial q_0^\sigma} - \frac{\partial b_\sigma}{\partial q_0^\nu}\right)q_1^\nu = A_\sigma + B_{\sigma\nu}q_1^\nu, \tag{6.24}$$

where

$$B_{\sigma\nu} = \frac{\partial b_\nu}{\partial q_0^\sigma} - \frac{\partial b_\sigma}{\partial q_0^\nu} = \frac{\partial^2 L}{\partial q_1^\nu \partial q_0^\sigma} - \frac{\partial^2 L}{\partial q_0^\nu \partial q_1^\sigma}.$$

The corresponding mechanical system is generated by the representative (see Eq. 3.29)

$$\alpha = A_\sigma \, dq_0^\sigma \wedge dt + \frac{1}{2} B_{\sigma\nu} \, dq_0^\sigma \wedge dq_0^\nu, \tag{6.25}$$

and the dynamical distribution has the annihilator

$$\Delta_\alpha^0 = \text{span}\,\{A_\sigma \, dt + B_{\sigma\nu} \, dq_0^\nu, \}, \quad 1 \le \sigma, \nu \le m. \tag{6.26}$$

Definition 6.7 (*Concept of a regular Lagrangian revisited*) The Euler-Lagrange form of a 1st order Lagrangian is called *regular*, if the matrix $B = (B_{\sigma\nu})$ is regular. A 1st order Lagrangian is called *regular*, if its Euler-Lagrange form is regular.

Recall that the matrix B in Definition 6.7 is

$$B = (B_{\sigma\nu}) = \left(\frac{\partial^2 L}{\partial q_1^\sigma \partial q_1^\nu} \right), \quad 1 \le \sigma, \nu \le m,$$

for the Euler-Lagrange form affine in accelerations (regularity in the sense of condition (a) of Proposition 6.8), and

$$B = (B_{\sigma\nu}) = \left(\frac{\partial^2 L}{\partial q_0^\sigma \partial q_1^\nu} - \frac{\partial^2 L}{\partial q_1^\sigma \partial q_0^\nu} \right), \quad 1 \le \sigma, \nu \le m,$$

for the Lagrangian as well as its Euler-Lagrange form affine in velocities (regularity in the sense of condition (b) of Proposition 6.8). Express the matrix B directly with help of $E_\lambda \in \Lambda_Y^2(TJ^2Y)$, in the case (a) i.e. $E_\sigma(t, q_0^\nu, q_1^\nu, q_2^\nu)$, or $E_\lambda \in \Lambda_Y^2(TJ^1Y)$, in the case (b) i.e. $E_\sigma(t, q_0^\nu, q_1^\nu)$:

$$B_{\sigma\nu} = \frac{\partial E_\sigma}{\partial q_2^\nu}, \quad \text{or} \quad B_{\sigma\nu} = \frac{\partial E_\sigma}{\partial q_1^\nu}.$$

Thus, the concept of the generalized regularity is related to the concept of regular Hamilton extremals.

Theorem 6.1 *Let (Y, π, X) be a fibred manifold, (V, ψ), $\psi = (t, q_0^\nu)$ a fibred chart on Y, $(\pi(V), \varphi)$, $\varphi = (t)$, $(\pi_{1,0}^{-1}(V), \psi_1)$, $\psi_1 = (t, q_0^\nu, q_1^\nu)$ and $(\pi_{2,0}^{-1}(V), \psi_2)$, $\psi_2 = (t, q_0^\nu, q_1^\nu, q_2^\nu)$ the associated chart on X, associated fibred chart on J^1Y and associated fibred chart on J^2Y, respectively. Let $\lambda \in \Lambda_X^1(TJ^1Y)$ be a Lagrangian defined on an open set $W_1 \subset J^1Y$ where $W_1 \subset \pi_{1,0}^{-1}(V)$ or $W_1 \cap \pi_{1,0}^{-1}(V) \neq \emptyset$, and $E_\lambda = E_\sigma \omega_0^\sigma \wedge dt$ its Euler-Lagrange form. Let $[\alpha]$ be the corresponding mechanical system and Δ_α the corresponding dynamical distribution, and let B be the matrix specified in Proposition 6.8 (a) or (b) (according to the situations described there). Then the following conditions are equivalent:*

(a) The Lagrangian λ is regular.
(b) The form E_λ is regular.
(c) The matrix $B = (B_{\sigma\nu})$, $1 \le \sigma, \nu \le m$, in the chart expression of E_λ is regular.
(d) Dynamical and characteristic distributions are identical and their rank is 1.
(e) All Hamilton extremals are regular.

The question arises: what do mean the Legendre transformation and Hamilton equations for the case of the regularity in the sense of condition (b) of Proposition 6.8.

Definition 6.8 (*Generalized momenta and Hamilton function*) Let (Y, π, X) be a fibred manifold, $\dim X = 1$, $\dim Y = m + 1$. Let (π, λ) be a 1st order Lagrange structure with a regular Lagrangian $\lambda \in \Lambda^1_X(TJ^1Y)$ defined on an open set $W_1 \subset J^1Y$. Let (V, ψ), $\psi = (t, q_0^\sigma)$, $(\pi(V), \varphi)$, $\varphi = (t)$, $(\pi_{1,0}^{-1}(V), \psi_1)$, $\psi_1 = (t, q_0^\sigma, q_1^\sigma)$, and $(\pi_{2,0}^{-1}(V), \psi_2)$, $\psi_2 = (t, q_0^\sigma, q_1^\sigma, q_2^\sigma)$, $1 \le \sigma \le m$, a fibred chart on Y, associated chart on X and associated fibred charts on J^1Y and J^2Y, respectively. Suppose that $W_1 \subset \pi_{1,0}^{-1}(V)$, or, alternatively, $W_1 \cap \pi_{1,0}^{-1}(V) \neq \emptyset$ and consider the chart expressions on $W_1 \cap \pi_{1,0}^{-1}(V)$ (and corresponding subsets of X, Y and J^2Y). Define the *generalized momenta* and the *Hamilton function* as

$$p_\sigma^0 = \frac{\partial L}{\partial q_1^\sigma}, \quad H = -L + p_\sigma^0 q_1^\sigma.$$

Definition 6.8 looks formally the same as in relation (6.2) for the generalized momenta and in relation (6.5) for the Hamilton function. However, the important difference lies in the fact that Definition 6.8 is formulated for a Lagrangian regular in the generalized sense contrary to the corresponding concepts introduced at the beginning of the chapter. This will be also evident in various forms of the Legendre transformation and Hamilton equations for 1st and 2nd order Euler-Lagrange forms (This difference is especially important for the higher order, where the definitions od regularity and then the corresponding Hamilton equations are different for even and odd orders—see the brief overview in the following section).

Definition 6.9 (*Generalized Legendre transformation*) Let us start from the same assumptions as in Definition 6.8. Define the mapping called the *generalized Legendre transformation*

(a) for a 1st order regular Euler-Lagrange form

$$\mathcal{L} : (t, q_0^\sigma, q_1^\sigma) \longrightarrow (t, p_\sigma^0), \quad p_\sigma^0 = b_\sigma, \quad H = -a,$$

(b) for a 2nd order regular Euler-Lagrange form

$$\mathcal{L} : (t, q_0^\sigma, q_1^\sigma) \longrightarrow (t, q_0^\sigma, p_\sigma^0).$$

Situation (b) is standard one because in this case the regularity in the generalized sense is fully equivalent to that given by the standard regularity condition (6.1). Let us discuss situation (a) in detail. Note that the relations defining the generalized momenta and the Hamilton function are formally identical for cases (a) and (b), but the consequences will be different.

On the other hand, it is clear that for a 1st order Euler-Lagrange form the mapping $\mathcal{L}$ is not a coordinate transformation on J^1Y. It represents only the coordinate transformation on fibres over points $t \in \pi(V), \pi(V) \subset X$. For this reason, we do not get Hamilton equations in their usual form. Let us consider this situation in more detail. For a regular 1st order Euler-Lagrange form $E_\lambda = E_\sigma(t, q_0^\mu, q_1^\mu)\, dq_0^\sigma \wedge dt$, (equivalently for a 1st order Lagrangian affine in velocities $L = a(t, q_0^\mu) + b_\sigma(t, q_0^\mu) q_1^\sigma$) it holds

$$E_\sigma = \left(\frac{\partial a}{\partial q_0^\sigma} - \frac{\partial b_\sigma}{\partial t}\right) + \left(\frac{\partial b_\nu}{\partial q_0^\sigma} - \frac{\partial b_\sigma}{\partial q_0^\nu}\right) q_1^\nu,$$
$$p_\sigma^0 = \frac{\partial L}{\partial q_1^\sigma} = b_\sigma(t, q_0^\mu), \quad H = -L + p_\sigma^0 q_1^\sigma = -a(t, q_0^\mu). \tag{6.27}$$

For the corresponding representative of the mechanical system we obtain—see (6.24) and (6.25)

$$\alpha_E = \left(\frac{\partial a}{\partial q_0^\sigma} - \frac{\partial b_\sigma}{\partial t}\right) dq_0^\sigma \wedge dt + \frac{1}{2}\left(\frac{\partial b_\nu}{\partial q_0^\sigma} - \frac{\partial b_\sigma}{\partial q_0^\nu}\right) dq_0^\sigma \wedge dq_0^\nu. \tag{6.28}$$

We can also express it with help of the generalized momenta and the Hamilton function. In calculations we use the formulas (knowing that $\mathcal{L}$ is in the discussed case the transformation on fibres)

$$dq_0^\sigma = \frac{\partial q_0^\sigma}{\partial p_\nu^0} dp_\nu^0, \quad \frac{\partial H}{\partial q_0^\sigma} = \frac{\partial H}{\partial p_\nu^0}\frac{\partial p_\nu^0}{\partial q_0^\sigma}, \quad \frac{\partial p_\nu^0}{\partial q_0^\sigma}\frac{\partial q_0^\sigma}{\partial p_\mu^0} = \delta_\nu^\mu,$$

and obtain

$$\alpha_E = \left(-\frac{\partial H}{\partial q_0^\sigma} - \frac{\partial p_\sigma^0}{\partial t}\right)\frac{\partial q_0^\sigma}{\partial p_\nu^0} dp_\nu^0 \wedge dt + \frac{1}{2}\left(\frac{\partial p_\nu^0}{\partial q_0^\sigma} - \frac{\partial p_\sigma^0}{\partial q_0^\nu}\right)\frac{\partial q_0^\sigma}{\partial p_\mu^0}\frac{\partial q_0^\nu}{\partial p_\lambda^0} dp_\mu^0 \wedge dp_\lambda^0,$$
$$\alpha_E = -\left(\frac{\partial H}{\partial p_\sigma^0} + \frac{\partial q_0^\sigma}{\partial t}\right) dp_\sigma^0 \wedge dt + \frac{1}{2}\left(\frac{\partial q_0^\sigma}{\partial p_\nu^0} - \frac{\partial q_0^\nu}{\partial p_\sigma^0}\right) dp_\sigma^0 \wedge dp_\nu^0.$$

Let $\zeta = \zeta_\mu \frac{\partial}{\partial p_\mu^0}$ be a π-vertical vector field. The annihilator of the dynamical distribution is then

$$\Delta_\alpha^0 = \text{span}\left\{ i_\zeta \alpha_E \mid \zeta = \zeta_\mu \frac{\partial}{\partial p_\mu^0} \right\},$$

$$\Delta_\alpha^0 = \text{span}\,\{\eta^\sigma \mid 1 \le \sigma \le m\}, \quad \eta^\sigma = -\left(\frac{\partial H}{\partial p_\sigma^0} + \frac{\partial q_0^\sigma}{\partial t}\right) \mathrm{d}t + \left(\frac{\partial q_0^\sigma}{\partial p_\nu^0} - \frac{\partial q_0^\nu}{\partial p_\sigma^0}\right) \mathrm{d}p_\nu^0.$$

So, we get the single set of "Hamilton equations" (the equations for a Hamilton extremal δ) as $\delta^* \eta^\sigma = 0$, i.e.

$$\left[\left(-\frac{\partial H}{\partial p_\sigma^0} - \frac{\partial q_0^\sigma}{\partial t}\right) + \left(\frac{\partial q_0^\sigma}{\partial p_\nu^0} - \frac{\partial q_0^\nu}{\partial p_\sigma^0}\right) \frac{\mathrm{d}p_\nu^0}{\mathrm{d}t}\right] \circ \delta = 0, \quad 1 \le \sigma \le m. \tag{6.29}$$

We can also express the closed representative α_E with help of the generalized momenta and the Hamilton function (it holds $\alpha_E = \mathrm{d}\theta_\lambda$). This expression is formally the same for situations (a) and (b),

$$\alpha_E = -\mathrm{d}H \wedge \mathrm{d}t + \mathrm{d}p_\sigma^0 \wedge \mathrm{d}q_0^\sigma. \tag{6.30}$$

We derive it for the situation (a), i.e. for a 1st order Euler-Lagrange form. This corresponds to a Lagrangian $\lambda = L\,\mathrm{d}t$, $L = a(t, q_0^\mu) + b_\sigma(t, q^\mu)\, q_1^\sigma$. The generalized momenta and the Hamilton function are expressed by relations (6.27). Then

$$-\mathrm{d}H \wedge \mathrm{d}t + \mathrm{d}p_\sigma^0 \wedge \mathrm{d}q_0^\sigma = \frac{\partial a}{\partial q_0^\sigma} \mathrm{d}q_0^\sigma \wedge \mathrm{d}t + \left(\frac{\partial b_\sigma}{\partial t} \mathrm{d}t + \frac{\partial b_\sigma}{\partial q_0^\nu} \mathrm{d}q_0^\nu\right) \wedge \mathrm{d}q_0^\sigma$$
$$= \left(\frac{\partial a}{\partial q_0^\sigma} - \frac{\partial b_\sigma}{\partial t}\right) \mathrm{d}q_0^\sigma \wedge \mathrm{d}t - \frac{1}{2}\left(\frac{\partial b_\sigma}{\partial q_0^\nu} - \frac{\partial b_\nu}{\partial q_0^\sigma}\right) \mathrm{d}q_0^\sigma \wedge \mathrm{d}q_0^\nu,$$

where we take into account that

$$\frac{\partial b_\sigma}{\partial q_0^\nu} \mathrm{d}q_0^\nu \wedge \mathrm{d}q_0^\sigma = -\left(\frac{\partial b_\sigma}{\partial q_0^\nu}\right)_{\text{alt}\,(\sigma,\nu)} \mathrm{d}q_0^\sigma \wedge \mathrm{d}q_0^\nu.$$

This is just the expression (6.28) for α_E.

Example 6.6 Let $(Y, \pi, X) = (\mathbf{R} \times \mathbf{R}^2, p, \mathbf{R}$ be the fibred manifold with global charts $(\mathbf{R} \times \mathbf{R}^2, \psi)$, $\psi = (t, x, y)$, on Y, $(\mathbf{R}, \varphi)$, $\varphi = (t)$, on and $(\mathbf{R} \times \mathbf{R}^2 \times \mathbf{R}^2, \psi_1)$, $\psi_1 = (t, x, y, \dot{x}, \dot{y})$ on J^1Y. (Here x, y, $\dot{x}$ and $\dot{y}$ mean $q_0^1 = q^1$, $q_0^2 = q^2$, $q_1^1 = \dot{q}^1$, $q_1^2 = \dot{q}^2$, respectively.) Consider a "model" 1st order Lagrangian

$$\lambda = L(t, x, y, \dot{x}, \dot{y})\,\mathrm{d}t, \quad L = \frac{m}{2}(\dot{x}^2 + \dot{y}^2) + \frac{b}{2}(\dot{x}y - x\dot{y}) - \frac{k}{2}(x^2 + y^2),$$

where m, b and k are positive constants. (Without the middle term, i.e. for $b = 0$, it would be the Lagrangian of a two-dimensional harmonic oscillator.) This Lagrangian is regular in the sense of the standard regularity (6.1). The corresponding matrix $\mathcal{M}$ has nonzero determinant $\det \mathcal{M} = m^2$. We will formulate and solve both the Euler-Lagrange and Hamilton equations of this standard example in Problem 6.14. Here we will focus on the case $m = 0$ for which we obtain again the 1st order Lagrangian, but not regular in the standard sense,

$$\lambda = L(t, x, y, \dot{x}, \dot{y})\, dt, \quad L = \frac{b}{2}(\dot{x}y - x\dot{y}) - \frac{k}{2}(x^2 + y^2),$$

$$L = a(t, x, y) + b_x(t, x, y)\dot{x} + b_y(t, x, y)\dot{y},$$
$$a = -\frac{k}{2}(x^2 + y^2), \quad b_x = \frac{by}{2}, \quad b_y = -\frac{bx}{2}.$$

However, this Lagrangian is regular in the sense of the "new" regularity, because the determinant of the matrix

$$B = (B_{\sigma\nu}) = \left(\frac{\partial^2 L}{\partial \dot{q}^\nu \partial q^\sigma} - \frac{\partial^2 L}{\partial \dot{q}^\sigma \partial q^\nu} \right) = \begin{pmatrix} 0 & \frac{\partial^2 L}{\partial \dot{y} \partial x} - \frac{\partial^2 L}{\partial \dot{x} \partial y} \\ \frac{\partial^2 L}{\partial \dot{x} \partial y} - \frac{\partial^2 L}{\partial \dot{y} \partial x} & 0 \end{pmatrix} = \begin{pmatrix} 0 & -b \\ b & 0 \end{pmatrix}$$

is nonzero, concretely $\det B = b^2 \neq 0$. The (regular) Euler-Lagrange form and the corresponding representative of the mechanical system $[\alpha]$ (of course, regular as well) are

$$E_\lambda = E_x\, dx \wedge dt + E_y\, dy \wedge dt = (-b\dot{y} - kx)\, dx \wedge dt + (b\dot{x} - ky)\, dy \wedge dt,$$

$$\begin{aligned} \alpha_E &= \left(\frac{\partial a}{\partial x} - \frac{\partial b_x}{\partial t} \right) dx \wedge dt + \left(\frac{\partial a}{\partial y} - \frac{\partial b_y}{\partial t} \right) dy \wedge dt - \left(\frac{\partial b_x}{\partial y} - \frac{\partial b_y}{\partial x} \right) dx \wedge dy \\ &= -kx\, dx \wedge dt - ky\, dy \wedge dt - b\, dx \wedge dy. \end{aligned}$$

(for obtaining α_E see relation (6.28)). The solution of the corresponding 1st order Euler-Lagrange equations $E_x \circ J^1\gamma = 0$, $E_y \circ J^1\gamma = 0$ is

$$x\gamma(t) = C_1 \sin\frac{kt}{b} - C_2 \cos\frac{kt}{b}, \quad y\gamma(t) = C_1 \cos\frac{kt}{b} + C_2 \sin\frac{kt}{b},$$

where C_1 and C_2 are arbitrary real constants.

Let us now solve this problem within the framework of the generalized Hamilton theory for 1st order regular Euler-Lagrange form. The generalized momenta and the Hamilton function are

$$p_x = \frac{by}{2}, \quad p_y = -\frac{bx}{2}, \quad H = \frac{2k}{b^2}(p_x^2 + p_y^2).$$

The generalized Legendre transformation (transformation of coordinates on fibres $\mathbf{R}^2$ of $Y = \mathbf{R} \times \mathbf{R}^2$) is

$$\mathcal{L} : (t, x, y) \longrightarrow (t, p_x, p_y),$$

i.e. the Jacobi matrix of this mapping is regular, $\det D\mathcal{L} = b^2/4$:

$$D\mathcal{L} = \begin{pmatrix} 1 & 0 & 0 \\ 0 & 0 & \frac{b}{2} \\ 0 & -\frac{b}{2} & 0 \end{pmatrix}, \quad \det D\mathcal{L} = \frac{b^2}{4}.$$

Expressing α_E (see above) with help of momenta, we obtain

$$\alpha_E = -\frac{4}{b}\left(\frac{k}{b}p_x\,\mathrm{d}p_x \wedge \mathrm{d}t + \frac{k}{b}p_y\,\mathrm{d}p_y \wedge \mathrm{d}t + \mathrm{d}p_x \wedge \mathrm{d}p_y\right).$$

Contracting α_E by π-vertical vector fields $\frac{\partial}{\partial p_x}$ and $\frac{\partial}{\partial p_y}$ we obtain the annihilator of the dynamical distribution

$$\Delta_\alpha^0 = \mathrm{span}\left\{-\frac{4k}{b^2}p_x\,\mathrm{d}t - \frac{4}{b}\,\mathrm{d}p_y,\ -\frac{4k}{b^2}p_y\,\mathrm{d}t + \frac{4}{b}\,\mathrm{d}p_x\right\}.$$

Let $\zeta = \zeta^0\frac{\partial}{\partial t} + \zeta_x\frac{\partial}{\partial p_x} + \zeta_y\frac{\partial}{\partial p_y}$ be a vector field on Y. The condition $i_\zeta\eta = 0$ for every $\eta \in \Delta_\alpha^0$ specifies the generator of the dynamical distribution

$$\Delta_\alpha = \mathrm{span}\left\{\frac{\partial}{\partial t} + \frac{kp_y}{b}\frac{\partial}{\partial p_x} - \frac{kp_x}{b}\frac{\partial}{\partial p_y}\right\}.$$

So, we can see that rank $\Delta_\alpha = 1$. Using the set of (two) Eq. (6.29) we finally obtain the Hamilton equations (equations for Hamilton extremals δ) corresponding to the generalized regularity:

$$\left[\left(-\frac{\partial H}{\partial p_x} - \frac{\partial x}{\partial t}\right) + \left(\frac{\partial x}{\partial p_y} - \frac{\partial y}{\partial p_x}\right)\frac{\mathrm{d}p_y}{\mathrm{d}t}\right] \circ \delta = 0,$$
$$\left[\left(-\frac{\partial H}{\partial p_y} - \frac{\partial y}{\partial t}\right) + \left(\frac{\partial y}{\partial p_x} - \frac{\partial x}{\partial p_y}\right)\frac{\mathrm{d}p_x}{\mathrm{d}t}\right] \circ \delta = 0.$$

Finally we obtain the set of two 1st order equations for momenta p_x and p_y as follows:

$$\frac{k}{b}p_x + \frac{\mathrm{d}p_y}{\mathrm{d}t} = 0, \quad \frac{k}{b}p_y - \frac{\mathrm{d}p_x}{\mathrm{d}t} = 0.$$

Its solution is

$$p_x\gamma(t) = P_1\cos\frac{kt}{b} + P_2\sin\frac{kt}{b}, \quad p_y\gamma(t) = -P_1\sin\frac{kt}{b} + P_2\cos\frac{kt}{b},$$

$$x_\gamma(t) = -\frac{2}{b}\left(-P_1 \sin\frac{kt}{b} + P_2 \cos\frac{kt}{b}\right) = C_1 \sin\frac{kt}{b} - C_2 \cos\frac{kt}{b},$$
$$y_\gamma(t) = \frac{2}{b}\left(P_1 \cos\frac{kt}{b} + P_2 \sin\frac{kt}{b}\right) = C_1 \cos\frac{kt}{b} + C_2 \sin\frac{kt}{b},$$

where we have denoted $C_1 = \frac{2}{b}P_1$ and $C_2 = \frac{2}{b}P_2$. We have confirmed the results obtained by solving the Euler-Lagrange equations.

6.4 A Note to Higher-Order Hamilton Theory

The different properties of generalized regularity and the corresponding generalized Hamiltonian theory for the case of even and odd order variational dynamical forms (Euler-Lagrange forms of a certain Lagrangians) stand out better in the higher-order theory. Although we have not dealt in detail with the basic geometric structures for higher order problems (fibred manifolds and their prolongations, sections and their prolongations, local diffeomorphisms and vector fields, differential forms, etc.), the basic ideas of formulating these concepts are quite analogous for the higher-order cases as for those of first or second order. For first-order and second-order situations, the concepts are thoroughly explained in Chaps. 2–5. In Chap. 8, which represents the extending of geometrical concepts to the field theory with regard to their applications in physics, higher-order structures are explained and prepared to the practical use in variational physics theories. Thus, the reader can gain a deeper understanding there as well. Here we will only summarize the basic results of higher-order Hamilton mechanics.

Definition 6.10 (*Standard regularity*) Let (Y, π, X) be a fibred manifold with one-dimensional base X, $\dim Y = m + 1$, and $(J^r Y, \pi_r, X)$, $r \geq 1$ its rth prolongation. Let (π, λ) be a rth order Lagrange structure with the Lagrangian $\lambda \in \Lambda^1_X(TJ^rY)$, defined on an open set $W_r \subset J^rY$. Let (V, ψ), $\psi = (t, q^\sigma)$, be a fibred chart on Y such that $W_r \subset V_r$, $V_r = \pi^{-1}_{r,0}(V)$ (or at least $W_r \cap \pi^{-1}_{r,0}(V) \neq \emptyset$), $(\pi(V), \varphi)$, $\varphi = (t)$ the associated chart on X and (V_s, ψ_s), $V_s = \pi^{-1}_{s,0}(V)$, $\psi_s = (t, q^\sigma_0, \dots, q^\sigma_s)$, $1 \leq \sigma \leq m$, the associated fibred charts on J^sY, $1 \leq s \leq r$. The Lagrangian $\lambda = L(t, q^\sigma_0, \dots, q^\sigma_r)$ is called *regular in the standard sense* if the matrix

$$\mathcal{M} = \left(\frac{\partial^2 L}{\partial q^\sigma_r \partial q^\nu_r}\right) \tag{6.31}$$

is regular.

(In the following, we will work with the structures specified in Definition 6.10, unless otherwise stated). It is known (see Chap. 8) that the Lepage equivalent θ_λ of an rth order Lagrangian is in general defined on $J^{2r-1}Y$ (for mechanics it is unique) and the corresponding Euler-Lagrange form E_λ on $J^{2r}Y$. It holds

$$\theta_\lambda = L\,\mathrm{d}t + \sum_{j=0}^{r-1}\left[\sum_{\ell=0}^{r-\ell-1}(-1)^\ell \frac{\mathrm{d}^\ell}{\mathrm{d}t^\ell}\left(\frac{\partial L}{\partial q^\sigma_{\ell+j+1}}\right)\right]\omega^\sigma_j, \quad \omega^\sigma_j = \mathrm{d}q^\sigma_j - q^\sigma_{j+1}\,\mathrm{d}t,$$

$$E_\lambda = p_1\,\mathrm{d}\theta_\lambda = \sum_{j=0}^{r}(-1)^j \frac{\mathrm{d}^j}{\mathrm{d}t^j}\left(\frac{\partial L}{\partial q^\sigma_j}\right)\mathrm{d}q^\sigma_0 \wedge \mathrm{d}t. \tag{6.32}$$

Definition 6.11 (*Hamilton extremals, regular Hamilton extremals*) Let (π, λ) be a rth order Lagrange structure specified in Definition 6.10 and let $\theta_\lambda \in \Lambda^1(TJ^{2r-1}Y)$ be the Lepage equivalent of the Lagrangian λ. A section $\delta \in \Gamma_{\pi_r(W_r)}(\pi_{2r-1})$ is called a *Hamilton extremal* of the Lagrangian λ (or alternatively of the given Lagrange structure), if for every π_{2r-1}-vertical vector field ζ on $J^{2r-1}Y$ (defined on $\pi^{-1}_{2r-1,r}(W_r)$) it holds

$$\delta^* i_\zeta\,\mathrm{d}\theta_\lambda = 0. \tag{6.33}$$

A Hamilton extremal δ is called *regular* if there exists an extremal $\gamma \in \Gamma_{\pi_r(W_r)}(\pi)$ such that $\delta = J^{2r-1}\gamma$.

Proposition 6.9 *Let (π, λ) be a rth order Lagrange structure on (Y, π, X) and $\delta \in \Gamma_{\pi_r(W_r)}(\pi_{2r-1})$ its Hamilton extremal. If the Lagrangian λ is regular in the sense of relation (6.31), then the Hamilton extremal δ is regular.*

Similarly as for 1st order Lagrange structures (Lepage equivalent of the Lagrangian is defined also on J^1Y, because for $r = 1$ we have $2r - 1 = 1$) we can formulate the equivalent definition of Hamilton extremals. Consider the fibred manifold $(\tilde{Y}, \tilde{\pi}, X) = (J^{2r-1}Y, \pi_{2r-1}, X)$ as the basic underlying space and denote

$$(\tilde{J}^1\tilde{Y}, \tilde{\pi}_1, X) = (\tilde{J}^1(J^{2r-1}Y), \tilde{\pi}_1, X), \quad \tilde{\pi}_1 = (\pi_{2r-1})_1.$$

Let δ be a section of π_{2r-1} and ζ a π_{2r-1}-vertical vector field on $J^{2r-1}Y$. Then the condition

$$\tilde{J}^1\delta^*\tilde{h} i_\zeta\,\mathrm{d}\theta_\lambda = 0, \tag{6.34}$$

where $\tilde{h}$ is the horizontalization mapping

$$\tilde{h} : \Lambda^k(TJ^{2r-1}Y) \ni \eta \longrightarrow \tilde{h}(\eta) \in \Lambda^k_X\big(T\tilde{J}^1(J^{2r-1}Y)\big).$$

The equivalence of definition relations (6.33) and (6.34) is the consequence of the generally valid relation $\delta^* = \tilde{J}^1\delta^* \circ \tilde{h}$.

Definition 6.12 (*Generalized momenta, Hamilton function, Hamilton vector field*) Let (π, λ) be a rth order Lagrange structure regular in the sense of condition (6.31), $\lambda = L\,dt$. Define the *generalized momenta, Hamilton function* and *Hamilton vector field* as follows:

$$p_\sigma^j = \sum_{\ell=0}^{r-j-1} (-1)^\ell \frac{d^\ell}{dt^\ell} \left(\frac{\partial L}{\partial q_{\ell+j+1}^\sigma} \right), \quad 0 \leq j \leq r-1, \tag{6.35}$$

(note that due to the repetition of the total derivative operator, these are functions on $J^{2r-1}Y$),

$$H = -L + \sum_{j=0}^{r-1} p_\sigma^j q_{j+1}^\sigma, \tag{6.36}$$

$$\zeta = \frac{\partial}{\partial t} + \sum_{j=1}^{r-1} \left(\frac{\partial H}{\partial p_\sigma^j} \frac{\partial}{\partial q_j^\sigma} - \frac{\partial H}{\partial q_j^\sigma} \frac{\partial}{\partial p_\sigma^j} \right).$$

Suppose that the rth order Lagrange structure is regular in the sense of (6.31). Since the Lepage equivalent of this Lagrangian is in general defined on $J^{2r-1}Y$, we introduce the *Legendre transformation* just on $J^{2r-1}Y$:

$$\mathcal{L} : (t, q_0^\sigma, \ldots, q_{2r-1}^\sigma) \longrightarrow (t, q_0^\sigma, \ldots, q_{r-1}^\sigma, p_0^\sigma, \ldots, p_\sigma^{r-1}), \quad 1 \leq \sigma \leq m.$$

The Jacobi matrix of the mapping $\mathcal{L}$ is

$$D\mathcal{L} = \left(\begin{array}{c|c|c} 1 & 0 & \cdots \\ \hline 0 & E_{rm} & \cdots \\ \hline 0 & 0 & \frac{\partial^2 L_\sigma}{\partial q_r^\sigma \partial q_r^\nu} \end{array} \right), \quad 1 \leq \sigma, \nu \leq m.$$

The Lepage equivalent (6.32) of the Lagrangian obtains in Legendre coordinates the following form:

$$\theta_\lambda = L\,dt + \sum_{j=0}^{r-1} p_\sigma^j \omega_j^\sigma = -H\,dt + \sum_{j=0}^{r-1} p_\sigma^j\, dq_j^\sigma.$$

Let ζ be a π_{2r-1}-vertical vector field on $J^{2r-1}Y$,

$$\zeta = \sum_{j=0}^{r-1} \zeta_j^\sigma \frac{\partial}{\partial q_j^\sigma} + \tilde{\zeta}_\sigma^j \frac{\partial}{\partial p_\sigma^j}.$$

Taking into account that $\tilde{h}\,\mathrm{d}q_j^\sigma = q_{j,1}^\sigma\,\mathrm{d}t$, $\tilde{h}\,\mathrm{d}p_\sigma^j = p_\sigma^{j,1}\,\mathrm{d}t$ we get by step-by-step calculations

$$\begin{aligned} \mathrm{d}\theta_\lambda &= \sum_{j=0}^{r-1}\left(-\frac{\partial H}{\partial q_j^\sigma}\,\mathrm{d}q_j^\sigma - \frac{\partial H}{\partial p_\sigma^j}\,\mathrm{d}p_\sigma^j\right)\wedge \mathrm{d}t + \sum_{j=0}^{r-1}\mathrm{d}p_\sigma^j\wedge \mathrm{d}q_j^\sigma,\\ i_\zeta\,\mathrm{d}\theta_\lambda &= \sum_{j=0}^{r-1}\left(-\frac{\partial H}{\partial q_j^\sigma}\,\zeta_j^\sigma - \frac{\partial H}{\partial p_\sigma^j}\,\tilde{\zeta}_\sigma^j\right)\mathrm{d}t + \sum_{j=0}^{r-1}(\tilde{\zeta}_\sigma^j\,\mathrm{d}q_j^\sigma - \zeta_j^\sigma\,\mathrm{d}p_\sigma^j),\\ \tilde{h}(i_\zeta\,\mathrm{d}\theta_\lambda) &= \sum_{j=0}^{r-1}\left[\left(-\frac{\partial H}{\partial q_j^\sigma} - p_\sigma^{j,1}\right)\zeta_j^\sigma + \left(-\frac{\partial H}{\partial p_\sigma^j} + q_{j,1}^\sigma\right)\tilde{\zeta}_\sigma^j\right]\mathrm{d}t. \end{aligned} \tag{6.37}$$

The equation for Hamilton extremals δ read $\tilde{J}^1\delta^*\tilde{h}i_\zeta\,\mathrm{d}\theta_\lambda = 0$ for arbitrary components ζ_j^σ and $\tilde{\zeta}_\sigma^j$, $1 \le \sigma \le m$, $0 \le j \le r-1$. We have obtained the *Hamilton equations*.

Theorem 6.2 *Let (π, λ) be a regular rth order Lagrange structure specified in Definition 6.10. A section δ of the projection π_{2r-1} is a Hamilton extremal of the given Lagrange structure if and only if it obeys the* Hamilton equations

$$-\left(\frac{\partial H}{\partial q_j^\sigma} + p_\sigma^{j,1}\right)\circ \tilde{J}^1\delta = 0, \quad \left(-\frac{\partial H}{\partial p_\sigma^j} + q_{j,1}^\sigma\right)\circ \tilde{J}^1\delta = 0,$$

for all $1 \le \sigma \le m$, $0 \le j \le r-1$.

Definition 6.13 (*Euler-Lagrange distribution*) Let (π, λ) be a regular rth order Lagrange structure specified in Definition 6.10. The one-dimensional distribution Δ_λ generated by the Hamilton vector field is called *Euler-Lagrange distribution*.

Proposition 6.10 *A section δ of the projection π_{2r-1} is a Hamilton extremal if and only if it is the integral section of the Euler-Lagrange distribution.*

Proof Let $\eta \in \Delta_\lambda^0$, in coordinates $\eta = \eta_0\,\mathrm{d}t + \sum_{j=0}^{r-1}\left(\eta_\nu^j\,\mathrm{d}q_j^\nu + \tilde{\eta}_j^\nu\,\mathrm{d}p_\nu^j\right)$. Then the requirement $i_\zeta\eta = 0$ for a form $\eta \in \Delta_\alpha^0$ and the Hamilton vector field ζ gives

$$\eta_0 + \sum_{j=0}^{r-1}\left(\frac{\partial H}{\partial p_\sigma^j}\eta_\sigma^j - \frac{\partial H}{\partial q_j^\sigma}\tilde{\eta}_j^\sigma\right) = 0 \implies \eta_0 = -\sum_{j=0}^{r-1}\left(\frac{\partial H}{\partial p_\sigma^j}\eta_\sigma^j - \frac{\partial H}{\partial q_j^\sigma}\tilde{\eta}_j^\sigma\right)$$

$$\implies \eta = \sum_{j=0}^{r-1}\left[\eta_\sigma^j\left(\mathrm{d}q_j^\sigma - \frac{\partial H}{\partial p_\sigma^j}\,\mathrm{d}t\right) + \tilde{\eta}_j^\sigma\left(\mathrm{d}p_\sigma^j + \frac{\partial H}{\partial q_j^\sigma}\,\mathrm{d}t\right)\right].$$

A section δ is a Hamilton extremal if and only if it obeys the Hamilton equations. This is equivalent to the condition $\delta^*\eta - 0$, i.e. δ is the integral section of Δ_α. □

Now we will present concepts leading to the generalized regularity of a Lagrangian, in analogy with the corresponding 1st and 2nd order. The following definition includes, of course, the "standard" regularity as well.

Definition 6.14 (*Dynamical forms, mechanical system. Regularity*) Let (Y, π, X) be a fibred manifold in mechanics ($\dim X = 1$) and $(J^s Y, \pi_s, X)$ its sth prolongation, (V, ψ), $\psi = (t, q^\nu)$, $(\pi(V), \varphi)$, $\varphi = (t)$, $(\pi_{\ell,0}^{-1}(V), \psi_\ell)$, $\psi_\ell = (t, q_0^\nu, \dots, q_\ell^\nu)$ a fibred chart on Y, the associated chart on X and the associated fibred charts on $J^\ell Y$, $1 \le \ell \le s$.

(a) Every 1-contact $\pi_{s,0}$-projectable form $E \in \Lambda_Y^2(TJ^sY)$, expressed in charts as $E = E_\sigma \, dq_0^\sigma \wedge dt$, $1 \le \sigma \le m$, where functions E_σ are affine in variables q_s^σ i.e.

$$E_\sigma = A_\sigma(t, q_0^\mu, \dots, q_{s-1}^\mu) + B_{\sigma\nu}(t, q_0^\mu, \dots, q_{s-1}^\mu) q_s^\nu$$

is called the *dynamical form*.

(b) A dynamical form E is called *regular*, if the matrix $B = (B_{\sigma\nu})$, $1 \le \sigma, \nu \le m$, is regular. This mechanical system is called *regular* if the corresponding dynamical form is regular.

(c) The class $[\alpha]$ of all forms α such that $\alpha = p_1 E$ is called the *mechanical system* corresponding to the dynamical form E.

(d) The mechanical system $[\alpha]$ corresponding to a dynamical form E is called *regular*, if E is regular.

(e) A Lagrangian λ is called *regular* if its Euler-Lagrange form is regular.

Proposition 6.11 *Let $\alpha \in [\alpha]$ be a representative of the mechanical system corresponding to the dynamical form E. Then α is $\pi_{s,s-1}$-projectable and $\alpha = E + F$, where F is a 2-contact form on $J^{s-1}Y$.*

There are two distributions connected with a representative $\alpha \in [\alpha]$ corresponding to the dynamical form E, called the *dynamical distribution* and *characteristic distribution*. Their annihilators are

$$\Delta_\alpha^0 = \operatorname{span}\{i_\zeta \alpha\},$$

where ζ is an arbitrary π_{s-1}-vertical vector field on $J^{s-1}Y$.

$$D_\alpha^0 = \operatorname{span}\{i_\zeta \alpha\},$$

where ζ is an arbitrary vector field on $J^{s-1}Y$. It is evident that the characteristic distribution D_α is a subdistribution of the dynamical distribution Δ_α.

For our considerations concerning variational physics variational dynamical forms are relevant above all.

Theorem 6.3 *Let (Y, π, X) be a fibred manifold and $(J^s Y, \pi_s, X)$ its sth prolongation. Let E be a variational dynamical form defined on an open set $W_s \subset J^s Y$ and (V, ψ), $(\pi(V), \varphi)$, $(\pi_{\ell,0}^{-1}(V), \psi_\ell)$ a fibred chart on Y, the associated chart on X and the associated fibred charts on $J^\ell Y$, $1 \le \ell \le s$, (as described above),*

$$E = \left(A_\sigma + B_{\sigma\nu} q_s^\nu\right) \mathrm{d}q_0^\sigma \wedge \mathrm{d}t, \quad 1 \le \sigma, \nu \le m,$$

where A_σ and $B_{\sigma\nu}$ are functions on $J^{s-1}Y$. The following conditions are equivalent:

(a) The form E is regular.
(b) The matrix $B = (B_{\sigma\nu})$ is regular.
(c) $\Delta_\alpha = D_\alpha$.
(d) The dynamical and the characteristic distribution are of the constant rank, $\operatorname{rank} \Delta_\alpha = \operatorname{rank} D_\alpha = 1$.
(e) All Hamilton extremals are regular.

The generalization of the concept of regularity leads to a difference between the Legendre transformation and the set of Hamilton equations for Euler-Lagrange forms of the even ($s = 2r$) and odd ($s = 2r - 1$) order.

Definition 6.15 (*Generalized Legendre transformation*) Let the underlying structures are specified as above, including the generalized momenta and the Hamilton function. Let E be a sth order regular variational dynamical form. Then we define the *generalized Legendre transformation* as follows:

(a) For $s = 2r$

$$\mathcal{L} : (t, q_0^\nu, \ldots, q_{2r-1}^\nu) \longrightarrow (t, q_0^\nu, \ldots, q_{r-1}^\nu, p_\nu^0, \ldots, p_\nu^{r-1}),$$

(b) and for $s = 2r - 1$

$$\mathcal{L} : (t, q_0^\nu, \ldots, q_{2r-1}^\nu) \longrightarrow (t, q_0^\nu, \ldots, q_{r-2}^\nu, p_\nu^0, \ldots, p_\nu^{r-1}).$$

The generalized momenta and the Hamilton function are defined by formally the same relations (6.35) and (6.36) for both, the even and odd orders. For both these situations it is possible to express the (uniquely defined closed) representative α_E of the mechanical system $[\alpha]$ of the form E,

$$\alpha_E = -\mathrm{d}H \wedge \mathrm{d}t + \sum_{j=0}^{r-1} \mathrm{d}p_\sigma^j \wedge \mathrm{d}q_j^\sigma .$$

Theorem 6.4 *Let the underlying structures be specified as above and let E be a sth order regular variational dynamical form defined on an open set $W_s \subset J^s Y$. Then the so called Hamilton canonical equations have the form*

(a) *For even order of the form E, $s = 2r$*

$$-\frac{dq_j^\sigma}{dt} + \frac{\partial H}{\partial p_\sigma^j} = 0, \quad -\frac{dp_\sigma^j}{dt} - \frac{\partial H}{\partial q_j^\sigma} = 0, \quad 1 \le \sigma \le m, \quad 0 \le j \le r-1. \tag{6.38}$$

This means that a section $\delta\colon \pi_{2r}(W_{2r}) \to J^{2r-1}Y$ obeying these equations is a regular Hamilton extremal, i.e. $\delta = J^{2r-1}\gamma$, where γ is an extremal of the Lagrangian corresponding to the form E.

(b) *For odd order of the form E, $s = 2r - 1$*

$$-\frac{dq_j^\sigma}{dt} + \frac{\partial H}{\partial p_\sigma^j} = 0,$$

$$-\frac{dp_\sigma^j}{dt} - \frac{\partial H}{\partial q_j^\sigma} - \frac{\partial q_{r-1}^\nu}{\partial q_j^\sigma}\frac{dp_\nu^{r-1}}{dt} = 0, \quad 1 \le \sigma \le m, \quad 0 \le j \le r-1.$$

$$\frac{\partial q_{r-1}^\sigma}{\partial t} + \sum_{j=0}^{r-2} \frac{\partial q_{r-1}^\sigma}{\partial q_j^\nu}\frac{dq_j^\nu}{dt} + \left(\frac{\partial q_{r-1}^\sigma}{\partial p_\nu^{r-1}} - \frac{\partial q_{r-1}^\nu}{\partial p_\sigma^{r-1}}\right)\frac{dp_\nu^{r-1}}{dt} - \frac{\partial H}{\partial p_\sigma^{r-1}} = 0, \tag{6.39}$$

for $1 \le \sigma, \nu \le m$, $0 \le j \le r-2$. This means that a section δ : $\pi_{2r-1}(W_{2r-1}) \to J^{2r-2}Y$ is a regular Hamilton extremal, i.e. $\delta = J^{2r-2}\gamma$, where γ is an extremal of the Lagrangian corresponding to the form E.

Verify that the case (b) for $r = 1$ is in agreement with results obtained by us in Sect. 6.4. For better understanding we give one more example for the odd-order of E higher than 1. This example will clarify the difference between even-order and odd-order situations, even if it has no particular relevance to physics.

Example 6.7 Consider the fibred manifold $(Y, \pi, X) = (\mathbf{R} \times \mathbf{R}^2, p, \mathbf{R})$ and the 3-rd order dynamical form E defined (globally) on J^3Y

$$E = \left(q_0^2 - q_3^2\right) dq_0^1 \wedge dt + \left(q_0^1 + q_3^1\right) dq_0^2 \wedge dt.$$

This dynamical form is variational and regular, a minimal Lagrangian is $\lambda = L\,dt$, where

$$L = \frac{1}{2}\left(q_1^1 q_2^2 - q_1^2 q_2^1\right) + q_0^1 q_0^2, \quad (A_\sigma) = \left(q_0^2 \;\; q_0^1\right),$$

$$B = (B_{\sigma\nu}) = \begin{pmatrix} 0 & -1 \\ 1 & 0 \end{pmatrix}, \quad 1 \le \sigma, \nu \le 2.$$

$$\mathcal{L} : \left(t\, q_0^1, q_0^2, q_1^1, q_1^2, q_2^1, q_2^2\right) \longrightarrow \left(t\, q_0^1, q_0^2, p_1^0, p_2^0, p_1^1, p_2^1\right),$$

$$p_1^0 = q_2^2,\ p_2^0 = -q_2^1,\ p_1^1 = -\frac{1}{2}q_1^2,\ p_2^1 = \frac{1}{2}q_1^1,\ H = -2p_2^0 p_1^1 + 2p_1^0 p_2^1 - q_0^1 q_0^2.$$

Using (6.39) for $1 \leq \sigma, \nu \leq 2$ we obtain the Hamilton equations:

$$\frac{\mathrm{d}p_1^0}{\mathrm{d}t} = q_0^2,\ \frac{\mathrm{d}p_2^0}{\mathrm{d}t} = q_0^1,\ \frac{\mathrm{d}q_0^1}{\mathrm{d}t} = 2p_2^1,\ \frac{\mathrm{d}q_0^2}{\mathrm{d}t} = -2p_1^1,\ \frac{\mathrm{d}p_1^1}{\mathrm{d}t} = -\frac{1}{2}p_1^0,$$
$$\frac{\mathrm{d}p_2^1}{\mathrm{d}t} = -\frac{1}{2}p_2^0.$$

Continue according to Problem 6.15.

6.5 A Note to Hamilton-Jacobi Theory

The Hamilton-Jacobi theory is very extensive and its presentation in this text has rather a supplementary informative role. Therefore, in this section we will again limit ourselves to mechanics. We will very briefly note the basic idea of Hamilton-Jacobi theory in 1st order mechanics, based on the use of fibred structures introduced in previous chapters. The task is, shortly speaking, to find differential equations for stationary values of functionals and to find these stationary values out without knowing the corresponding extremals in advance. Extremals can be then found with help of the solution of equations for stationary values. The results for the 1st order mechanics will be then generalized to higher odder mechanics.

Example 6.8 This example serves as a certain motivation problem for introducing key concepts of the theory. Consider a fibred manifold $(Y, \pi, X) = (\mathbf{R} \times \mathbf{R}^m, p, \mathbf{R})$, a 1st order Lagrange structure (π, λ), with the Lagrangian $\lambda = L\,\mathrm{d}t$. Suppose that the Lagrangian is regular in the sense of relation (6.1). Consider the action function as it was introduced in Definition 4.1 for a generally chosen section of (Y, π, X) defined on a compact set $\Omega \subset X$

$$\Gamma_\Omega(\pi) \ni \gamma \longrightarrow \int_a^b J^1\gamma^*\lambda \in \mathbf{R}, \quad \Omega = [a, b].$$

Consider two sections $\gamma, \bar{\gamma} \in \Gamma_\Omega(\pi)$ and action functions (variational integrals) with so called *free ends*, see Fig. 6.2,

$$\gamma\colon [\tau, t + \delta t] \longrightarrow \gamma(t) = \big(t, q^\sigma\gamma(t)\big),\quad \bar{\gamma}\colon [\tau + \delta\tau,\ t] \to \bar{\gamma}(t) = (t,\ q^\sigma\bar{\gamma}(t)),$$

$$S[\gamma] = \int_\tau^{t+\delta t} J^1\gamma^*\lambda,\quad S[\bar{\gamma}] = \int_{\tau+\delta\tau}^{t} J^1\bar{\gamma}^*\lambda,\quad [\tau, t + \delta t], [\tau + \delta\tau,\ t] \subset \Omega.$$

Denote

$$u^\sigma(t) = q^\sigma J^1\bar{\gamma}(t) - q^\sigma J^1\gamma(t) \ \text{ for } \ t \in [a, b].$$

Calculating the difference of values $S[\bar{\gamma}]$ and $S[\gamma]$ in the linear approximation we obtain

$$\delta S = S[\bar{\gamma}] - S[\gamma] = \int_{\tau}^{t+\delta t} L \circ J^1\bar{\gamma}(t)\,\mathrm{d}t - \int_{\tau+\delta\tau}^{t} L \circ J^1\gamma(t)\,\mathrm{d}t$$

$$= \int_{\tau}^{t} \Big[L\big(t, q^\sigma J^1\gamma(t) + u^\sigma(t), \dot{q}^\sigma J^1\gamma(t) + \dot{u}^\sigma(t)\big)$$

$$- L\big(t, q^\sigma J^1\gamma(t), \dot{q}^\sigma J^1\gamma(t)\big) \Big] \mathrm{d}t$$

$$+ \int_{t}^{t+\delta t} L\big(t, q^\sigma J^1\bar{\gamma}(t), \dot{q}^\sigma J^1\bar{\gamma}(t)\big)\,\mathrm{d}t - \int_{\tau+\delta\tau}^{\tau} L\big(t, q^\sigma J^1\gamma(t), \dot{q}^\sigma J^1\gamma(t)\big)\,\mathrm{d}t.$$

Applying the standardly used per partes calculation on the first integral (within limits $[\tau, t]$) we obtain in the above mentioned linear approximation

$$\delta S \doteq \int_{\tau}^{t} \left[\frac{\partial L}{\partial q^\sigma} - \frac{\mathrm{d}}{\mathrm{d}t}\frac{\partial L}{\partial \dot{q}^\sigma} \right] \circ J^2\gamma(t) u^\sigma(t)\,\mathrm{d}t$$

$$+ \left(\frac{\partial L}{\partial \dot{q}^\sigma} \circ J^1\gamma(t) \right) u^\sigma(t) \Big]_{\tau}^{t} + L \circ J^1\gamma(t)(\delta t + \delta\tau).$$

In this approximation we also express $u^\sigma(\tau)$ and $u^\sigma(t)$ with help of sections γ and $\bar{\gamma}$,

$$u^\sigma(t) = q^\sigma J^1\bar{\gamma}(t) - q^\sigma J^1\gamma(t)$$

$$\doteq [q^\sigma J^1\bar{\gamma}(t+\delta t) - \dot{q}^\sigma \bar{\gamma}(t+\delta t)\,\delta t] - q^\sigma J^1\gamma(t) \doteq \delta q^\sigma(t) - \dot{q}^\sigma J^1\gamma(t),$$

where we have denoted $q^\sigma J^1\bar{\gamma}(t+\delta t) - q^\sigma J^1\gamma(t) \doteq \delta q^\sigma(t)$. And similarly for $u^\sigma(\tau)$. For expressing $q^\sigma J^1\bar{\gamma}(t)$ we have used the linear extrapolation to the point t. Then, in the linear approximation,

$$\delta S \doteq \int_{\tau}^{t} \left[\frac{\partial L}{\partial q^\sigma} - \frac{\mathrm{d}}{\mathrm{d}t}\left(\frac{\partial L}{\partial \dot{q}^\sigma}\right) \right] \circ J^2\gamma(t) u^\sigma(t)\,\mathrm{d}t$$

$$+ \left[\left(\frac{\partial L}{\partial \dot{q}} \circ J^1\gamma(t) \right) \delta q^\sigma(t) + \left(L - \frac{\partial L}{\partial \dot{q}^\sigma}\dot{q}^\sigma \right) \circ J^1\gamma(t)\,\delta t \right]_{\tau}^{t}.$$

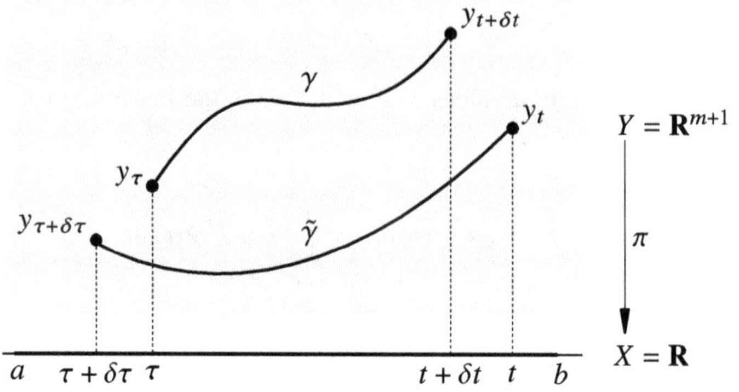

Fig. 6.2 Variational problem with free ends

Furthermore, suppose that γ is an extremal of the Lagrangian λ, i.e. a stationary point of the variational integral, $\delta S = 0$, i.e. (because $\delta t(\tau) = -\delta\tau$)

$$\delta S = \left[\frac{\partial L}{\partial \dot{q}^\sigma} \delta q^\sigma + \left(L - \frac{\partial L}{\partial \dot{q}^\sigma} \dot{q}^\sigma \right) \delta t \right]_\tau^t = 0. \tag{6.40}$$

Since the Lagrangian is supposed to be regular in the sense of standard regularity, the Legendre transformation reads

$$\mathcal{L} : (t, q^\sigma, \dot{q}^\sigma) \longrightarrow (t, q^\sigma, p_\sigma), \quad 1 \le \sigma \le m.$$

Using the Hamilton function $H = L + p_\sigma \dot{q}^\sigma$, and the generalized momenta $p_\sigma = \partial L/\partial \dot{q}^\sigma$, $1 \le \sigma \le m$, we can rewrite Eq. (6.40) as

$$\delta S = \left[-H(t, q^\nu, p_\nu)\, \delta t + p_\sigma\, \delta q^\sigma \right] \circ J^1\gamma = 0.$$

On the base of motivation ideas presented in Example 6.8 we introduce geometrical concepts of the Hamilton-Jacobi theory in 1st order mechanics.

Definition 6.16 Let (Y, π, X), $\dim X = 1$, $\dim Y = m + 1$ be a fibred manifold and (π, λ) a 1st order Lagrange structure with the Lagrangian λ regular in the sense of (6.1) and defined on an open set $W_1 \subset J^1Y$, expressed in charts as $\lambda = L(t, q^\sigma, \dot{q}^\sigma)\, dt$, $1 \le \sigma \le m$. Let p_σ are generalized momenta and $H(t, q^\sigma, p_\sigma)$ the Hamilton function. Define the 1-form by the relation

$$\eta = -H\, dt + p_\sigma\, dq^\sigma.$$

Proposition 6.12 *The form* $\eta = -H(t, q^\nu, p_\nu)\, dt + p_\sigma\, dq^\sigma$ *is closed.*

Proof The exterior derivative of the form η is

$$\mathrm{d}\eta = -\mathrm{d}H \wedge \mathrm{d}t + \mathrm{d}p_\sigma \wedge \mathrm{d}q^\sigma,$$

which is just the representative α_E of the mechanical system $[\alpha]$ corresponding to the Euler-Lagrange form E_λ, see (6.30). We have formerly proved that this representative is closed. So, there locally exist a function S of variables (t, q^σ), $1 \leq \sigma \leq m$, such that $\eta = \mathrm{d}S$. □

On the base of the previous proposition and its proof we have

$$\mathrm{d}S = -H\,\mathrm{d}t + p_\sigma\,\mathrm{d}q^\sigma \implies -H = \frac{\partial S}{\partial t}, \quad p_\sigma = \frac{\partial S}{\partial q^\sigma}. \tag{6.41}$$

Equations (6.41) called the *Hamilton-Jacobi equations* can be interpreted as the partial differential equation for the function $S = S(t, q^1, \ldots, q^m)$, where we put $H(t, q^\sigma, p_\sigma) = H(t, q^\sigma, \frac{\partial S}{\partial q_\sigma})$,

$$\frac{\partial S}{\partial t} + H\left(t, q^\sigma, \frac{\partial S}{\partial q^\sigma}\right) = 0. \tag{6.42}$$

Solving the set of Hamilton-Jacobi equations can be, of course, difficult, or even impossible without numerical procedures. Nevertheless, suppose that we have found its complete integral, i.e. general solution. It will contain m parameters $u^1, \ldots, u^m$ (For given initial conditions these parameters take concrete values and we get the corresponding particular solution).

$$S = S(t, q^1, \ldots, q^m, u^1, \ldots, u^m), \quad \text{where regularity of the matrix} \quad \left(\frac{\partial^2 S}{\partial q^\sigma \partial u^\nu}\right)$$

for $1 \leq \sigma, \nu \leq m$ is the necessary condition.

The behavior of the function S along extremals, i.e. along solutions of Hamilton equations, is characterized by the following proposition.

Proposition 6.13 *Let the assumptions formulated in Definition 6.16 are fulfilled and let $S = S(t, q^1, \ldots, q^m, u^1, \ldots, u^m)$ be the solution of the Hamilton-Jacobi equation (6.42). Then a section*

$$\delta : (t) \longrightarrow \delta(t) = (t, q^\sigma\delta(t), u^\sigma\delta(t)), \quad 1 \leq \sigma \leq m,$$

is a Hamilton extremal if and only if the quantities

$$v_\sigma = \frac{\partial S}{\partial u^\sigma}$$

are constant along δ, i.e. they are first integrals of the set of Hamilton equations (Equations $v_\sigma - \frac{\partial S}{\partial u^\sigma} = 0$ are the explicit equations of extremals).

Proof Calculating step-by-step the total derivative of the function v_σ with the use of the Hamilton-Jacobi equation we obtain

$$\begin{aligned}\frac{\mathrm{d}v_\sigma}{\mathrm{d}t} &= \frac{\mathrm{d}}{\mathrm{d}t}\left(\frac{\partial S}{\partial u^\sigma}\right) = \frac{\partial^2 S}{\partial t \partial u^\sigma} + \frac{\partial^2 S}{\partial q^\nu \partial u^\sigma}\dot{q}^\nu = \frac{\partial^2 S}{\partial q^\nu \partial u^\sigma}\dot{q}^\nu - \frac{\partial H}{\partial u^\sigma} \\ &= \frac{\partial^2 S}{\partial q^\nu \partial u^\sigma}\dot{q}^\nu - \frac{\partial H}{\partial p_\nu}\frac{\partial p_\nu}{\partial u^\sigma} = \frac{\partial^2 S}{\partial q^\nu \partial u^\sigma}\dot{q}^\nu - \frac{\partial H}{\partial p_\nu}\frac{\partial^2 S}{\partial q^\nu \partial u^\sigma} \\ &= \left(\dot{q}^\nu - \frac{\partial H}{\partial p_\nu}\right)\frac{\partial^2 S}{\partial q^\nu \partial u^\sigma}.\end{aligned}$$

Along a section δ it holds

$$\left[\frac{\mathrm{d}}{\mathrm{d}t}\left(\frac{\partial S}{\partial u^\sigma}\right)\right] \circ \delta = \left[\left(\frac{\mathrm{d}q^\nu}{\mathrm{d}t} - \frac{\partial H}{\partial p_\nu}\right)\frac{\partial^2 S}{\partial q^\nu \partial u^\sigma}\right] \circ \delta, \quad 1 \le \sigma \le m.$$

If the section δ is a Hamilton extremal, then the Hamilton equations hold. At the same time, the expressions $\left(\frac{\mathrm{d}q^\nu}{\mathrm{d}t} - \frac{\partial H}{\partial p_\nu}\right)$ for $1 \le \nu \le m$ represent the left-hand-sides of their first set. So, if δ is a Hamilton extremal, both sets of Hamilton equations are fulfilled along it and thus quantities $v_\sigma = \frac{\partial S}{\partial u^\sigma}$, $1 \le \sigma \le m$, are constants along Hamilton extremals. Because for regular Lagrangians all Hamilton extremals are regular, i.e. $\delta = J^1\gamma$ for a certain extremal γ, then quantities v_σ are constant along extremals.

Conversely, suppose that the quantities v_σ are constant along a Hamilton extremal δ. The matrix $(\frac{\partial^2 S}{\partial q^\nu \partial u^\sigma})$, $1 \le \sigma, \nu \le m$, is supposed to be regular, then the set of equations

$$\left[\left(\frac{\mathrm{d}q^\nu}{\mathrm{d}t} - \frac{\partial H}{\partial p_\nu}\right)\frac{\partial^2 S}{\partial q^\nu \partial u^\sigma}\right] \circ \delta = 0$$

has only the trivial solution. Thus $\left(\frac{\mathrm{d}q^\nu}{\mathrm{d}t} - \frac{\partial H}{\partial p_\nu}\right) \circ \delta = 0$, so, the first set of Hamilton equations is fulfilled for the section δ. As for the second set of Hamilton equations, let us calculate $\frac{\mathrm{d}p_\nu}{\mathrm{d}t} \circ \delta$ realizing that $H = H(t, q^\nu, p_\nu)$. Quantities p_ν are here to be considered as functions of q^μ, $1 \le \nu, \mu \le m$.

$$\begin{aligned}\frac{\mathrm{d}p_\nu}{\mathrm{d}t} &= \frac{\mathrm{d}}{\mathrm{d}t}\left(\frac{\partial S}{\partial q^\nu}\right) = \frac{\partial^2 S}{\partial t \partial q^\nu} + \frac{\partial^2 S}{\partial q^\sigma \partial q^\nu}\dot{q}^\sigma = \left(-\frac{\partial H}{\partial q^\nu} - \frac{\partial H}{\partial p_\sigma}\frac{\partial p_\sigma}{\partial q^\nu}\right) + \frac{\partial^2 S}{\partial q^\sigma \partial q^\nu}\dot{q}^\sigma, \\ \frac{\mathrm{d}p_\nu}{\mathrm{d}t} &= -\frac{\partial H}{\partial q^\nu} - \frac{\partial H}{\partial p_\sigma}\frac{\partial^2 S}{\partial q^\nu \partial q^\sigma} + \dot{q}^\sigma\frac{\partial^2 S}{\partial q^\nu \partial q^\sigma} = -\frac{\partial H}{\partial q^\nu} - \left(\frac{\partial H}{\partial p_\sigma} - \dot{q}^\sigma\right)\frac{\partial^2 S}{\partial q^\nu \partial q^\sigma}.\end{aligned}$$

Along a Hamilton extremal it holds $\dot{q}^\sigma \delta(t) = \frac{dq^\sigma \delta(t)}{dt}$ and

$$\left(\frac{\partial H}{\partial p_\sigma} - \dot{q}^\sigma\right) \circ \delta = 0 \text{ and then } \left(\frac{dp_\nu}{dt} + \frac{\partial H}{\partial q^\nu}\right) \circ \delta = 0.$$

The second set of Hamilton equations is then fulfilled as well. □

Summarizing the results leads to the following assertion.

Proposition 6.14 *Let* (Y, π, X), dim, $X = 1$, dim $Y = m + 1$, *be a fibred manifold and* (π, λ) *a* 1*st order Lagrange structure with the Lagrangian* $\lambda \in \Lambda^1_X(TJ^1Y)$ *defined on an open set* $W_1 \subset J^1Y$. *Suppose that* λ *is regular in the sense of relation (6.1). Let* $S(t, q^1, \ldots, q^m, u^1, \ldots, u^m)$ *be the complete integral (general solution) of Hamilton equations ((6.7), (6.22) for* 1*st order, (6.38) for* r*th order putting* $r = 1$*) depending on parameters* $u^1, \ldots, u^m$. *Then it holds*

(a) Functions $\partial S/\partial u^\sigma$, $1 \leq \sigma \leq m$, *are first integrals of the set of Hamilton equations, i.e. first integrals of the corresponding mechanical system* $[\alpha]$.
(b) The complete system of Hamilton extremals (their general solution) is determined by implicit equations

$$v_\sigma = \frac{\partial S}{\partial u^\sigma}, \quad p_\sigma = \frac{\partial S}{\partial q^\sigma}, \quad 1 \leq \sigma \leq m.$$

Example 6.9 Let $(Y, \pi, X) = (\mathbf{R} \times \mathbf{R}, p, \mathbf{R})$ be the fibred manifold with global fibred chart on Y $(\mathbf{R} \times \mathbf{R}, \psi)$, $\psi = (t, x)$, associated chart on X $(\mathbf{R}, \varphi)$, $\varphi = (t)$, and associated fibred chart on J^1Y $(\mathbf{R} \times \mathbf{R} \times \mathbf{R}, \psi_1)$, $\psi_1 = (t, x, \dot{x})$. Let (π, λ) be the 1st order Lagrange structure with the Lagrangian $\lambda = L(t, x, \dot{x})\, dt$, where

$$L = \sqrt{t^2 + x^2}\sqrt{1 + \dot{x}^2},$$

see also Problem 6.3. The Lagrangian is regular in the sense of (6.1), because

$$\frac{\partial^2 L}{\partial \dot{x}^2} = \frac{(t^2 + x^2)^{1/2}}{(1 + \dot{x}^2)^{3/2}} \neq 0.$$

Let us construct the Hamilton-Jacobi equation and find the explicit equations of extremals. The generalized momentum p is

$$p = \frac{\partial L}{\partial \dot{x}} = \sqrt{t^2 + x^2}\frac{\dot{x}}{\sqrt{1 + \dot{x}^2}} \implies \dot{x} = \frac{p}{\sqrt{t^2 + x^2 - p^2}},$$

the Hamilton function and the Hamilton-Jacobi equation (6.42) are then

$$H(t, x, p) = -L + p\dot{x} = -\sqrt{t^2 + x^2 - p^2}, \quad \frac{\partial S}{\partial t} - \sqrt{t^2 + x^2 - \left(\frac{\partial S}{\partial x}\right)^2} = 0,$$

$$\left(\frac{\partial S}{\partial t}\right)^2 + \left(\frac{\partial S}{\partial x}\right)^2 = t^2 + x^2.$$

The general solution (complete integral) is—see Problem 6.17

$$S(t, x, u) = -ut^2 + \sqrt{1 - 4u^2}\, tx + ux^2, \quad u \in \left[-\frac{1}{2}, \frac{1}{2}\right], \quad \frac{\partial^2 S}{\partial x \partial u} \neq 0.$$

The first integral of the corresponding Hamilton equation (alternatively of the mechanical system $[\alpha]$) is

$$v = \frac{\partial S}{\partial u} = -t^2 - \frac{4u}{\sqrt{1 - 4u^2}}\, tx + x^2. \tag{6.43}$$

This quantity is constant along extremals and (6.43) gives the implicit expression of all extremals, i.e. it is the general solution of the given variational problem. Parameters u and v have the meaning of integration constants. In addition, calculating $f(x, p) = p^2 - x^2$ (see Problem 6.3) we can see that this quantity is constant along extremals as well,

$$p^2 - x^2 = \left(\frac{\partial S}{\partial x}\right)^2 - x^2 = -v(1 - 4u^2).$$

(Perform the previous calculation).

We have dealt with the idea of Hamilton-Jacobi theory for the case of the first order Lagrange structure regular in the sense of the regularity (6.1), i.e. for second order Euler-Lagrange forms (even order). In the case of an odd order, there will be deviations, similar as it was in the case of Hamilton equations. Further considerations, including the generalization of Hamilton-Jacobi theory to general-order mechanics or field theory, brings new theoretical aspects and require the introduction of additional concepts that are beyond the scope of this text. The reader can find them, for example, through monograph [2] and papers [1,3].

6.6 Problems

Problem 6.1 Consider a 1st order Lagrangian λ regular in the sense of relation (6.1). Prove that every Lagrangian $\tilde{\lambda}$ of the 1st order equivalent with λ (i.e. $\lambda_0 = \tilde{\lambda} - \lambda$ being a trivial Lagrangian) is also regular in the sense of (6.1).

Problem 6.2 Make in detail calculations leading to Eqs. (6.8) and (6.9).

Problem 6.3 Consider the fibred manifold $(Y, \pi, X) = (\mathbf{R} \times \mathbf{R}, p, \mathbf{R})$ with global fibred chart on Y $(\mathbf{R} \times \mathbf{R}, \psi)$, $\psi = (t, x)$, associated chart on X $(\mathbf{R}, \varphi)$, $\varphi = (t)$, and associated fibred chart on J^1Y $(\mathbf{R} \times \mathbf{R} \times \mathbf{R}, \psi_1)$, $\psi_1 = (t, x, \dot{x})$. Let (π, λ) be the Lagrange structure with

$$\lambda = \sqrt{t^2 + x^2}\sqrt{1 + \dot{x}^2}\, dt.$$

(a) Construct the corresponding Lepage equivalent (Cartan form) and the Euler-Lagrange form. Write the Euler-Lagrange equation.
(b) Check the regularity of λ.
(c) Construct the generalized momentum and the Hamilton function.
(d) Write the Hamilton equations and solve them (Note the simplicity of the Hamiltonian approach compared to Lagrangan one).

Problem 6.4 First-order Lagrangians in mechanics on (Y, π, X), dim $X = 1$, dim $Y = m + 1$ have usually the form

$$\lambda = L\, dt, \quad L = \frac{1}{2} g_{\sigma\nu}(q^1, \ldots, q^m)\, \dot{q}^\sigma \dot{q}^\nu - U(t, q^1, \ldots, q^m), \quad g_{\sigma\nu} = g_{\nu\sigma},$$

for all $1 \leq \sigma, \nu \leq m$.

(a) Express the generalized momenta and the Hamilton function.
(b) Specify the conditions for the Hamilton function (corresponds to the mechanical energy) to be constant along extremals of the Lagrangian.

Problem 6.5 Perform in detail all calculations leading to Proposition 6.3.

Problem 6.6 Perform all calculations omitted in Example 6.3, as well as calculations leading to the resulting solution (finding Hamilton extremals).

Problem 6.7 Perform all necessary calculations leading to relation (6.21).

Problem 6.8 Let E be a 1st order regular variational dynamical form

$$E = (A_\sigma + B_{\sigma\nu} q_1^\nu)\, \omega^\sigma \wedge dt, \quad A_\sigma = A_\sigma(t, q_0^\mu), \quad B_{\sigma\nu} = B_{\sigma\nu}(t, q_0^\mu),$$
$$1 \leq \sigma, \nu, \mu \leq m.$$

The minimal 1st order Lagrangian corresponding to this form is $L = \big(a(t, q_0^\mu) + b_\nu(t, q_0^\mu)\, \dot{q}^\nu\big)\, dt$. Solve the following tasks:

(a) Find the relations between coefficients A_σ and $B_{\sigma\nu}$ following from the variationality of the form E and express the coefficients A_σ and $B_{\sigma\nu}$ with help of a and b_σ, $1 \leq \sigma, \nu \leq m$.
(b) Construct the representative $\alpha \in [\alpha]$ of the corresponding mechanical system $[\alpha]$.
(c) Find the vector fields and 1-forms generating the dynamical distribution Δ_α and its annihilator Δ_α^0, respectively.

Problem 6.9 Solve the same questions as in Problem 6.8 for a 2nd order variational dynamical form

$$E = (A_\sigma + B_{\sigma\nu} q_2^\nu)\, \omega^\sigma \wedge dt, \quad A_\sigma = A_\sigma(t, q_0^\mu, q_1^\mu), \quad B_{\sigma\nu} = B_{\sigma\nu}(t, q_0^\mu, q_1^\mu),$$

for all $1 \leq \sigma, \nu, \mu \leq m$.

Problem 6.10 Prove relation (6.27) for a 1st order Euler-Lagrange form.

Problem 6.11 Let λ be a 1st order Lagrangian regular in the sense of condition (6.1) and θ_λ its Lepage equivalent. Let ξ be the corresponding Hamilton vector field (6.23). Prove that it holds $i_\xi\, d\theta_\lambda \in \Delta_\lambda^0$.

Problem 6.12 Let $E_\lambda = (A_\sigma + B_{\sigma\nu} q_1^\nu)\, dq_0^\sigma \wedge dt$ be the regular 1st order Euler-Lagrange form corresponding to the Lagrangian $\lambda = \left(a(t, q_0^\nu) + b_\sigma(t, q_0^\nu) q_1^\sigma\right) dt$. Express vector fields generating the dynamical distribution of α_E.

Problem 6.13 Prove that for the representative $\alpha_E \in [\alpha]$ (uniquely determined closed representative) of a variational dynamical form E_λ it holds $\alpha_E = d\theta_\lambda$. Derive relation (6.30).

Problem 6.14 Consider the regular Lagrangian given in Example 6.6 (with positive constants m, b and k).

(a) Formulate and solve the corresponding Euler-Lagrange equations.
(b) Write the Legendre transformation, formulate and solve the corresponding Hamilton equations.

Problem 6.15 Perform all calculations in Example 6.7 in detail and solve the obtained Hamilton equations. Formulate and solve Euler-Lagrange equations and show that both solutions are identical.

Problem 6.16 Construct the Hamilton-Jacobi equation for a linear harmonic oscillator with $\lambda = L(t, x, \dot{x})\, dt$, $L = \frac{m}{2}\dot{x}^2 - \frac{k}{2}x^2$, m and k positive constants.

Problem 6.17 Determine the complete integral $S(t, x, u)$ of the Hamilton-Jacobi equation formulated in Example 6.9.

References

1. O. Krupková, Hamiltonian field theory. J. Geom. Phys. **43**, 93–132 (2002)
2. O. Krupková, *The Geometry of Ordinary Variational Equations*. Lectures in Mathematical Notes, vol. 1678 (Springer, Berlin, Heidelberg, 1997)
3. O. Krupková, Lepagean 2-forms in higher order Hamiltonian mechanics. I. Regularity, Arch. Math. (Brno) **22**, 97–120 (1986)

Elements of Variational Sequences 7

In previous sections we solved two important problems of calculus of variations—the *trivial problem* and the *inverse problem*. We have recognized that in both these problems the exterior derivative operator plays an important role. Summarize main facts:

- Every trivial Lagrangian on $J^r Y$, $r = 1, 2$, is the horizontal component of the exterior derivative of a function on $J^{r-1}Y$ (the same holds also for higher order problems).
- Every (locally) variational mechanical system contains a representative which is the exterior derivative of a 1-form (Lepage equivalent of Lagrangian).
- The Euler-Lagrange mapping assigns to every Lagrangian its Euler-Lagrange form.
- The set of all first-order trivial Lagrangians is the kernel of the Euler-Lagrange mapping.
- The set of Euler-Lagrange forms (a subset of the set of dynamical forms) is the image of the Euler-Lagrange mapping.

In this chapter, we generalize the concept of Euler-Lagrange mapping and show that it is a first (nontrivial) term of an exact sequence of classes of forms arising as the factor sequence of the de Rham exact sequence of exterior operators which assign to k-forms their exterior derivatives ($(k+1)$-forms). The classes of forms arise by factorization of spaces of k-forms with respect to their subspaces of forms of a certain type of contactness forming the differential ideal. The mentioned sequence of mappings is called the *finite order variational sequence*. For the calculus of variations also the second term of this sequence, called the *Helmholtz-Sonin mapping*, is important. It assigns to dynamical forms special 3-forms (called *Helmholtz-Sonin forms*) and the set of variational dynamical forms is its kernel. (Note that the relationship of

J. Musilová et al., *Calculus of Variations on Fibred Manifolds and Variational Physics*, Lecture Notes in Physics 1033, https://doi.org/10.1007/978-3-031-77408-9_7

these mappings and the exterior derivative operator is intuitively clear: the kernel of exterior derivative operator is formed by exact forms, equivalently closed forms in the local version.) The variational sequence was mostly studied as so called *variational bicomplex* working on infinite jets of an underlying fibred structure—see, e.g. [1–5]. Nevertheless, the variational sequence on *finite order* jet prolongations of fibred manifolds as well as its representation by differential forms is an effective tool for the study of both the local and global problems of calculus of variations especially for physical applications: We can solve the trivial variational problem and the inverse problem of the calculus of variations by studying kernels and images of mappings in the variational sequence.

The general concept of the finite order variational sequence was first introduced by Krupka [6] and intensively studied in numerous works of him and his co-workers and other researchers—see, e.g. [8–10]. The problem of the representation of the finite order variational sequence by forms was completely solved in [11,12], detailed calculations in coordinates for lower order problems see, e.g. [13]. Application of the finite order variational sequence for general solution of trivial Lagrangian problem see in [7].

Our considerations concern the first-order variational sequence with the representation by forms lifted in general up to the third-order jet prolongation of the underlying fibred manifold. (Notes to higher order calculus of variations in mechanics see in Sect. 5.1 and for field theory in Chap. 8). So, the definitions and propositions are formulated and proved for these "low-order" situations. Nevertheless, they are valid for an arbitrary finite order r of the underlying fibred manifold. The most consistent presentation of this problem discussed in full generality can be found in [14].

Prerequisites exact sequence, short exact sequence, de Rham sequence, factorization of a group with respect to a normal subgroup, kernel and image of a mapping, basic concepts of sheaf theory, elements of cohomology theory. □

7.1 Preliminary Considerations

In this section we present or summarize the concepts required especially for understanding of the variational sequence and for practical calculations leading to solution of concrete problems.

Definition 7.1 Total derivative of a form. Let $\eta \in \Lambda^k(TJ^rY), r = 0, 1, 2, (V, \psi)$, $\psi = (t, q^\sigma)$, a fibred chart on Y and (U, φ), $\varphi = (t)$. Let (V_s, ψ_s), $s = 1, 2, 3$ (or higher, as needed), be the associated fibred charts on J^sY, where $\psi_s = (t, q^\sigma, \dot{q}^\sigma, \ddot{q}^\sigma, q^\sigma_{(3)})$ for $s = 3$, or $\psi_s = (t, q^\sigma_0, q^\sigma_1, \dots, q^\sigma_s)$ in general, as needed, $1 \leq \sigma \leq m$. Consider vector fields on $J^{r+1}Y$

$$\Xi = \frac{\partial}{\partial t} + \dot{q}^\sigma \frac{\partial}{\partial q^\sigma} \quad \text{for} \quad r = 0,$$

$$\Xi = \frac{\partial}{\partial t} + \dot{q}^\sigma \frac{\partial}{\partial q^\sigma} + \ddot{q}^\sigma \frac{\partial}{\partial \dot{q}^\sigma} \quad \text{for} \quad r = 1,$$

$$\Xi = \frac{\partial}{\partial t} + \dot{q}^\sigma \frac{\partial}{\partial q^\sigma} + \ddot{q}^\sigma \frac{\partial}{\partial \dot{q}^\sigma} + q^\sigma_{(3)} \frac{\partial}{\partial \ddot{q}^\sigma} \quad \text{for} \quad r = 2.$$

These vector fields are called the *total derivative vector fields*. We will denote them as d_t or $\frac{\mathrm{d}}{\mathrm{d}t}$. The form

$$\mathrm{d}_t \eta = \partial_{\mathrm{d}_t} \pi^*_{r+1,r} \eta = i_{\mathrm{d}_t} \mathrm{d}(\pi^*_{r+1,r} \eta) + \mathrm{d} i_{\mathrm{d}_t} \pi^*_{r+1,r} \eta \in \Lambda^k(T J^{r+1} Y) \tag{7.1}$$

is called the *total derivative of a form* $\eta \in \Lambda^k(T J^r Y)$. So, the total derivative of a form is in fact the Lie derivative of this form with respect to the corresponding total derivative vector field.

Note that for a function f on $J^r Y$, $r = 0, 1, 2$, the expression (7.1) is the same as for the total derivative of this function in the standard sense. (The same is valid for general r.)

Example 7.1 We calculate the total derivatives of coordinate 1-forms on $J^r Y$. Using (7.1) we obtain

$$\begin{aligned}
\mathrm{d}_t f &= \frac{\mathrm{d}f}{\mathrm{d}t}, \\
\mathrm{d}_t \,\mathrm{d}t &= i_{\mathrm{d}_t} \mathrm{d}^2 t + \mathrm{d} i_{\mathrm{d}_t} \mathrm{d}t = 0, \\
\mathrm{d}_t \,\mathrm{d}q^\sigma &= i_{\mathrm{d}_t} \mathrm{d}^2 q^\sigma + \mathrm{d} i_{\mathrm{d}_t} \mathrm{d}q^\sigma = \mathrm{d}\dot{q}^\sigma, \\
\mathrm{d}_t \,\mathrm{d}\dot{q}^\sigma &= i_{\mathrm{d}_t} \mathrm{d}^2 \dot{q}^\sigma + \mathrm{d} i_{\mathrm{d}_t} \mathrm{d}\dot{q}^\sigma = \mathrm{d}\ddot{q}^\sigma, \\
\mathrm{d}_t \,\mathrm{d}\ddot{q}^\sigma &= i_{\mathrm{d}_t} \mathrm{d}^2 \ddot{q}^\sigma + \mathrm{d} i_{\mathrm{d}_t} \mathrm{d}\ddot{q}^\sigma = \mathrm{d}q^\sigma_{(3)}, \\
\mathrm{d}_t \,\omega^\sigma &= \mathrm{d}_t(\mathrm{d}q^\sigma - \dot{q}^\sigma \,\mathrm{d}t) = \mathrm{d}\dot{q}^\sigma - \ddot{q}^\sigma \,\mathrm{d}t = \dot{\omega}^\sigma \\
\mathrm{d}_t \,\dot{\omega}^\sigma &= \mathrm{d}_t(\mathrm{d}\dot{q}^\sigma - \ddot{q}^\sigma \,\mathrm{d}t) = \mathrm{d}\ddot{q}^\sigma - q^\sigma_{(3)} \,\mathrm{d}t = \ddot{\omega}^\sigma .
\end{aligned}$$

The construction of the total derivative operator of forms is independent of the order of the jet prolongation of the underlying fibred manifold. This means that it can be used in principle for an arbitrary order r, including all properties. The action of this operator on forms forming the base of 1-forms was presented in Example 7.1 up to the second order. The obtained relations can be, of course, extrapolated to higher orders.

Now let us study the decomposition of the contact components of the exterior derivative of a k-form $\eta \in \Lambda^k(T J^r Y)$, $r = 0, 1$. Following relation (3.20) we obtain

$$\begin{aligned}
\pi^*_{r+1,r} \,\eta = p_{k-1} \,\eta + p_k \,\eta \implies \pi^*_{r+2,r} \,\mathrm{d}\eta &= \pi^*_{r+2,r+1} \left(\mathrm{d} p_{k-1} \,\eta + \mathrm{d} p_k \,\eta\right) \\
&= p_k \,\mathrm{d} p_{k-1} \,\eta + p_{k+1} \,\mathrm{d} p_{k-1} \,\eta + p_k \,\mathrm{d} p_k \,\eta + p_{k+1} \,\mathrm{d} p_k \,\eta.
\end{aligned}$$

Because $p_{k+1}\, \mathrm{d} p_{k-1}\, \eta = 0$ we obtain

$$\pi^*_{r+2,r}\, \mathrm{d}\eta = p_k\, \mathrm{d} p_k\, \eta + (p_k\, \mathrm{d} p_{k-1}\, \eta + p_{k+1}\, \mathrm{d} p_k\, \eta)\,.$$

So the form $\pi^*_{r+2,r}\, \mathrm{d}\eta$ can be decomposed as

$$\pi^*_{r+2,r}\, \mathrm{d}\eta = \mathrm{d}_h\, \eta + \mathrm{d}_c\, \eta, \quad \mathrm{d}_h \eta = p_k\, \mathrm{d} p_k\, \eta, \quad \mathrm{d}_c\, \eta = p_k\, \mathrm{d} p_{k-1}\, \eta + p_{k+1}\, \mathrm{d} p_k\, \eta. \tag{7.2}$$

For a strongly contact form η (i.e. for $p_{k-1}\, \eta = 0$) we have

$$\mathrm{d}_h\, \eta = p_k\, \mathrm{d} p_k\, \eta, \quad \mathrm{d}_c\, \eta = p_{k+1}\, \mathrm{d} p_k\, \eta.$$

By direct calculations we can easily prove the following properties of operators d_h and d_c:

$$\begin{aligned} \mathrm{d}_h \circ \mathrm{d}_h = 0, \;\; \mathrm{d}_c \circ \mathrm{d}_c = 0, \;\; \mathrm{d}_h \circ \mathrm{d}_c &= -\mathrm{d}_c \circ \mathrm{d}_h, \\ \mathrm{d}_h\, (\eta \wedge \varrho) = \mathrm{d}_h\, \eta \wedge (\pi^*_{r+2,r}\, \varrho) &+ (-1)^k (\pi^*_{r+2,r}\, \eta \wedge \mathrm{d}_h\, \varrho), \\ \mathrm{d}_c\, (\eta \wedge \varrho) = \mathrm{d}_c\, \eta \wedge (\pi^*_{r+2,r}\, \varrho) &+ (-1)^k (\pi^*_{r+2,r}\, \eta \wedge \mathrm{d}_c\, \varrho), \end{aligned} \tag{7.3}$$

where $\eta \in \Lambda^k(TJ^rY)$ and $\varrho \in \Lambda^l(TJ^rY)$, $r = 0, 1$.

The following proposition summarizes the main properties of total derivatives of forms.

Proposition 7.1 *Let (Y, π, X) be a fibred manifold, (V, ψ), $\psi = (t, q^\sigma)$, a fibred chart on Y and (U, φ), $\varphi = (t)$, the associated chart on X. Let (V_s, ψ_s), $s = 1, 2, \ldots$, be the associated fibred charts on J^sY. Let $\eta \in \Lambda^k(TJ^rY)$ and $\varrho \in \Lambda^l(TJ^rY)$ be differential forms. Then the following holds:*

(a) Total derivatives of exterior products of forms obey the Leibniz rule

$$\mathrm{d}_t (\eta \wedge \varrho) = \mathrm{d}_t\, \eta \wedge \varrho + \eta \wedge \mathrm{d}_t\, \varrho. \tag{7.4}$$

(b) The form $\mathrm{d}_h\, \eta$ can be locally expressed as

$$\mathrm{d}_h\, \eta = (-1)^k\, \mathrm{d}_t\, (\pi^*_{r+1,r} \eta) \wedge \mathrm{d}t \tag{7.5}$$

(c) Let $(\bar{V}, \bar{\psi})$, $\bar{\psi} = (\bar{t}, \bar{q}^\sigma)$, be another fibred chart on Y, $(\bar{U}, \bar{\varphi})$ the corresponding associated chart on X and $(\bar{V}_s, \bar{\psi}_s)$, $s = 1, 2, 3, \ldots$, the corresponding associated fibred charts on J^sY. On nonempty intersections of charts the transformation rule holds

$$\mathrm{d}_t\, \eta = \left(\frac{\mathrm{d}\bar{t}}{\mathrm{d}t}\right) \bar{\mathrm{d}}_{\bar{t}}\, \eta. \tag{7.6}$$

Proof We will prove the proposition for k-forms on J^1Y. (Even though we are interested only in first-order problems and the formulation of the proposition is adapted to this fact, the proof shows that it is valid for an arbitrary order of the underlying fibred manifold.)

(a) The Leibniz rule follows directly from the definition relation (7.1) as the consequence of the corresponding property of the Lie derivative:

$$\partial_\xi (\eta \wedge \varrho) = \partial_\xi \eta \wedge \varrho + \eta \wedge \partial_\xi \varrho.$$

(b) We proceed by induction with respect to k. First we prove relation (7.5) for a 1-form on J^1Y calculating and comparing both sides of this relation. For calculating the right-hand side we use the just proved assertion (a). The chart expression of $\pi_{2,1}^* \eta$ reads

$$\pi_{2,1}^* \eta = A \, \mathrm{d}t + B_\sigma \, \omega^\sigma + C_\sigma \, \dot{\omega}^\sigma,$$

where B_σ and C_σ are functions on J^1Y and A is a function on J^2Y affine in variables $\ddot{q}^\sigma$ (see Sect. 3.3). Using (7.4) and results of Example 7.1 we obtain

$$\begin{aligned}
-\mathrm{d}_t (\pi_{2,1}^* \eta) \wedge \mathrm{d}t &= -\mathrm{d}_t \left(A \, \mathrm{d}t + B_\sigma \, \omega^\sigma + C_\sigma \, \dot{\omega}^\sigma\right) \wedge \mathrm{d}t \\
&= -\left(\mathrm{d}_t A \, \mathrm{d}t + \mathrm{d}_t B_\sigma \, \omega^\sigma + B_\sigma \, \mathrm{d}_t \omega^\sigma + \mathrm{d}_t C_\sigma \, \dot{\omega}^\sigma + C_\sigma \, \mathrm{d}_t \dot{\omega}^\sigma\right) \wedge \mathrm{d}t \\
&= -\left(\frac{\mathrm{d}A}{\mathrm{d}t} \mathrm{d}t + \frac{\mathrm{d}B_\sigma}{\mathrm{d}t} \omega^\sigma + B_\sigma \dot{\omega}^\sigma + \frac{\mathrm{d}C_\sigma}{\mathrm{d}t} \dot{\omega}^\sigma + C_\sigma \ddot{\omega}^\sigma\right) \wedge \mathrm{d}t \\
&= -\left(\frac{\mathrm{d}B_\sigma}{\mathrm{d}t} \omega^\sigma + B_\sigma \dot{\omega}^\sigma + \frac{\mathrm{d}C_\sigma}{\mathrm{d}t} \dot{\omega}^\sigma + C_\sigma \ddot{\omega}^\sigma\right) \wedge \mathrm{d}t.
\end{aligned}$$

On the other hand, it holds $p_1 \eta = B_\sigma \omega^\sigma + C_\sigma \dot{\omega}^\sigma$ and

$$\begin{aligned}
\pi_{3,2}^* \mathrm{d} p_1 \eta = &-\frac{\mathrm{d}B_\sigma}{\mathrm{d}t} \omega^\sigma \wedge \mathrm{d}t + \left\{\frac{\partial B_\sigma}{\partial q^\nu}\right\}_{\mathrm{alt}\,(\sigma,\nu)} \omega^\nu \wedge \omega^\sigma + \frac{\partial B_\sigma}{\partial \dot{q}^\nu} \dot{\omega}^\nu \wedge \omega^\sigma - B_\sigma \dot{\omega}^\sigma \wedge \mathrm{d}t \\
&- \frac{\mathrm{d}C_\sigma}{\mathrm{d}t} \dot{\omega}^\sigma \wedge \mathrm{d}t + \frac{\partial C_\sigma}{\partial q^\nu} \omega^\nu \wedge \dot{\omega}^\sigma + \left\{\frac{\partial C_\sigma}{\partial \dot{q}^\nu}\right\}_{\mathrm{alt}\,(\sigma,\nu)} \dot{\omega}^\nu \wedge \dot{\omega}^\sigma - C_\sigma \ddot{\omega}^\sigma \wedge \mathrm{d}t,
\end{aligned}$$

Then for $\mathrm{d}_h \eta$ we obtain

$$\mathrm{d}_h \eta = p_1 \pi_{3,2}^* \mathrm{d} p_1 \eta = -\left(\frac{\mathrm{d}B_\sigma}{\mathrm{d}t} \omega^\sigma + B_\sigma \dot{\omega}^\sigma + \frac{\mathrm{d}C_\sigma}{\mathrm{d}t} \dot{\omega}^\sigma + C_\sigma \ddot{\omega}^\sigma\right) \wedge \mathrm{d}t$$

The equality (7.5) for a 1-form η is now evident. Now let us suppose that (7.5) holds for a k-form. Let η be a $(k+1)$-form. So it is a sum of products of the type $\varrho \wedge \chi$, where ϱ is a k-form and χ is a 1-form. Following the Leibniz rule (7.4) we obtain

$$(-1)^{k+1} \mathrm{d}_t (\varrho \wedge \chi) \wedge \mathrm{d}t = (-1)^{k+1} \left(\mathrm{d}_t \varrho \wedge \chi + \varrho \wedge \mathrm{d}_t \chi\right) \wedge \mathrm{d}t.$$

Using relation (7.5) proved for 1-forms for the form χ and the same relation supposed for k-forms for the form ϱ we obtain

$$(-1)^{k+1}\mathrm{d}_t(\varrho\wedge\chi)\wedge\mathrm{d}t = (-1)^k\mathrm{d}_t\varrho\wedge\mathrm{d}t\wedge\chi + (-1)^k\varrho\wedge(-\mathrm{d}_t\chi\wedge\mathrm{d}t)$$
$$= \mathrm{d}_h\varrho\wedge\chi + (-1)^k\varrho\wedge\mathrm{d}_h\chi.$$

Taking into account the second relation of (7.3) we immediately obtain the desired result.

(c) For a general k-form the decomposition (7.5) is not invariant in general. On the other hand we have the invariant decomposition of the $(k+1)$-form $\mathrm{d}\eta$

$$\pi^*_{r+2,r}\,\mathrm{d}\eta = p_k(\mathrm{d}p_{k-1}\eta + \mathrm{d}p_k\eta) + p_{k+1}\,\mathrm{d}p_k\eta$$

For a k-contact k-form (i.e. strongly contact form) this gives

$$\pi^*_{r+2,r}\eta = p_k\,\mathrm{d}p_k\eta + p_{k+1}\,\mathrm{d}p_k\eta,$$

where the first summand is $\mathrm{d}_h\eta$ and the second one is $\mathrm{d}_c\eta$. This means that for a k-contact k-form the decomposition (7.5) is invariant. Thus for such a case it holds

$$(-1)^k\,\mathrm{d}_t\eta\wedge\mathrm{d}t = (-1)^k\,\bar{\mathrm{d}}_t\eta\wedge\mathrm{d}\bar{t} \implies \mathrm{d}_t\eta\wedge\mathrm{d}t = \bar{\mathrm{d}}_t\eta\wedge\left(\frac{\mathrm{d}\bar{t}}{\mathrm{d}t}\right)\mathrm{d}t,$$

and then

$$\mathrm{d}_t\eta = \left(\frac{\mathrm{d}\bar{t}}{\mathrm{d}t}\right)\bar{\mathrm{d}}_t\eta.$$

For a general k-form its $(k-1)$-contact component is nonzero. On the other hand it is invariant and it can be expressed as $p_{k-1}\eta = \varrho\wedge\mathrm{d}t$, where ϱ is a $(k-1)$-contact $(k-1)$-form and then the transformation rule (7.6) holds. Using the Leibniz rule (7.4) and the relation $\mathrm{d}_h\,\mathrm{d}t = 0$ we obtain

$$\mathrm{d}_h(\varrho\wedge\mathrm{d}t) = \mathrm{d}_h\varrho\wedge\mathrm{d}t + \varrho\wedge\mathrm{d}_h\,\mathrm{d}t = \left(\frac{\mathrm{d}\bar{t}}{\mathrm{d}t}\right)\bar{\mathrm{d}}_t\varrho\wedge\mathrm{d}t.$$

This confirms the validity of the transformation rule for a general k-form. □

Example 7.2 Let η be a k-form on J^1Y. We express the action of the total derivative operator on this form in charts. The chart expression of the form $\pi^*_{2,1}\eta$ is

$$\sum_{a=0}^{k-1} A^a_{\sigma_1\ldots\sigma_a\sigma_{a+1}\ldots\sigma_{k-1}}\omega^{\sigma_1}\wedge\ldots\wedge\omega^{\sigma_a}\wedge\dot{\omega}^{\sigma_{a+1}}\wedge\ldots\wedge\dot{\omega}^{\sigma_{k-1}}\wedge\mathrm{d}t$$
$$+\sum_{b=0}^{k} B^b_{\sigma_1\ldots\sigma_b\sigma_{b+1}\ldots\sigma_k}\omega^{\sigma_1}\wedge\ldots\wedge\omega^{\sigma_b}\wedge\dot{\omega}^{\sigma_{b+1}}\wedge\ldots\wedge\dot{\omega}^{\sigma_k},$$

where $A^a_{\sigma_1\ldots\sigma_a\sigma_{a+1}\ldots\sigma_{k-1}}$ and $B^b_{\sigma_1\ldots\sigma_b\sigma_{b+1}\ldots\sigma_k}$ are functions on J^2Y. Functions $A^a_{\sigma_1\ldots\sigma_a\sigma_{a+1}\ldots\sigma_{k-1}}$ are affine in variables $\ddot{q}^\nu$ and antisymmetric in indices $\{\sigma_1,\ldots,\sigma_a\}$ and in indices $\{\sigma_{a+1},\ldots,\sigma_{k-1}\}$, $0\le a\le k-1$. Functions $B^b_{\sigma_1\ldots\sigma_b\sigma_{b+1}\ldots\sigma_k}$ are projectable on J^1Y and antisymmetric in indices $\{\sigma_1,\ldots,\sigma_b\}$ and in indices $\{\sigma_{b+1},\ldots,\sigma_k\}$, $0\le b\le k$. Using the Leibniz formula, results of Example 7.1 and the antisymmetry properties of function coefficients of the form we obtain

$$
\begin{aligned}
\mathrm{d}_t\eta = {} & \sum_{a=0}^{k-1}\frac{\mathrm{d}A^a_{\sigma_1\ldots\sigma_a\sigma_{a+1}\ldots\sigma_{k-1}}}{\mathrm{d}t}\omega^{\sigma_1}\wedge\ldots\wedge\omega^{\sigma_a}\wedge\dot{\omega}^{\sigma_{a+1}}\wedge\ldots\wedge\dot{\omega}^{\sigma_{k-1}}\wedge\mathrm{d}t\\
&+\sum_{b=0}^{k}\frac{\mathrm{d}B^b_{\sigma_1\ldots\sigma_b\sigma_{b+1}\ldots\sigma_k}}{\mathrm{d}t}\omega^{\sigma_1}\wedge\ldots\wedge\omega^{\sigma_b}\wedge\dot{\omega}^{\sigma_{b+1}}\wedge\ldots\wedge\dot{\omega}^{\sigma_k}\\
&+\sum_{a=0}^{k-1}aA^a_{\sigma_1\ldots\sigma_{a-1}\{\sigma_a\sigma_{a+1}\ldots\sigma_{k-1}\}}\omega^{\sigma_1}\wedge\ldots\wedge\omega^{\sigma_{a-1}}\wedge\dot{\omega}^{a}\wedge\dot{\omega}^{\sigma_{a+1}}\wedge\ldots\wedge\dot{\omega}^{\sigma_{k-1}}\wedge\mathrm{d}t\\
&+\sum_{b=0}^{k}bB^b_{\sigma_1\ldots\sigma_{b-1}\{\sigma_b\sigma_{b+1}\ldots\sigma_k\}}\omega^{\sigma_1}\wedge\ldots\wedge\omega^{\sigma_{b-1}}\wedge\dot{\omega}^{\sigma_b}\wedge\dot{\omega}^{\sigma_{b+1}}\wedge\ldots\wedge\dot{\omega}^{\sigma_k}\\
&+\sum_{a=0}^{k-1}(k-1-a)A^a_{\sigma_1\ldots\sigma_a\sigma_{a+1}\ldots\sigma_{k-1}}\omega^{\sigma_1}\wedge\ldots\wedge\omega^{\sigma_a}\wedge\dot{\omega}^{\sigma_{a+1}}\wedge\ldots\wedge\dot{\omega}^{\sigma_{k-2}}\wedge\ddot{\omega}^{\sigma_{k-1}}\wedge\mathrm{d}t\\
&+\sum_{b=0}^{k}(k-b)B^b_{\sigma_1\ldots\sigma_b\sigma_{b+1}\ldots\sigma_k}\omega^{\sigma_1}\wedge\ldots\wedge\omega^{\sigma_b}\wedge\dot{\omega}^{\sigma_{b+1}}\wedge\ldots\wedge\dot{\omega}^{k-1}\wedge\ddot{\omega}^{\sigma_k},
\end{aligned}
$$

where $\{\sigma_a\sigma_{a+1}\ldots\sigma_{k-1}\}$ and $\{\sigma_b\sigma_{b+1}\ldots\sigma_k\}$ denote the antisymmetrization in indices in brackets $\{\ldots\}$.

7.2 Variational Sequence

In this section, we apply the general concept of a finite order variational sequence on the first-order case. The deep understanding of the problem requires basic knowledge of the theory of shaves (see, e.g. [15]). For our purposes, this is not strictly necessary.

In following sections, we adopt the notation standardly used in the theory of finite order variational sequences. Denote by Ω^r_k, $k\ge 0$, $1\le r$, the space of k forms on J^rY, i.e. $\Omega^r_k\sim\Lambda^k(TJ^rY)$. More precisely Ω^r_k is the *direct image* of the sheaf of (smooth) k-forms on J^rY by the projection $\pi^*_{r,0}$. Functions are considered as 0-forms. The spaces of strongly contact k-forms (i.e. k-contact k-forms) are denoted by

$$\Omega^r_{k,c}=\ker \mathrm{p}_{k-1},$$

where the mapping p_{k-1} assigning to a k-form its $(k-1)$-contact component has the meaning of the morphism of sheaves induced by this mapping. (Recall that the mapping p_0 is the same as the horizontalization mapping.) Denote by $\mathrm{d}\Omega^r_{k-1,c}$ the space of exterior derivatives of strongly contact $(k-1)$-forms, i.e. the *image sheaf* of $\Omega^r_{k-1,c}$ by the exterior derivative operator d, and

$$\Theta^r_k = \Omega^r_{k,c} + \mathrm{d}\Omega^r_{k-1,c}.$$

Consider the sequence of mappings

$$\{0\} \to \Theta^r_1 \to \Theta^r_2 \to \ldots \to \Theta^r_{mr+1} \to \{0\}, \tag{7.7}$$

where $k_{\max} = mr + 1$ is the maximal order of nontrivial forms of the type (7.7). The arrows except the first one denote the exterior derivative. Let us study the properties of the sequence (7.7). We will prove the following proposition.

Proposition 7.2 *Let $W \subset Y$ be an open set. Denote $\Omega^r_0 W$ the ring of smooth functions defined on $W_r = \pi^{-1}_{r,0}(W)$, $\Omega^r_k W$ the $\Omega^r_0 W$-module of (smooth) k-forms defined on W_r and $\Theta^r_k W = \Omega^r_{k,c} W + \mathrm{d}\Omega^r_{k-1,c} W$. Let $\eta \in \Theta^r_k W$, $r = 1, 2, 3$, be a k-form, $1 \le k \le r(m+1) = \dim J^r Y$. Then the decomposition $\eta = \eta_c + \mathrm{d}\chi_c$, where $\eta_c \in \Omega^r_{k,c} W$ is the strongly contact k-form and $\chi_c \in \Omega^r_{k-1,c} W$ is the strongly contact $(k-1)$-form is unique.*

Proof The proof will be made in fibred charts. Because we will study primarily the first-order variational sequence, we prove the proposition for $r = 1$. The proof for higher order situation is quite analogous. Suppose that there is an other decompositions of the form η, $\eta = \bar{\eta}_c + \mathrm{d}\bar{\chi}$. This means that

$$(\eta - \bar{\eta}_c) + \mathrm{d}(\chi_c - \bar{\chi}_c) = 0 \implies \mathrm{d}(\eta_c - \bar{\eta}_c) = 0.$$

The form $\varrho_c = \eta_c - \bar{\eta}_c$ is closed strongly contact k-form, i.e. k-contact k-form. Its chart expression then reads

$$\pi^*_{2,1}\varrho_c = \sum_{b=0}^{k} B^b_{\sigma_1\ldots\sigma_b\sigma_{b+1}\ldots\sigma_k}\, \omega^{\sigma_1} \wedge \ldots \wedge \omega^{\sigma_b} \wedge \dot{\omega}^{\sigma_{b+1}} \wedge \ldots \wedge \dot{\omega}^{\sigma_k},$$

where coefficients $B^b_{\sigma_1\ldots\sigma_b\sigma_{b+1}\ldots\sigma_k}$ are projectable on J^1Y and antisymmetric in indices $\{\sigma_1, \ldots, \sigma_b\}$ and also in indices $\{\sigma_{b+1}, \ldots, \sigma_k\}$. Calculating $\mathrm{d}\varrho_c$, rearranging the order of factors in wedge products, and taking into account these antisymmetry relations we obtain

$$\pi^*_{3,2}\mathrm{d}\pi^*_{2,1}\varrho_c = \sum_{b=0}^{k} \frac{\mathrm{d}B^b_{\sigma_1\ldots\sigma_b\sigma_{b+1}\ldots\sigma_k}}{\mathrm{d}t}\, \mathrm{d}t \wedge \omega^{\sigma_1} \wedge \ldots \wedge \omega^{\sigma_b} \wedge \dot{\omega}^{\sigma_{b+1}} \wedge \ldots \wedge \dot{\omega}^{\sigma_k}$$

$$+ \sum_{b=0}^{k} \frac{\partial B^b_{\sigma_1\ldots\sigma_b\sigma_{b+1}\ldots\sigma_k}}{\partial q^\sigma_0}\, \omega^{\sigma_0} \wedge \omega^{\sigma_1} \wedge \ldots \wedge \omega^{\sigma_b} \wedge \dot{\omega}^{\sigma_{b+1}} \wedge \ldots \wedge \dot{\omega}^{\sigma_k}$$

$$+\sum_{b=0}^{k}\frac{\partial B^b_{\sigma_1\ldots\sigma_b\sigma_{b+1}\ldots\sigma_k}}{\partial\dot{q}^\sigma_0}\,\dot{\omega}^{\sigma_0}\wedge\omega^{\sigma_1}\wedge\ldots\wedge\omega^{\sigma_b}\wedge\dot{\omega}^{\sigma_{b+1}}\wedge\ldots\wedge\dot{\omega}^{\sigma_k}$$

$$+(-1)^k\sum_{b=0}^{k} b B^b_{\sigma_1\ldots\sigma_{b-1}\{\sigma_b\ldots\sigma_k\}}\,\omega^{\sigma_1}\wedge\ldots\wedge\omega^{\sigma_{b-1}}\wedge\dot{\omega}^{\sigma_b}\wedge\dot{\omega}^{\sigma_{b+1}}\wedge\ldots\wedge\dot{\omega}^{\sigma_k}\wedge \mathrm{d}t$$

$$+(-1)^k\sum_{b=0}^{k}(k-b)B^b_{\sigma_1\ldots\sigma_b\sigma_{b+1}\ldots\sigma_k}\omega^\sigma_1\wedge\ldots\wedge\omega^{\sigma_b}\wedge\dot{\omega}^{\sigma_{b+1}}\wedge\ldots\wedge\dot{\omega}^{\sigma_{k-1}}\wedge\ddot{\omega}^{\sigma_k}\wedge \mathrm{d}t.$$

This form vanishes if all coefficients vanish. Especially considering the last term containing $\ddot{\omega}^{\sigma_k}$ together with the antisymmetric properties of coefficients we finally obtain $B^b_{\sigma_1\ldots\sigma_b\sigma_{b+1}\ldots\sigma_k}=0$ and thus $\varrho_c=0$ and $\eta_c-\bar{\eta}_c=0$. Then $\mathrm{d}(\chi_c-\bar{\chi}_c)=0$. The analogous procedure applied to this condition leads to the conclusion $\chi_c-\bar{\chi}_c=0$. This finishes the proof. (General considerations concerning strongly contact forms see in [15].) □

Proposition 7.2 ensures that the sequence (7.7) is exact. (Recall that for the exact sequence of mappings $f_1, f_2, \ldots$ it holds $\operatorname{Im} f_{i-1}\subset \operatorname{Ker} f_i, i=2,3,\ldots$) Especially, it is an *exact subsequence* of the *de Rham sequence* of forms

$$\{0\}\to\mathbf{R}\to\Omega^r_0\to\Omega^r_1\to\Omega^r_2\to\ldots\to\Omega^r_{(r+1)m+1}\to\{0\}. \tag{7.8}$$

Arrows in (7.8) represent the *exterior derivative operator* $\mathrm{d}_k:\Omega^r_k\ni\eta\to\mathrm{d}\eta\in\Omega^r_{k+1}$, $k=0,1,\ldots,r(m+1)+1$, with the exception of first two of them. The first arrow denotes the inclusion $\{0\}\ni 0\to 0\in\mathbf{R}$ and the second one assigns to every real number the corresponding constant function (0-form). The exactness of the sequence follows from the fact that every exact form is also closed, i.e. $\mathrm{d}_k(\mathrm{d}_{k-1}\eta)=0$ (indices of the operators d are in fact redundant, they serve only for identification of the concrete arrow). This means that $\operatorname{Im}\mathrm{d}_{k-1}\subseteq\operatorname{Ker}\mathrm{d}_k$. The opposite inclusion need not be fulfilled because it depends on the underlying fibred manifold as the definition domain (a closed form defined on certain domain need not be necessarily exact).

Denote Ω^r_k/Θ^r_k the quotient of Ω^r_k with respect to Θ^r_k. We obtain the *short exact sequence*

$$\{0\}\to\Theta^r_k\to\Omega^r_k\to\Omega^r_k/\Theta^r_k\to\{0\}. \tag{7.9}$$

Arrows in this sequence denote the following mappings of groups of forms

$$\begin{aligned}
f_{1,k}&:\{0\}\ni 0\longrightarrow f_1(0)=0_{\Theta^r_k},\ \operatorname{Ker} f_{1,k}=\{0\},\ \operatorname{Im} f_{1,k}=0_{\Theta^r_k},\\
f_{2,k}&:\Theta^r_k\ni\eta\longrightarrow f_{2,k}(\eta)=\eta\in\Omega^r_k,\ \operatorname{Ker} f_{2,k}=0_{\Theta^r_k}=0_{\Omega^r_k},\ \operatorname{Im} f_{2,k}=0_{\Omega^r_k},\\
f_{3,k}&:\Omega^r_k\ni\eta\longrightarrow f_{3,k}([\eta])\in\Omega^r_k/\Theta^r_k,\ \operatorname{Ker} f_{3,k}=\Theta^r_k,\ \operatorname{Im} f_{3,k}=0_{\Omega^r_k/\Theta^r_k},\\
f_{4,k}&:\Omega^r_k/\Theta^r_k\ni[\eta]\longrightarrow f_{4,k}([\eta])=0\in\{0\},\ \operatorname{Ker} f_{4,k}=\Omega^r_k/\Theta^r_k.
\end{aligned}$$

The exactness of this sequence is evident: $\operatorname{Im} f_{\alpha,k}\subseteq\operatorname{Ker} f_{\alpha+1,k}$, $\alpha=1,2,3$.

Definition 7.2 Variational sequence. Define mappings

$$E_k^r : \Omega_k^r/\Theta_k^r \ni [\eta] \to E_k^r([\eta]) = [\mathrm{d}\eta] \in \Omega_{k+1}^r/\Theta_{k+1}^r,$$

where $[\eta]$ are classes of equivalent forms. The mappings E_1^r and E_2^r are called *Euler-Lagrange mapping* and *Helmholtz-Sonin mapping*. The quotient sequence for $r = 1$

$$\{0\} \to \Omega_0^1 \to \Omega_1^1/\Theta_1^r \to \ldots \to \Omega_{m+1}^1/\Theta_{m+1}^r \to \Omega_{m+2}^1 \to \ldots \to \Omega_{2m+1}^1 \to \{0\}, \tag{7.10}$$

where (for $r = 1$) arrows for $k = 0, \ldots, 2m + 1$ represent the mappings E_k^1, is called the *first-order variational sequence*.

The complete scheme of the variational sequence is presented on Fig. 7.1. (Putting $r = 1$ we obtain the scheme for the first-order variational sequence.) The following (locally valid) proposition characterizes the relations between variational sequences of different orders.

Proposition 7.3 *Let W be an open set. Let $\eta \in \Theta_k^{r+1}W$, $r = 1, 2$, be a $\pi_{r+1,r}$-projectable form, i.e. $\eta = \pi_{r+1,r}^*\varrho$, where $\varrho \in \Omega_k^r W$. Then ϱ is an element of $\Theta_k^r W$.*

Proof We prove the proposition for $r = 1$. For $r = 2$ (and higher orders in general) the proof is analogous. If the form $\eta \in \Theta_k^2 W$ is $\pi_{2,1}^*$-projectable, then there exists a form $\varrho \in \Omega_k^1 W$ such that $\eta = \pi_{2,1}^*\varrho = \eta_c + \mathrm{d}\chi_c$, $\eta_c \in \Omega_{k,c}^2 W$, $\chi_c \in \Omega_{k-1,c}^2 W$. This decomposition is unique by Proposition 7.2. We prove that both forms η_c and χ_c are $\pi_{2,1}$-projectable individually, i.e. $\eta_c = \pi_{2,1}^*\varrho_c$, $\chi_c = \pi_{2,1}^*\mu_c$, where $\varrho_c \in \Omega_{k,c}^1 W$ and $\mu_c \in \Omega_{k-1,c}^1 W$. The exterior derivative of the equation

$$\pi_{2,1}^*\varrho = \eta = \eta_c + \mathrm{d}\chi_c$$

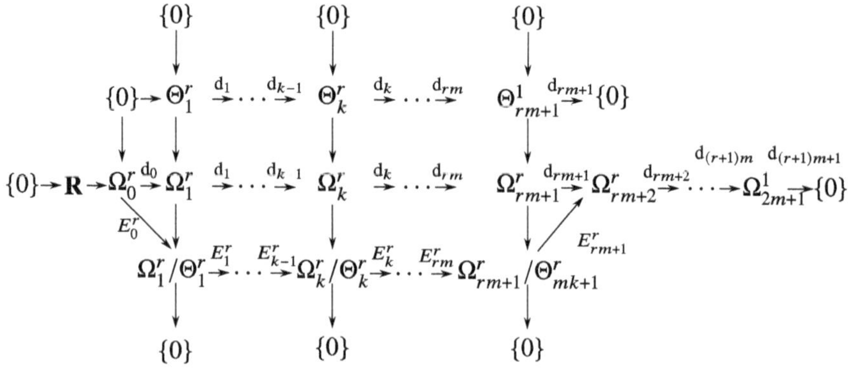

Fig. 7.1 Scheme of the variational sequence

gives $\mathrm{d}\eta_c = \mathrm{d}\pi^*_{2,1}\varrho = \pi^*_{2,1}\,\mathrm{d}\varrho$. This means that the exterior derivative of the k-contact k-form η_c is projectable onto J^1Y. The chart expression of the k-contact k-form η_c on J^2Y is

$$\eta_c = \sum_{b=0}^{k} B^b_{\sigma_1\ldots\sigma_b\sigma_{b+1}\ldots\sigma_k}\,\omega^\sigma_1 \wedge \ldots \omega^{\sigma_b} \wedge \dot\omega^{\sigma_{b+1}} \wedge \ldots \wedge \dot\omega^{\sigma_k},$$

where the coefficients $B^b_{\sigma_1\ldots\sigma_b\sigma_{b+1}\ldots\sigma_k}$ are functions on J^2Y (or, more precisely on $W_2 = \pi^{-1}_{2,1}(W)$) antisymmetric in indices $\{\sigma_1,\ldots,\sigma_b\}$ and in indices $\{\sigma_{b+1},\ldots,\sigma_k\}$. We calculate $\mathrm{d}\eta_c$ with the use of these antisymmetry properties similarly as in Proposition 7.2. However, we must take into account that the coefficients are functions on W_2. We obtain

$$\begin{aligned}
\pi^*_{3,2}\mathrm{d}\eta_c = {} & \sum_{b=0}^{k} \frac{\mathrm{d}B^b_{\sigma_1\ldots\sigma_b\sigma_{b+1}\ldots\sigma_k}}{\mathrm{d}t}\,\mathrm{d}t \wedge \omega^{\sigma_1} \wedge \ldots \wedge \omega^{\sigma_b} \wedge \dot\omega^{\sigma_{b+1}} \wedge \ldots \wedge \dot\omega^{\sigma_k} \\
& + \sum_{b=0}^{k} \frac{\partial B^b_{\sigma_1\ldots\sigma_b\sigma_{b+1}\ldots\sigma_k}}{\partial q^\sigma_0}\,\omega^{\sigma_0} \wedge \omega^{\sigma_1} \wedge \ldots \wedge \omega^{\sigma_b} \wedge \dot\omega^{\sigma_{b+1}} \wedge \ldots \wedge \dot\omega^{\sigma_k} \\
& + \sum_{b=0}^{k} \frac{\partial B^b_{\sigma_1\ldots\sigma_b\sigma_{b+1}\ldots\sigma_k}}{\partial \dot q^\sigma_0}\,\dot\omega^{\sigma_0} \wedge \omega^{\sigma_1} \wedge \ldots \wedge \omega^{\sigma_b} \wedge \dot\omega^{\sigma_{b+1}} \wedge \ldots \wedge \dot\omega^{\sigma_k} \\
& + \sum_{b=0}^{k} \frac{\partial B^b_{\sigma_1\ldots\sigma_b\sigma_{b+1}\ldots\sigma_k}}{\partial \ddot q^\sigma_0}\,\ddot\omega^{\sigma_0} \wedge \omega^{\sigma_1} \wedge \ldots \wedge \omega^{\sigma_b} \wedge \dot\omega^{\sigma_{b+1}} \wedge \ldots \wedge \dot\omega^{\sigma_k} \\
& + (-1)^k \sum_{b=0}^{k} bB^b_{\sigma_1\ldots\sigma_{b-1}\{\sigma_b\ldots\sigma_k\}}\,\omega^{\sigma_1} \wedge \ldots \wedge \omega^{\sigma_{b-1}} \wedge \dot\omega^{\sigma_b} \wedge \dot\omega^{\sigma_{b+1}} \wedge \ldots \wedge \dot\omega^\sigma_k \wedge \mathrm{d}t \\
& + (-1)^k \sum_{b=0}^{k} (k-b)B^b_{\sigma_1\ldots\sigma_b\sigma_{b+1}\ldots\sigma_k}\,\omega^\sigma_1 \wedge \ldots \wedge \omega^{\sigma_b} \wedge \dot\omega^{\sigma_{b+1}} \wedge \ldots \wedge \dot\omega^{\sigma_{k-1}} \wedge \ddot\omega^{\sigma_k} \wedge \mathrm{d}t.
\end{aligned}$$

Because of the projectability of this form onto J^1Y all components containing $\ddot\omega^\nu$ must vanish. This means that

$$B^b_{\sigma_1\ldots\sigma_b\sigma_{b+1}\ldots\sigma_k} = 0 \quad \text{for} \quad 1 \le b \le k-1, \quad \text{and} \quad \frac{\partial B^b_{\sigma_1\ldots\sigma_k}}{\partial \ddot q^{\sigma_0}} = 0.$$

Finally we obtain

$$\eta_c = \pi^*_{2,1}\left(B_{\sigma_1\ldots\sigma_k}\,\omega^{\sigma_1} \wedge \ldots \wedge \omega^{\sigma_k}\right),$$

with coefficients $B_{\sigma_1\ldots\sigma_k}$ antisymmetric in all indices and defined on W_1. So we have proved that the form η_c is projectable on J^1Y. This implies that also $\mathrm{d}\chi_c$ is projectable on J^1Y. The same procedure as for η_c leads to the conclusion that also the form χ_c is projectable on J^1Y. This finishes the proof. □

$$\begin{array}{ccccccccc}
\{0\} & \to & \Theta_k^r & \to & \Theta_k^{r+1} & \to & \Theta_k^{r+1}/\Theta_k^r & \to & \{0\} \\
 & & \downarrow & & \downarrow & & \downarrow & & \\
\{0\} & \to & \Omega_k^r & \to & \Omega_k^{r+1} & \to & \Omega_k^{r+1}/\Omega_k^r & \to & \{0\}
\end{array}$$

$$\begin{array}{ccccccccc}
\{0\} & \to & \Theta_k^{r+1} & \to & \Omega_k^{r+1} & \to & \Omega_k^{r+1}/\Theta_k^{r+1} & \to & \{0\} \\
 & & \uparrow & & \uparrow & & \uparrow & & \\
\{0\} & \to & \Theta_k^r & \to & \Omega_k^r & \to & \Omega_k^r/\Theta_k^r & \to & \{0\}
\end{array}$$

Fig. 7.2 Injectivity of the quotient mappings

Consider two diagrams on Fig. 7.2 (for $r = 1, 2$ for our purposes). In the first of them the first two vertical arrows denote the inclusions of sets and the last one represents the quotient mapping. Second horizontal arrows (from the left) represent the immersions by pullbacks $\pi^*_{r+1,r}$. The meaning of remaining horizontal arrows is clear. The just proved Proposition 7.3 says that the quotient mapping $\Theta_k^{r+1}/\Theta_k^r \to \Omega_k^{r+1}/\Omega_k^r$ is injective. In the second diagram the horizontal lines represent the short exact sequences (7.9). The first two vertical arrows denote the immersions by pullbacks $\pi^*_{r+1,r}$, the third one defines the quotient mapping

$$Q_k^{r+1,r} : \Omega_k^r/\Theta_k^r \ni [\eta] \longrightarrow [\pi^*_{r+1,r}\eta] \in \Omega_k^{r+1}/\Theta_k^{r+1}.$$

The mapping $Q_k^{r+1,r}$, $r = 1, 2$, is injective by Proposition 7.3.

$$Q_k^{s,r} : \Omega_k^r/\Theta_k^r \ni [\eta] \longrightarrow [\pi^*_{s,r}\eta] \in \Omega_k^s/\Theta_k^s, \quad s > r.$$

These mappings are injective as well. Proposition 7.3 ensures also the exactness of the sequence

$$\{0\} \longrightarrow \Omega_k^r/\Theta_k^r \longrightarrow \Omega_k^{r+1}/\Theta_k^{r+1} \longrightarrow (\Omega_k^{r+1}/\Theta_k^{r+1})/(\Omega_k^r/\Theta_k^r).$$

Recall that the mappings E_1^r and E_2^r in the variational sequence are called the *Euler-Lagrange* and *Helmholtz-Sonin* mapping, respectively. We will study their properties in detail in following sections.

7.3 Representation of the Variational Sequence by Forms

Let us give a brief characterization of the representation problem. We wish to represent every class $[\eta]$ of differential forms in the first-order variational sequence by an appropriately chosen differential form. We obtain a *representative* $\mathcal{I}(\eta)$ by applying a specific operator $\mathcal{I}$ called the *interior Euler operator* to an arbitrary element of the class $[\eta]$. By this operator the *representation mapping* assigns to every class of forms its specific representative. It is possible that such a representative will be defined on higher jet prolongation of the underlying manifold $J^s Y$, closely related to the starting order r of the variational sequence (in our presentation $r = 1$ and $1 \leq s \leq 3$). There are the following specific but very natural requirements for the properties of the representation.

- **R**-linearity: Linear combinations of classes must correspond to linear combinations of their representatives.
- Invariance: The representative of a class is to be a differential form, i.e. it must obey the corresponding transformation rules.
- Exactness: The sequence of mappings of spaces of representatives induced by the representation mappings is required to be exact as is the starting first-order variational sequence. This will have important practical consequences in calculus of variations, especially for solving the trivial variational problem and the inverse variational problem.
- Projection property: The mapping assigning to a class of forms its representative should be a projection, i.e. $\mathcal{I} \circ \mathcal{I} = \mathcal{I}$ up to a possible jet projection.

The general representation problem (for rth order calculus of variations in the field theory) as a partial problem of the variational sequence theory was completely solved [12].

Let W be an open subset of Y. We present the (local) construction of the representation sequence for the first-order variational sequence, i.e. $r = 1$. It will be given by mappings assigning to every class $[\eta] \in \Omega_k^1 W/\Theta_k^1 W$ a k-form $R_k(\eta) \in \Omega_k^s W$, where $s \geq 1$ in general,

$$R_k : \Omega_k^1 W/\Theta_k^1 W \ni [\eta] \longrightarrow R_k\big([\eta]\big) \in \Omega_k^s W$$

satisfying the above specified requirements. Because for $k = 1$ it holds $\Theta_1^1 W = \Omega_{1,c}^1$ ($\mathrm{d}\Omega_{0,c}^1$ contains only the zero 0-form) we define R_1 with help of the horizontalization mapping, i.e.

$$R_1 : \Omega_1^1 W/\Theta_1^1 W \ni [\eta] \longrightarrow R_1\big([\eta]\big) = h\eta \in \Omega_1^2 W.$$

For strongly contact k-forms (i.e. with $1 < k \leq m + 1$) the representation mapping is constructed by different ways leading to the interior Euler operator. For $m + 1 < k \leq 2m + 1$ the module $\Theta_k^1 W$ contains only the trivial (zero) form. For these forms the representation mapping is defined as

$$R_k : \Omega_k^1 W/\Theta_k^1 W \ni [\eta] \longrightarrow R_k\big([\eta]\big) = \eta \in \Omega_k^1 W.$$

Let us construct the interior Euler operator. Let $\eta \in \Omega_k^1 W$, $1 < k \leq m + 1$. Then $\pi_{2,1}^* \eta = p_{k-1}\eta + p_k\eta$. Because $p_k\eta \in \Omega_{k,c}^2 W \subset \Theta_k^2 W$ the operator will act on $p_{k-1}\eta$. This is the $(k-1)$-contact k-form, i.e. it can be locally (not uniquely, of course) expressed as

$$p_{k-1}\eta = \omega^\sigma \wedge \varrho_\sigma + \dot{\omega}^\sigma \wedge \tilde{\varrho}_\sigma,$$

where ϱ_σ and $\tilde{\varrho}_\sigma$ are some $(k-2)$-contact $(k-1)$-forms on J^2Y in general. If we express forms $\dot{\omega}^\sigma$ using the total derivative operator (7.1) (see Example 7.2) as $\dot{\omega}^\sigma = \mathrm{d}_t\omega^\sigma$ and using the Leibniz rule (7.4) we can write

$$\pi^*_{3,2}p_{k-1}\eta = \omega^\sigma \wedge [\pi^*_{3,2}\varrho_\sigma - \mathrm{d}_t\tilde{\varrho}_\sigma] + \mathrm{d}_t(\omega^\sigma \wedge \tilde{\varrho}_\sigma).$$

Here $\omega^\sigma \wedge \tilde{\varrho}_\sigma$ is the $(k-1)$-contact k-form. Thus it can be written as $\omega^\sigma \wedge \tilde{\varrho}_\sigma = \chi \wedge \mathrm{d}t$, where χ is a $(k-1)$-contact $(k-1)$-form. Taking into account relation (7.5) in Proposition 7.1 we can write

$$\mathrm{d}_t(\omega^\sigma \wedge \tilde{\varrho}_\sigma) = \mathrm{d}_t(\chi \wedge \mathrm{d}t) = \mathrm{d}_t\chi \wedge \mathrm{d}t = (-1)^{k-1}\,\mathrm{d}_h\chi = (-1)^{k-1}p_{k-1}\,\mathrm{d}p_{k-1}\chi.$$

We obtained the local decomposition (up to a projection)

$$p_{k-1}\eta = \omega^\sigma \wedge [\varrho_\sigma - \mathrm{d}_t\tilde{\varrho}_\sigma] + p_{k-1}\,\mathrm{d}p_{k-1}\chi,$$

more precisely

$$\pi^*_{3,2}p_{k-1}\eta = \omega^\sigma \wedge [\pi^*_{3,2}\varrho_\sigma - \mathrm{d}_t\tilde{\varrho}_\sigma] + p_{k-1}\,\mathrm{d}p_{k-1}\chi,$$

where χ is a $(k-1)$-contact $(k-1)$-form on J^2Y in general. Keeping in mind that this decomposition is defined on $J^3Y = J^{2r+1}Y$ we can use the above simplified expression.

Proposition 7.4 *Let W be an open subset of Y. Let $\eta \in \Omega^1_k W$, $1 < k \leq m+1$, be a k-form. Consider the decomposition*

$$p_{k-1}\eta = I(\eta) + p_{k-1}\,\mathrm{d}p_{k-1}\chi = \omega^\sigma \wedge [\varrho_\sigma - \mathrm{d}_t\tilde{\varrho}_\sigma] + p_{k-1}\,\mathrm{d}p_{k-1}\chi. \qquad (7.11)$$

There exists the unique decomposition of this type such that the mapping I is $\mathbf{R}-$*linear. The corresponding decomposition is defined globally.*

Proof We will prove the above mentioned assertions successively.

(a) For a form $\eta \in \Omega^1_k W$ it obviously holds

$$p_{k-1}\eta = \frac{1}{k-1}\left(\omega^\sigma \wedge i_{\partial/\partial q^\sigma}\,p_{k-1}\eta + \dot{\omega}^\sigma \wedge i_{\partial/\partial \dot{q}^\sigma}\,p_{k-1}\eta\right).$$

This represents the special choice of forms ϱ_σ, $\tilde{\varrho}_\sigma$,

$$\varrho_\sigma = \frac{1}{k-1}\,i_{\partial/\partial q^\sigma}\,p_{k-1}\eta, \quad \tilde{\varrho}_\sigma = \frac{1}{k-1}\,i_{\partial/\partial \dot{q}^\sigma}\,p_{k-1}\eta.$$

For such a choice the operator I has the form

$$I : \Omega^1_k W \ni \eta \longrightarrow I(\eta) \in \Omega^3_k W,$$

$$\mathcal{I}(\eta) = \frac{1}{k-1}\,\omega^\sigma \wedge \left[i_{\partial/\partial q^\sigma}\, p_{k-1}\eta - \mathrm{d}_t \left(i_{\partial/\partial \dot{q}^\sigma}\, p_{k-1}\eta \right) \right]. \tag{7.12}$$

It is evident that this operator is **R**-linear.

(b) We prove the uniqueness of the decomposition (7.11) with the operator (7.12). The primary decomposition of $p_{k-1}\eta = \omega^\sigma \wedge \varrho_\sigma + \dot{\omega}^\sigma \wedge \tilde{\varrho}_\sigma$ is, of course, not unique. So, the mapping $\mathcal{I}$ in the decomposition (7.11) is not unique in general. Suppose that there are two decompositions

$$p_{k-1}\eta = \omega^\sigma \wedge \varrho_\sigma + \dot{\omega}^\sigma \wedge \tilde{\varrho}_\sigma, \quad p_{k-1}\eta = \omega^\sigma \wedge \varrho'_\sigma + \dot{\omega}^\sigma \wedge \tilde{\varrho}'_\sigma,$$

i.e.

$$\omega^\sigma \wedge \left(\varrho_\sigma - \varrho'_\sigma\right) + \dot{\omega}^\sigma \wedge \left(\tilde{\varrho}_\sigma - \tilde{\varrho}'_\sigma\right) = 0.$$

Because of the required **R**-linearity of the operator $\mathcal{I}$ it holds $\mathcal{I}(0_{\Omega^1_k W}) = 0_{\Omega^3_k W}$. This means that

$$\omega^\sigma \wedge \left[\left(\varrho_\sigma - \mathrm{d}_t\tilde{\varrho}_\sigma\right) - \left(\varrho'_\sigma - \mathrm{d}_t\tilde{\varrho}'_\sigma\right) \right] = 0.$$

This proves the uniqueness of the **R**-*linear* operator $\mathcal{I}$.

(c) Using the partition of unity we can construct a global **R**-linear operator $\mathcal{I}$. Because this operator is unique due to **R**-linearity then different partitions of unity lead to the same decomposition.

Denote the form χ in the decomposition (7.11) as $\mathcal{R}(\eta)$. The mapping $\mathcal{R} : \eta \to \mathcal{R}(\eta)$ need not be unique. This means that it need not be global. □

Definition 7.3 Interior Euler operator. The mapping

$$\mathcal{I} : \Omega^1_k W \ni \eta \longrightarrow \mathcal{I}(\eta) \in \Omega^3_k W,$$

$$\mathcal{I}(\eta) = \frac{1}{k-1}\,\omega^\sigma \wedge \left[i_{\partial/\partial q^\sigma}\, p_{k-1}\eta - \mathrm{d}_t \left(i_{\partial/\partial \dot{q}^\sigma}\, p_{k-1}\eta \right) \right]$$

is called the *interior Euler operator.*

Proposition 7.5 *The properties of the interior Euler operator $\mathcal{I}$. Let $\eta \in \Omega^1_k W$, $1 < k \le m+1$, $W \subset Y$ being an open set. Then it holds*

$$\pi^*_{3,1}\eta - \mathcal{I}(\eta) \in \Theta^3_k W, \tag{7.13}$$

$$\mathcal{I}\left(p_{k-1}\,\mathrm{d}p_{k-1}\,\mathcal{R}(\eta)\right) = 0, \tag{7.14}$$

$$\mathcal{I}^2(\eta) = \mathcal{I}(\eta), \tag{7.15}$$

$$\operatorname{Ker} \mathcal{I} = \Theta^1_k W.$$

Note that relation (7.15) is satisfied up to a projection.

Proof We prove the previous relations by direct calculations. Because of the uniqueness of the interior Euler operator we chose the arbitrary version of its expression.

(a) Calculate the difference $\pi^*_{3,1}\eta - \mathcal{I}(\eta)$ taking into account that $\mathcal{I}(\eta) = \mathcal{I}(p_{k-1}\eta)$ and using relation (7.11):

$$\pi^*_{3,1}\eta - \mathcal{I}(\eta) = \pi^*_{3,2}(p_{k-1}\eta + p_k\eta) - \mathcal{I}(p_{k-1}\eta) = p_{k-1}\,\mathrm{d}p_{k-1}\mathcal{R}(\eta) + \pi^*_{3,2}p_k\eta$$

$$\Rightarrow \pi^*_{3,1}\eta - \mathcal{I}(\eta) = \pi^*_{3,2}\,\mathrm{d}p_{k-1}\mathcal{R}(\eta) - p_k\,\mathrm{d}p_{k-1}\mathcal{R}(\eta) + \pi^*_{3,2}p_k\eta. \quad (7.16)$$

The right-hand side is evidently the element of $\Theta^3_k W$.

(b) Expressing the form $p_{k-1}\,\mathrm{d}p_{k-1}\mathcal{R}(\eta)$ from (7.11) we obtain

$$\begin{aligned} p_{k-1}\,\mathrm{d}p_{k-1}\mathcal{R}(\eta) &= p_{k-1}\eta - \mathcal{I}(\eta) = p_{k-1}\eta - \omega^\sigma \wedge (\varrho_\sigma - \mathrm{d}_t\tilde{\varrho}_\sigma) \\ &= (\omega^\sigma \wedge \varrho_\sigma + \dot{\omega}^\sigma \wedge \tilde{\varrho}_\sigma) - \omega^\sigma \wedge (\varrho_\sigma - \mathrm{d}_t\tilde{\varrho}_\sigma) = \omega^\sigma \wedge \mathrm{d}_t\tilde{\varrho}_\sigma + \dot{\omega}^\sigma \wedge \tilde{\varrho}_\sigma. \end{aligned}$$

Now we apply the operator $\mathcal{I}$ on both sides of the obtained equality:

$$\mathcal{I}\big(p_{k-1}\,\mathrm{d}p_{k-1}\mathcal{R}(\eta)\big) = \mathcal{I}\left(\omega^\sigma \wedge \mathrm{d}_t\tilde{\varrho_\sigma} + \dot{\omega}^\sigma \wedge \tilde{\varrho}_\sigma\right) = \omega^\sigma \wedge (\mathrm{d}_t\tilde{\varrho}_\sigma - \mathrm{d}_t\tilde{\varrho}_\sigma) = 0.$$

(c) For proving (7.15) we apply the operator $\mathcal{I}$ to relation (7.16). We use relation (7.14) and the fact that the action of the interior Euler operator on the k-contact component of a k-form vanishes by definition. Then it holds

$$\mathcal{I}\big(\pi^*_{3,1}\eta - \mathcal{I}(\eta)\big) = \mathcal{I}\big(\pi^*_{3,2}\,\mathrm{d}p_{k-1}\mathcal{R}(\eta)\big) - \mathcal{I}\big(p_k\,\mathrm{d}p_{k-1}\mathcal{R}(\eta)\big) + \mathcal{I}(\pi^*_{3,2}p_k\eta) = 0.$$

Note that the relation $\mathcal{I} = \mathcal{I}^2$ holds up to a projection because every application of the operator $\mathcal{I}$ raises the order of the image $\mathcal{I}(\eta)$ with respect to the initial order of the form η.

(d) Suppose that $\mathcal{I}(\eta) = 0$. Then property (7.13) gives $\pi^*_{3,1}\eta \in \Theta^3_k W$. Then $\eta \in \Theta^1_k W$ by Proposition 7.3. This means that $\Theta^1_k W \supset \operatorname{Ker}\mathcal{I}$. Now let $\eta \in \Theta^1_k W$. Then $\eta = \eta_c + \mathrm{d}\chi_c$, where $\eta_c \in \Omega^1_{k,c}W$ and $\chi_c \in \Omega^1_{k-1,c}W$. For the $(k-1)$-contact component of η it holds, because of the facts that $p_{k-1}\eta_c = 0$ and $\pi^*_{3,2}p_{k-1}\,\mathrm{d}\chi_c = p_{k-1}\,\mathrm{d}p_{k-1}\chi_c$,

$$\pi^*_{3,2}p_{k-1}\eta = \pi^*_{3,2}\,(p_{k-1}\eta_c + p_{k-1}(\mathrm{d}\chi_c) = p_{k-1}\,\mathrm{d}p_{k-1}\chi_c.$$

This means that

$$\mathcal{I}(\eta) = \mathcal{I}(p_{k-1}\,\mathrm{d}p_{k-1}\chi_c).$$

Due to the uniqueness of the decomposition (7.11) and the property (7.14) we obtain $\mathcal{I}(\eta) = 0$. So, $\Theta^1_k W \subset \operatorname{Ker}\mathcal{I}$. (Recall that the uniqueness of the decomposition (7.11) is related to $p_{k-1}\,\mathrm{d}p_{k-1}\chi_c$, not to χ_c itself.)

□

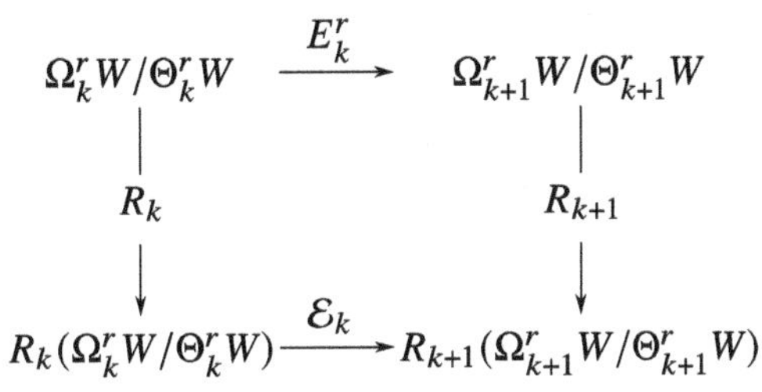

Fig. 7.3 Mappings of the representation sequence

Now we are ready to define the representation of the variational sequence by differential forms.

Definition 7.4 Representation of the variational sequence by forms. Consider the variational sequence of the first order. Let $W \subset Y$ be an open set. We define a family of mappings

$$R_k : \Omega_k^1 W/\Theta_k^1 W \ni [\eta] \longrightarrow R_k([\eta]) \in \Omega_k^s W, \quad 1 \le s \le 3, \tag{7.17}$$

$$\begin{aligned} R_k([\eta]) &= h\eta = p_0\eta \quad \text{for} \quad k = 0, 1, \quad s = 2, \\ R_k([\eta]) &= \mathcal{I}(\eta) \quad \text{for} \quad 2 \le k \le m+1, \quad s = 3, \\ R_k([\eta]) &= \eta \quad \text{for} \quad m+2 \le k \le 2m+1, \quad s = 1. \end{aligned} \tag{7.18}$$

In these relations η is an arbitrarily chosen element of the class $[\eta] \in \Omega_k^1 W/\Theta_k^1 W$. The sequence

$$\{0\} \to R_0(\Omega_0^1 W) \to R_1(\Omega_1^1 W/\Theta_1^1 W) \to \ldots \to R_{m+1}(\Omega_{m+1}^1 W/\Theta_{m+1}^1 W)$$

$$\to R_{m+2}(\Omega_{m+2}^1 W) \to \ldots \to R_{2m+1}(\Omega_{2m+1}^1 W) \to \{0\},$$

in which the arrows denote the mappings

$$\mathcal{E}_k : R_k(\Omega_k^1 W/\Theta_k^1 W) \ni R_k([\eta]) \longrightarrow \mathcal{E}_k\big(R_k([\eta])\big) \in R_{k+1}(\Omega_{k+1}^1 W/\Theta_{k+1}^1 W)$$

induced by the diagram on Fig. 7.3 by the requirement of its commutativity. (For our situations $r = 1, 2$.) This means that

$$\mathcal{E}_k \circ R_k([\eta]) = R_{k+1} \circ E_k^1([\eta]) = R_{k+1}([\mathrm{d}\eta]).$$

This sequence is called the *representation sequence* of the first-order variational sequence.

Note: For obtaining the formal uniformity of the order s in the previous definition we could write for $k = 0, 1$ $R_k([\eta]) = \pi_{3,2}^* h\eta$ instead of $h\eta$ and for $m + 2 \le k \le 2m + 1$ $R_k([\eta]) = \pi_{3,1}^* \eta$ instead of η.

Theorem 7.1 *The (first-order) representation sequence is exact.*

Proof The exactness of the representation sequence is the consequence of two properties. First of them is the fact that

$$\{0\} \longrightarrow \Theta^1_k W \longrightarrow \Omega^1_k W \longrightarrow R_k(\Omega^1_k W/\Theta^1_k W) \longrightarrow \{0\}$$

is the short exact sequence. This follows directly from the properties of the representation mappings (7.18).

The second property is the commutativity of the diagram on Fig. 7.4. □

Now we present the representatives of classes of 1-forms, 2-forms and 3-forms, because the special forms of these orders occur in various variational problems as Lagrangians (horizontal 1-forms), dynamical forms ($\pi_{2,0}$)-horizontal 2-forms) and exterior derivatives of dynamical forms (3-forms). Recall that the class $[\eta] \in \Omega^1_k W/\Theta^1_k W$ contains all k-forms which differ from η by a sum $\eta_c + \mathrm{d}\chi_c$, where η_c is a strongly contact k-form and χ_c is a strongly contact $(k-1)$-form.

Example 7.3 Let $\eta \in \Omega^1_1 W$ be a 1-form. Its chart expression in the basis of 1-forms adapted to the contact structure reads

$$\eta = A\,\mathrm{d}t + B_\sigma\,\omega^\sigma + C_\sigma\,\mathrm{d}\dot{q}^\sigma,$$

where coefficients A, B_σ and C_σ are functions on J^1Y. Then the representative of the class $[\eta]$ is

$$\Lambda_\eta = R_1\big([\eta]\big) = h\eta = (A + C_\sigma \ddot{q}^\sigma)\,\mathrm{d}t,$$

i.e. it is the second-order Lagrangian. Notice that the class $[\lambda]$ of a first-order Lagrangian λ is represented by the Lagrangian itself, i.e. up to a projection it holds, $\Lambda_\lambda = R_1([\lambda]) = \pi^*_{2,1}\lambda$. (The Lepage equivalent of the first-order Lagrangian is represented by the Lagrangian as well, because for such a case $C_\sigma = 0$, $1 \le \sigma \le m$.)

Example 7.4 Let $\eta \in \Omega^1_2 W$ be a 2-form. Let us express it again in the basis of 1-forms adapted to the contact structure. We have

$$\eta = A_\sigma\,\omega^\sigma \wedge \mathrm{d}t + B_{\sigma\nu}\,\omega^\sigma \wedge \mathrm{d}\dot{q}^\nu + C_\sigma\,\mathrm{d}\dot{q}^\sigma \wedge \mathrm{d}t + D_{\sigma\nu}\,\mathrm{d}\dot{q}^\sigma \wedge \mathrm{d}\dot{q}^\nu + F_{\sigma\nu}\,\omega^\sigma \wedge \omega^\nu,$$

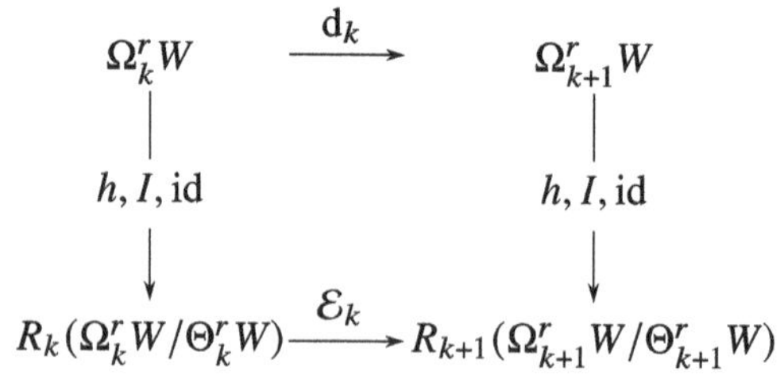

Fig. 7.4 Properties of the representative sequence

where coefficients A_σ, C_σ, $B_{\sigma\nu}$, $D_{\sigma\nu}$ and $F_{\sigma\nu}$ are functions on J^1Y, $D_{\sigma\nu}$ and $F_{\sigma\nu}$ being antisymmetric. The 1-contact component of this form is

$$p_1\eta = \left(A_\sigma + B_{\sigma\nu}\ddot{q}^\nu\right)\omega^\sigma \wedge \mathrm{d}t + C_\sigma\,\dot{\omega}^\sigma \wedge \mathrm{d}t + 2D_{\sigma\nu}\ddot{q}^\nu\,\dot{\omega}^\sigma \wedge \mathrm{d}t\,.$$

The representative of the class $[\eta]$ reads

$$\begin{aligned} E_\eta &= R_2\big([\eta]\big) = \omega^\sigma \wedge \left(i_{\partial/\partial q^\sigma}\,p_1\eta - \mathrm{d}_t\left(i_{\partial/\partial \dot{q}^\sigma}\,p_1\eta\right)\right) \\ &= \left(A_\sigma + \left(B_{\sigma\nu} - 2\frac{\mathrm{d}D_{\sigma\nu}}{\mathrm{d}t}\right)\ddot{q}^\nu - \frac{\mathrm{d}C_\sigma}{\mathrm{d}t} - 2D_{\sigma\nu}q^\sigma_{(3)}\right)\omega^\sigma \wedge \mathrm{d}t\,. \end{aligned}$$

As a special case let us consider a 2-form η for which $p_1\eta$ is a dynamical form, i.e. $p_1\eta = E = E_\sigma\,\omega^\sigma \wedge \mathrm{d}t$, where coefficients E_σ are defined on J^2Y and they are affine in variables $\ddot{q}^\nu$. Then we trivially obtain

$$E_\eta = R_2\big([\eta]\big) = R_2([E]) = E_\sigma\,\omega^\sigma \wedge \mathrm{d}t\,.$$

This means that the class $[\eta]$ containing such a form is represented by $p_1\eta$, or, up to a projection, the class containing a dynamical form is represented by this dynamical form itself. (Notice that the above equation is valid up to a projection, because $R_2\big([\eta]\big) \in \Omega^3_2 W$, but $p_1\eta = E \in \Omega^2_2 W$ and thus formally $R_2(p_1\eta) \in \Omega^5_2 W$.)

Example 7.5 Let $\eta \in \Omega^1_3 W$ be a 3-form. Its expression in the basis of 1-forms adapted to the contact structure is

$$\begin{aligned} \eta &= A_{\sigma\nu}\,\omega^\sigma \wedge \omega^\nu \wedge \mathrm{d}t + B_{\sigma\nu}\,\omega^\sigma \wedge \mathrm{d}\dot{q}^\nu \wedge \mathrm{d}t + C_{\sigma\nu}\,\mathrm{d}\dot{q}^\sigma \wedge \mathrm{d}\dot{q}^\nu \wedge \mathrm{d}t \\ &\quad + D_{\sigma\nu\lambda}\,\omega^\sigma \wedge \omega^\nu \wedge \omega^\lambda + E_{\sigma\nu\lambda}\omega^\sigma \wedge \omega^\nu \wedge \mathrm{d}\dot{q}^\lambda + F_{\sigma\nu\lambda}\omega^\sigma \wedge \mathrm{d}\dot{q}^\nu \wedge \mathrm{d}\dot{q}^\lambda \\ &\quad + G_{\sigma\nu\lambda}\,\mathrm{d}\dot{q}^\sigma \wedge \mathrm{d}\dot{q}^\nu \wedge \mathrm{d}\dot{q}^\lambda, \end{aligned}$$

the function coefficients being defined on J^1Y. Functions $A_{\sigma\nu}$, $C_{\sigma\nu}$ and $E_{\sigma\nu\lambda}$ are antisymmetric in indices σ and ν, coefficients $F_{\sigma\nu\lambda}$ are antisymmetric in ν and λ and coefficients $D_{\sigma\nu\lambda}$ and $G_{\sigma\nu\lambda}$ are antisymmetric in all three indices. The 2-contact component of the form η is

$$\begin{aligned} p_2\eta &= \left(A_{\sigma\nu} + E_{\sigma\nu\lambda}\ddot{q}^\lambda\right)\omega^\sigma \wedge \omega^\nu \wedge \mathrm{d}t + \left(B_{\sigma\nu} + 2F_{\sigma\nu\lambda}\ddot{q}^\lambda\right)\omega^\sigma \wedge \dot{\omega}^\nu \wedge \mathrm{d}t \\ &\quad + \left(C_{\sigma\nu} + 3G_{\sigma\nu\lambda}\ddot{q}^\lambda\right)\dot{\omega}^\sigma \wedge \dot{\omega}^\nu \wedge \mathrm{d}t\,. \end{aligned}$$

Applying the operator $\mathcal{I}$ on this expression we obtain the representative

$$\begin{aligned} H_\eta &= R_3\big([\eta]\big) \\ &= \left[\left(A_{\sigma\nu} + E_{\sigma\nu\lambda}\ddot{q}^\lambda\right) - \left\{\frac{1}{2}\frac{\mathrm{d}B_{\sigma\nu}}{\mathrm{d}t} + \frac{\mathrm{d}F_{\sigma\nu\lambda}}{\mathrm{d}t}\ddot{q}^\lambda + F_{\sigma\nu\lambda}q^\lambda_{(3)}\right\}_{\mathrm{alt}\,(\sigma,\nu)}\right]\omega^\sigma \wedge \omega^\nu \wedge \mathrm{d}t \\ &\quad - \left(\frac{\mathrm{d}C_{\sigma\nu}}{\mathrm{d}t} + 3\frac{\mathrm{d}G_{\sigma\nu\lambda}}{\mathrm{d}t}\ddot{q}^\lambda + 3G_{\sigma\nu\lambda}q^\lambda_{(3)}\right)\omega^\sigma \wedge \dot{\omega}^\nu \wedge \mathrm{d}t \\ &\quad + \left(C_{\sigma\nu} + 3G_{\sigma\nu\lambda}\ddot{q}^\lambda\right)\omega^\sigma \wedge \ddot{\omega}^\nu \wedge \mathrm{d}t\,. \end{aligned}$$

Summarizing the results of previous three examples we can see that representatives of classes of 1-forms are *Lagrangians*, representatives of classes of 2-forms are *dynamical forms*. Representatives of classes of 3-forms are called the *Helmholtz-Sonin* forms.

An important generalizing note: In this text we concentrate to first-order variational problems. For this reason we have introduced the concept of the variational sequence and the interior Euler operator on $J^1 Y$. On the other hand we have seen that the application of the interior Euler operator raises the order of forms. Even though we consider the first-order variational sequence as mentioned, there are some exceptional situations concerning its representation in which we need to apply the interior Euler operator to a special type of forms on $J^2 Y$ (concretely to dynamical forms $E = E_\sigma \, \omega^\sigma \wedge \mathrm{d}t$, $E_\sigma = E_\sigma(t, q^\nu, \dot{q}^\nu, \ddot{q}^\nu)$, which are $\pi_{2,0}$-horizontal and moreover affine in second-order variables). Therefore we present the higher order version of Propositions 7.4 and 7.5.

Proposition 7.6 *The rth order interior Euler operator. Let W be an open subset of Y. Let $\eta \in \Omega^r_k W$, $1 < k \leq m + 1$, be a form. Then there exists the unique decomposition*

$$p_{k-1}\eta = \mathcal{I}(\eta) + p_{k-1}\,\mathrm{d}p_{k-1}\mathcal{R}(\eta) = \left(\omega^\sigma \wedge \sum_{j=0}^{r} (-1)^j \, \mathrm{d}_t^j \, \varrho_j^\sigma\right) + p_{k-1}\,\mathrm{d}p_{k-1}\mathcal{R}(\eta),$$

such that the mapping $\mathcal{I}$ is **R***-linear. Here $\mathcal{R}(\eta) \in \Omega^r_{k-1,c} W$ and $\varrho_0, \ldots, \varrho_r$ are some $(k-1)$-forms. The mapping $\mathcal{I}$ (called the interior Euler operator) is given as*

$$\mathcal{I} : \Omega^r_k W \ni \eta \longrightarrow \mathcal{I}(\eta) \in \Omega^{2r+1}_k W,$$

$$\mathcal{I}(\eta) = \frac{1}{k-1} \omega^\sigma \wedge \sum_{j=0}^{r} (-1)^j \, \mathrm{d}_t^j \left(i_{\partial/\partial q_j^\sigma} \, p_{k-1}\eta\right).$$

and satisfies the relations

$$\begin{aligned} \pi^*_{2r+1,1}\eta - \mathcal{I}(\eta) &\in \Theta^{2r+1}_k W, \\ \mathcal{I}\big(p_{k-1}\,\mathrm{d}p_{k-1}\,\mathcal{R}(\eta)\big) &= 0, \\ \mathcal{I}^2(\eta) &= (\pi^{4r+3}_{2r+1})^* \mathcal{I}(\eta), \\ \operatorname{Ker} \mathcal{I} &= \Theta^r_k W. \end{aligned}$$

We do not present the proof of this proposition because it is completely analogous as for the first order, and its complete version can be found in [12]. Using the interior Euler operator we can define the representation sequence for the rth order variational sequence. The representation sequence is of the order $(2r + 1)$ and it is, of course, exact.

7.4 Variational Sequence in Calculus of Variations

The results of the previous section, especially the form of the representation sequence, indicate that there should be a correspondence of the first three mappings of the variational sequence E_0, E_1 and E_2 with the calculus of variations. It is also to be expected that there is the close correlation of the exactness of the variational sequence and its representation sequence with the solution of the trivial variational problem and the inverse variational problem. In this section, we will study the above mentioned three mappings and the analytic structure of its kernels using the exactness of the representative sequence. This gives a possibility to obtain the local solution of both the trivial and the inverse problems of calculus of variations. As an important parallel consequence of properties of the variational sequence we introduce the concept of *Lepage forms*.

Let $\eta \in \Omega_0^1 W$ be a 0-form, i.e. a function. The class of this 0-form is formed only by η itself, and $R_0\big([\eta]\big) = \eta$. Calculating the representative of $\mathrm{d}\eta$ we obtain

$$R_1\big([\mathrm{d}\eta]\big) = p_0 \mathrm{d}\eta = \frac{\mathrm{d}\eta}{\mathrm{d}t}\,\mathrm{d}t.$$

The mapping $\mathcal{E}_0$ in the representative sequence assigns to a first-order function η the second-order Lagrangian $\frac{\mathrm{d}\eta}{\mathrm{d}t}\,\mathrm{d}t$. These Lagrangians form the image of the mapping $\mathcal{E}_0$. The kernel of this mapping is formed by constant functions. We have

$$\operatorname{Ker}\mathcal{E}_0 = \{\eta \in \Omega_0^1 W \mid \eta(t, q^\sigma, \dot{q}^\sigma) = \text{const.}\}, \quad \operatorname{Im}\mathcal{E}_0 = \left\{\frac{\mathrm{d}\eta}{\mathrm{d}t}\,\mathrm{d}t\right\}.$$

The set $\Omega_0^1 W$ contains also functions $\pi_{1,0}^* f$, where $f \in \Omega_0^0 W$. Their representatives $h\,\mathrm{d}f \in \operatorname{Im}\mathcal{E}_0$ are first-order Lagrangians.

Let $\eta \in \Omega_1^1 W$ be a 1-form. Then in the basis of 1-forms adapted to the contact structure it holds

$$\begin{aligned}
\eta &= L\,\mathrm{d}t + B_\sigma \omega^\sigma + C_\sigma\,\mathrm{d}\dot{q}^\sigma, \\
p_1 \mathrm{d}\eta &= \left(\frac{\partial L}{\partial q^\sigma} - \frac{\mathrm{d}B_\sigma}{\mathrm{d}t} + \frac{\partial C_\nu}{\partial q^\sigma}\ddot{q}^\nu\right)\omega^\sigma \wedge \mathrm{d}t \\
&\quad + \left(\frac{\partial L}{\partial \dot{q}^\sigma} - B_\sigma - \frac{\mathrm{d}C_\sigma}{\mathrm{d}t} + \frac{\partial C_\nu}{\partial \dot{q}^\sigma}\ddot{q}^\nu\right)\dot{\omega}^\sigma \wedge \mathrm{d}t, \qquad (7.19) \\
i_{\partial/\partial q^\sigma}\, p_1\,\mathrm{d}\eta &= \left(\frac{\partial L}{\partial q^\sigma} - \frac{\mathrm{d}B_\sigma}{\mathrm{d}t} + \frac{\partial C_\nu}{\partial q^\sigma}\ddot{q}^\nu\right)\mathrm{d}t,
\end{aligned}$$

$$
\begin{aligned}
i_{\partial/\partial\dot{q}^\sigma}\, p_1\, \mathrm{d}\eta &= \left(\frac{\partial L}{\partial \dot{q}^\sigma} - B_\sigma - \frac{\mathrm{d}C_\sigma}{\mathrm{d}t} + \frac{\partial C_\nu}{\partial \dot{q}^\sigma}\ddot{q}^\nu\right)\mathrm{d}t,\\
\mathcal{I}(\mathrm{d}\eta) &= \left[\left(\frac{\partial L}{\partial q^\sigma} - \frac{\mathrm{d}}{\mathrm{d}t}\frac{\partial L}{\partial \dot{q}^\sigma}\right) + \left(\frac{\partial C_\nu}{\partial q^\sigma} - \frac{\mathrm{d}}{\mathrm{d}t}\frac{\partial C_\nu}{\partial \dot{q}^\sigma}\right)\ddot{q}^\nu\right.\\
&\quad \left. + \frac{\mathrm{d}^2 C_\sigma}{\mathrm{d}t^2} - \frac{\partial C_\nu}{\partial \dot{q}^\sigma} q^\nu_{(3)}\right]\omega^\sigma \wedge \mathrm{d}t\\
&= \left(\frac{\partial \mathcal{L}}{\partial q^\sigma} - \frac{\mathrm{d}}{\mathrm{d}t}\frac{\partial \mathcal{L}}{\partial \dot{q}^\sigma} + \frac{\mathrm{d}^2}{\mathrm{d}t^2}\frac{\partial \mathcal{L}}{\partial \ddot{q}^\sigma}\right)\omega^\sigma \wedge \mathrm{d}t,
\end{aligned}
\tag{7.20}
$$

where we denoted $\mathcal{L} = L + C_\nu \ddot{q}^\nu$. This is the Lagrange function of the special type second-order Lagrangian $\tilde{\lambda} = \mathcal{L}\,\mathrm{d}t$. Notice that this Lagrangian is equal to the horizontal component of the form η, i.e. it is the representative of the class $[\eta]$. The representative $\mathcal{I}(\mathrm{d}\eta)$ of the class $[\mathrm{d}\eta]$ is an element of $\Omega^3_1 W$. It is the Euler-Lagrange form of the Lagrangian $\tilde{\lambda} = \mathcal{L}\,\mathrm{d}t$. In a special case of a first-order Lagrangian $\lambda = L\,\mathrm{d}t$, $L \in \Omega^1_0 W$, i.e. for $C_\sigma = 0$, the representative obtains the form

$$
\mathcal{I}(\mathrm{d}\lambda) = \left(\frac{\partial L}{\partial q^\sigma} - \frac{\mathrm{d}}{\mathrm{d}t}\frac{\partial L}{\partial \dot{q}^\sigma}\right)\omega^\sigma \wedge \mathrm{d}t
$$

which is the Euler-Lagrange form of the Lagrangian λ. This means that $E_1([\lambda]) = [E_\lambda]$, or $R_2(\mathrm{d}\lambda) = E_\lambda$.

Finally, let $\eta \in \Omega^1_2 W$ be a 2-form. Its chart expression is

$$
\eta = A_\sigma\, \omega^\sigma \wedge \mathrm{d}t + B_{\sigma\nu}\, \omega^\sigma \wedge \mathrm{d}\dot{q}^\nu + F_{\sigma\nu}\, \omega^\sigma \wedge \omega^\nu + C_\sigma\, \mathrm{d}\dot{q}^\sigma \wedge \mathrm{d}t + D_{\sigma\nu}\, \mathrm{d}\dot{q}^\sigma \wedge \mathrm{d}\dot{q}^\nu,
\tag{7.21}
$$

where A_σ, $B_{\sigma\nu}$, C_σ, $F_{\sigma\nu}$ and $D_{\sigma\nu}$ are functions on $J^1 Y$. Functions $F_{\sigma\nu}$ and $D_{\sigma\nu}$ are antisymmetric in indices σ and ν. We calculate the representative as $\mathcal{I}(\eta)$ (see also Problem 7.2):

$$
\begin{aligned}
p_2\, \mathrm{d}\eta &= \left(\left\{\frac{\partial A_\nu}{\partial q^\sigma} + \frac{\partial B_{\nu\lambda}}{\partial q^\sigma}\ddot{q}^\lambda\right\}_{\mathrm{alt}\,(\sigma,\nu)} + \frac{\mathrm{d}F_{\sigma\nu}}{\mathrm{d}t}\right)\omega^\sigma \wedge \omega^\nu \wedge \mathrm{d}t\\
&\quad + \left(-\frac{\partial A_\sigma}{\partial \dot{q}^\nu} + \frac{\mathrm{d}B_{\sigma\nu}}{\mathrm{d}t} - \frac{\partial B_{\sigma\lambda}}{\partial \dot{q}^\nu}\ddot{q}^\lambda + 2F_{\sigma\nu} + \frac{\partial C_\nu}{\partial q^\sigma}\right.\\
&\quad \left. + 2\frac{\partial D_{\nu\lambda}}{\partial q^\sigma}\ddot{q}^\lambda\right)\omega^\sigma \wedge \dot{\omega}^\nu \wedge \mathrm{d}t\\
&\quad + \left(\left\{B_{\sigma\nu} + \frac{\partial C_\nu}{\partial \dot{q}^\sigma} + 2\frac{\partial D_{\nu\lambda}}{\partial \dot{q}^\sigma}\ddot{q}^\lambda\right\}_{\mathrm{alt}\,(\sigma,\nu)} + \frac{\mathrm{d}D_{\sigma\nu}}{\mathrm{d}t}\right)\dot{\omega}^\sigma \wedge \dot{\omega}^\nu \wedge \mathrm{d}t.
\end{aligned}
\tag{7.22}
$$

The representative of the class $[\mathrm{d}\eta]$ is

$$R_2([\mathrm{d}\eta]) = \mathcal{I}(\mathrm{d}\eta) = \frac{1}{2}\omega^\sigma \wedge \left[i_{\partial/\partial q^\sigma}\, p_2\, \mathrm{d}\eta - \mathrm{d}_t(i_{\partial/\partial \dot{q}^\sigma}\, p_2\, \mathrm{d}\eta)\right] \tag{7.23}$$

$$\begin{aligned}
&= \left\{\frac{\partial A_\nu}{\partial q^\sigma} + \frac{\partial B_{\nu\lambda}}{\partial q^\sigma}\ddot{q}^\lambda - \frac{1}{2}\frac{\mathrm{d}}{\mathrm{d}t}\left(\frac{\partial A_\nu}{\partial \dot{q}^\sigma} - \frac{\mathrm{d}B_{\nu\sigma}}{\mathrm{d}t} + \frac{\partial B_{\nu\lambda}}{\partial \dot{q}^\sigma}\ddot{q}^\lambda - \frac{\partial C_\sigma}{\partial q^\nu} - 2\frac{\partial D_{\sigma\lambda}}{\partial q^\nu}\ddot{q}^\lambda\right)\right\}_{\mathrm{alt}\,(\sigma,\nu)} \\
&\qquad \omega^\sigma \wedge \omega^\nu \wedge \mathrm{d}t \\
&+ \left(\left[-\frac{\partial A_\sigma}{\partial \dot{q}^\nu} + \frac{\mathrm{d}B_{\sigma\nu}}{\mathrm{d}t} - \frac{\partial B_{\sigma\lambda}}{\partial \dot{q}^\nu}\ddot{q}^\lambda + \frac{\partial C_\nu}{\partial q^\sigma} + 2\frac{\partial D_{\nu\lambda}}{\partial \dot{q}^\sigma}\ddot{q}^\lambda\right]_{\mathrm{sym}\,(\sigma,\nu)}\right. \\
&\left. - \frac{\mathrm{d}}{\mathrm{d}t}\left\{B_{\sigma\nu} + \frac{\partial C_\nu}{\partial \dot{q}^\sigma} + \frac{\mathrm{d}D_{\sigma\nu}}{\mathrm{d}t} + 2\frac{\partial D_{\nu\lambda}}{\partial \dot{q}^\sigma}\ddot{q}^\lambda\right\}_{\mathrm{alt}\,(\sigma,\nu)}\right)\omega^\sigma \wedge \dot{\omega}^\nu \wedge \mathrm{d}t \\
&- \left\{B_{\sigma\nu} + \frac{\partial C_\nu}{\partial \dot{q}^\sigma} + \frac{\mathrm{d}D_{\sigma\nu}}{\mathrm{d}t} + 2\frac{\partial D_{\nu\lambda}}{\partial \dot{q}^\sigma}\ddot{q}^\lambda\right\}_{\mathrm{alt}\,(\sigma,\nu)} \omega^\sigma \wedge \ddot{\omega}^\nu \wedge \mathrm{d}t.
\end{aligned} \tag{7.24}$$

As a special situation consider the form $\alpha \in \Omega_2^1 W$ for which $C_\sigma = 0$ and $D_{\sigma\nu} = 0$, $1 \le \sigma, \nu \le m$. This is the case of forms α generating the dynamical system of a dynamical form $E = E_\sigma\, \omega^\sigma \wedge \mathrm{d}t \in \Omega_2^2 W$. (Recall that the form E is $\pi_{2,0}$-horizontal.) The representative of the class $[\mathrm{d}\alpha]$ is

$$\begin{aligned}
R_2([\mathrm{d}\alpha]) &= \mathcal{I}(\mathrm{d}\alpha) \\
&= \left\{\frac{\partial A_\nu}{\partial q^\sigma} + \frac{\partial B_{\nu\lambda}}{\partial q^\sigma}\ddot{q}^\lambda - \frac{1}{2}\frac{\mathrm{d}}{\mathrm{d}t}\left(\frac{\partial A_\nu}{\partial \dot{q}^\sigma} - \frac{\mathrm{d}B_{\nu\sigma}}{\mathrm{d}t} + \frac{\partial B_{\nu\lambda}}{\partial \dot{q}^\sigma}\ddot{q}^\lambda\right)\right\}_{\mathrm{alt}\,(\sigma,\nu)} \omega^\sigma \wedge \omega^\nu \wedge \mathrm{d}t \\
&+ \left(\left[-\frac{\partial A_\sigma}{\partial \dot{q}^\nu} + \frac{\mathrm{d}B_{\sigma\nu}}{\mathrm{d}t} - \frac{\partial B_{\sigma\lambda}}{\partial \dot{q}^\nu}\ddot{q}^\lambda\right]_{\mathrm{sym}\,(\sigma,\nu)} - \left\{\frac{\mathrm{d}B_{\sigma\nu}}{\mathrm{d}t}\right\}_{\mathrm{alt}\,(\sigma,\nu)}\right)\omega^\sigma \wedge \dot{\omega}^\nu \wedge \mathrm{d}t \\
&- \{B_{\sigma\nu}\}_{\mathrm{alt}\,(\sigma,\nu)}\, \omega^\sigma \wedge \ddot{\omega}^\nu \wedge \mathrm{d}t.
\end{aligned}$$

We will discuss this expression. Let $E_\sigma\, \omega^\sigma \wedge \mathrm{d}t$ be a dynamical form. It is (locally) variational if and only if there exists a Lagrangian λ such that E is its Euler-Lagrange form, $E = E_\lambda$. (Recall that for a second-order variational dynamical form there always exists a first-order Lagrangian—see Sect. 5.4.) So, a variational dynamical form is an element of the image of the Euler-Lagrange mapping and simultaneously an element of the kernel of the Helmholtz-Sonin mapping (due to the exactness of the variational sequence), i.e. $E_2^1([\alpha]) = 0_{\Omega_3^1 W/\Theta_3^1 W}$. Thus, there must exist an element $\alpha \in [\alpha]$ such that $\varepsilon_2(\alpha) = \mathcal{I}(\mathrm{d}\alpha) = 0$. Putting all coefficients of the form $\mathcal{I}(\mathrm{d}\alpha)$ equal to zero we obtain Helmholtz conditions presented in Proposition 5.2 (Problem 7.7).

The schematic Fig. 7.5 illustrates the trivial variational problem and the inverse variational problem. In this figure only spaces of forms relevant for these problems are denoted for simplicity. The order increase is considered. Using the representative sequence we can directly derive relations (5.14), (5.15) and (5.16) for components E_σ of a (locally) variational dynamical form $E = E_\sigma\, \omega \wedge \mathrm{d}t \in \Omega_{2,Y}^2 W$. For a 2-form $\eta \in \Omega_2^1 W$ it holds $\mathcal{I}(\mathrm{d}\eta) = \mathcal{I}(p_2\, \mathrm{d}p_1\eta)$ (up to a formal projection, of course, $\pi_{7,3}^*\mathcal{I}(\mathrm{d}\eta) = \mathcal{I}(p_2\, \mathrm{d}p_1\eta)$) in agreement with Proposition 7.6 and Fig. 7.5. As the form

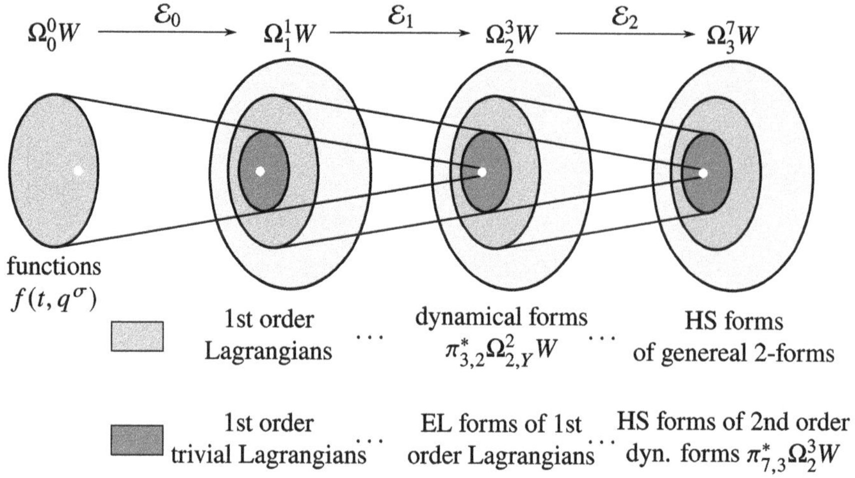

Fig. 7.5 Representation sequence in the calculus of variations

η we consider an arbitrary representative α of the corresponding dynamical systems, i.e. $p_1\alpha = E$. Then the representative $R_2\big([\mathrm{d}\alpha]\big)$ reads

$$R_2\left([\mathrm{d}\alpha]\right) = \mathcal{I}(p_2 \mathrm{d}E) = \frac{1}{2}\omega^\sigma \wedge \Big[i_{\partial/\partial q^\sigma}\, p_2\, \mathrm{d}E - \mathrm{d}_t(i_{\partial/\partial \dot{q}^\sigma}\, p_2\, \mathrm{d}E) + \mathrm{d}_t^2(i_{\partial/\partial \ddot{q}^\sigma}\, p_2\, \mathrm{d}E) \Big].$$

By successive calculations we obtain

$$p_2\, \mathrm{d}E = \{\frac{\partial E_\nu}{\partial q^\sigma}\}\omega^\sigma \wedge \omega^\nu \wedge \mathrm{d}t - \frac{\partial E_\sigma}{\partial \dot{q}^\nu}\omega^\sigma \wedge \dot{\omega}^\nu \wedge \mathrm{d}t - \frac{\partial E_\sigma}{\partial \ddot{q}^\nu}\omega^\sigma \wedge \ddot{\omega}^\nu \wedge \mathrm{d}t,$$

where the brackets $\{X_{\sigma\nu}\}$ denote the antisymmetrization of quantities $X_{\sigma\nu}$ in their two indices,

$$\begin{aligned}
i_{\partial/\partial q^\sigma}\, p_2\, \mathrm{d}E &= 2\{\frac{\partial E_\nu}{\partial q^\sigma}\}\, \omega^\nu \wedge \mathrm{d}t - \frac{\partial E_\sigma}{\partial \dot{q}^\nu}\, \dot{\omega}^\nu \wedge \mathrm{d}t - \frac{\partial E_\sigma}{\partial \ddot{q}^\nu}\, \ddot{\omega}^\nu \wedge \mathrm{d}t, \\
i_{\partial/\partial \dot{q}^\sigma}\, p_2\, \mathrm{d}E &= \frac{\partial E_\nu}{\partial \dot{q}^\sigma}\, \omega^\nu \wedge \mathrm{d}t, \\
i_{\partial/\partial \ddot{q}^\sigma}\, p_2\, \mathrm{d}E &= \frac{\partial E_\nu}{\partial \ddot{q}^\sigma}\, \omega^\nu \wedge \mathrm{d}t, \\
\mathrm{d}_t(i_{\partial/\partial \dot{q}^\sigma}\, p_2\, \mathrm{d}E) &= \left(\frac{\mathrm{d}}{\mathrm{d}t}\frac{\partial E_\nu}{\partial \dot{q}^\sigma}\right)\, \omega^\nu \wedge \mathrm{d}t + \frac{\partial E_\nu}{\partial \dot{q}^\sigma}\, \dot{\omega}^\nu \wedge \mathrm{d}t, \\
\mathrm{d}_t^2(i_{\partial/\partial \ddot{q}^\sigma}\, p_2\, \mathrm{d}E) &= \left(\frac{\mathrm{d}^2}{\mathrm{d}t^2}\frac{\partial E_\nu}{\partial \ddot{q}^\sigma}\right)\, \omega^\nu \wedge \mathrm{d}t + 2\left(\frac{\mathrm{d}}{\mathrm{d}t}\frac{\partial E_\nu}{\partial \ddot{q}^\sigma}\right)\, \dot{\omega}^\nu \wedge \mathrm{d}t + \frac{\partial E_\nu}{\partial \ddot{q}^\sigma}\, \ddot{\omega}^\nu \wedge \mathrm{d}t,
\end{aligned}$$

$$I(\mathrm{d}E) = \left(\left\{ \frac{\partial E_\nu}{\partial q^\sigma} \right\} - \frac{1}{2} \frac{\mathrm{d}}{\mathrm{d}t} \frac{\partial E_\nu}{\partial \dot{q}^\sigma} + \frac{1}{2} \frac{\mathrm{d}^2}{\mathrm{d}t^2} \frac{\partial E_\nu}{\partial \ddot{q}^\sigma} \right) \omega^\sigma \wedge \omega^\nu \wedge \mathrm{d}t$$
$$+ \left[-\frac{1}{2} \left(\frac{\partial E_\sigma}{\partial \dot{q}^\nu} + \frac{\partial E_\nu}{\partial \dot{q}^\sigma} \right) + \frac{\mathrm{d}}{\mathrm{d}t} \frac{\partial E_\nu}{\partial \ddot{q}^\sigma} \right] \omega^\sigma \wedge \dot{\omega}^\nu \wedge \mathrm{d}t$$
$$+ \frac{1}{2} \left(\frac{\partial E_\nu}{\partial \ddot{q}^\sigma} - \frac{\partial E_\sigma}{\partial \ddot{q}^\nu} \right) \omega^\sigma \wedge \ddot{\omega}^\nu \wedge \mathrm{d}t.$$

Because of the fact that classes of variational dynamical forms are elements of the kernel of Helmholtz-Sonin mapping the representatives of their exterior derivatives must vanish, i.e. $I(\mathrm{d}E) = 0$. We obtain the system of equations

$$\left\{ \frac{\partial E_\nu}{\partial q^\sigma} \right\} - \frac{1}{2} \frac{\mathrm{d}}{\mathrm{d}t} \frac{\partial E_\nu}{\partial \dot{q}^\sigma} + \frac{1}{2} \frac{\mathrm{d}^2}{\mathrm{d}t^2} \frac{\partial E_\nu}{\partial \ddot{q}^\sigma} = 0,$$
$$-\frac{1}{2} \left(\frac{\partial E_\sigma}{\partial \dot{q}^\nu} + \frac{\partial E_\nu}{\partial \dot{q}^\sigma} \right) + \frac{\mathrm{d}}{\mathrm{d}t} \frac{\partial E_\nu}{\partial \ddot{q}^\sigma} = 0,$$
$$\frac{\partial E_\nu}{\partial \ddot{q}^\sigma} - \frac{\partial E_\sigma}{\partial \ddot{q}^\nu} = 0,$$

which is equivalent with conditions (5.14)–(5.16).

Finally we notice the fact that the representation sequence of the variational sequence enables us to introduce the generalized concept of *Lepage forms* by quite natural way.

Definition 7.5 Generalized Lepage forms. Let η be a k-form on $J^r Y$, $r = 1, 2$, $k \geq 1$. The form is called the *generalized Lepage form* if it satisfies the relation

$$\pi^*_{2r+1,r+1} p_k \, \mathrm{d}\eta = I(\mathrm{d}\eta), \quad \text{or, equivalently} \quad p_k \, \mathrm{d}\mathcal{R}(p_k \, \mathrm{d}\eta) = 0.$$

Example 7.6 Let $\eta \in \Omega_1^1 W$. We derive the conditions for this form to be a generalized Lepage form. We of course expect that Lepage equivalents θ_λ of Lagrangians λ satisfy this definition. For the form $\eta = L \, \mathrm{d}t + B_\sigma \, \omega^\sigma + C_\sigma \, \mathrm{d}\dot{q}^\sigma$ (L, B_σ and C_σ being functions on $J^1 Y$) we compare expressions (7.19) and (7.20) for $p_1 \, \mathrm{d}\eta$ and $I(\mathrm{d}\eta)$, respectively. We obtain two conditions. First of them declares that the $(\dot{\omega}^\sigma \wedge \mathrm{d}t)$-component of $p_1 \, \mathrm{d}\eta$ must vanish, i.e.

$$\frac{\partial L}{\partial \dot{q}^\sigma} - B_\sigma - \frac{\mathrm{d}C_\sigma}{\mathrm{d}t} + \frac{\partial C_\nu}{\partial \dot{q}^\sigma} \ddot{q}^\nu = \left(\frac{\partial L}{\partial \dot{q}^\sigma} - B_\sigma - \frac{\mathrm{d}' C_\sigma}{\mathrm{d}t} \right) + \left(\frac{\partial C_\nu}{\partial \dot{q}^\sigma} - \frac{\partial C_\sigma}{\partial \dot{q}^\nu} \right) \ddot{q}^\nu = 0$$

$$\Longrightarrow \quad \frac{\partial C_\nu}{\partial \dot{q}^\sigma} - \frac{\partial C_\sigma}{\partial \dot{q}^\nu} = 0, \quad B_\sigma = \frac{\partial L}{\partial \dot{q}^\sigma} - \frac{\mathrm{d}' C_\sigma}{\mathrm{d}t}.$$

The second condition resulting from comparison of $(\omega^\sigma \wedge dt)$-components of the forms $p_1\,\mathrm{d}\eta$ and $I(\mathrm{d}\eta)$ is then fulfilled identically. Thus, all first-order Lepage 1-forms are of the form

$$\theta = L\,\mathrm{d}t + \left(\frac{\partial L}{\partial \dot{q}^\sigma} - \frac{\mathrm{d}' C_\sigma}{\mathrm{d}t}\right)\omega^\sigma + C_\sigma\,\mathrm{d}\dot{q}^\sigma, \quad \frac{\partial C_\nu}{\partial \dot{q}^\sigma} - \frac{\partial C_\sigma}{\partial \dot{q}^\nu} = 0.$$

A special case is represented by $C_\sigma = 0$. Then $B_\sigma = \frac{\partial L}{\partial \dot{q}^\sigma}$ and we have the unique type of Lepage 1-forms completely derived from (first order) Lagrangians, their Lepage equivalents

$$\theta_\lambda = L\,\mathrm{d}t + \frac{\partial L}{\partial \dot{q}}\,\omega^\sigma, \quad \text{where } \lambda = L\,\mathrm{d}t, \quad L \in \Omega_0^1 W.$$

For the Lepage form θ_λ we finally have the self-evident relation $p_1\,\mathrm{d}\theta_\lambda = I(\mathrm{d}\theta_\lambda) = E_\lambda$.

For the calculus of variations Lepage 2-forms are important as well. In the following example we find the general expression for generalized Lepage 2-forms on J^1Y.

Example 7.7 Let $\eta \in \Omega_2^1 W$ be a form. Its chart expression in the basis of 1-forms adapted to the contact structure is given by (7.21). Comparing the expression (7.22) for the 2-contact component of $\mathrm{d}\eta$ with the expression (7.23) for $I(\mathrm{d}\eta)$ we obtain conditions for η to be a Lepage 2-form. After some technical manipulations with them we obtain

$$\begin{aligned}
F_{\sigma\nu} - \frac{1}{2}\left\{\frac{\partial A_\sigma}{\partial \dot{q}^\nu} - \frac{\mathrm{d}' B_{\sigma\nu}}{\mathrm{d}t} + \frac{\partial C_\sigma}{\partial q^\nu}\right\}_{\mathrm{alt}\,(\sigma,\nu)} &= 0,\\
\left\{B_{\sigma\nu} + \frac{\partial C_\nu}{\partial \dot{q}^\sigma}\right\}_{\mathrm{alt}\,(\sigma,\nu)} + \frac{\mathrm{d}' D_{\sigma\nu}}{\mathrm{d}t} &= 0,\\
\left\{\frac{\partial B_{\sigma\nu}}{\partial \dot{q}^\lambda} - \frac{\partial B_{\sigma\lambda}}{\partial \dot{q}^\nu} + 2\frac{\partial D_{\nu\lambda}}{\partial q^\sigma}\right\}_{\mathrm{alt}\,(\sigma,\nu)} &= 0,\\
\left[\frac{\partial D_{\sigma\lambda}}{\partial q^\nu} - \frac{\partial D_{\sigma\lambda}}{\partial \dot{q}^\nu}\right]_{\mathrm{sym}\,(\sigma,\nu)} &= 0,\\
\frac{\partial D_{\nu\lambda}}{\partial \dot{q}^\sigma} + \frac{\partial D_{\lambda\sigma}}{\partial \dot{q}^\nu} + \frac{\partial D_{\sigma\nu}}{\partial \dot{q}^\lambda} &= 0.
\end{aligned} \tag{7.25}$$

These conditions define Lepage 2-forms.

The context of Lepage 2-forms with the calculus of variations gives the following definition.

Definition 7.6 Let $E = E_\sigma\,\omega^\sigma \wedge \mathrm{d}t = (A_\sigma + B_{\sigma\nu}\ddot{q}^\nu)\,\omega^\sigma \wedge \mathrm{d}t$, $1 \leq \sigma, \nu \leq m$, be a dynamical form on J^2Y. A 2-form α is called the *Lepage equivalent of* E if it is a Lepage form and $p_1\alpha = E$ (up to a possible projection).

It is evident that the necessary condition for a 2-form α to be a Lepage equivalent of the given dynamical form $E = (A_\sigma + B_{\sigma\nu}\ddot{q}^\nu)\,\omega^\sigma \wedge dt$ is $p_1\alpha = (A_\sigma + B_{\sigma\nu}\ddot{q}^\nu)\,\omega^\sigma \wedge dt$. This means that coefficients C_σ and $D_{\sigma\nu}$ in Example 7.7 are zero. Conditions (7.25) then give

$$B_{\sigma\nu} = B_{\nu\sigma}, \quad \frac{\partial B_{\sigma\lambda}}{\partial \dot{q}^\nu} = \frac{\partial B_{\sigma\nu}}{\partial \dot{q}^\lambda}, \quad F_{\sigma\nu} = \frac{1}{4}\left(\frac{\partial A_\sigma}{\partial \dot{q}^\nu} - \frac{\partial A_\nu}{\partial \dot{q}^\sigma}\right). \tag{7.26}$$

So, we can see that there are dynamical forms for which their dynamical system contains no Lepage equivalent. If for the given dynamical form the Lepage equivalent exists, then it is unique. Dynamical forms having the Lepage equivalent are not necessarily variational. On the other hand, variational dynamical form has obviously the Lepage equivalent, which besides conditions (7.26) satisfies the additional conditions (5.11) and (5.12).

7.5 A Note to Local and Global Aspects

In the preceding chapters and sections we found the solutions of trivial variational problem (Sect. 4.3, Definition 4.7, Proposition 4.4) and the inverse variational problem (Sect. 5.2, Definition 5.2, Theorem 5.1) in their *local* version: Trivial Lagrangians are locally given by total derivatives of functions. For a variational dynamical form E there exists locally a Lagrangian λ. There arises a question under what conditions these object exist globally, i.e. on the whole underlying fibred manifold. The answer is connected with properties of the variational sequence, especially with its *cohomology*. In the preceding section we saw that trivial Lagrangians are elements of the image of the mapping E_0^1 and simultaneously elements of the kernel of the Euler-Lagrange mapping E_1^1, because $\mathrm{Im}\, E_0^1 \subseteq \mathrm{Ker}\, E_1^1$ due to the exactness of the variational sequence. Variational dynamical forms are elements of the image of the Euler-Lagrange mapping and then elements of the kernel of the Helmholtz-Sonin mapping, because $\mathrm{Im}\, E_1^1 \subseteq \mathrm{Ker}\, E_2^1$. A function $f \in \Omega_0^1$ leading to the given trivial Lagrangian $\lambda = h\, df$ exists globally if the image of the mapping E_0^1 coincides with the kernel of the Euler-Lagrange mapping E_1^1. A Lagrangian λ leading to the given locally variational dynamical form E exists globally if the image of the Euler-Lagrange mapping E_1^1 coincides with the kernel of the Helmholtz-Sonin mapping E_2^1. The cohomology of the variational sequence plays the important role in solving this problem.

The global properties of the variational sequence follow from the abstract de Rham theorem. We summarize them in the following proposition. (Its deeper understanding requires, of course, a more detailed study of original works, especially at least elements of the sheaf theory—a very detailed approach see, e.g. in [15].)

Proposition 7.7 *Consider the variational sequence of the rth order (for our purposes $r = 1, 2$). It has the following properties.*

(a) Each of the sheaves Ω_k^r is fine.
(b) The variational sequence is an acyclic resolution of the constant sheaf $\mathbf{R}_Y$ over $\mathcal{V}$ (the shortened notation $\{0\} \to \mathbf{R}_Y \to \mathcal{V}$ used for (7.10)).
(c) Let $\Gamma(Y, \Omega_0^r)$ be a cochain complex of global sections

$$\Gamma(Y, \mathcal{V}) : \{0\} \longrightarrow \Gamma(Y, \mathbf{R}_Y) \longrightarrow \Gamma(Y, \Omega_0^r) \longrightarrow \cdots \longrightarrow \Gamma(Y, \Omega_{(r+1)m+1}^r) \longrightarrow \{0\}$$

and denote $H^k(\Gamma(\mathbf{R}_Y, \mathcal{V})$ its kth cohomology group. Then for every $k \geq 0$ it holds $H^k(\Gamma(\mathbf{R}_Y, \mathcal{V})) = H^k(\Gamma(Y, \mathbf{R}))$, where $H^k(\Gamma(Y, \mathbf{R}))$ is the standard cohomology group of the manifold Y.

Proof We present only comments to the proof. Details can we found elsewhere (see, e.g. [15]).

(a) The property (a) follows from the fact that Θ_k^r admits a sheaf partition of unity. This is the immediate consequence of Proposition 7.2.
(b) The variational sequence is exact and thus it is a resolution of the constant sheaf $\mathbf{R}_Y$. Because of the property (a) stating that each of the shaves Θ_k^r is fine and then soft. The sheaves Ω_k^r are soft as well. Then the same holds for the quotient sheaves Ω_k^r / Θ_k^r. Thus the resolution is acyclic.
(c) The assertion is the immediate consequence of the general de Rham theorem applied to the variational sequence $\{0\} \to \mathbf{R}_Y \to \mathcal{V}$. □

7.6 Problems

Problem 7.1 In Sect. 7.4 we have obtained the representative of a class $[\mathrm{d}\eta] \in \Omega_2^1 W / \Theta_2^1 W$, $\eta \in \Omega_1^1 W$, by the direct use of Definition 7.4, i.e. as the image of $\mathrm{d}\eta$ (or any form belonging to the class $[\mathrm{d}\eta]$) by the interior Euler operator $\mathcal{I}$. Prove that we can obtain this representative also as the action of the operator $\mathcal{I}$ on the form $p_1\, \mathrm{d}h\eta$. Explain this possibility by generally valid arguments. Use the same arguments for 2-forms, i.e. for explaining the relation $\mathcal{I}(\mathrm{d}\eta) = \mathcal{I}(p_2\, \mathrm{d}p_1\eta)$ for $\eta \in \Omega_2^1 W$.

Problem 7.2 The same general arguments as in Problem 7.1 are valid for 2-forms, i.e. it holds $\mathcal{I}(\mathrm{d}\eta) = \mathcal{I}(\mathrm{d}\mathcal{I}(\eta))$ for $\eta \in \Omega_2^1 W$. Calculate $\mathcal{I}(\mathrm{d}\mathcal{I}\eta)$ and compare the result with the representative of the class $[\mathrm{d}\eta]$, $\eta \in \Omega_2^1 W$, obtained in Sect. 7.4.

Problem 7.3 Prove relations (7.3) for operators d_h and d_c defined by relations (7.2).

Problem 7.4 For a 1-form η and a 2-form ϱ on J^1Y derive the chart expressions for $\mathrm{d}_t\eta$ and $\mathrm{d}_t\varrho$. The obtained results compare with the general expression in Example 7.2.

Problem 7.5 Calculate in detail the chart expression of the form $\mathrm{d}\varrho_c$ in the proof of Proposition 7.2. First make the calculations for a 1-form and for a 2-form.

Problem 7.6 Make all calculations in Examples 7.3, 7.4 and 7.5 in detail.

Problem 7.7 Derive Helmholtz conditions of variationality presented in Proposition 5.2 from the equation $\mathcal{I}(\mathrm{d}\alpha) = 0$, where $\alpha \in \Omega_2^1 W$ is an element of the dynamical system of a dynamical form $E \in \Omega_2^2 W$.

Problem 7.8 Let $\eta \in \Omega_1^1 W$ be a generalized Lepage form. Prove that $p_1\,\mathrm{d}\eta$ is obviously $\pi_{2,0}$-horizontal.

Problem 7.9 Let $\eta \in \Omega_2^1 W$ be a generalized Lepage form. Prove that $p_2\,\mathrm{d}\eta$ is obviously of the form $p_2\,\mathrm{d}\eta = A_{\sigma\nu}\,\omega^\sigma \wedge \omega^\nu \wedge \mathrm{d}t + B_{\sigma\nu}\,\omega^\sigma \wedge \dot{\omega}^\nu \wedge \mathrm{d}t$, i.e. the component $C_{\sigma\nu}\,\dot{\omega}^\sigma \wedge \dot{\omega}^\nu \wedge \mathrm{d}t$ vanishes.

References

1. A.M. Vinogradov, A spectral sequence associated with a non-linear differential equation, and algebro-geometric foundations of Lagrangian field theory with constraints. Soviet Math. Dokl. **19**, 144–148 (1978)
2. A.M. Vinogradov, The C-spectral Sequence, Lagrangian Formalism and Conservation Laws I and II. Jour. Math. Anal. Appl. **100**(1), 1–129 (1984)
3. I. Anderson, T. Duchamp, On the existence of global variational principles. Am. J. Math. **102**, 781–867 (1980)
4. I.M. Anderson, Introduction to the variational bicomplex. Contemp. Math. **132**, 51–73 (1992)
5. I.M. Anderson, *The Variational Bicomplex*, Technical report of the Utah State University (1989)
6. D. Krupka, Variational sequences on finite order jet spaces. In: *Differential Geometry and its Applications*, ed. J. Janyška, D. Krupka (Brno, 1989) (World Science Publication, Teaneck, NJ, 1990), pp. 236–254
7. D. Krupka, J. Musilová, Trivial Lagrangians in field theory. Diff. Geom. Appl. **9**, 293–305 (1998)
8. D.R. Grigore, The variational sequence on finite jet bundle extensions and the Lagrange formalism. Diff. Geom. Appl. **10**, 43–77 (1999)
9. R. Vitolo, Finite order Lagrangian bicomplexes. Math. Proc. Cambridge Phil. Soc. **125**, 321–333 (1999)
10. R. Vitolo, Finite order formulation of Vinogradov's C-spectral sequence, Acta Applicandae Mathematicae, 133–154 (2002)
11. M. Krbek, J. Musilová, Representation of the variational sequence by differential forms. Rep. Math. Phys. **51**, 251–258 (2003)
12. M. Krbek, J. Musilová, Representation of the variational sequence by differential forms. Acta Appl. Math. **88**, 177–199 (2005)

13. D. Krupka, J. Šedĕnková, Variational sequences and Lepage forms, in *Differential Geometry and its Applications*. ed. by J. Bureš, O. Kowalski, D. Krupka, J. Slovák (Matfyzpress, Prague, 2005), pp.617–627
14. D. Krupka, *Global variational theory in fibred spaces, in Handbook of Global Analysis* (Elsevier Sci. B. V, Amsterdam, 2008), pp.773–836
15. D. Krupka, *Introduction to Global Variational Geometry*. Atlantis Studies in Variational Geometry, vol. 1 (Atlantis Press, Paris, 2015)

Extension: Geometrical Structures for Field Theories

8

With regard to applications of the calculus of variations in physics (Chap. 9), we will now briefly extend the mathematical tools for field theories, especially with respect to the inverse problem in field theories in physics. The needed geometrical tools differ from mechanics only by the dimension of the base X of the underlying fibred manifold (Y, π, X) (and thus also its prolongations), $\dim X = n > 1$. (We focus on the practical use of the presented concepts which will be also explained with the help of some physical examples.) For a more detailed explanation of concepts and proofs from the mathematical point of view, namely, for 1st and 2nd prolongations of fibred manifolds, the reader can also follow Chap. 2, with the fact that the general (coordinate free) properties of the presented mappings are the same in field theory as in mechanics, only the calculations in charts differ in the number of variables and respect the interchangeability of mixed partial derivatives of functions of several variables. Even though for applications in physical field theories in Chap. 9 only 1st and 2nd prolongations of relevant geometrical objects (fibred manifolds, vector fields, differential forms) are needed, we present here the general coordinate free definitions for correctness and completeness. They are then illustrated by examples of first and second order.

Prerequisites concepts explained for mechanics in Chaps. 2–5 (including prerequisites listed there), tangent bundles on differential manifolds (vectors as classes of equivalent parametrized curves on differential manifolds), differential and integral calculus of functions of several variables including practical calculations, symmetrization and antisymmetrization operators, vector fields and differential forms in Euclidean spaces, one-parameter groups of vector fields, Lie derivatives of vector fields and differential forms. □

J. Musilová et al., *Calculus of Variations on Fibred Manifolds and Variational Physics*, Lecture Notes in Physics 1033, https://doi.org/10.1007/978-3-031-77408-9_8

8.1 Fibred Manifolds with n-Dimensional Bases, Sections, Jet Prolongations

In Chap. 2, we explained in detail concepts mentioned in the title of this chapter for the case of mechanics, i.e. for fibred manifolds with a one-dimensional base. Now we extend the definitions and assertions to n-dimensional bases with $n > 1$, including calculations in coordinates. The concepts for fibred manifolds with multidimensional bases are quite analogous to those with one-dimensional bases. Therefore, we will deal with them only briefly. (In this section one can follow also schematic figures in Chap. 2 for illustration.)

The following definition is the same as Definitions 2.1 and 2.2 differing only by the dimension of the base.

Definition 8.1 (*Fibred manifolds, fibration*) Let X be a n-dimensional manifold, $n > 1$, Y a manifold of the dimension $m + n$ and $\pi : Y \to X$ surjective submersion. Then (Y, π, X) is called a *fibred manifold with n-dimensional base* X, π being the *projection* and Y the *total space*. The set $\pi^{-1}(x)$ for a point $x \in X$ is the *fibre over the point* x. (It has the structure of m-dimensional submanifold of Y. The dimension of the total space is $m + n$.) A fibred manifold is called the *fibration* if it has a connected base and their fibres are diffeomorphic from each other.

(Instead of *a fibred manifold* (Y, π, X) one often use shortened names *a fibred manifold* π), or *fibred manifold* Y.

Let (V, ψ), be a map on Y. Then the map (U, φ) on X, such that $U = \pi(V)$ and

$$p \circ \psi = \varphi \circ \pi,$$

where $p : \mathbf{R}^n \times \mathbf{R}^m \to \mathbf{R}^n$ is the Cartesian projection on the first factor, is the *associated chart on* X. We denote $\varphi = (x^1, \ldots, x^n)$ and $\psi = (x^1, \ldots, x^n, y^1, \ldots, y^m)$. Using these maps we can write for $y \in V$

$$\psi(y) = \big(x^i \psi(y), y^\sigma \psi(y)\big), \quad x^i \psi(y) = x^i \varphi\big(\pi(y)\big), \quad 1 \le i \le n, 1 \le \sigma \le m.$$

The chart expression of π is $p = \varphi \circ \pi \circ \psi^{-1} : \psi(V) \to \varphi(U)$, i.e. (in shortened notation)

$$\varphi \circ \pi \circ \psi^{-1} : \mathbf{R}^{m+n} \ni \big(x^i(y), y^\sigma(y)\big) = (x^i) \in \mathbf{R}^n.$$

Coordinate transformations for (V, ψ) and $(\bar{V}, \bar{\psi})$, $V \cap \bar{V} \neq \emptyset$, are given by one-to-one mappings $\psi \circ \bar{\psi}^{-1}$ on $V \cap \bar{V}$ and $\varphi \circ \bar{\varphi}^{-1}$ on $U \cap \bar{U}$ with relations

$$\bar{x}^i = \bar{x}^i(\bar{\varphi} \circ \varphi^{-1})(x^j), \quad \bar{y}^\sigma = \bar{y}^\sigma(\bar{\psi} \circ \psi^{-1})(x^j, y^\nu). \tag{8.1}$$

The definition of a *section of the projection* π is quite analogous as in the case of mechanics (one-dimensional base).

Definition 8.2 (*Sections*) Let (Y, π, X) be a fibred manifold, dim $X = n > 1$, dim $Y = m + n$. Let $U \subset X$ be an open set. A smooth mapping

$$\gamma : U \ni x \longrightarrow \gamma(x) \in Y, \quad \text{such that} \quad \pi \circ \gamma = \mathrm{id}_U, \tag{8.2}$$

is called the *(local) section* of the projection π (alternatively the section of (Y, π, X)). If $U = X$, then the section is *global*.

Chart expressions of sections are as follows: Let (V, ψ), $\psi = (x^i, y^\sigma)$, $1 \le i \le n$, $1 \le \sigma \le m$, be a fibred chart on Y and (U, φ) the associated chart on X, $U \subset \mathrm{dom}\, \gamma$. For a point $x \in U$ it holds

$$\begin{aligned} \psi \circ \gamma(x) &= (\psi \circ \gamma \circ \varphi^{-1})(\varphi(x)) \\ &= \left(x^i(\psi \circ \gamma \circ \varphi^{-1})(\varphi(x)), y^\sigma(\psi \circ \gamma \circ \varphi^{-1})(\varphi(x))\right), \end{aligned} \tag{8.3}$$

or in shortened notation

$$\gamma = \left(x^1\gamma, \ldots, x^n\gamma, y^1\gamma, \ldots, y^m\gamma\right) = \left(x^i\gamma, y^\sigma\gamma\right), \tag{8.4}$$

$1 \le i \le n$, $1 \le \sigma \le m$. Similarly, as in mechanics, we can say that sections are in fact "graphs" of vector functions with components $y^1(x^1, \ldots, x^n), \ldots, y^m(x^1, \ldots, x^n)$.

Example 8.1 (Motion of an ideal liquid) The motion of an ideal liquid in the three-dimensional Euclidean space is described by the vector field of velocity $\boldsymbol{v}(t, x, y, z)$. The corresponding fibred manifold is then $(Y, \pi, X) = (\mathbf{R} \times \mathbf{R}^3 \times \mathbf{R}^3, p, \mathbf{R} \times \mathbf{R}^3)$ and the form of sections is

$$\begin{aligned} \gamma : &\mathbf{R} \times \mathbf{R}^3 \ni (t, x, y, z) \longrightarrow \gamma(t, x, y, z) \\ &= \left(t, x, y, z, v_x(t, x, y, z), v_y(t, x, y, z), v_z(t, x, y, z)\right). \end{aligned}$$

Example 8.2 (Electromagnetic field) Let $(Y, \pi, X) = (\mathbf{M} \times \mathbf{R}^6, p, \mathbf{M})$, where $\mathbf{M}$ is four-dimensional Minkowski time-space. Electromagnetic field is described by the antisymmetric tensor with components given by vector fields of electric intensity and magnetic induction $F = (F_{\alpha\beta})$, $F_{\alpha\beta} + F_{\beta\alpha} = 0$, $\alpha, \beta \in \{0, 1, 2, 3\}$. The sections of this fibred manifold are of the form

$$\begin{aligned} \gamma : &\mathbf{M} \ni (x^0, x^1, x^2, x^3) \longrightarrow \gamma(x^0, x^1, x^2, x^3) \\ &= \left(x^0, x^1, x^2, x^3, F_{\alpha\beta}(x^0, x^1, x^2, x^3)\right), \quad 0 \le \alpha, \beta \le 3. \end{aligned}$$

Definition 8.3 (*Local isomorphisms*) Consider fibred manifolds (Y, π, X) and $(\bar{Y}, \bar{\pi}, \bar{X})$ and the mapping α: $\bar{Y} \rightarrow Y$ such that there exists a mapping α_0: $\bar{X} \rightarrow X$ such that it holds

$$\alpha_0 \circ \pi = \bar{\pi} \circ \alpha. \tag{8.5}$$

If condition (8.5) is fulfilled, then the mapping α is called the *homomorphism of fibred manifolds* and α_0 its *projection*. If the mappings α and α_0 are diffeomorphisms, then α is called the *isomorphism* of mentioned fibred manifolds. (In the equivalent notation (α, α_0).)

(For illustration see the commuting diagram for mechanics in Fig. 2.12.) It can be easily proved that for a local homomorphism α of fibred manifolds (as well as for its special case—local isomorphism) its projection is unique, see Problem 8.2.

Proposition 8.1 *Let (α, α_0) be a local isomorphism of fibred manifolds (Y, π, X) and $(\bar{Y}, \bar{\pi}, \bar{X})$. Then for every section $\gamma \in \Gamma_U(\pi)$ the mapping $\bar{\gamma} = \alpha \circ \gamma \circ \alpha_0^{-1}$ is the section of the fibred manifold $(\bar{Y}, \bar{\pi}, \bar{X})$, $\bar{\gamma} \in \Gamma_{\bar{U}}(\bar{\pi})$, where $\bar{U} = \alpha(U)$.*

The proof of this proposition is based on the same argumentation as in mechanics.

In the following, we introduce the concept of *jet prolongations* of a fibred manifold. It is again analogous as in mechanics, only a small "technical" problem can be connected with interchangeability of mixed partial derivatives. For simplicity we will suppose that the mappings we will work with are of the necessary class C^r, i.e. they are differentiable into the order of r inclusive. By the even more simplified notation $\Gamma_W(\pi)$ for open subset $W \subset X$ or $\Gamma(\pi)$ for $W = X$ we will understand the set of all *smooth* sections, i.e. differentiable into any order. (In practice, however, this requirement may not be fulfilled.)

Definition 8.4 (*Equivalent sections*) Let $\Gamma^{(r)}(\pi)$ be the set of all C^r-class local sections of the fibred manifold (Y, π, X), and let $W \subset X$ be an open set. Denote as

$$\Gamma_W^{(r)}(\pi) = \left\{ \gamma \in \Gamma^{(r)}(\pi) \mid W \subset \operatorname{dom} \gamma \right\}.$$

We say that the sections $\gamma_1, \gamma_2 \in \Gamma_W^{(r)}(\pi)$ have the rth *order contact* (or they are *r-equivalent*) at the point $x \in W$, if $\gamma_1(x) = \gamma_2(x)$ and there exists a fibred chart (V, ψ), $\psi = (x^i, y^\sigma)$, $1 \le i \le n$, $1 \le \sigma \le m$, on Y and the corresponding associated chart $\big(\pi(V), \varphi\big)$ on X such that following conditions are fulfilled at the above mentioned point $x \in W$ (shortened notation (8.4)):

$$y^\sigma \gamma_1(x^1, \dots, x^n) = y^\sigma \gamma_2(x^1, \dots, x^n), \quad \frac{\partial y^\sigma \gamma_1(x^1, \dots, x^n)}{\partial x^j} = \frac{\partial y^\sigma \gamma_2(x^1, \dots, x^n)}{\partial x^j},$$
$$\dots\dots, \quad \frac{\partial^r y^\sigma \gamma_1(x^1, \dots, x^n)}{\partial (x^{j_1})^{k_1} \dots \partial (x^{j_p})^{k_p}} = \frac{\partial^r y^\sigma \gamma_2(x^1, \dots, x^n)}{\partial (x^{j_1})^{k_1} \dots \partial (x^{j_p})^{k_p}}, \tag{8.6}$$

for all $1 \le \sigma \le m$, $1 \le j \le n$, $1 \le j_1, \dots, j_p \le n$, $j_1 \ne \dots \ne j_p \ne j_1$, $k_1 + \dots + k_p = r$. Or equivalently

$$\dots, \quad \frac{\partial^r y^\sigma \gamma_1(x^1, \dots, x^n)}{\partial x^{\ell_1} \dots \partial x^{\ell_r}} = \frac{\partial^r y^\sigma \gamma_2(x^1, \dots, x^n)}{\partial x^{\ell_1} \dots \partial x^{\ell_r}} \tag{8.7}$$

for all $1 \le \ell_1, \dots, \ell_r \le n$ and all $1 \le \sigma \le m$, where some of the variables $x^{\ell_1}, \dots, x^{\ell_r}$ can be identical.

Proposition 8.2 *Let $\gamma_1, \gamma_2 \in \Gamma^{(r)}_W(\pi)$ be r-equivalent sections. Then conditions (8.6) and equivalent conditions (8.7) are fulfilled in every fibred chart $(\bar{V}, \bar{\psi})$, $\bar{V} \subset W$, $\bar{\psi} = (\bar{x}^i, \bar{y}^\sigma)$, $\bar{U} = \pi(\bar{V})$, $1 \le i \le n$, $1 \le \sigma \le m$.*

The proof is rather technical (see Problem 8.3 or follow the proof of Proposition 2.1).

Definition 8.5 (*r-jets of sections*) The class $J^r_x\gamma$ of all sections r-equivalent at the point x with a section $\gamma \in \Gamma_W(\pi)$ is called *r-jet* of the section γ at x.

We now introduce *jet prolongations of a fibred manifold* (Y, π, X) with help of prolongations of its sections (follow Definition 2.7).

Definition 8.6 (*Jet prolongations of fibred manifolds*) Let $\Gamma^{(r)}(\pi)$ be the set of all C^r-class sections of (Y, π, X) defined on open subsets of the base X. Denote as J^rY the set of all r-jets of sections $\gamma \in \Gamma^{(r)}(\pi)$, i.e.

$$J^rY = \left\{J^r_x\gamma \mid x \in X, \gamma \in \Gamma^{(r)}(\pi)\right\}$$

and the projection

$$\pi_r : J^rY \ni J^r_x\gamma \longrightarrow \pi_r\left(J^r_x\gamma\right) = x \in X.$$

The fibred manifold (J^rY, π_r, X) is called the rth *jet prolongation of the fibred manifold* $(Y, \pi\ X)$.

It is evident that the dimension of the total space J^rY is

$$\dim J^rY = n + m\sum_{s=0}^{r}\binom{n+s-1}{s}.$$

(The number of independent sth order partial derivatives of a function of n variables, with consideration of interchangeability, is $\binom{n+s-1}{s}$.) Denote $J^0Y = Y, \pi_0 = \pi$. By the natural way there arise projections and corresponding fibred manifolds with the base X for $1 \le s \le r, 0 \le p < s$

$$\pi_{s,p} : J^sY \ni J^s_x\gamma \longrightarrow \pi_{s,p}\left(J^s_x\gamma\right) = J^p_x\gamma \in J^pY.$$

(see also schematic Fig. 2.14). There arise also sections of projections π_s, i.e. mappings

$$\delta : W \ni x \longrightarrow \delta(x) \in J^sY, \quad \text{such that } \pi \circ \delta = \mathrm{id}_W.$$

Proposition 8.3 *Let $\gamma \in \Gamma_W^{(r)}(\pi)$ be a section of the projection π. The mapping*

$$J^r\gamma \, : \, W \ni x \longrightarrow J^r\gamma(x) = J_x^r\gamma \in J^rY$$

is the section of the projection π_r.

A section δ of the projection π_r, $r \geq 1$, is called *holonomic* if there exists a section γ of π such that $\delta = J^r\gamma$.

For physics the calculations in coordinates are important. So, we introduce *associated fibred charts* on jet prolongations of a fibred manifold (Y, π, X).

Definition 8.7 (*Associated fibred charts*) Let (J^rY, π_r, X) be the rth prolongation of a fibred manifold (Y, π, X). Let (V, ψ) be a fibred chart on Y and (U, φ), $U = \pi(V)$, the associated chart on X. Denote $V_r = \pi_{r,0}^{-1}(V)$. For a section $\gamma \in \Gamma_{\pi(V)}(\pi)$ and $1 \leq s \leq r$, $1 \leq j, j_1, \ldots, j_s \leq n$, $1 \leq \sigma \leq m$, denote

$$y^\sigma_{j_1 \ldots j_s} J^r_x\gamma = \frac{\partial^s y^\sigma \gamma\left(x^1\varphi(x), \ldots, x^n\left(\varphi(x)\right)\right)}{\partial x^{j_1} \ldots \partial x^{j_s}}, \quad \psi_r = \left(x^i, y^\sigma, y^\sigma_j, \ldots, y^\sigma_{j_1 \ldots j_r}\right),$$

taking into account that the expressions $y^\sigma_{j_1 \ldots j_s} J^r_x\gamma$ are symmetric in indices $j_1, \ldots, j_s$, so that each $y^\sigma_{j_1 \ldots j_s}$ contains only independent choices of indices. Then by the *associated fibred chart* on (J^rY, π_r, X) we mean the pair (V_r, ψ_r).

Note—summation conventions: Consider formal operator expressions as, e.g.

$$Z^{i_1 \ldots i_s}\partial^s_{i_1 \ldots i_s} = Z_{i_1 \ldots i_s}\frac{\partial^s}{\partial x^{i_1} \ldots \partial x^{i_s}}, \quad \Xi^\sigma_{i_1 \ldots i_s}\frac{\partial}{\partial y^\sigma_{i_1 \ldots i_s}}$$

where $1 \leq i_1, \ldots, i_s \leq n$ are arbitrary indices, some of them may be identical, $y^\sigma_{i_1 \ldots i_s}$ are symmetric in indices $i_1, \ldots, i_s$, and $Z^{i_1 \ldots i_s}$, or $\Xi^\sigma_{i_1 \ldots i_s}$ are some coefficients (e.g. functions). No permutation of indices changes the operators $\partial^s_{i_1 \ldots i_s}$, $\frac{\partial}{\partial y^\sigma_{i_1 \ldots i_s}}$. If the summations of independent expressions are needed, we write

$$\sum_{1 \leq j_1 \leq \cdots \leq j_s \leq n} \tilde{Z}^{j_1 \ldots j_s}\frac{\partial^s}{\partial x^{j_1} \ldots \partial x^{j_s}}, \quad \sum_{1 \leq j_1 \leq \cdots \leq j_s \leq n} \tilde{\Xi}^\sigma_{j_1 \ldots j_s}\frac{\partial}{\partial y^\sigma_{j_1 \ldots j_s}}. \tag{8.8}$$

In these expressions, the Einstein summation convention cannot be used. On the other hand, if we do not want resign to the Einstein summation, we use symmetrized coefficients as follows: Let $1 \leq j_1 \leq \cdots \leq j_s \leq n$, and let $p(j_1 \ldots j_s)$ be an arbitrary arrangement of indices $j_1, \ldots, j_s$ (e.g. for $(j_1, j_2, j_3, j_4, j_5, j_6) = (1, 3, 3, 3, 4, 4)$ there are possibilities as $p(j_1 \ldots j_6) = (3, 1, 3, 4, 4, 3)$ and others).

The number of such arrangements denote $N(j_1 \dots j_s)$ (for the mentioned example $N(1,3,3,3,4,4) = 6 \cdot \binom{5}{2} = 60$.) Then using Einstein summation convention

$$Z^{i_1 \dots i_s} \frac{\partial^s}{\partial x^{i_1} \dots \partial x^{i_s}}, \quad \Xi^{\sigma}_{i_1 \dots i_s} \frac{\partial}{\partial y^{\sigma}_{i_1 \dots i_s}},$$

we must take into account that for $(i_1 \dots i_s) = p(j_1 \dots j_s)$

$$Z^{i_1 \dots i_s} = Z^{p(j_1 \dots j_s)} = \frac{1}{N(j_1 \dots j_s)} \tilde{Z}^{j_1 \dots j_s},$$

$$\Xi^{\sigma}_{i_1 \dots i_s} = \Xi^{\sigma}_{p(j_1 \dots j_s)} = \frac{1}{N(j_1 \dots j_s)} \tilde{\Xi}^{\sigma}_{j_1 \dots j_s}.$$

The choice of one of two mentioned summation conventions follows the context of a concrete problem.

For a function $f(x^i, y^\sigma, \dots, y^\sigma_{j_1 \dots j_r})$ on $J^r Y$ denote the so called *total derivative operator* with respect to the variable x^j as the mapping

$$\begin{aligned} \mathrm{d}_j = \frac{\mathrm{d}}{\mathrm{d}x^j} &: f(x^i, y^\sigma, \dots, y^\sigma_{j_1 \dots j_r}) \longrightarrow \frac{\mathrm{d} f(x^i, y^\sigma, \dots, y^\sigma_{j_1 \dots j_r})}{\mathrm{d}x^j} \\ &= \frac{\partial f(x^i, y^\sigma, \dots, y^\sigma_{j_1 \dots j_r})}{\partial x^j} + \sum_{s=0}^{r} \frac{\partial f(x^i, y^\sigma, \dots, \dots, y^\sigma_{j_1 \dots j_r})}{\partial y^\nu_{\ell_1 \dots \ell_s}} y^\sigma_{\ell_1 \dots \ell_s j}. \end{aligned}$$

For the purposes of physics applications in Chap. 9 the first and second prolongations of the underlying fibred manifold are important, where the fibred charts associated with (V, ψ), $\psi = (x^i, y^\sigma)$, $1 \le i \le n$, $1 \le \sigma \le m$, are

$$\left((\pi^{-1}_{1,0}(V), \psi_1\right), \quad \psi_1 = (x^i, y^\sigma, y^\sigma_j), \quad \left(\pi^{-1}_{2,0}(V), \psi_2\right), \quad \psi_2 = \left(x^i, y^\sigma, y^\sigma_j, y^\sigma_{j_1 j_2}\right).$$

At the end of this section, we define the prolongations of local isomorphisms of a fibred manifold (Y, π, X) needed for the definition of vector fields on this manifold.

Proposition 8.4 *Let $\alpha : W \to Y$ be a local isomorphism of a fibred manifold (Y, π, X) defined on an open set $W \subset Y$ and $\alpha_0 : \pi(W) \to X$ its projection. Let U be an open set containing $\pi(W)$ and $\gamma \in \Gamma_U(\pi)$ a smooth section. Then the mapping*

$$J^r \alpha : \pi^{-1}_{r,0}(W) \ni J^r_x \gamma \longrightarrow J^r \alpha \left(J^r_x \gamma\right) = J^r_{\alpha_0(x)} \left(\alpha \circ \gamma \circ \alpha_0^{-1}\right) \in J^r Y$$

is a local isomorphism of the fibred manifold $(J^r Y, \pi_r, X)$. Its projection is α_0.

Proof It holds

$$\begin{aligned}\pi_r \circ J^r\alpha\left(J^r_x\gamma\right) &= \pi_r\left[J^r_{\alpha_0(x)}\left(\alpha\circ\gamma\circ\alpha_0^{-1}\right)\right]\\ &= \left[\pi_r\circ J^r\left(\alpha\circ\gamma\circ\alpha_0^{-1}\right)\right]\left(\alpha_0(x)\right) = \alpha_0\circ\pi_r\left(J^r_x\gamma\right).\end{aligned}$$

The last equation follows from the fact that the mapping $\alpha\circ\gamma\circ\alpha_0^{-1}$ is the section of π. Thus, $J^r\left(\alpha\circ\gamma\circ\alpha_0^{-1}\right)$ is the section of the projection π_r, and thus $\pi_r\circ J^r\left(\alpha\circ\gamma\circ\alpha^{-1}\right) = \mathrm{id}_U$. On the other hand it holds $\alpha_0\left[\pi_r\left(J^r_x\gamma\right)\right] = \alpha_0(x)$. Thus for the mapping $J^r\alpha$ holds $\pi_r\circ J^r\alpha = \alpha_0\circ\pi_r$, i.e. the pair $(J^r\alpha, \alpha_0)$ is the local isomorphism of (J^rY, π_r, X). □

8.2 Vector Fields and Differential Forms

In this section we introduce the concept of vector fields and differential forms on fibred manifolds and their prolongations adapted to the fibred structure of these manifolds.

Let (Y, π, X) be a fibred manifold, $\dim X = n$, $\dim Y = m + n$. Let (V, ψ), $\psi = (x^i, y^\sigma)$ and $\left(\pi(V), \varphi\right)$, $\varphi = (x^i)$, $1 \le i \le n$, $1 \le \sigma \le m$, be a fibred chart on Y and associated chart on X. Let (TY, Π, Y), $TY = \cup_{y\in Y} T_yY$, be the tangent bundle on Y (for a general differential manifold see Problem 2.3). Let $W \subset Y$ be an open set. The mapping

$$\xi : W \ni y \to \xi(y) \in T_yY \subset TY$$

is called a *local vector field* on Y. The definitions of vector fields on X and on jet-prolongations of Y are quite analogous. We express vector fields on Y, X and jet prolongations of Y in charts. Let $\left(\pi_{r,0}^{-1}(V), \psi_r\right)$, $\psi_r = (x^i, y^\sigma, y^\sigma_j, \ldots, y^\sigma_{j_1\ldots j_r})$. Then for vector fields on X, Y, J^1Y, ..., J^rY we can write

$$\begin{aligned}
\xi_0 &= \xi^i(x^j)\frac{\partial}{\partial x^i},\\
\xi &= \xi^i(x^j, y^\nu)\frac{\partial}{\partial x^j} + \Xi^\sigma(x^j, y^\nu)\frac{\partial}{\partial y^\sigma},\\
\xi_1 &= \xi^i(x^j, y^\nu, y^\nu_j)\frac{\partial}{\partial x^i} + \Xi^\sigma(x^j, y^\nu, y^\nu_j)\frac{\partial}{\partial y^\sigma} + \Xi^\sigma_j(x^j, y^\nu, y^\nu_j)\frac{\partial}{\partial y^\sigma_j},\\
&\ldots,\\
\xi_r &= \xi^i(x^j, y^\nu, y^\nu_j, \ldots, y^\nu_{j_1\ldots j_r})\frac{\partial}{\partial x^i} + \Xi^\sigma(x^j, y^\nu, y^\nu_j, \ldots, y^\nu_{j_1\ldots j_r})\frac{\partial}{\partial y^\sigma}\\
&\quad + \Xi^\sigma_j(x^j, y^\nu, y^\nu_j, \ldots, y^\nu_{j_1\ldots j_r})\frac{\partial}{\partial y^\sigma_j}\\
&\quad + \cdots + \Xi^\sigma_{j_1\ldots j_r}(x^j, y^\nu, y^\nu_j, \ldots, y^\nu_{j_1\ldots j_r})\frac{\partial}{\partial y^\sigma_{j_1\ldots j_r}}.
\end{aligned} \tag{8.9}$$

General expression for a vector field on J^rY:

$$\xi = \xi^i(x^j, y^\nu, \ldots, y^\nu_{j_1\ldots j_r})\frac{\partial}{\partial x^i} + \sum_{s=0}^{r} \Xi^\sigma_{j_1\ldots j_s}(x^j, y^\nu, \ldots, y^\nu_{\ell_1\ldots\ell_s})\frac{\partial}{\partial y^\sigma_{\ell_1\ldots\ell_s}},$$

where $1 \le i, j, j_1, \ldots, j_r, \ell_1, \ldots, \ell_r \le n$, $1 \le \sigma, \nu \le m$. In the following we will suppose that all considered vector fields are smooth, (or at least differentiable up to the required order). Similarly as in mechanics there are some special vector fields on fibred manifolds that respect their fibred structure. Their definitions are quite analogous as these in mechanics.

Definition 8.8 (*Projectable vector fields*) Let (Y, π, X) be a fibred manifold, dim $Y = m + n$, and (J^r, π_r, X) its rth jet prolongation. Denote $J^0Y = Y$, $\pi = \pi_0$ as in Chap. 2. Let $0 \le s \le r$.

(a) A vector field ξ on J^sY is called π_s-*projectable* if there exists a vector field ξ_0 on X such that it holds

$$T\pi_s \circ \xi = \xi_0 \circ \pi_s.$$

The vector field ξ_0 is called the π_s-*projection of* ξ.

(b) Let $s < r$. A vector field ξ_r on J^rY is called $\pi_{r,s}$ *projectable* if there exists a vector field ξ_s on J^sY such that it holds

$$T\pi_{r,s} \circ \xi_r = \xi_s \circ \pi_{r,s}.$$

The vector field ξ_s is called the $\pi_{r,s}$-*projection of* ξ_r.

Proposition 8.5 *Consider the notation presented in Definition 8.8. Let ξ be a π-projectable vector field on Y. Let (V, ψ), $\psi = (x^i, y^\sigma)$, be an arbitrarily chosen fibred chart on Y and $(\pi(V), \varphi)$, $\varphi = (x^i)$, the associated chart on X, and $(\pi^{-1}_{s,0}(V), \psi_s)$, $\psi_s = (x^i\ y^\sigma, y^\sigma_j, \ldots, y^\sigma_{j_1\ldots j_s})$, the associated fibred charts on J^sY, $0 \le s \le r$, $1 \le i, j, j_1, \ldots, j_s \le n$, $1 \le \sigma \le m$.*

(a) Let ξ_r be a π_r-projectable vector field on J^rY. Then its π_r-projection ξ_0 is unique and its chart expression reads

$$\xi_0 = \xi^i_0(x^1, \ldots, x^n)\frac{\partial}{\partial x^i}.$$

(b) Let ξ_r be a $\pi_{r,s}$-projectable vector field on J^rY. Then its $\pi_{r,s}$-projection ξ_s is unique and its chart expression reads

$$\xi_s = \xi^i_0(x^j, y^\nu, y^\nu_\ell, \ldots, y^\nu_{\ell_1\ldots\ell_s})\frac{\partial}{\partial x^i} + \sum_{p=0}^{s} \Xi^\sigma_{\ell_1\ldots\ell_p}(x^j, y^\nu, y^\nu_\ell, \ldots, y^\nu_{\ell_1\ldots\ell_s})\frac{\partial}{\partial y^\nu_{\ell_1\ldots\ell_p}}.$$

Proof Prove the assertion (a). We use charts specified in the proposition. Let ξ_r be a vector field defined on an open set $W_r \subset J^rY$ such that $W_r \subset V_r$, $V_r = \pi_{r,0}^{-1}(V)$. The chart expression of π_r is

$$\varphi \circ \pi_r \circ \psi_r^{-1} : \psi_r(V_r) \ni (x^j, y^\nu, y^\nu_\ell, \dots, y^\nu_{\ell_1 \dots \ell_r}) \longrightarrow (x^1, \dots, x^n) \in \varphi\big(\pi(V)\big).$$

The tangent mapping $T\pi_r$ is represented by the Jacobi matrix

$$D(\varphi \circ \pi_r \circ \psi_r^{-1})(x^j, y^\nu, y^\nu_\ell, \dots, y^\nu_{\ell_1 \dots \ell_r}) = \begin{pmatrix} E_n \\ 0_{M,n} \end{pmatrix},$$

where E_n is the unit matrix of the order n and $0_{M,n}$ is the zero matrix of the type M/n (M rows and n columns), $M = \dim J^rY - n$. So the tangent image of the vector

$$\xi_r(z) = \xi^i(z) \left.\frac{\partial}{\partial x^i}\right|_z + \sum_{s=0}^{r} \Xi^\nu_{\ell_1 \dots \ell_s}(z) \left.\frac{\partial}{\partial y^\nu_{\ell_1 \dots \ell_s}}\right|_z \quad \text{is}$$

$$\xi_0\big(\pi_r(z)\big) = T\pi_r\big(\xi_r(z)\big) = \xi^i\big(\pi_r(z)\big) \left.\frac{\partial}{\partial x^i}\right|_{\pi_r(z)}$$

Thus the vector field ξ_r is π_r-projectable if and only if $\xi_0 = \xi^i \frac{\partial}{\partial x^i}$ is a vector field on X. In such a case ξ_0 is unique and its components depend only on coordinates on X. The proof of the assertion (b) is quite analogous. □

Definition 8.9 (*Vertical vector fields*)

(a) vector field ξ_r on J^rY is called π_r*-vertical* on an open set $W_r \subset J^rY$ if $T\pi_r \circ \xi_r = 0$ on W_r.
(b) A vector field ξ_r on J^rY is called $\pi_{r,s}$*-vertical* on an open set $W_r \subset J^rY$ for $0 \le s < r$ if $T\pi_{r,s} \circ \xi_r = 0$ on W_r.

Similarly as in mechanics we need the procedure of prolongations of vector fields based on onc-parameter groups of vector fields. This concerns projectable vector fields. Their one-parameter groups are formed by local isomorphisms of underlying fibred manifolds. The procedure of prolongations of vector fields was in detail explained in Chap. 3. Here we present a brief recapitulation of general considerations and focus on our attention to chart expressions of prolonged vector fields.

Proposition 8.6 *Let ξ be a π-projectable vector field defined on an open set $W \subset Y$, and ξ_0 its projection on $\pi(W) \subset X$. Let $\mathcal{G}_W = \{\alpha_u \mid u \in (-\varepsilon, \varepsilon)\}$ and $\mathcal{G}_{\pi(W)} = \{\alpha_{0u} \mid u \in (-\varepsilon, \varepsilon)\}$ be one-parameter groups of ξ and ξ_0, respectively. Then for each value of the parameter u the mappings α_u and α_{0u} are local isomorphisms of Y and X, respectively, and α_{0u} is the projection of α_u, i.e. $\pi \circ \alpha_u = \alpha_{0u} \circ \pi$.*

Proof The proof is based on the properties of one-parameter groups of vector fields,

$$T\alpha_u \circ \xi = \xi \circ \alpha_{0u}, \quad \text{and} \quad T\alpha_{0u} \circ \xi_0 = \xi_0 \circ \alpha_{0u}$$

One can follow the proof of Proposition 3.2. □

Definition 8.10 (*Prolongations of vector fields*) Let ξ be a (smooth) π-projectable vector field on an open set $W \subset Y$ of a fibred manifold (Y, π, X), $\dim Y = m + \dim X = m + n$ and let the set $\{(\alpha_u, \alpha_{0u}) \mid u \in (-\varepsilon, \varepsilon)\}$ of local isomorphisms of the fibred manifold be its local one-parameter group. The vector field $J^r\xi$ on J^rY having the local one-parameter group $\{(J^r\alpha_u, \alpha_{0u}) \mid u \in (-\varepsilon, \varepsilon)\}$ is called *the* rth *prolongation of the vector field* ξ.

Let ξ be a π-projectable vector field defined on an open set $W \subset Y$. Let (V, ψ), $\psi = (x^i, y^\sigma)$, be a fibred chart on Y such that $\alpha_u(W) \subset V$ for some $u \in (-\varepsilon, \varepsilon)$, $(\pi(V), \varphi)$, $\varphi = (x^i)$, the associated chart on X, and $(\pi^{-1}_{s,0}(V), \psi_s)$, $\psi_s = (x^i\, y^\sigma, y^\sigma_j, \dots, y^\sigma_{j_1 \dots j_s})$, the associated fibred charts on J^sY, $1 \le s \le r$, $1 \le i, j, j_1, \dots, j_s \le n$, $1 \le \sigma \le m$.

For a point $z \in \pi^{-1}_{r,0}(W) \subset J^rY$, $z = (x^l, y^\nu, y^\nu_\ell, \dots, y^\nu_{\ell_1 \dots \ell_r})$, it holds

$$J^r\xi(z) = \xi^i(x^l)\frac{\partial}{\partial x^i} + \Xi^\sigma(x^l, y^\nu)\frac{\partial}{\partial y^\sigma} + \sum_{s=1}^{r} \Xi^\sigma_{j_1 \dots j_s}(x^l, y^\nu, y^\nu_\ell, \dots, y^\nu_{\ell_1 \dots \ell_s})\frac{\partial}{\partial y^\sigma_{j_1 \dots j_s}},$$

$$\Xi^\sigma_{j_1 \dots j_s}(x^l, y^\nu, y^\nu_\ell, \dots, y^\nu_{\ell_1 \dots \ell_s}) = \left. \frac{\mathrm{d}y^\sigma_{j_1 \dots j_s} J^s\alpha_u(x^l, y^\nu, y^\nu_\ell, \dots, y^\nu_{\ell_1 \dots \ell_s})}{\mathrm{d}u} \right|_{u=0},$$

shortly expressed

$$J^r\xi = \xi^i\frac{\partial}{\partial x^i} + \Xi^\sigma\frac{\partial}{\partial y^\sigma} + \sum_{s=1}^{r}\left[\frac{\mathrm{d}y^\sigma_{j_1 \dots j_s} J^s\alpha_u}{\mathrm{d}u}\right]_{u=0}\frac{\partial}{\partial y^\sigma_{j_1 \dots j_s}}. \tag{8.10}$$

Alternatively, for expressing $J^r\xi$ we can use jets of sections putting $z = J^r_x\gamma$. Then

$$J^r\xi(J^r_x\gamma) = \left.\frac{\mathrm{d}J^r\alpha_u(J^r_x\gamma)}{\mathrm{d}u}\right|_{u=0}. \tag{8.11}$$

Definition 8.10 and relations (8.10) enable us to express the components of vector fields arising as prolongations of π-projectable vector fields on Y explicitly using the prolongations of their one-parameter groups. Using relations (8.10), we can derive the recursive formulas for these components as follows:

$$\begin{aligned}
\xi &= \xi^i(x^j)\frac{\partial}{\partial x^i} + \Xi^\sigma(x^j, y^\nu)\frac{\partial}{\partial y^\sigma},\\
J^r\xi &= \xi^i(x^j)\frac{\partial}{\partial x^i} + \Xi^\sigma(x^j, y^\nu)\frac{\partial}{\partial y^\sigma} + \sum_{s=1}^{r}\Xi^\sigma_{i_1\ldots i_s}(x^j, y^\nu, \ldots, y^\nu_{\ell_1\ldots\ell_s})\frac{\partial}{\partial y^\sigma_{i_1\ldots i_s}},\\
\Xi^\sigma_{i_1\ldots i_s} &= \frac{\mathrm{d}\Xi^\sigma_{i_1\ldots i_{s-1}}}{\mathrm{d}x^{j_s}} - y^\sigma_{i_1\ldots i_{s-1}\ell}\frac{\mathrm{d}\xi^\ell}{\mathrm{d}x^{i_s}}.
\end{aligned} \tag{8.12}$$

Example 8.3 (Prolongations of vector fields) Let $(Y, \pi, X) = (\mathbf{R}^2 \times \mathbf{R}, p, \mathbf{R}^2)$ be a fibred manifold (fibred Euclidean space). Such a situation is relevant in physics, e.g. for the description of scalar waves propagating along a fixed line, where the wave amplitude depends on time t and the coordinate x. Fibred charts on this manifold are of the form (V, ψ), $\psi = (t, x, y)$, associated charts on X are $(p(V), \varphi)$, $\varphi = (t, x)$, associated fibred charts on J^1Y and J^2Y are $(\pi^{-1}_{1,0}(V), \psi_1)$, $\psi_1 = (t, x, y, y_t, y_x)$ and $(\pi^{-1}_{2,0}(V), \psi_2)$, $\psi_2 = (t, x, y, y_t, y_x, y_{tt}, y_{tx}, y_{xx})$. (The identical coordinates $y_{tx} = y_{xt}$ are represented by the unique independent coordinate y_{tx}). Let ξ be a (smooth) p-projectable vector field defined on an open set $W \subset V$,

$$\xi = \xi^t(t, x)\frac{\partial}{\partial t} + \xi^x(t, x)\frac{\partial}{\partial x} + \Xi(t, x, y)\frac{\partial}{\partial y}.$$

Using (8.12), we express the components of $J^1\xi$ and $J^2\xi$ explicitly through components of ξ and their derivatives. (These calculations and results will be important in Chap. 9 especially for formulating and solving the Noether equation, or Noether-Bessel-Hagen equation.)

$$\begin{aligned}
\Xi_t &= \frac{\mathrm{d}\Xi}{\mathrm{d}t} - y_t\frac{\partial\xi^t}{\partial t} - y_x\frac{\partial\xi^x}{\partial t} = \frac{\partial\Xi}{\partial t} + \frac{\partial\Xi}{\partial y}y_t - y_t\frac{\partial\xi^t}{\partial t} - y_x\frac{\partial\xi^x}{\partial t},\\
\Xi_x &= \frac{\mathrm{d}\Xi}{\mathrm{d}x} - y_t\frac{\partial\xi^t}{\partial x} - y_x\frac{\partial\xi^x}{\partial x} = \frac{\partial\Xi}{\partial x} + \frac{\partial\Xi}{\partial y}y_x - y_t\frac{\partial\xi^t}{\partial x} - y_x\frac{\partial\xi^x}{\partial x},\\
\Xi_{tt} &= \frac{\mathrm{d}\Xi_t}{\mathrm{d}t} - y_{tt}\frac{\partial\xi^t}{\partial t} - y_{tx}\frac{\partial\xi^x}{\partial t} = \frac{\partial\Xi_t}{\partial t} + \frac{\partial\Xi_t}{\partial y}y_t + \frac{\partial\Xi_t}{\partial y_t}y_{tt} + \frac{\partial\Xi_t}{\partial y_x}y_{tx}\\
&\quad - y_{tt}\frac{\partial\xi^t}{\partial t} - y_{tx}\frac{\partial\xi^x}{\partial t}
\end{aligned}$$

$$
\begin{aligned}
&= \frac{\partial^2 \Xi}{\partial t^2} + 2\frac{\partial^2 \Xi}{\partial t \partial y} y_t + \frac{\partial^2 \Xi}{\partial y^2} y_t^2 + \frac{\partial \Xi}{\partial y} y_{tt} - 2y_{tt}\frac{\partial \xi^t}{\partial t} - 2y_{tx}\frac{\partial \xi^x}{\partial t} \\
&- y_t \frac{\partial^2 \xi^t}{\partial t^2} - y_x \frac{\partial^2 \xi^x}{\partial t^2},
\end{aligned}
$$

$$
\begin{aligned}
\Xi_{tx} &= \frac{\mathrm{d}\Xi_t}{\mathrm{d}x} - y_{tt}\frac{\partial \xi^t}{\partial x} - y_{tx}\frac{\partial \xi^x}{\partial x} = \frac{\partial \Xi_t}{\partial x} + \frac{\partial \Xi_t}{\partial y} y_x + \frac{\partial \Xi_t}{\partial y_t} y_{tx} + \frac{\partial \Xi_t}{\partial y_x} y_{xx} \\
&- y_{tt}\frac{\partial \xi^t}{\partial x} - y_{tx}\frac{\partial \xi^x}{\partial x} \\
&= \frac{\partial^2 \Xi}{\partial t \partial x} + \frac{\partial^2 \Xi}{\partial t \partial y} y_x + \frac{\partial^2 \Xi}{\partial x \partial y} y_t + \frac{\partial^2 \Xi}{\partial y^2} y_t y_x + \frac{\partial \Xi}{\partial y} y_{tx} \\
&- y_{tt}\frac{\partial \xi^t}{\partial x} - y_{tx}\left(\frac{\partial \xi^t}{\partial t} + \frac{\partial \xi^x}{\partial x}\right) - y_{xx}\frac{\partial \xi^x}{\partial t} - y_t\frac{\partial^2 \xi^t}{\partial t \partial x} - y_x \frac{\partial^2 \xi^x}{\partial x^2},
\end{aligned}
$$

$$
\begin{aligned}
\Xi_{xx} &= \frac{\mathrm{d}\Xi_x}{\mathrm{d}x} - y_{tx}\frac{\partial \xi^t}{\partial x} - y_{xx}\frac{\partial \xi^x}{\partial x} = \frac{\partial \Xi_x}{\partial t} + \frac{\partial \Xi_x}{\partial y} y_x + \frac{\partial \Xi_x}{\partial y_t} y_{tx} + \frac{\partial \Xi_x}{\partial y_x} y_{xx} \\
&- y_{tx}\frac{\partial \xi^t}{\partial x} - y_{xx}\frac{\partial \xi^x}{\partial x} = \\
&= \frac{\partial^2 \Xi}{\partial x^2} + 2\frac{\partial^2 \Xi}{\partial x \partial y} y_x + \frac{\partial^2 \Xi}{\partial y^2} y_x^2 + \frac{\partial \Xi}{\partial y} y_{xx} - 2y_{tx}\frac{\partial \xi^t}{\partial x} - 2y_{xx}\frac{\partial \xi^x}{\partial x} \\
&- y_t \frac{\partial^2 \xi^t}{\partial x^2} - y_x \frac{\partial^2 \xi^x}{\partial x^2}.
\end{aligned}
$$

Definition 8.11 *(Distributions)* Let (Y, π, X) be a fibred manifold. Denote $T_y Y = T(\mathbf{R}^n \times \mathbf{R}^m)$ the tangent space to Y at the point $y \in Y$, and $TY = \cup_{y \in Y} T_y Y$. Similarly $T_z J^r Y$ and $T J^r Y$. *Distribution* on $J^r Y$, resp. on an open set $W_r \subset J^r Y$ is the mapping assigning to every point $z \in W_r$ certain vector subspace $\mathcal{D}(z)$ of $T_z J^r Y$, $r \geq 0$, i.e.

$$\mathcal{D} : (J^r Y \text{ resp. } W_r) \ni z \longrightarrow \mathcal{D}(z) \subset T_z J^r Y.$$

By the *rank of the distribution* we mean the mapping rank $\mathcal{D} : z \to \dim D(z)$. If $\dim \mathcal{D}(z)$ is the same in all points $z \in W_r$ the distribution is *of constant rank.*

The distribution is generated by vector fields forming in every point $z \in W$ the basis of the vector subspace $\mathcal{D}(z)$, $(\zeta_1, \ldots, \zeta_p)$, $1 \leq p \leq \dim \mathcal{D}(z)$.

Definition 8.12 *(Differential forms)* By a *differential k-form*, $k \geq 1$, on a fibred manifold $(J^r Y, \pi, X)$, $r \geq 0$, we mean a C^r-differentiable (or smooth) antisymmetric covariant k-tensor field, i.e. (C^r-differentiable, or smooth) mapping defined on an open set $W_r \subset J^r Y$

$$\omega : W_r \ni y \longrightarrow \omega(y) \in \Lambda^k(T_y J^r Y) \subset \Lambda^k(T J^r Y).$$

By 0-*form we mean a function* $f : W_r \to \mathbf{R}$. (Note that for $k > \dim J^r Y$ k-forms are in principle zero.)

The structure $(\Lambda^k(TY), \Pi, Y)$, where $\Pi : \Lambda^k(T_yY) \ni \omega(y) \to y \in Y$, is the fibred manifold. Let us express k-forms in fibred charts chosen as usual, (V, ψ), $\psi = (x^i, y^\sigma)$, $(\pi_{r,0}^{-1}(V), \psi_r)$, $\psi_r = (x^i, y^\sigma, y^\sigma_j, \ldots, y^\sigma_{j_1 \ldots j_r})$, $(\pi(V), \varphi)$, $\varphi = (x^i)$, $1 \leq i, j, j_1, \ldots, j_r \leq n$, $1 \leq \sigma \leq m$, $0 \leq r$. Let $\omega \in \Lambda^k(TY)$. Then we can write

(a) Chart expression of a k-form on the base X:

$$\omega = A_{i_1 \ldots i_k}(x^\ell)\, \mathrm{d}x^{i_1} \wedge \ldots \wedge \mathrm{d}x^{i_k}, \quad 1 \leq i_1, \ldots, i_k \leq n,$$

$A_{i_1 \ldots i_k}(x^\ell)$ are functions on X, antisymmetric in all indices. An n-form

$$\omega_0 = \mathrm{d}x^1 \wedge \mathrm{d}x^2 \wedge \ldots \wedge \mathrm{d}x^n$$

is the *volume element* on X.

(b) Chart expressions of $(n-1)$-forms, $(n-2)$-forms, ..., $(n-p)$-forms on X:

$$\begin{aligned} \omega_i &= i_{\partial/\partial x^i} \omega_0 = (-1)^{i-1}\, \mathrm{d}x^1 \wedge \ldots \wedge \mathrm{d}x^{i-1} \wedge \mathrm{d}x^{i+1} \wedge \ldots \wedge \mathrm{d}x^n, \\ \omega_{ij} &= i_{\partial/\partial x^j} \omega_i, \quad j \neq i, \quad \text{etc.} \end{aligned} \tag{8.13}$$

(c) Chart expression of a k-form, $1 \leq k \leq \dim Y$, on Y:

$$\omega = \sum_{p=0}^{k} A_{\sigma_1 \ldots \sigma_p, i_{p+1} \ldots i_k}(x^\ell, y^\nu)\, \mathrm{d}y^{\sigma_1} \wedge \ldots \wedge \mathrm{d}y^{\sigma_p} \wedge \mathrm{d}x^{i_{p+1}} \wedge \ldots \wedge \mathrm{d}x^{i_k},$$

$1 \leq i_{p+1}, \ldots, i_k \leq n$, $1 \leq \sigma_1, \ldots, \sigma_p \leq m$, coefficients are functions on Y and they are antisymmetric in indices $i_{p+1}, \ldots, i_k$, as well as in indices $\sigma_1, \ldots, \sigma_p$.

(d) A k form, $1 \leq k \leq \dim J^rY$, on J^rY has the chart expression

$$\omega = \sum_{p=0}^{k} \sum_{|J_q|=0}^{r} A^{J_1}_{\sigma_1} \cdots ^{J_p}_{\sigma_p, i_{p+1} \ldots i_k}\, \mathrm{d}y^{\sigma_1}_{J_1} \wedge \ldots \wedge \mathrm{d}y^{\sigma_p}_{J_p} \wedge \mathrm{d}x^{i_{p+1}} \wedge \ldots \wedge \mathrm{d}x^{i_k}, \tag{8.14}$$

$1 \leq i_{p+1}, \ldots, i_k \leq n$, $1 \leq \sigma_1, \ldots, \sigma_p \leq m$, where $J_1, \ldots, J_p$ are *multiindices*. For example, $J = (j_1 \ldots j_s)$, $1 \leq j_1, \ldots, j_s \leq n$, $0 \leq s \leq r$. The symbol $|J|$ is the *length of the multiindex*, i.e. the number of indices in the multiindex. Coefficients are functions on J^rY. They are antisymmetric in indices $i_{p+1}, \ldots, i_k$ and in multiindices $^{J_1}_{\sigma_1}, \ldots, ^{J_p}_{\sigma_p}$.

Similarly as in Sect. 3.2 we now define properties of differential forms connected with the fibred structure of manifolds.

Definition 8.13 (*Projectable forms*) Let (Y, π, X) be a fibred manifold and (J^rY, π_r, X) its r-jet prolongation, $r \geq 1$.

(a) A form $\omega \in \Lambda^k(TY)$ is called π*-projectable*, if there exists a form $\eta \in \Lambda^k(TX)$ such that $\omega = \pi^*\eta$.
(b) A form $\omega \in \Lambda^k(TJ^rY)$, $r \geq 1$, is called π_r*-projectable*, if there exists a form $\eta \in \Lambda^k(TX)$ so that $\omega = \pi_r^*\eta$.
(c) A form $\omega \in \Lambda^k(TJ^rY)$, $r \geq 1$, is called $\pi_{r,s}$*-projectable*, $0 \leq s < r$, if there exists a form $\eta \in \Lambda^k(TJ^sY)$ such that$\omega = \pi_{r,s}^*\eta$.

The form η is the *projection* of the form ω on the base X in cases (a) and (b) and the projection on J^sY in the case (c).

Definition 8.14 (*Horizontal forms*) Let (Y, π, X) be a fibred manifold and (J^rY, π_r, X) its r-jet prolongation, $r \geq 0$. Let $k \geq 1$.

(a) A k-form on J^rY is called π_r*-horizontal* , if it vanishes whenever any of its vector arguments is a π_r-vertical vector field, i.e.

$$i_\xi \omega = 0 \quad \text{for every} \quad \pi_r\text{-vertical vector field } \xi.$$

π_r-horizontal n forms on J^rY are called rth *order Lagrangians*.
(b) A k-form ω is called $\pi_{r,s}$*-horizontal*, if it vanishes whenever any of its vector argument is $\pi_{r,s}$-vertical vector field, i.e.

$$i_\xi \omega = 0 \quad \text{for every} \quad \pi_{r,s} \text{ vertical vector field } \xi.$$

Expressions of just defined forms in charts chosen above are as follows:

(a) π_r-horizontal k forms, $1 \leq k \leq n$:

$$\omega = \omega_{j_1 \dots j_k}\, \mathrm{d}x^{j_1} \wedge \dots \wedge \mathrm{d}x^{j_k}, \quad 1 \leq j_1, \dots, j_k \leq n,$$

where $\omega_{j_1 \dots j_k}$ are functions on J^rY, and $\omega_{j_1 \dots j_k} = \operatorname{sgn}(\sigma)\, \omega_{\sigma(j_1),\dots,\sigma(j_k)}$, $(\sigma(j_1), \dots, \sigma(j_k))$ is a permutation of indices $j_1, \dots, j_k$ (functions $\omega_{j_1 \dots j_k}$ are antisymetric in all indices).
(b) $\pi_{r,s}$-horizontal k-forms, $1 \leq k \leq n$:

$$\omega = \sum_{p=0}^{k} \sum_{|J_q|=0}^{r} A^{J_1}_{\sigma_1} \cdots {}^{J_p}_{\sigma_p, i_{p+1} \dots i_k}\, \mathrm{d}y^{\sigma_1}_{J_1} \wedge \dots \wedge \mathrm{d}y^{\sigma_p}_{J_p} \wedge \mathrm{d}x^{i_{p+1}} \wedge \dots \wedge \mathrm{d}x^{i_k},$$

$1 \leq i_{p+1}, \dots, i_k \leq n$, $1 \leq \sigma_1, \dots, \sigma_p \leq m$, where $J_1, \dots, J_p$ are *multiindices* for every of which it holds $0 \leq |J| \leq s$. This means that the chart expression of $\pi_{r,s}$-horizontal form contain any terms with $\mathrm{d}y^\sigma_{j_1 \dots j_s, j_{s+1} \dots j_{s+q}}$ for $q \geq 1$. On the other hand, coefficients of such a form are function defined on J^rY.

Another type of differential forms connected with the fibred structure of underlying manifolds are contact forms, similarly as in mechanics (see Sect. 3.3).

Definition 8.15 (*Contact forms*) Let (Y, π, X), $\dim X = n > 1$, be a fibred manifold, and $k \geq 1$.

(a) A k-form ω on $J^r Y$ is called *contact* if for every section γ of the projection π it holds $J^r\gamma^*\omega = 0$.
(b) A contact 1-form ω on $J^r Y$ is called 1-*contact*. A k-form ω on $J^r Y$, where $k > 1$, is called 1-*contact*, if its contraction by an arbitrary π_r-vertical vector field on $J^r Y$ is a horizontal form. A k-form ω on $J^r Y$, where $k > 1$, is called q-*contact*, $2 \leq q \leq k$, if its contraction by an arbitrary π_r-vertical vector field on $J^r Y$ is a $(q-1)$-contact $(k-1)$-form. Horizontal forms are called 0-*contact*.

On the base of Definition 8.15 one can formulate the following simple (no less important) proposition:

Proposition 8.7 *Let $\eta \in \Lambda^k(TJ^rY)$ be a contact k-form defined on an open set $W_r \subset J^rY$. Then it holds*

(a) The form $\eta \wedge \omega$ is contact for and arbitrary ℓ-form $\omega \in \Lambda^\ell(TJ^rY)$ defined on W_r.
(b) The exterior derivative $\mathrm{d}\eta$ of the form η is contact.

Proof The proof of both parts of the proposition is very simple. Let $\gamma \in \Gamma_U(\pi)$ be an arbitrary section. Then

$$J^r\gamma^*(\eta \wedge \omega) = J^r\gamma^*\eta \wedge J^r\gamma^*\omega = 0, \quad J^r\gamma^*\,\mathrm{d}\eta = \mathrm{d}J^r\gamma^*\eta = 0$$

because η is contact and then $J^r\gamma^*\eta = 0$ by Definition 8.15. □

As the consequence of Proposition 8.7 we can state that all contact forms on $W_r \subset J^rY$ form the ideal, called *the contact ideal*. (See also Sect. 3.3 for mechanics.)

Example 8.4 (The contact ideal) Let $\dim X = 2$, $\dim Y = 3$. Let (V, ψ), $\psi = (x^1, x^2, y)$, $(\pi(V), \varphi)$, $\varphi = (x^1, x^2)$, $(\pi_{1,0}^{-1}(V), \psi_1)$, $\psi_1 = (x^1, x^2, y, y_1, y_2)$, $(\pi_{2,0}^{-1}(V), \psi_2)$, $\psi_2 = (x^1, x^2, y, y_1, y_2, y_{11}, y_{12}, y_{22})$, be a fibred chart on Y, associated chart on X, and associated fibred charts on J^1Y and J^2Y.

(a) Let $\omega \in \Lambda^1(TY)$ be defined on an open set $W \subset Y$, and let $\gamma \in \Gamma_{\pi(V)}(\pi)$ be an arbitrarily chosen section of the projection π. The chart expression of ω is

$$\omega = \omega_y\,\mathrm{d}y^1 + \omega_1\,\mathrm{d}x^1 + \omega_2\,\mathrm{d}x^2$$

where ω_y, ω_1 and ω_2 are functions on W (the components of the form). The form ω is contact if and only if $\gamma^*\omega = 0$, i.e.

$$(\omega_y \circ \gamma)\gamma^*\,\mathrm{d}y^1 + (\omega_1 \circ \gamma)\gamma^*\,\mathrm{d}x^1 + (\omega_2 \circ \gamma)\gamma^*\,\mathrm{d}x^2 = 0,$$

$$\gamma^* \mathrm{d}x^1 = \mathrm{d}x^1, \quad \gamma^* \mathrm{d}x^2 = \mathrm{d}x^2, \quad \gamma^* \mathrm{d}y = \frac{\partial y\gamma(x^1, x^2)}{\partial x^1} \mathrm{d}x^1 + \frac{\partial y\gamma(x^1, x^2)}{\partial x^2} \mathrm{d}x^2.$$

Putting $\gamma^*\omega = 0$ we obtain

$$\left((\omega_y \circ \gamma) \frac{\partial y\gamma}{\partial x^1} + (\omega_1 \circ \gamma) \right) \mathrm{d}x^1 + \left((\omega_y \circ \gamma) \frac{\partial y\gamma}{\partial x^2} + (\omega_2 \circ \gamma) \right) \mathrm{d}x^2 = 0$$

$$\Longrightarrow (\omega_y \circ \gamma) \frac{\partial y\gamma}{\partial x^1} + \omega_1 \circ \gamma = 0, \quad (\omega_y \circ \gamma) \frac{\partial y\gamma}{\partial x^2} + \omega_2 \circ \gamma = 0$$

for an arbitrary section γ. This condition can be fulfilled in only situation, when all components of ω are zero. This means that no contact 1-forms exist on the considered fibred manifold.

(b) Let $\omega \in \Lambda^2(TY)$,

$$\omega = \omega_{y,1} \, \mathrm{d}y \wedge \mathrm{d}x^1 + \omega_{y,2} \, \mathrm{d}y \wedge \mathrm{d}x^2 + \omega_{12} \, \mathrm{d}x^1 \wedge \mathrm{d}x^2,$$

where $\omega_{y,1}$, $\omega_{y,2}$ and ω_{12} are again functions on Y, and let $\gamma \in \Gamma_{\pi(W)}$ be an arbitrary section. Then

$$\begin{aligned} \gamma^*\omega &= (\omega_{y,1} \circ \gamma) \left(\frac{\partial y\gamma}{\partial x^1} \mathrm{d}x^1 + \frac{\partial y\gamma}{\partial x^2} \mathrm{d}x^2 \right) \wedge \mathrm{d}x^1 \\ &+ (\omega_{y,2} \circ \gamma) \left(\frac{\partial y\gamma}{\partial x^1} \mathrm{d}x^1 + \frac{\partial y\gamma}{\partial x^2} \mathrm{d}x^2 \right) \wedge \mathrm{d}x^2 + (\omega_{12} \circ \gamma) \, \mathrm{d}x^1 \wedge \mathrm{d}x^2 \\ &= \left((-\omega_{y,1} \circ \gamma) \frac{\partial y\gamma}{\partial x^2} + (\omega_{y,2} \circ \gamma) \frac{\partial y\gamma}{\partial x^1} + (\omega_{12} \circ \gamma) \right) \mathrm{d}x^1 \wedge \mathrm{d}x^2. \end{aligned}$$

The required condition

$$\gamma^*\omega = 0 \Longrightarrow -(\omega_{y,1} \circ \gamma) \frac{\partial y\gamma}{\partial x^2} + (\omega_{y,2} \circ \gamma) \frac{\partial y\gamma}{\partial x^1} + (\omega_{12} \circ \gamma) = 0$$

again can not be fulfilled for an arbitrary section.

(c) Consider a form $\omega \in \Lambda^1(TJ^1Y)$ defined on $W_1 \subset J^1Y$

$$\omega = \omega_y \, \mathrm{d}y + \omega_{y1} \, \mathrm{d}y_1 + \omega_{y2} \, \mathrm{d}y_2 + \omega_1 \, \mathrm{d}x^1 + \omega_2 \, \mathrm{d}x^2,$$

where coefficients ω_y, ω_{y1}, ω_{y2}, ω_1 and ω_2 are now functions defined on the subset W_1 of J^1Y, and let $\gamma \in \Gamma_U(\pi)$, $U = \pi_1(W_1)$, be an arbitrary section.

$$\begin{aligned} J^1\gamma^*\omega &= (\omega_y \circ J^1\gamma) \left(\frac{\partial y J^1\gamma}{\partial x^1} \mathrm{d}x^1 + \frac{\partial y J^1\gamma}{\partial x^2} \mathrm{d}x^2 \right) \\ &+ (\omega_{y1} \circ J^1\gamma) \left(\frac{\partial y_1 J^1\gamma}{\partial x^1} \mathrm{d}x^1 + \frac{\partial y_1 J^1\gamma}{\partial x^2} \mathrm{d}x^2 \right) \end{aligned}$$

$$
\begin{aligned}
&+ (\omega_{y2} \circ J^1\gamma)\left(\frac{\partial y_2 J^1\gamma}{\partial x^1}\,\mathrm{d}x^1 + \frac{\partial y_2 J^1\gamma}{\partial x^2}\,\mathrm{d}x^2\right) \\
&+ (\omega_1 \circ J^1\gamma)\,\mathrm{d}x^1 + (\omega_2 \circ J^1\gamma)\,\mathrm{d}x^2 \\
&= \Bigg((\omega_y \circ J^1\gamma)\frac{\partial y J^1\gamma}{\partial x^1} + (\omega_{y1} \circ J^1\gamma)\frac{\partial y_1 J^1\gamma}{\partial x^1} \\
&+ (\omega_{y2} \circ J^1\gamma)\frac{\partial y_2 J^1\gamma}{\partial x^1} + (\omega_1 \circ J^1\gamma)\Bigg)\,\mathrm{d}x^1 \\
&+ \Bigg((\omega_y \circ J^1\gamma)\frac{\partial y J^1\gamma}{\partial x^2} + (\omega_{y1} \circ J^1\gamma)\frac{\partial y_1 J^1\gamma}{\partial x^2} \\
&+ (\omega_{y2} \circ J^1\gamma)\frac{\partial y_2 J^1\gamma}{\partial x^2} + (\omega_2 \circ J^1\gamma)\Bigg)\,\mathrm{d}x^2.
\end{aligned}
$$

Putting $J^1\gamma^*\omega = 0$ for every section γ and taking into account that

$$
\frac{\partial y J^1\gamma}{\partial x^1} = y_1 J^1\gamma, \quad \text{and} \quad \frac{\partial y J^1\gamma}{\partial x^2} = y_2 J^1\gamma
$$

we can see that the condition for the contactness of the considered form reads

$$
\omega_{y1} = \omega_{y2} = 0, \quad \omega_1 = -y_1\omega_y, \quad \omega_2 = -y_2\omega_y,
$$

$$
\omega = \omega_y(\mathrm{d}y - y_1\,\mathrm{d}x^1 - y_2\,\mathrm{d}x^2).
$$

(d) A 2-form on J^1Y

$$
\begin{aligned}
\omega &= \omega_{y,1}\,\mathrm{d}y \wedge \mathrm{d}x^1 + \omega_{y,2}\,\mathrm{d}y \wedge \mathrm{d}x^2 + \omega_{y_1,1}\,\mathrm{d}y_1 \wedge \mathrm{d}x^1 + \omega_{y_1,2}\,\mathrm{d}y_1 \wedge \mathrm{d}x^2 \\
&+ \omega_{y_2,1}\,\mathrm{d}y_2 \wedge \mathrm{d}x^1 + \omega_{y_2,2}\,\mathrm{d}y_2 \wedge \mathrm{d}x^2 + \omega_{12}\,\mathrm{d}x^1 \wedge \mathrm{d}x^2 \\
&+ \omega_{y,y_1}\,\mathrm{d}y \wedge \mathrm{d}y_1 + \omega_{y,y_2}\,\mathrm{d}y \wedge \mathrm{d}y_2 + \omega_{y_1,y_2}\,\mathrm{d}y_1 \wedge \mathrm{d}y_2
\end{aligned}
$$

The form ω is contact if and only if its coefficients obey following relations:

$$
\begin{aligned}
&\omega_{12} - y_2\omega_{y,1} + y_1\omega_{y,2} = 0, \quad \omega_{y_1,y_2} = 0, \\
&\omega_{y_1,2} = y_2\omega_{y,y_1}, \quad \omega_{y_2,1} = y_1\omega_{y,y_2}, \\
&\omega_{y_1,1} - \omega_{y_2,2} = y_1\omega_{y,y_1} - y_2\omega_{y,y_2}.
\end{aligned}
$$

Let (V, ψ), $\psi = (x^i, y^\sigma)$, $(\pi(V), \varphi)$, $\varphi = (x^i)$, $1 \le i \le n$, $1 \le \sigma \le m$, be a fibred chart on Y and the associated chart on X, respectively, and $\left(\pi_{r,0}^{-1}(V), \psi_r\right)$ the associated fibred chart on J^rY. The chart expression of every k-form on J^rY, $1 \le k \le \dim J^rY$, $r \ge 0$, i.e. k-form defined on an open set $W_r \subset J^rY$ is expressed as linear combination (with functions on W_r as coefficients) of terms formed by a products of k 1-forms chosen from the set

$$
\{\mathrm{d}x^i, \mathrm{d}y^\nu, \mathrm{d}y^\nu_i, \mathrm{d}y^\nu_{ij}, \dots, \mathrm{d}y^\nu_{j_1\dots j_s}, \dots, \mathrm{d}y^\nu_{j_1\dots j_r}\}, \tag{8.15}
$$

where $1 \leq i, j, j_1, \dots, j_r \leq n$, $1 \leq \sigma \leq m$, $r \geq 0$, and

$$y^{\nu}_{j_1 \dots j_s} = y^{\nu}_{\sigma(j_1)\dots\sigma(j_s)}, \quad 1 \leq s \leq r,$$

for every permutation $\{\sigma(j_1)\dots\sigma(j_s)\}$ of indices $j_1, \dots, j_s$. We recall this set *basis of* 1-*forms* on $J^r Y$.

Reproducing the procedure used in Example 8.4 on can relatively easily prove the following proposition.

Proposition 8.8 *Let* (V, ψ), $\psi = (x^i, y^\sigma)$, $\big(\pi(V), \varphi\big)$, $\varphi = (x^i)$, $1 \leq i \leq n$, $1 \leq \sigma \leq m$, *be a fibred chart on* Y *and the associated chart on* X, *respectively, and* $\big(\pi^{-1}_{r,0}(V), \psi_r\big)$, $\psi_r = (x^i, y^\sigma, y^\sigma_j, \dots, y^\sigma_{j_1\dots j_r})$, *the associated fibred chart on* $J^r Y$. *The* 1-*forms* $\omega^\sigma, \omega^\sigma_j, \omega^\sigma_{ij}, \dots, \omega^\sigma_{j_1\dots j_{r-1}}$ *defined on an open set* $W_r \subset J^r Y$ *contained in* $V_r = \pi^{-1}_{r,0}(V)$, *where (in Einstein summation convention)*

$$\begin{aligned}
\omega^\sigma &= \mathrm{d}y^\sigma - y^\sigma_i \,\mathrm{d}x^i \\
\omega^\sigma_i &= \mathrm{d}y^\sigma_i - y_{ij} \,\mathrm{d}x^j \\
&\dots \\
\omega^\sigma_{j_1\dots j_s} &= \mathrm{d}y^\sigma_{j_1\dots j_s} - y^\sigma_{j_1\dots j_s \ell} \,\mathrm{d}x^\ell, \\
\omega^\sigma_{j_1\dots j_{r-1}} &= \mathrm{d}y^\sigma_{j_1\dots j_{r-1}} - y^\sigma_{j_1\dots j_{r-1}\ell} \,\mathrm{d}x^\ell,
\end{aligned} \tag{8.16}$$

are contact. (Alternatively, the above presented summations can be expressed with independent terms only, see (8.8).)

Proof Let $\gamma \in \Gamma_W(\pi)$, where $W \subset \pi(V)$ be an open set. Then by the direct calculation we obtain for $1 \leq s \leq r - 1$:

$$\begin{aligned}
J^s\gamma^* \omega^\sigma_{j_1\dots j_s} &= J^s\gamma^* \,\mathrm{d}y^\sigma_{j_1\dots j_s} - (y^\sigma_{j_1\dots j_{s-1}\ell} \circ J^s\gamma) J^s\gamma^* \,\mathrm{d}x^\ell \\
&= \frac{\partial (y^\sigma_{j_1\dots j_{s-1}} \circ J^s\gamma)}{\partial x^\ell} \,\mathrm{d}x^\ell - (y^\sigma_{j_1\dots j_{s-1}\ell} \circ J^s\gamma) \,\mathrm{d}x^\ell = 0.
\end{aligned}$$

As a section γ was chosen arbitrarily, this finishes the proof. □

Using relations (8.16) we can write

$$\mathrm{d}y^\sigma = \omega^\sigma + y^\sigma_i \,\mathrm{d}x^i, \quad \dots, \quad \mathrm{d}y^\sigma_{j_1\dots j_{s-1}} = \omega^\sigma_{j_1\dots j_{s-1}} + y^\sigma_{j_1\dots j_{s-1}\ell} \,\mathrm{d}x^\ell, \quad \dots$$

and instead of the basis of 1-forms given by (8.15) we can use contact forms given by (8.16) as follows.

Definition 8.16 (*Basis of 1-forms adapted to the contact structure*) The basis of 1-forms defined on $J^r Y$, $r \geq 1$,

$$\{dx^i, \omega^\sigma, \omega^\sigma_i, \omega^\sigma_{ij}, \ldots, \omega^\sigma_{j_1 \ldots j_{r-1}}, dy^\nu_{j_1 \ldots j_r}\} \tag{8.17}$$

is called the *basis of* 1-*forms adapted to the contact structure.*

Example 8.5 (Chart expressions of forms) Let $(Y, \pi, X) = (\mathbf{R}^n \times \mathbf{R}^m, p, \mathbf{R}^n)$. Let η_r be a 1-form defined on an open set $W_r \subset J^r Y$ for $r = 1, 2$. The chart expressions of η_1 and η_2 in the basis of 1 forms (8.15) is

$$\eta_1 = A_i \, dx^i + B_\sigma \, dy^\sigma + C^i_\sigma \, dy^\sigma_i ,$$

coefficients being functions defined on W_r, and

$$\eta_2 = A_i \, dx^i + B_\sigma \, dy^\sigma + C^i_\sigma \, dy^\sigma_i + D^{ij}_\sigma \, dy^\sigma_{ij}, \quad D^{ij}_\sigma = D^{ji}_\sigma .$$

Putting $dy^\sigma = \omega^\sigma + y^\sigma_j \, dx^j$ and $dy^\sigma_i = \omega^\sigma_i + y^\sigma_{ij} \, dx^j$ we can express the forms η_1 and η_2 with help of the basis adapted to the contact structure (8.17):

$$\eta_1 = (A_i + B_\sigma y^\sigma_i) \, dx^i + B_\sigma \, \omega^\sigma + C^i_\sigma \, dy^\sigma_i ,$$
$$\eta_2 = (A_i + B_\sigma y^\sigma_i + C^j_\sigma y^\sigma_{ji}) \, dx^i + B_\sigma \, \omega^\sigma + C^i_\sigma \omega^\sigma_i + D^{ij}_\sigma \, dy^\sigma_{ij} .$$

The procedure leading to chart expressions of forms in bases adapted to the contact structure can be in principle easily generalized for arbitrary order r of prolongation of the underlying fibred manifold for arbitrary k-form, $1 \leq k \leq \dim J^r Y$, using relations $dy^\sigma_{j_1 \ldots j_{s-1}} = \omega^\sigma_{j_j \ldots j_{s-1}} + y^\sigma_{j_1 \ldots j_{s-1} \ell} \, dx^\ell$. (However, explicit transitions between bases (8.15) and (8.17), especially for general k, are tedious from the point of view of technical calculations. For theoretical considerations, it is advantageous to express differential forms primarily in the adapted bases.)

Example 8.6 (Decomposition of forms) Consider the 1-form $\eta_1 \in \Lambda^1(T J^1 Y)$ studied in Example 8.5. This form is defined on an open set $W_1 \subset J^1 Y$,

$$\eta_1 = A_i \, dx^i + B_\sigma \, dy^\sigma + C^i_\sigma \, dy^\sigma_i ,$$

coefficients A_i, B_σ, C^i_σ are functions on W_1, i.e. functions of variables (x^j, y^ν, y^ν_j), $1 \leq i, j \leq n$, $1 \leq \sigma, \nu \leq m$. Using lift on the space $J^2 Y$ we can express the lifted form completely with help of contact 1-forms on $J^2 Y$:

$$\pi^*_{2,1} \eta_1 = (A_i + B_\sigma y^\sigma_i + C^j_\sigma y^\sigma_{ij}) \, dx^i + B_\sigma \, \omega^\sigma + C^i_\sigma \omega^\sigma_i .$$

We can see that the form

$$h\eta_1 = (A_i + B_\sigma y^\sigma_i + C^j_\sigma y^\sigma_{ij}) \, dx^i$$

is horizontal, while the form

$$p\eta_1 = B_\sigma\, \omega^\sigma + C^i_\sigma \omega^\sigma_i$$

is contact. So,

$$\pi^*_{2,1}\eta_1 = h\eta_1 + p\eta_1.$$

Analogously, for a 1-form η_2 on J^2Y we can write

$$\pi^*_{3,2}\eta_2 = h\eta_2 + p\eta_2,$$

where

$$h\eta_2 = (A_i + B_\sigma y^\sigma_i + C^j_\sigma y^\sigma_{ji} + D^{\ell j}_\sigma\, y^\sigma_{\ell ji})\, \mathrm{d}x^i\,, \quad p\eta_2 = B_\sigma\, \omega^\sigma + C^i_\sigma \omega^\sigma_i + D^{ij}_\sigma\, \omega^\sigma_{ij}.$$

This simple example implies that it will be possible to decompose lifts of forms to horizontal and contact components (see Sect. 3.3 in Chap. 3).

Let us now examine the structure of contact forms on J^rY, (Y, π, X) with $\dim X = n > 1$ in general. Denote the specific types of k-forms defined, e.g. on an open set $W_r \subset J^rY$ as follows:

$$\begin{aligned}
\omega_0 &= \mathrm{d}x^{i_1} \wedge \ldots \wedge \mathrm{d}x^{i_k}, \quad 1 \le k \le n,\\
\omega_1 &= \omega^\sigma_{j_1\ldots j_s} \wedge \mathrm{d}x^{i_1} \wedge \ldots \wedge \mathrm{d}x^{i_{k-1}}, \quad 0 \le s \le r-1,\\
\omega_2 &= \omega^\sigma_{j_1\ldots j_s} \wedge \omega^\nu_{l_1\ldots l_p} \wedge \mathrm{d}x^{i_1} \wedge \ldots \wedge \mathrm{d}x^{i_{k-2}}, \quad 0 \le s, p \le r-1,\\
&\ldots,\\
\omega_{k-1} &= \omega^{\sigma_1}_{J_1} \wedge \omega^{\sigma_2}_{J_2} \wedge \ldots \wedge \omega^{\sigma_{k-1}}_{J_{k-1}} \wedge \mathrm{d}x^i,\\
\omega_k &= \omega^{\sigma_1}_{J_1} \wedge \ldots \wedge \omega^{\sigma_k}_{J_k},
\end{aligned} \tag{8.18}$$

with multiindices $J_1, \ldots, J_k$ of various lengths $0 \le |J| \le r-1$. The form ω_0 is evidently horizontal, remaining forms ω_q, $1 \le q \le k-1$ are contact, because $J^r\gamma^*\omega_q = 0$ for an arbitrary section $\gamma \in \Gamma_U(\pi)$, $U = \pi_r(W_r)$. (Notice that the form ω_q contains just q 1-contact factors of the type ω^σ_J.) Let

$$\xi = \sum_{s=0}^{r} \Xi^\nu_{j_1\ldots j_s} \frac{\partial}{\partial y^\nu_{j_1\ldots j_s}}, \quad 1 \le j_1, \ldots, j_s \le n, \quad 1 \le \nu \le m,$$

be a π_r-vertical vector field defined on W. Contraction of forms ω_q by ξ gives

$$\begin{aligned}
\chi_0 &= i_\xi\omega_0 = 0,\\
\chi_1 &= i_\xi\omega_1 = \Xi^\sigma_{j_1\ldots j_s}\, \mathrm{d}x^{i_1} \wedge \ldots \wedge \mathrm{d}x^{i_{k-1}},\\
\chi_2 &= i_\xi\omega_2 = \Xi^\sigma_{j_1\ldots j_s} \omega^\nu_{l_1\ldots l_p} \wedge \mathrm{d}x^{i_1} \wedge \ldots \wedge \mathrm{d}x^{i_{k-2}} - \Xi^\nu_{l_1\ldots l_p} \omega^\sigma_{j_1\ldots j_s}\, \mathrm{d}x^{i_1} \wedge \ldots \wedge \mathrm{d}x^{i_{k-2}},\\
&\ldots,\\
\chi_k &= i_\xi\omega_k = \Xi^{\sigma_1}_{J_1}\, \omega^{\sigma_2}_{J_2} \wedge \ldots \wedge \omega^{\sigma_k}_{J_k} + \cdots + (-1)^{k-1}\, \Xi^{\sigma_k}_{J_k}\, \omega^{\sigma_1}_{J_1} \wedge \ldots \wedge \omega^{\sigma_{k-1}}_{J_{k-1}}.
\end{aligned}$$

As the consequence of these relations and from the point of view of Definition 8.15, we can formulate the assertion on the decomposition of forms into their contact components of various degrees.

Theorem 8.1 *Let $\omega_q \in \Lambda^k(TJ^rY)$, $0 \le q \le k$, be k-forms defined by relations (8.18). Let $\eta \in \Lambda^k(TJ^rY)$ be a k-form defined on an open set $W_s \subset J^sY$, $0 \le s \le r-1$. Then*

(a) The form ω_q is q-contact.
(b) The form

$$\pi^*_{s+1,s}\eta = \sum_{p=0}^{k} p_k\eta = p_0\eta + \cdots + p_k\eta, \tag{8.19}$$

where $p_0\eta = h\eta$, $p\eta = p_1\eta + \cdots + p_k\eta$ and forms $p_1\eta, \dots, p_k\eta$ are 1-contact, ..., k-contact, respectively. The decomposition (8.19) is unique.
(c) Denote as $\Lambda^k_{J^{r-1}Y}(TJ^rY)$ the module of all $\pi_{r,r-1}$-horizontal k-forms and $\Lambda^{(k,q)}_{J^{r-1}Y}(TJ^rY)$, $0 \le q \le k$, a submodule of q-contact $\pi_{r,r-1}$-horizontal k-forms on $W_r \subset J^rY$. Then

$$\Lambda^k_{J^{r-1}Y}(TJ^rY) = \oplus^k_{q=0}\Lambda^{(k,q)}_{J^{r-1}Y}(TJ^rY).$$

Definition 8.17 (*Horizontalization, contactization*) Denote $\Lambda^k_X(TJ^rY)$ the space of all π_r-horizontal k-forms defined on J^rY (or on an open set $W_r \subset J^rY$). Then we use the following terminology:

(a) The mapping

$$h : \Lambda^k(TJ^rY) \ni \omega \longrightarrow h\omega \in \Lambda^k_X(TJ^{r+1}Y)$$

is called *the horizontalization*.
(b) The mapping

$$p : \Lambda^k(TJ^rY) \ni \omega \longrightarrow \pi^*_{r+1,r}\omega - h\omega \in \Lambda^k(TJ^{r+1}Y)$$

is called *the contactization*.
(c) The mapping

$$p_q : \Lambda^k(TJ^rY) \ni \omega \longrightarrow p_q\omega \in \Lambda^k(TJ^{r+1}Y)$$

is called *the q-contactization*.
(d) A form $\omega \in \Lambda^k(TJ^rY)$ is called *at most q-contact*, if $p_{q+1}\omega = \cdots = p_k\omega = 0$.
(e) A form $\omega \in \Lambda^k(TJ^rY)$ is called *at least q-contact*, if $p_0 = \ldots = p_{q-1} = 0$.
(f) A form $\omega \in \Lambda^k(TJ^rY)$ is called *strongly contact* , if $p_0 = \ldots = p_{k-n} = 0$.

(g) A vector field ξ defined on $W_r \subset J^rY$ is called *the contact symmetry* if for every contact form $\omega \in \Lambda^k(TJ^rY)$ on W_r is its Lie derivative contact too. (This means that for a contact symmetry on J^rY it holds

$$\omega \in \mathcal{I}_c(J^rY) \implies \partial_\xi \omega \in \mathcal{I}_c(J^rY),$$

where $\mathcal{I}_c(J^rY)$ is the contact ideal on J^rY.)

Notice that all operators mentioned in Definition 8.17 are linear. Now consider the properties of horizontalization and contactization with respect to other algebraic and analytic operations.

Proposition 8.9 *Let $\omega \in \Lambda^k(TJ^rY)$ and $\eta \in \Lambda^\ell(TJ^rY)$ are differential forms defined on an open set $W_r \subset J^rY$. Let ξ be a π-projectable vector field on $W = \pi_{r,0}(W_r)$. Then the following relations hold:*

(a) $h(\omega \wedge \eta) = h\omega \wedge h\eta$,
(b) The Lie derivative of every π_r-horizontal form $\omega \in \Lambda^k(TJ^rY)$ with respect to every vector field $J^r\xi$ is horizontal. Moreover

$$\partial_{J^{r+1}\xi} h\omega = h\partial_{J^r\xi}\omega.$$

(c) For every π-projectable vector field ξ on $W \subset Y$ the prolongation $J^r\xi$ is the contact symmetry on J^rY, i.e. $\partial_{J^r\xi}\omega$ is contact for every contact form defined on $\pi^{-1}_{r,0}(W)$.
(d) If a π_r-projectable vector field ζ on $W_r \subset J^rY$ is the contact symmetry on J^rY, then there exists a π-projectable vector field ξ on $\pi_{r,0}(W_r)$ such that $\zeta = J^r\xi$.

Proof Let ω and η be forms specified in the proposition.

(a) Calculating $\pi^*_{r+1,r}(\omega \wedge \eta)$ we obtain

$$\begin{aligned}\pi^*_{r+1,r}(\omega \wedge \eta) &= \pi^*_{r+1,r}\omega \wedge \pi^*_{r+1,r}\eta = (h\omega + p\omega) \wedge (h\eta + p\eta)\\ &= h\omega \wedge h\eta + (h\omega \wedge p\eta + p\omega \wedge h\eta + p\omega \wedge p\eta),\\ h(\omega \wedge \eta) &= h\omega \wedge h\eta, \quad p(\omega \wedge \eta) = h\omega \wedge p\eta + p\omega \wedge h\eta + p\omega \wedge p\eta.\end{aligned}$$

(b) Let ω be a π_r-horizontal k form on $W_r \subset J^rY$, $1 \leq k \leq n$,

$$\omega = A_{i_1 \ldots i_k}\,\mathrm{d}x^{i_1} \wedge \ldots \wedge \mathrm{d}x^{i_k}$$

and let $J^r\xi$ be the rth prolongation of a π-projectable vector field ξ defined on $\pi_{r,0}(W_r)$ with the one-parameter group (α_u, α_{0u}). The Lie derivative of ω with respect of $J^r\xi$ is by definition

$$\partial_{J^r\xi}\omega = \lim_{u\to 0}\frac{J^r\alpha_u^*\omega - \omega}{u}. \text{ Then}$$

$$J^r\alpha_u^*\omega = (A_{i_1\dots i_k}\circ J^r\alpha_u)\, J^r\alpha_u^*\,\mathrm{d}x^{i_1}\wedge\dots\wedge J^r\alpha_u^*\,\mathrm{d}x^{i_k}$$
$$= (A_{i_1\dots i_k}\circ J^r\alpha_u)\,\alpha_{0u}^*\,\mathrm{d}x^{i_1}\wedge\dots\wedge\alpha_{0u}^*\,\mathrm{d}x^{i_k},$$

which is the horizontal form. This finishes the proof of (b).

(c) Let ξ be a π-projectable vector field on $W\subset Y$. Let $\omega\in\Lambda^k(TJ^rY)$ be a contact form defined on $\pi_{r,0}^{-1}(W)$ and let $\gamma\in\Gamma_{\pi(W)}(\pi)$. Then it holds

$$J^r\gamma^* J^r\alpha_u^*\omega = (J^r\alpha_u\circ J^r\gamma)^*\omega = (J^r\alpha_u\circ J^r\gamma\circ\alpha_{0u}^{-1}\circ\alpha_{0u})^*\omega$$
$$= \alpha_{0u}^*(J^r\alpha_u\circ J^r\gamma\circ\alpha_{0u}^{-1})^*\omega = \alpha_{0u}^* J^r(\alpha_u\circ\gamma\circ\alpha_{0u}^{-1})^*\,\omega = 0.$$

The last equality results from the facts that the mapping $\alpha_u\circ\gamma\circ\alpha_{0u}^{-1}$ is the section of π and ω is contact. As a result we obtain

$$J^r\gamma^*\frac{J^r\alpha_u^*\omega - \omega}{u} = 0$$

for all $u\in(-\varepsilon,\varepsilon)$ for admissible ε. This finishes the proof of (c). □

For completeness let us present the Poincaré lemma for contact forms without proof (the proof for mechanics see in detail in Chap. 3, Theorem 3.2, the complete proof of quite general case see, e.g. in [1,2]).

Theorem 8.2 *Poincaré lemma. Let* (Y,π,X), $\dim X = n > 1$, $\dim Y = m+n$, *be a fibred manifold. Let* $\eta\in\Lambda^k(TJ^rY)$, $k < \dim J^rY$ *be a closed* q *contact form (i.e.* $1\le q\le k$, $\mathrm{d}\eta = 0$*) defined on an simply connected open set* $W_r\subset J^rY$. *Then there exists a* $(q-1)$*-contact form* $\varrho\in\Lambda^{k-1}(TJ^rY)$ *defined on* W_r *such that* $\eta = \mathrm{d}\varrho$.

For practical purposes important in applications in physics we now present, instead of the proof, the general technical procedure for obtaining the form ϱ, where $\mathrm{d}\varrho = \eta$ for a closed form η. The reader can find the complete proof in [1] or [2]. For mechanics of first-order one can follow the proof of Theorem 3.2. The $(q-1)$-contactness of the $(k-1)$-form $\varrho = \mathrm{d}\eta$ for a closed q-contact k-form η results from the construction of ϱ.

Consider the fibred manifold $(Y,\pi,X) = (\mathbf{R}^n\times\mathbf{R}^m, p, \mathbf{R}^n)$. Let W_r be an open simply connected subset of J^rY (e.g. an open ball). By the analogy with the proof of Theorem 3.2 define the mapping

$$\chi : [0,1]\times W_r \ni (u, x^i, y^\sigma, y^\sigma, y^\sigma_{j_1},\dots, y^\sigma_{j_1\dots j_r})$$
$$\longrightarrow (x^i, uy^\sigma, uy^\sigma_{j_1},\dots, uy^\sigma_{j_1\dots j_r})\in J^rY, \tag{8.20}$$

$1\le i, j_1,\dots, j_r\le n,\quad 1\le\sigma\le m$. Then it holds

$$\chi^*\,\mathrm{d}x^i = \mathrm{d}x^i,\quad \chi^*\,\mathrm{d}y^\sigma_{j_1\dots j_s} = u\,\mathrm{d}y^\sigma_{j_1\dots j_s} + y^\sigma_{j_1\dots j_s}\,\mathrm{d}u,\quad 0\le s\le r. \tag{8.21}$$

Taking into account the chart expression of a general k-form (8.14), we can decompose the form $\chi^*\eta$ into two terms

$$\chi^*\eta = \mathrm{d}u \wedge \mu + \tilde{\eta}, \tag{8.22}$$

where neither the $(k-1)$-form μ nor the k-form $\tilde{\eta}$ contain the factor $\mathrm{d}u$. The coefficients of these forms depend, of course, on the variable u.

Introduce the operator $\mathcal{A}$ formally defined as follows.

$$\mathcal{A} : \Lambda^k(TJ^rY) \ni \eta \longrightarrow \mathcal{A}\eta = \int\limits_{u=0}^{u=1} (\mathrm{d}u \wedge \mu). \tag{8.23}$$

(In fact, this denotes a form with coefficients obtained as integrals of coefficients of the form μ with respect to the variable u. See the following Example 8.7 showing the action of the operator $\mathcal{A}$ on a form η.) By the direct and rather technical calculations in coordinates, one can prove that

$$\eta = \mathrm{d}\mathcal{A}\eta + \mathcal{A}\,\mathrm{d}\eta.$$

If η is a closed form ($\mathrm{d}\eta = 0$), then

$$\eta = \mathrm{d}(\mathcal{A}\eta) = \mathrm{d}\varrho, \quad \varrho = \mathcal{A}\eta = \int\limits_{u=0}^{u=1} (\mathrm{d}u \wedge \mu).$$

Example 8.7 (Operator $\mathcal{A}$ explicitly) Let $\eta = E_\sigma\,\mathrm{d}y^\sigma \wedge \omega_0 \in \Lambda^{n+1}(TJ^rY)$ be a dynamical form defined on an open simply connected set $W_r \subset J^rY$, $E_\sigma = E_\sigma(x^i, y^\nu, y^\nu_j, \ldots, y^\nu_{j_1\ldots j_r})$. Then

$$\begin{aligned} \chi^*\eta &= (E_\sigma \circ \chi)\,\chi^*\,\mathrm{d}y^\sigma \wedge \chi^*\omega_0 = (E_\sigma \circ \chi)(y^\sigma\,\mathrm{d}u + u\,\mathrm{d}y^\sigma) \wedge \omega_0 \\ &= y^\sigma(E_\sigma \circ \chi)\,\mathrm{d}u \wedge \omega_0 + u(E_\sigma \circ \chi)\,\mathrm{d}y^\sigma \wedge \omega_0 \\ &= \mathrm{d}u \wedge \mu + \tilde{\eta}, \quad \mu = y^\sigma(E_\sigma \circ \chi)\,\omega_0, \quad \tilde{\eta} = u(E_\sigma \circ \chi)\,\mathrm{d}y^\sigma \wedge \omega_0, \\ \mathcal{A}\eta &= \left[y^\sigma \int\limits_{u=0}^{u=1} E_\sigma(x^i\, uy^\nu, uy^\nu_j, \ldots, uy^\nu_{j_1\ldots j_r})\,\mathrm{d}u \right] \omega_0. \end{aligned}$$

Definition 8.18 (*Dynamical forms*) Let (Y, π, X) be a fibred manifold with $\dim X = n > 1$. A form $E \in \Lambda^{n+1}(TJ^rY)$ defined on an open set $W_r \subset J^rY$ is called *the dynamical form* if it is $\pi_{r,0}$-horizontal and 1-contact. (For mechanics see Definition 3.6.)

Chart expressions of dynamical forms are as follows:

$$E = E_\sigma(x^i, y^\nu, \dots, y^\nu_{j_1\dots j_r})\,\mathrm{d}y^\sigma \wedge \omega_0 = E_\sigma(x^i, y^\nu, \dots, \omega^\nu_{j_1\dots j_r})\,\omega^\sigma \wedge \omega_0.$$

A dynamical form important for general field theories in calculus of variations is *affine in highest coordinates*, which means that

$$E_\sigma = A_\sigma(x^i, y^\mu, \dots, y^\mu_{j_1\dots j_{r-1}}) + B^{j_1\dots j_r}_{\sigma\nu}(x^i, y^\mu, \dots, y^\mu_{j_1\dots j_{r-1}})y^\nu_{j_1\dots j_r}.$$

For applications in physics the case $r = 2$ is relevant in most of the situations. In such a case, the dynamical form affine in higher coordinates is called *affine in accelerations*.

The remaining part of this section is devoted to distributions. This is important for the purpose of following sections concerning calculus of variations on fibred manifolds with multidimensional bases ($\dim X > 1$—field theory) which we finally apply to field theories in physics.

The concept of distributions was presented in Definition 3.11 for mechanics ($\dim X = 1$) and Definition 8.11 for field theory ($\dim X > 1$). In accordance with Definition 8.11 a concrete distribution was generated by vector fields. Similarly as in mechanics there is an alternative possibility to generate distributions: with help of annihilating differential 1-forms.

Definition 8.19 (*Annihilator of the distribution*) Integral sections of the distribution.

(a) Let $\mathcal{D}$ be a distribution on an open set $W_r \subset J^rY$. Let $\mathcal{D}^0$ be the set of all 1-forms $\eta \in \Lambda^1(TJ^rY)$ defined on W_r such that $i_\xi\eta = 0$ for every vector field $\xi \in \mathcal{D}$. $\mathcal{D}^0$ is called *the annihilator of the distribution* $\mathcal{D}$.
(b) The mapping assigning to every point $z \in W_r$ the dimension of $\mathcal{D}^0(z)$, i.e.

$$J^rY \supset W_r \ni z \longrightarrow \dim J^rY - \dim \mathcal{D} \in \mathbf{N} \cup 0,$$

is called *the corank of the distribution* $\mathcal{D}$.
(c) A section δ of π_r is called *the integral section of the distribution* $\mathcal{D}$ if for every form $\chi \in \mathcal{D}^0$ it holds $\delta^*\chi = 0$.

8.3 Lagrange Structures, First Variational Formula, Euler-Lagrange Equations

In this section, we follow the structure of first three sections of Chap. 4. Definitions and assertions (propositions and theorems) as well as examples formulated in Chap. 4 for mechanics are modified for the field theory, i.e. on fibred manifolds with multidimensional bases (Y, π, X), $\dim X = n > 1$, and their prolongations. In the following, we will suppose for simplicity that all underlying manifolds as well as other geometrical objects (functions, vector fields, differential forms, etc.) are smooth (or at least C^r-class necessary with respect to the context).

In the following, we denote as $\Omega \subset X$ an n-dimensional connected compact submanifold of the base X and as $\Gamma_\Omega(\pi)$ the set of all sections of π for which $\Omega \subset \mathrm{Dom}\,\gamma$.

Definition 8.20 (*Variational integral*) Let (Y, π, X), $\dim X = n > 1$, $\dim Y = n + m$, be a fibred manifold and $\lambda \in \Lambda^n_X(TJ^rY)$ a horizontal n-form (it is called *the* rth *order Lagrangian*). The pair (π, λ) is *the Lagrange structure*. The mapping

$$S : \Gamma_\Omega(\pi) \ni \gamma \longrightarrow S[\gamma] = \int_\Omega J^r\gamma^*\lambda \in \mathbf{R},$$

for λ defined on an open set $W_r \subset J^rY$ for which $\Omega \subset \pi_r(W_r)$ is called the *variational integral* or, alternatively, the *action function* of the Lagrange structure (π, λ) on Ω.

The concepts presented in Definition 8.20 are introduced similarly as in mechanics, see Definition 4.1. Note that similarly as in mechanics, the first-order variational integral is important for most field theories in physics. And, again similarly as in mechanics, the concept of the variational integral can be generalized for an arbitrarily chosen form $\varrho \in \Lambda^n(TJ^rY)$ by the relation

$$\int_\Omega J^r\gamma^*\varrho = \int_\Omega J^{r+1}\gamma^*\pi^*_{r+1,r}\varrho = \int_\Omega J^{r+1}\gamma^*\, h\varrho.$$

Moreover, just on the rth prolongation (J^rY, π_r, X) of (Y, π, X) it holds for a form $\varrho = \lambda + \eta$, where $\eta \in \Lambda^n(TJ^rY)$ is a contact n-form,

$$\int_\Omega J^r\gamma^*\varrho = \int_\Omega J^r\gamma^*(\lambda + \eta) = \int_\Omega J^r\gamma^*\lambda.$$

Definition 8.21 (*Variations of sections*) Let ξ be a π-projectable vector field on Y and $\{\alpha_u\}$ its local one-parameter group with projection $\{\alpha_{0u}\}$. Consider a section $\gamma \in \Gamma_\Omega(\pi)$. The one-parameter family $\{\gamma_u\}$ of sections such that $\gamma_u = \alpha_u \circ \gamma \circ \alpha_{0u}^{-1} \in \Gamma_{\alpha_u(\Omega)}(\pi)$ is called *the variation* or *deformation of the section* γ *induced by the vector field* ξ. The vector field ξ itself is called *the variation.* A π-vertical variation ξ such that $\xi(y) = 0$ for every point $y \in \pi^{-1}(\partial\Omega)$ is called *variation with fixed ends.*

Calculating the derivative of the mapping

$$(-\varepsilon, \varepsilon) \ni u \longrightarrow S[\gamma_u] = \int_{\alpha_{0u}(\Omega)} J^r\gamma_u^*\lambda \in \mathbf{R}$$

we obtain (perform in detail)

$$\left.\frac{\mathrm{d}S[\gamma_u]}{\mathrm{d}u}\right|_{u=0} = \int_\Omega J^r\gamma^* \partial_{J^r\xi}\lambda.$$

Definition 8.22 (*Variational derivative*) The mapping

$$\Gamma_\Omega(\pi) \ni \gamma \longrightarrow \left[\frac{\mathrm{d}}{\mathrm{d}u}\int_{\alpha_{0u}(\Omega)} J^r\gamma^*\lambda\right]_{u=0} = \int_\Omega J^r\gamma^*\partial_{J^r\xi}\lambda$$

is called *the variational derivative* of the action function (of the variational integral).

Note that $\partial_{J^r\xi}\lambda$ is the horizontal form on J^rY, i.e. the Lagrangian. We can introduce also 2nd, 3rd and higher order variational derivatives of the action function with respect to various π-projectable vector fields,

$$\int_\Omega J^r\gamma^*\partial_{J^r\xi_1}\partial_{J^r\xi_2}\dots\partial_{J^r\xi_p}\lambda.$$

Taking into account the well-known general formula for Lie derivatives of forms

$$\partial_\xi\eta = i_\xi\,\mathrm{d}\eta + \mathrm{d}i_\xi\eta$$

and the general Stokes theorem we obtain

$$\int_\Omega J^r\gamma^*\partial_{J^r\xi}\lambda = \int_\Omega J^r\gamma^*\left(i_{J^r\xi}\,\mathrm{d}\lambda + \mathrm{d}i_{J^r\xi}\lambda\right) = \int_\Omega J^r\gamma^* i_{J^r\xi}\,\mathrm{d}\lambda + \int_{\partial\Omega} J^r\gamma^* i_{J^r\xi}\lambda. \tag{8.24}$$

To arrive at the first variational formula, we need to introduce the concept of Lepagean forms and Lepagean equivalents of Lagrangians.

Definition 8.23 (*Lepagean forms, Lepage equivalent and Euler-Lagrange form of a Lagrangian*) Let (Y, π, X) be a fibred manifold with n-dimensional base $\dim X = n > 1$ and let (π, λ) be the rth order Lagrange structure.

(a) A form $\theta \in \Lambda^n(TJ^rY)$ defined on an open set $W_r \subset J^rY$ is called *Lepagean* if the 1-contact component of its exterior derivative $p_1\,\mathrm{d}\varrho$ is $\pi_{r+1,0}$-horizontal, i.e. $p_1\,\mathrm{d}\varrho$ is the dynamical form.
(b) A Lepagean form $\theta_\lambda \in \Lambda^n(TJ^sY)$, $s \geq r$ in general, is called the *Lepage equivalent of the Lagrangian* λ if it holds $h\theta_\lambda = \pi^*_{s+1,r}\lambda$.

(c) A dynamical form $E_\lambda = p_1\, \mathrm{d}\theta_\lambda$ is called the *Euler-Lagrange form* corresponding to the Lagrangian λ.

The item (b) in Definition 8.23 means that the horizontal component of the Lepage equivalent of a Lagrangian equals to this Lagrangian up to a possible projection as it will be explained below. (In contradiction to the case of mechanics the Lepage equivalent of a Lagrangian in field theories is not unique for $r > 2$. On the other hand, situations with $r = 1$ and $r = 2$ are the most relevant for physics applications.)

Example 8.8 (First order: Poincaré-Cartan form, Euler-Lagrange form) Let (π, λ) be a first-order Lagrange structure on $(Y, \pi, X) = (\mathbf{R}^n \times \mathbf{R}^m, p, \mathbf{R}^n)$,

$$\lambda = L(x^i, y^\sigma, y^\sigma_j)\, \mathrm{d}x^1 \wedge \mathrm{d}x^2 \wedge \ldots \wedge \mathrm{d}x^n, \quad 1 \le \sigma \le m, \quad 1 \le i, j \le n,$$

and $\mathrm{Dom}\, L = W_1 \subset J^1 Y$. Looking for the first-order Lepage equivalent of λ, we follow the procedure used in Sect. 4.2 and leading to Proposition 4.1. We modify relation (8.24) by adding to λ a contact form

$$\eta = B^i_\sigma \omega^\sigma \wedge \omega_i \quad \text{defined on } W_1$$

and requiring the first integral on the right-hand side of this relation to be independent of the first-order components Ξ^σ_i of the π-vertical vector field ξ

$$J^1\xi = \Xi^\sigma \frac{\partial}{\partial y^\sigma} + \Xi^\sigma_i \frac{\partial}{\partial y^\sigma_i}.$$

This means that the horizontal terms of the form $i_{J^1\xi}\, \mathrm{d}(\lambda + \eta)$ (containing components Ξ^σ_i) must vanish. Taking into account that $\mathrm{d}\omega^\sigma = -\mathrm{d}y^\sigma_j \wedge \mathrm{d}x^j$ and $\mathrm{d}x^j \wedge \omega_i = \delta^j_i \omega_0$, expressing these terms and putting them equal to zero we obtain

$$i_{J^1\xi}\left(\frac{\partial L}{\partial y^\sigma_i} \mathrm{d}y^\sigma_i \wedge \omega_0 - B^i_\sigma\, \mathrm{d}y^\sigma_j \wedge \mathrm{d}x^j \wedge \omega_i \right) = \left(\frac{\partial L}{\partial y^\sigma_i} - B^i_\sigma \right) \Xi^\sigma_i\, \omega_0 = 0.$$

for $\le i \le n$, $1 \le \sigma \le m$. So $B^i_\sigma = \frac{\partial L}{\partial y^\sigma_i}$. For the Lepage equivalent of the first-order Lagrangian, we have the resulting relation

$$\varrho = L\, \omega_0 + \frac{\partial L}{\partial y^\sigma_i} \omega^\sigma \wedge \omega_i + p_2 \varrho = \theta_\lambda + \nu \tag{8.25}$$

the form θ_λ is called the *Poincaré-Cartan form*. The Euler-Lagrange form of the Lagrangian λ then reads

$$E_\lambda = p_1\, \mathrm{d}\theta_\lambda = \left[\frac{\partial L}{\partial y^\sigma} - \frac{\mathrm{d}}{\mathrm{d}x^i} \left(\frac{\partial L}{\partial y^\sigma_i} \right) \right] \omega^\sigma \wedge \omega_0. \tag{8.26}$$

Notice that for $r > 2$ the Lepage equivalent of a Lagrangian λ is not unique, in general. The general expressions of the specific Lepage equivalent (called the *principal Lepage equivalent*) of a Lagrangian and the corresponding Euler-Lagrange dynamical form can be obtained from their definitions by direct calculations in charts. The results are given in the following proposition.

Proposition 8.10 *Let (Y, π, X) be a fibred manifold,* $\dim X = n > 1$ *and (π, λ) Lagrange structure of the rth order. Then there exists an integer $s \leq 2r - 1$ and a Lepage equivalent $\theta_\lambda \in \Lambda^n(TJ^sY)$ of λ such that its decomposition to components of various degrees of contactness contains only the horizontal and 1-contact components. This Lepage equivalent is uniquely determined by the Lagrangian and it is of the form*

$$\theta_\lambda = L\,\omega_0 + \sum_{q=0}^{r-1}\left[\sum_{p=0}^{r-q-1}(-1)^p \mathrm{d}_{l_1}\mathrm{d}_{l_2}\cdots\mathrm{d}_{l_p}\left(\frac{\partial L}{\partial y^\sigma_{j_1\ldots j_q l_1\ldots l_p i}}\right)\right]\omega^\sigma_{j_1\ldots j_q}\wedge\omega_i\,.$$

The corresponding Euler-Lagrange form is uniquely determined by the relation $E_\lambda = p_1\,\mathrm{d}\theta_\lambda = p_1\,\mathrm{d}\varrho$, where ϱ is arbitrary Lepage equivalent of λ. For the Euler-Lagrange form the following relation hold:

$$E_\lambda = \left[\sum_{l=0}^{r}(-1)^l\,\mathrm{d}_{j_1}\cdots\mathrm{d}_{j_l}\left(\frac{\partial L}{\partial y^\sigma_{j_1\ldots j_l}}\right)\right]\omega^\sigma\wedge\omega_0.$$

The Euler-Lagrange form of λ is defined on $J^{2r}Y$ in general.

The mapping assigning to Lagrangians their Euler-Lagrange forms is called the *Euler-Lagrange mapping.*

(As mentioned above, for $r > 2$ the Lepage equivalent of the Lagrangian is not unique. The principal Lepage equivalent is unfortunately not invariant with respect to coordinate transformations, i.e. it is not a differential form in general, except for the situation when on J^rY there exists a global associated fibred chart and when the Lagrangian is global too. Nevertheless for field theories in physics situations with $r = 1$ and $r = 2$ are mostly relevant. On the other hand, the Euler-Lagrange form is uniquely determined by the Lagrangian for arbitrary order r of the Lagrange structure.)

Now we are ready to formulate the general first variational formula in its integral as well as differential form.

Theorem 8.3 (First variational formula) Let (π, λ) be an rth order Lagrange structure with the Lagrangian λ defined on an open set $W_r \subset J^rY$ and ϱ its arbitrary Lepage equivalent defined on $\pi^{-1}_{2r-1,r}(W_r)$. Let ξ be a π-projectable vector field defined on $\pi_{r,0}(W_r)$ and let $\Omega \subset X$ be a connected compact set such that $\Omega \subset \pi_r(W_r)$. Then for every section $\gamma \in \Gamma_\Omega(\pi)$ the following equivalent relations hold:

$$\int_\Omega J^r\gamma^* \partial_{J^r\xi}\lambda = \int_\Omega J^{2r-1}\gamma^* i_{J^{2r-1}\xi}\, \mathrm{d}\varrho + \int_{\partial\Omega} J^{2r-1}\gamma^* i_{J^{2r-1}\xi}\varrho, \tag{8.27}$$

$$\int_\Omega J^r\gamma^* \partial_{J^r\xi}\lambda = \int_\Omega J^{2r}\gamma^* i_{J^{2r}\xi} E_\lambda + \int_{\partial\Omega} J^{2r-1}\gamma^* i_{J^{2r-1}\xi}\varrho, \tag{8.28}$$

$$\pi_{2r,r}^* \partial_{J^r\xi}\lambda = h i_{J^{2r-1}\xi}\, \mathrm{d}\varrho + h\, \mathrm{d} i_{J^{2r-1}\xi}\varrho. \tag{8.29}$$

Relations (8.27) and (8.28) represent the *first variational formula in the integral form*, relation (8.29) is the *first variational formula in the infinitesimal form*. These relations will be important in the following Sect. 8.4.

The last part of this section is devoted to *Euler-Lagrange equations*, i.e. equations leading to *stationary sections of a Lagrange structure*.

Definition 8.24 (*Extremals*) Let (Y, π, X) be a fibred manifold with $\dim X = n > 1$ and let (π, λ) be the Lagrange structure of the rth order, where $\lambda \in \Lambda_X^n(TJ^rY)$ is defined on an open set $W_r \subset J^rY$. Let $\Omega \subset X$ be a connected compact set such that $\Omega \subset \pi_r(W_r)$. A section $\gamma \in \Gamma_\Omega(\pi)$ is called the *stationary point of the variational functional on* Ω if for every variation ξ with fixed ends on Ω it holds

$$\int_\Omega J^r\gamma^* \partial_{J^r\xi}\lambda = 0. \tag{8.30}$$

Suppose that a section γ is defined on an open set $U \subset X$. Such a section is called the *stationary point, or critical section of the variational functional* if it is the stationary point of this functional on every $\Omega \subset U$. Stationary points of a functional are alternatively called *critical points of the variational functional* or *extremals of the Lagrangian* λ.

The schematic Fig. 8.1 (drawn for $n = 1$, $m = 1$ for better imagination—see also Fig. 4.2) demonstrates the concept of a variation with fixed ends on connected compact sets Ω_1, Ω_2, Ω_3 with help of corresponding sections. (The difference between definition of an *extremal on certain* Ω and that of an *extremal* as such is as follows: if a section γ is an extremal of λ for every $\Omega \subset U$, is stationarity is in fact independent of Ω.) The equivalent conditions for extremals are summarized in the following theorem.

Theorem 8.4 *Let* (Y, π, X), $\dim X = n > 1$, *be a fibred manifold and let* (π, λ) *be a corresponding* r*th order Lagrange structure with Lagrangian* λ *defined on an open set* $W_r \subset J^rY$. *Denote the Lepage equivalent of* λ *as* ϱ. *Let* γ *be a section of* π *defined on* $\pi_r(W_r)$. *The following conditions are equivalent:*

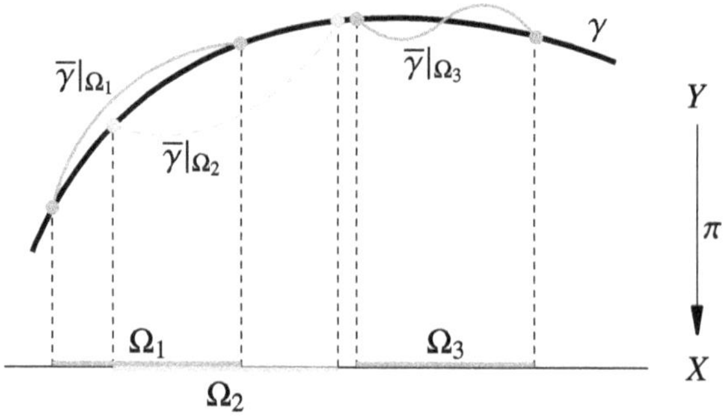

Fig. 8.1 Sections with fixed ends on various Ω

(a) The section γ is an extremal of the given Lagrange structure.
(b) For every π-vertical vector field ξ on $\pi_{r,0}(W_r)$ it holds

$$J^{2r-1}\gamma^* i_{J^{2r-1}\xi}\, \mathrm{d}\varrho = 0. \tag{8.31}$$

(c) Relation (8.31) is valid for every π-projectable vector field ξ on $\pi_{r,0}(W_r)$.
(d) For every π-projectable vector field ξ on $\pi_{r,0}(W_r)$ it holds

$$J^{2r}\gamma^* i_{J^{2r}\xi}\, E_\lambda = 0. \tag{8.32}$$

(e) The Euler-Lagrange form vanishes along $J^{2r}\gamma$, i.e. $E_\lambda \circ J^{2r}\gamma = 0$. Then

$$\left[\sum_{\ell=0}^{r}(-1)^\ell \mathrm{d}_{j_1}\cdots \mathrm{d}_{j_\ell}\left(\frac{\partial L}{\partial y^\sigma_{j_1\ldots j_\ell}}\right)\right] \circ J^{2r}\gamma = 0. \tag{8.33}$$

Relation (8.33) represents so called *Euler-Lagrange equations*. The proof of this theorem is based on the 1st variational formula and for its understanding is sufficient to follow the proof of Theorem 4.2.

Example 8.9 (Scalar waves) Let (π, λ) be a first-order Lagrange structure on $(Y, \pi, X) = (\mathbf{R}^4 \times \mathbf{R}, p, \mathbf{R}^4)$ with charts (V, ψ), $\psi = (x^1, x^2, x^3, x^4, y^1)$, $(\pi(V), \varphi)$, $\varphi = (x^1, x^2, x^3, x^4)$, $(\pi^{-1}_{1,0}(V), \psi_1)$, $\psi_1 = (x^i, y^1, y^1_i)$, $1 \le i \le 4$, $(\pi^{-1}_{2,0}(V), \psi_2)$, $\psi_2 = (x^i, y^1, y^1_i, y^1_{ij})$, $1 \le i, j \le 4$, and

$$\lambda = \frac{1}{2}\left[(y^1_1)^2 - \sum_{i=2}^{4}(y^1_i)^2\right]\omega_0.$$

Notice that this is the Lagrangian representing scalar waves in the four-dimensional time-space, x^1 representing the time variable and x^2, x^3, x^4 being the space variables. Its Lepage equivalent is (see Eq. (8.25))

$$\theta_\lambda = L\omega_0 + \frac{\partial L}{\partial y^1_i}\omega^1 \wedge \omega_i,$$

where

$$\begin{aligned}\omega_0 &= \mathrm{d}x^1 \wedge \mathrm{d}x^2 \wedge \mathrm{d}x^3 \wedge \mathrm{d}x^4,\\ \omega_1 &= \mathrm{d}x^2 \wedge \mathrm{d}x^3 \wedge \mathrm{d}x^4,\\ \omega_2 &= -\mathrm{d}x^1 \wedge \mathrm{d}x^3 \wedge \mathrm{d}x^4,\\ \omega_3 &= \mathrm{d}x^1 \wedge \mathrm{d}x^2 \wedge \mathrm{d}x^4,\\ \omega_4 &= -\mathrm{d}x^1 \wedge \mathrm{d}x^2 \wedge \mathrm{d}x^3.\end{aligned}$$

$$\begin{aligned}\theta_\lambda = {} & \frac{1}{2}\left[(y_1^1)^2 - \sum_{i=2}^{4}(y_i^1)^2\right]\omega_0 + y_1^1\omega^1 \wedge \mathrm{d}x^2 \wedge \mathrm{d}x^3 \wedge \mathrm{d}x^4\\ & + y_2^1\omega^1 \wedge \mathrm{d}x^1 \wedge \mathrm{d}x^3 \wedge \mathrm{d}x^4 - y_3^1\omega^1 \wedge \mathrm{d}x^1 \wedge \mathrm{d}x^2 \wedge \mathrm{d}x^4\\ & + y_4^1\omega^1 \wedge \mathrm{d}x^1 \wedge \mathrm{d}x^2 \wedge \mathrm{d}x^3,\\ E_\lambda = {} & \left[\frac{\partial L}{\partial y^1} - \frac{\mathrm{d}}{\mathrm{d}x^i}\left(\frac{\partial L}{\partial y_i^1}\right)\right]\omega^1 \wedge \omega_0 = -\left[y_{11}^1 - y_{22}^1 - y_{33}^1 - y_{44}^1\right]\omega^1 \wedge \omega_0.\end{aligned}$$

Euler-Lagrange equation then reads

$$y_{11}^1 - y_{22}^1 - y_{33}^1 - y_{44}^1 = 0, \quad \text{or in physics notation} \quad \frac{\partial^2 y^1}{\partial t^2} - \Delta y^1 = 0,$$

where Δ is the Laplace differential operator.

Similarly as in mechanics (Sect. 4.3) we will now discuss the *trivial variational problem*. The problem is: By the Euler-Lagrange mapping the unique Euler-Lagrange forms are assigned to a Lagrangian. However, the Euler-mapping is not one-to-one, i.e. there is the class of equivalent Lagrangians with the identical Euler-Lagrange form. This means that the Euler-Lagrange form of the difference of two arbitrarily chosen equivalent Lagrangians is identically zero. Such a difference makes a *trivial Lagrangian*. Trivial first-order Lagrangians in mechanics are characterized by Proposition 4.4.

Definition 8.25 (*Trivial Lagrange structure*) Let (π, λ) and $(\pi, \tilde{\lambda})$ be Lagrange structures of rth and sth order, respectively, on (Y, π, X), $s \geq r$ in general. Suppose that Lagrangians λ and $\tilde{\lambda}$ are defined on open sets $W_r \subset J^rY$ and $W_s \subset J^sY$, respectively, such that $W = \pi_{s,r}^{-1} W_r \cap W_s \neq \emptyset$. Then:

(a) Lagrangians λ and $\tilde{\lambda}$ are called *variationally equivalent*, if $E_\lambda = E_{\tilde{\lambda}}$ up to a formal projection.
(b) A rth order Lagrange structure on (Y, π, X) with the Lagrangian $\lambda \in \Lambda_X^n(TJ^rY)$ defined on an open set $W_r \subset J^rY$ is called *variationally trivial* if the Euler-Lagrange form of λ is identically zero. In such a case the Lagrangian λ is called *variationally trivial* as well.

Example 8.10 (Equivalent Lagrangians) Notice that two equivalent Lagrangians need not obviously to be of the same order. Let $(Y, \pi, X) = (\mathbf{R}^2 \times \mathbf{R}^2, p, \mathbf{R}^2)$. Let

$$(V, \psi), \quad \psi = (x^1, x^2, y^1, y^2), \quad (p(V), \varphi), \quad \varphi = (x^1, x^2),$$

be a fibred chart on Y and the associated chart on X, and

$$\left(\pi_{1,0}^{-1}(V), \psi_1\right), \quad \psi_1 = (x^1, x^2, y^1, y^2, y^1_1, y^1_2, y^2_1, y^2_2),$$

the associated fibred chart on J^1Y and finally

$$\left(\pi_{2,0}^{-1}(V), \psi_2\right), \quad \psi_2 = (x^1, x^2, y^1, y^2, y^1_1, y^1_2, y^2_1, y^2_2, y^1_{11}, y^1_{12}, y^1_{22}, y^2_{11}, y^2_{12}, y^2_{22})$$

the associated fibred chart on J^2Y. Let λ and $\tilde{\lambda}$ be the first- and second-order Lagrangians

$$\lambda = \left\{-\frac{1}{2}\left[\left(y^1_2\right)^2 + \left(y^2_2\right)^2\right] - \left(y^1 y^2_1 - y^2 y^1_1\right)\right\} \mathrm{d}x^1 \wedge \mathrm{d}x^2,$$

$$\tilde{\lambda} = \left[\frac{1}{2}(y^1 y^1_{22} + y^2 y^2_{22}) - (y^1 y^2_1 - y^2 y^1_1)\right] \mathrm{d}x^1 \wedge \mathrm{d}x^2.$$

It can be easily verified that

$$E_\lambda = E_{\tilde{\lambda}} = \left(y^1_{22} - 2y^2_1\right) \mathrm{d}y^1 \wedge \mathrm{d}x^1 \wedge \mathrm{d}x^2 + \left(y^2_{22} + 2y^1_1\right) \mathrm{d}y^2 \wedge \mathrm{d}x^1 \wedge \mathrm{d}x^2.$$

So, we have two equivalent Lagrangians of different orders with the same Euler-Lagrange form.

The following consideration leads to the general form for a trivial Lagrangian: Let λ and $\tilde{\lambda}$ be equivalent Lagrangians and ϱ and $\tilde{\varrho}$ their Lepage equivalents, respectively. Then $p_1\, \mathrm{d}\varrho = p_1\, \mathrm{d}\tilde{\varrho}$, up to a possible (nonsignificant) projection. Adding, e.g. to ϱ a form $\mathrm{d}\eta + \mu$, where η is an arbitrary $(n-1)$-form and μ is an arbitrary at least 2-contact n-form, we obtain

$$p_1\, \mathrm{d}\varrho = p_1\, \mathrm{d}(\varrho + \mathrm{d}\eta + \mu) \implies h(\varrho + \mathrm{d}\eta + \mu) = h\varrho + h\, \mathrm{d}\eta = \lambda + h\, \mathrm{d}\eta.$$

Proposition 8.11 *The rth order Lagrangian $\lambda_0 \in \Lambda^n_X(TJ^rY)$ defined on an open set $W_r \subset J^rY$ is variationally trivial if and only if for every fibred chart (V, ψ), $\psi = (x^i, y^\sigma)$, $1 \le i \le n$, $1 \le \sigma \le m$, $\pi_{r,0}^{-1}(V) \subset W_r$, there exists a form $\eta \in \Lambda^{n-1}$ $(TJ^{r-1}Y)$ such that on $\pi_{r,0}^{-1}(V)$ it holds*

$$\lambda_0 = h\, \mathrm{d}\eta.$$

The reader can find the precise and detailed proof of this Proposition in [3] or [1,2]. It is, of course, evident that the Proposition 8.11 is local. For practical use the following equivalent condition for trivial Lagrangians is useful ([2]).

Proposition 8.12 *The rth order Lagrangian $\lambda_0 \in \Lambda^n_X(TJ^rY)$ defined on an open set $W_r \subset J^rY$ is variationally trivial if and only if for every fibred chart (V, ψ), $\psi = (x^i, y^\sigma)$, $1 \le i \le n$, $1 \le \sigma \le m$, $\pi^{-1}_{r,0}(V) \subset W_r$, there exist functions (0-forms) f^i defined on $\pi^{-1}_{r-1,0}(V)$ such that for*

$$\lambda_0 = L_0(x^i, y^\sigma, y^\sigma_j, \dots, y^\sigma_{j_1\dots j_r})\,\mathrm{d}x^1 \wedge \dots \wedge \mathrm{d}x^n$$

it holds

$$L_0 = \mathrm{d}_i f^i = \frac{\mathrm{d}f^i}{\mathrm{d}x^i} = \frac{\partial f^i}{\partial x^i} + \sum_{\ell=0}^{r-1} \frac{\partial f^i}{\partial y^\sigma_{j_1\dots j_\ell}} y^\sigma_{j_1\dots j_\ell i}.$$

So, trivial Lagrangians have the form of "divergencies". (The existence of a form η on $J^{r-1}Y$ for which the rth order trivial Lagrangian λ_0 is expressed as $\lambda_0 = h\,\mathrm{d}\eta$ is proved, e.g. in [1–3].)

8.4 Symmetries, Noether Theorem, Generalized Noether Theorem

For symmetries and conservation laws in field theory the procedure is formally the same as in mechanics. General considerations will be made for the general order of jet prolongations of fibred manifolds, the applications are mainly related to the first and second order.

Definition 8.26 (*Invariance transformations and symmetries of Lagrangians and EL forms*) Let (Y, π, X) be a fibred manifold and (π, λ) the rth order Lagrange structure with the Lagrangian λ defined on an open set $W_r \subset J^rY$.

(a) A local isomorphism (α, α_0) with $\mathrm{Dom}\,\alpha = \pi_{r,0}(W_r)$ is called the *invariance transformation* of the given Lagrange structure if it holds

$$J^r\alpha^*\lambda = \lambda. \tag{8.34}$$

(b) A π-projectable vector field ξ defined on $\pi_{r,0}(W_r)$ with one-parameter group (α_u, α_{0u}), $u \in (-\varepsilon, \varepsilon)$, is called the *point symmetry*, or *symmetry of the Lagrange structure* if all α_u, $u \in (-\varepsilon, \varepsilon)$, are invariance transformations.
(c) A local isomorphism (α, α_0) with $\mathrm{Dom}\,\alpha = \pi_{r,0}(W_r)$ is called the *invariance transformation* of the Euler-Lagrange form E_λ if

$$J^{2r}\alpha^* E_\lambda = E_\lambda. \tag{8.35}$$

(d) Let ξ be a π-projectable vector field defined on $\pi_{r,0}(W_r)$ and (α_u, α_{0u}), $u \in (-\varepsilon, \varepsilon)$, its one-parameter group. Such a vector field is called the *point symmetry*, or *symmetry of the Euler-Lagrange form* E_λ if all α_u are its invariance transformations.

Conditions (8.34) and (8.35) for symmetries of the Lagrange structure and symmetries of E_λ, respectively, can be expressed as follows:

$$\begin{aligned} J^r\alpha_u^*\lambda - \lambda = 0 &\Longrightarrow \frac{J^r\alpha_u^*\lambda - \lambda}{u} = 0 \text{ for } u \in (-\varepsilon, \varepsilon), \quad u \neq 0 \\ &\Longrightarrow \partial_{J^r\xi}\lambda = 0 \text{ the \textit{Noether equation}}, \end{aligned} \tag{8.36}$$

$$\begin{aligned} J^{2r}\alpha_u^* E_\lambda - E_\lambda = 0 &\Longrightarrow \frac{J^{2r}\alpha_u^* E_\lambda - \lambda}{u} = 0 \text{ for } u \in (-\varepsilon, \varepsilon), \quad u \neq 0 \\ &\Longrightarrow \partial_{J^{2r}\xi} E_\lambda = 0 \text{ the \textit{Noether-Bessel-Hagen equation}}. \end{aligned} \tag{8.37}$$

The purpose of (8.36) and (8.37) can be twofold:

- for a given Lagrange structure (or a given dynamical form affine in highest variables) to find all symmetries ξ,
- for a given π-projectable vector field ξ to find all Lagrange structures (or dynamical forms affine in highest variables) for which ξ is the corresponding symmetry.

The practical use of (8.36) and (8.37) lies in their chart expressions:

$$J^r\xi = \xi^i \frac{\partial}{\partial x^i} + \Xi^\sigma \frac{\partial}{\partial y^\sigma} + \sum_{s=1}^{r} \Xi^\sigma_{j_1\cdots j_s} \frac{\partial}{\partial y^\sigma_{j_1\cdots j_s}},$$

$$\begin{aligned} \partial_{J^r\xi}\lambda = 0 &\Longrightarrow i_{J^r\xi}\, \mathrm{d}\lambda + \mathrm{d} i_{J^r\xi}\lambda = 0 \\ &\Longrightarrow \left(i_{J^r\xi}\, \mathrm{d}L\right)\omega_0 - \mathrm{d}L \wedge i_{J^r\xi}\omega_0 + \mathrm{d}L \wedge i_{J^r\xi}\omega_0 + L\, \mathrm{d} i_{J^r\xi}\omega_0 = 0 \\ &\Longrightarrow i_{J^r\xi}\, \mathrm{d}L + L\frac{\partial \xi^i}{\partial x^i} = 0, \end{aligned} \tag{8.38}$$

$$\begin{aligned} \partial_{J^{2r}\xi} E_\lambda &= i_{J^{2r}\xi}\, \mathrm{d}E_\lambda + \mathrm{d} i_{J^{2r}\xi} E_\lambda \\ &= i_{J^{2r}\xi}\, \mathrm{d}\left(E_\sigma\omega^\sigma \wedge \omega_0\right) + \mathrm{d}\left[i_{J^{2r}\xi}\left(E_\sigma\omega^\sigma \wedge \omega_0\right)\right]. \end{aligned}$$

After standard calculations we obtain (see Problem 8.25)

$$\partial_{J^{2r}\xi} E_\lambda = \left[\left(i_{J^{2r}\xi}\, \mathrm{d}E_\sigma\right) + E_\nu \frac{\mathrm{d}\Xi^\nu}{\mathrm{d}y^\sigma} + E_\sigma \frac{\mathrm{d}\xi^i}{\mathrm{d}x^i}\right] \omega^\sigma \wedge \omega_0.$$

The condition $\partial_{J^{2r}\xi} E_\lambda = 0$ leads to the Noether-Bessel-Hagen (NBH) equation

$$\left(i_{J^{2r}\xi}\, \mathrm{d}E_\sigma\right) + E_\nu \frac{\mathrm{d}\Xi^\nu}{\mathrm{d}y^\sigma} + E_\sigma \frac{\mathrm{d}\xi^i}{\mathrm{d}x^i} = 0. \tag{8.39}$$

Let α be an invariance transformation of the Euler-Lagrange form E_λ of a certain Lagrangian λ. We prove that the form $J^{2r}\alpha^* E_\lambda$ has the structure of Euler-Lagrange form as well and we will find out its Lagrangian.

Proposition 8.13 *Let (π, λ) be a rth order Lagrange structure on (Y, π, X) with Lagrangian λ defined on an open set $W_r \subset J^r Y$.*

(a) Suppose that the local isomorphism α is an invariance transformation of λ, i.e. $J^r\alpha^*\lambda = \lambda$. Then α is the invariance transformation of E_λ as well, i.e. $J^{2r}\alpha^* E_\lambda = E_\lambda$

(b) Suppose that the local isomorphism α is an invariance transformation of the Euler-Lagrange form E_λ of the Lagrangian λ, i.e. $J^{2r}\alpha^* E_\lambda = E_\lambda$. Then $J^r\alpha^*\lambda - \lambda$ is a trivial Lagrangian.

Proof Let $\varrho \in \Lambda^n(TJ^{2r-1}Y)$ defined on W_r be an arbitrary Lepage equivalent of λ. Taking into account the properties of the pullback mapping and the fact that the form $J^{2r-1}\alpha^*\varrho$ is Lepage, we obtain step by step

$$d\pi^*_{2r,2r-1}\varrho = \pi^*_{2r,2r-1}\, d\varrho = E_\lambda + F_\lambda,$$

where F_λ is at least 2-contact $(n+1)$-form. Then we have

$$J^{2r}\alpha^*\pi^*_{2r,2r-1}\, d\varrho = J^{2r}\alpha^* E_\lambda + J^{2r}\alpha^* F_\lambda.$$

On the other hand,

$$J^{2r}\alpha^*\pi^*_{2r,2r-1}\, d\varrho = \pi^*_{2r,2r-1}J^{2r-1}\alpha^*\, d\varrho = \pi^*_{2r,2r-1}\, dJ^{2r-1}\alpha^*\varrho.$$

Because ϱ is the Lepage form, then $\tilde{\varrho} = J^{2r-1}\alpha^*\varrho$ is the Lepage form as well. Denote $\tilde{\lambda} = h\tilde{\varrho}$. Then

$$\tilde{\lambda} = h\left(J^{2r-1}\alpha^*\varrho\right) = J^{2r}\alpha^*(h\varrho) = J^r\alpha^*\lambda,$$

so the form $\tilde{\varrho} = J^{2r-1}\alpha^*\varrho$ is a Lepage equivalent of the Lagrangian $\tilde{\lambda} = J^r\alpha^*\lambda$. This means that $E_{J^r\alpha^*\lambda} = E_\lambda$ which proves part (a) of the proposition.

Let α be an invariance transformation of E_λ, i.e. $J^{2r}\alpha^* E_\lambda = E_\lambda$. Then

$$J^{2r}\alpha^* E_\lambda - E_\lambda = E_{J^r\alpha^*\lambda - \lambda} = 0,$$

which means that $J^r\alpha^*\lambda - \lambda$ is a trivial Lagrangian. This proves part (b) of the proposition. □

Proposition 8.13 presents the fact that every symmetry ξ of the Lagrangian is obviously the symmetry of its Euler-Lagrange form, but the reverse is not true.

With symmetries of Lagrangians conservation laws are connected. This is specified in the following *Emmy Noether theorem*.

Theorem 8.5 *Let (π, λ) be a rth order Lagrange structure with the Lagrangian λ defined on an open set $W_r \subset J^r Y$, $\varrho \in \Lambda^n(T J^s Y)$ its arbitrary Lepage equivalent and $\theta \in \Lambda^n(T J^s Y)$ its principal equivalent, both defined on $\pi_{2r-1,r}^{-1}(W_r)$ (in general it holds, $r \leq s \leq 2r-1$). Let ξ be a symmetry of the Lagrangian defined on $\pi_{r,0}(W_r)$ and γ its extremal. Then it holds*

$$\mathrm{d}J^s\gamma^* i_{J^s\xi}\varrho = \mathrm{d}J^s\gamma^* i_{J^s\xi}\theta = 0,$$

resp. equivalently

$$\mathrm{d}\left(hi_{J^s\xi}\varrho \circ J^{s+1}\gamma\right) = \mathrm{d}\left(hi_{J^s\xi}\theta \circ J^{s+1}\gamma\right) = 0.$$

So, the form $\Phi(\xi) = hi_{J^s\xi}\varrho$ is closed along prolongations of extremals.

Proof Suppose for simplicity that the Lepage equivalents ϱ and θ are of the $(2r-1)$th order, as it is obvious. Consider the infinitesimal first variational formula taking into account that $\partial_{J^r\xi}\lambda = 0$ (ξ is a symmetry of the Lagrangian) we obtain

$$hi_{J^{2r-1}\xi}\,\mathrm{d}\varrho + h\,\mathrm{d}i_{J^{2r-1}\xi}\varrho = 0$$

$$\implies J^{2r}\gamma^* hi_{J^{2r-1}\xi}\,\mathrm{d}\varrho + J^{2r}\gamma^* h\,\mathrm{d}i_{J^{2r-1}\xi}\varrho = 0 \tag{8.40}$$

(and the same equation for θ). Consider a general expression $J^s\gamma^*\omega$ for an arbitrary form $\omega \in \Lambda^k(T J^s Y)$ and arbitrary section γ:

$$J^s\gamma^*\omega = J^{s+1}\gamma^*\pi_{s+1,s}^*\omega = J^{s+1}\gamma^* h\omega + J^{s+1}\gamma^* p\omega = J^{s+1}\gamma^* h\omega.$$

Then it follows from (8.40)

$$J^{2r-1}\gamma^* i_{J^{2r-1}\xi}\,\mathrm{d}\varrho + J^{2r-1}\gamma^*\,\mathrm{d}i_{J^{2r-1}\xi}\varrho = 0.$$

However, if γ is an extremal, then the first term of the preceding equation is zero and then

$$J^{2r-1}\gamma^*\,\mathrm{d}i_{J^{2r-1}\xi}\varrho = 0 \implies \mathrm{d}J^{2r-1}\gamma^* i_{J^{2r-1}\xi}\varrho = 0 \implies \mathrm{d}hi_{J^{2r-1}\xi}\varrho = 0.$$

In the last steps, we used the commutativity of the pullback and exterior derivative operators, as well as the fact that the pullback of the contact component of every form by $J^{2r-1}\gamma$ vanishes. As a result we obtain:

$$\Phi(\xi) = hi_{J^{2r-1}\xi}\varrho = \text{const.}\ \text{ along extremals.}$$

Briefly speaking, the quantities $\Phi(\xi)$ called *Noether currents* (where ξ is a symmetry of the Lagrangian) are horizontal $(n-1)$-forms closed along extremals.

Example 8.11 (Symmetries and currents—first order) For $r = 1$ we have

$$J^1\xi = \xi^i \frac{\partial}{\partial x^i} + \Xi^\sigma \frac{\partial}{\partial y^\sigma} + \Xi^\sigma_i \frac{\partial}{\partial y^\sigma_i}, \quad \Xi^\sigma_i = \frac{\mathrm{d}\Xi^\sigma}{\mathrm{d}x^i} - y^\sigma_j \frac{\mathrm{d}\xi^j}{\mathrm{d}x^i},$$

and from general relations (8.38) we obtain the Noether equation in the form

$$\frac{\partial L}{\partial x^i}\xi^i + \frac{\partial L}{\partial y^\sigma}\Xi^\sigma + \frac{\partial L}{\partial y^\sigma_i}\left(\frac{\mathrm{d}\Xi^\sigma}{\mathrm{d}x^i} - y^\sigma_j \frac{\mathrm{d}\xi^j}{\mathrm{d}x^i}\right) + L\frac{\mathrm{d}\xi^i}{\mathrm{d}x^i} = 0.$$

The corresponding Noether current calculated for the Poincaré-Cartan form θ_λ is

$$hi_{J^1\xi}\theta_\lambda = \left[L\xi^i + \frac{\partial L}{\partial y^\sigma_i}\left(\Xi^\sigma - y^\sigma_j \xi^j\right)\right]\omega_i.$$

8.5 Inverse Problem

The question in field theories is in principle the same as in mechanics: Is the given dynamical form $E \in \Lambda^{n+1}_Y(TJ^sY)$ (i.e. a $\pi_{s,0}$-horizontal form, in charts expressed as $E = E_\sigma \omega^\sigma \wedge \omega_0 \in \Lambda^{n+1}_Y(TJ^sY)$) variational or not? This means whether it is the Euler-Lagrange form of a certain Lagrangian or not. This question presents the so called *strong inverse problem* of calculus of variations. Notice what it means for the problem of variationality of equations of motion of a physical system: The left-hand sides of equations of motion, $E_\sigma = 0$, are considered as components of the given dynamical form, i.e. they must not be rearranged. (See, e.g. Example 5.2 for mechanics.) In the following we deal only with the problem of so called *local variationality*. (The problem of *global variationality* requires specific mathematical tools.)

Definition 8.27 (*Variational dynamical forms*) Let $E \in \Lambda^{n+1}_Y(TJ^sY)$ be a dynamical form defined on an open set $W_s \subset J^sY$. This form is called *variational* if there exists an integer r (usually $r \le s$) and a Lagrangian $\lambda \in \Lambda^n_X(TJ^rY)$ defined on $\pi^{-1}_{s,r}(W_s)$ such that $E = E_\lambda$ up to a possible projection. The mentioned form E is *locally variational* (on W_s), if every point $z \in W_s$ has an open surrounding $O(z)$ such that the form E is variational on $O(z)$.

Similarly as in mechanics the geometrical criterion of variationality of a dynamical form is as follows: A dynamical form is (locally) variational and associated with an rth order Lagrangian λ if and only if it is the 1-contact component of the exterior derivative of the corresponding Lepage equivalent ϱ, i.e.

$$\pi^*_{2r,s}E = p_1\,\mathrm{d}\varrho \implies \pi^*_{2r,s}E + \sum_{\ell=2}^{n+1} p_\ell\,\mathrm{d}\varrho = \pi^*_{2r,s}\,\mathrm{d}\varrho.$$

Denoting $F = \sum_{\ell=2}^{n+1} p_\ell \, \mathrm{d}\varrho$ (F is at least 2-contact) we can formulate the required criterion.

Proposition 8.14 *Let (Y, π, X) be a fibred manifold,* $\dim X = n \geq 1$, *and let $E \in \Lambda_Y^{n+1}(TJ^sY)$ be a dynamical form defined on an open subset $W_s \subset J^sY$. The form E is locally variational if and only if for every point $z \in W_s$ there exists its open surrounding $O(z) \subset W_s$ and a form $F \in \Lambda^{n+1}(TJ^sY)$ defined on $O(z)$ such that it holds*

(a) the form F is at least 2-contact,
(b) the form $E + F$ is closed, i.e. $\mathrm{d}(E + F) = 0$.

Proof First, suppose that the given form E is locally variational. As it was said before, then for every point $z \in W_s$ there exists its open surrounding $O(z)$ and a Lagrangian λ (in general of the rth order, $r \leq s$, defined on $\pi_{s,r}\big(O(z)\big)$) such that E is its Euler-Lagrange form, up to a possible projection, i.e. $\pi_{2r,s}^* E = E_\lambda$. Let $\varrho \in \Lambda_Y^n(TJ^{2r-1}Y)$ be an arbitrary Lepage equivalent of λ. This means that it holds

$$E = p_1 \, \mathrm{d}\varrho, \quad \text{and thus} \quad \mathrm{d}\varrho = E + p_2 \mathrm{d}\varrho + \cdots + p_{n+1} \, \mathrm{d}\varrho$$

(up to a possible projection). Denoting

$$F = p_2 \mathrm{d}\varrho + \cdots + p_{n+1} \, \mathrm{d}\varrho$$

we proved part (a) of the proposition. The assertion (b), $\mathrm{d}(E + F) = \mathrm{d}^2 \varrho = 0$ is then trivial.

Now suppose that assertions (a) and (b) are valid for a form E specified in the proposition. Let $z \in W_s$. The form $E + F$ is (at least) 1-contact and closed (on an open surrounding $O(z)$ of the point z, which can be chosen as convex). Following the Poincaré lemma there exists on $O(z)$ a form $\varrho \in \Lambda^n(TJ^sY)$ such that $E + F = \mathrm{d}\varrho$. The 1-contact component of $\mathrm{d}\varrho$ equals E and the horizontal component of ϱ is just a possible Lagrangian. □

The previous proposition gives a geometrical criterion of local variationality of a given dynamical form. However, for practical purposes it is important to have a possibility of how to decide if a given dynamical form is (locally) variational or not from chart expressions of the form E itself. Another important question is how to find corresponding Lagrangians of variational dynamical forms. We solved these problems in detail in Chap. 5 for mechanics. Here we present only results for field theories, without proof. The proof lies in coordinate calculations which are in principle based on the same ideas as in mechanics, but they are more technically complicated.

Theorem 8.6 *Let (Y, π, X) be a fibred manifold,* $\dim X = n \geq 1$, *and let $E \in \Lambda_Y^{n+1}(TJ^sY)$ be a dynamical form defined on an open subset $W_s \subset J^sY$. The form E is locally variational if and only if for every point $z \in W_s$ there exists a fibred* *chart* (V, ψ) *on* Y *such that* $z \in \pi_{s,0}^{-1}(V)$, *where* $\pi_{s,0}^{-1}(V) \subset W_s$, *and for components* E_σ *of the form* E *it holds for* $1 \leq l \leq s$

$$\frac{\partial E_\sigma}{\partial y^\nu_{p_1 \dots p_l}} - (-1)^l \frac{\partial E_\nu}{\partial y^\sigma_{p_1 \dots p_l}} - \sum_{j=l+1}^{s} (-1)^j \binom{j}{l} \mathrm{d}_{p_{l+1}} \dots \mathrm{d}_{p_j} \frac{\partial E_\nu}{\partial y^\sigma_{p_1 \dots p_l p_{l+1} \dots p_j}} = 0. \tag{8.41}$$

If we find a given dynamical form as variational, i.e. obeying Helmholtz conditions, there arises the problem: how to find corresponding Lagrangians, especially the *minimal order Lagrangian*. The solution lies in the following proposition. (For mechanics see relation (5.24).)

Proposition 8.15 *Let* $E \in \Lambda^{n+1}_Y(TJ^sY)$ *be a locally variational dynamical form defined on an open set* $W_s \subset J^sY$ *and* (V, ψ), $\psi = (x^i, y^\sigma)$, $(\pi(V), \varphi)$, $\varphi = (x^i)$ *and* $\pi_{s,0}^{-1}(V)$, $\psi = (x^i, y^\sigma, y^\sigma_{j_1}, \dots, y^\sigma_{j_1 \dots j_s})$, $1 \leq i, j_1, \dots, j_s \leq n$, $1 \leq \sigma \leq m$, *a fibred chart on* Y, *associated chart on* X *and associated fibred chart on* J^sY. *Suppose that* $\pi_{s,0}^{-1}(V) \subset W_s$. *(The chart expression of* E *is:* $E_\sigma \omega^\sigma \wedge \omega_0$. *Then the form* $\Lambda \in \Lambda^n_X(TJ^sY)$

$$\Lambda = \left(y^\sigma \int_0^1 (E_\sigma \circ \chi) \, \mathrm{d}u \right) \omega_0, \tag{8.42}$$

where the mapping χ *is defined by relation (8.20), is the* s*th order Lagrangian such that* E *is its Euler-Lagrange form.*

The Lagrangian (8.42) is called the *Vainberg-Tonti Lagrangian*, as in mechanics.

The analogous Proposition 5.4 for second order locally variational dynamical form in mechanics was proved with help of Theorem 5.1 in Sect. 5.2. There was also proved relation (5.25) for the first order (i.e. minimal) Lagrangian (see Proposition 5.5 and its proof). The problem of minimal Lagrangian in mechanics for a locally variational dynamical form of the general order s is completely solved in [4] as follows.

For a locally variational sth order dynamical form $E \in \Lambda^2_Y(TJ^sY)$ in mechanics there exists a Lagrangian of the order $s/2$ for even s, resp. $(s+1)/2$th order Lagrangian for odd s. The problem of so called order reduction (i.e. finding the Lagrangian of minimal order) is non-completely solved for field theories. Partial solution was presented in [5]. In concrete situations, we can obtain the minimal Lagrangian by subtracting an appropriate trivial Lagrangian from the Vainberg-Tonti one.

Example 8.12 (Helmholtz conditions for third-order dynamical form) Let $E \in \Lambda^{n+1}_Y(TJ^3Y)$ be a dynamical form

$$E = E_\sigma(x^i, y^\nu, y^\nu_j, y^\nu_{jk}, y^\nu_{jk\ell}) \, \omega^\sigma \wedge \omega_0, \; 1 \leq i, j, k, \ell \leq n, \; 1 \leq \sigma, \nu \leq m, \; s = 3.$$

The Helmholtz conditions for its variationality read:

$$\frac{\partial E_\sigma}{\partial y^\nu_{ijk}} + \frac{\partial E_\nu}{\partial y^\sigma_{ijk}} = 0,$$

$$\frac{\partial E_\sigma}{\partial y^\nu_{ij}} - \frac{\partial E_\nu}{\partial y^\sigma_{ij}} + 3\mathrm{d}_k \frac{\partial E_\nu}{\partial y^\sigma_{ijk}} = 0,$$

$$\frac{\partial E_\sigma}{\partial y^\nu_{i}} + \frac{\partial E_\nu}{\partial y^\sigma_{i}} - 2\mathrm{d}_j \frac{\partial E_\nu}{\partial y^\sigma_{ij}} + 3\mathrm{d}_j\,\mathrm{d}_k \frac{\partial E_\nu}{\partial y^\sigma_{ijk}} = 0,$$

$$\frac{\partial E_\sigma}{\partial y^\nu} - \frac{\partial E_\nu}{\partial y^\sigma} + \mathrm{d}_i \frac{\partial E_\nu}{\partial y^\sigma_{i}} - \mathrm{d}_i\,\mathrm{d}_j \frac{\partial E_\nu}{\partial y^\sigma_{ij}} + \mathrm{d}_i\,\mathrm{d}_j\,\mathrm{d}_k \frac{\partial E_\nu}{\partial y^\sigma_{ijk}} = 0.$$

The Vainberg-Tonti Lagrangian is

$$\Lambda = \left(y^\sigma \int_0^1 E_\sigma(x^i, uy^\nu, uy^\nu_j, uy^\nu_{ij}, y^\nu_{ijk})\,\mathrm{d}u \right) \omega_0, \quad 1 \le i, j \le n, \quad 1 \le \sigma, \nu \le m.$$

8.6 Problems

Problem 8.1 Write Jacobi matrices of mappings $\bar{\psi} \circ \psi^{-1}$, $\psi \circ \bar{\psi}^{-1}$, $\bar{\varphi} \circ \varphi^{-1}$ and $\varphi \circ \bar{\varphi}^{-1}$, (8.1).

Problem 8.2 Let $\alpha : \bar{Y} \to Y$ be a local isomorphism of fibred manifolds (Y, π, X) and $(\bar{Y}, \bar{\pi}, \bar{X})$.

(a) Prove that its projection α_0 is unique.
(b) Prove that (local) isomorphisms preserve fibres, i.e. if two different points $y_1, y_2 \in \pi^{-1}(\{x\})$ for every $x \in X$, then $\alpha(y_1), \alpha(y_2) \in \bar{\pi}^{-1}\{\alpha_0(x)\}$.
(c) Prove that (local) isomorphisms preserve sections. This means that for an arbitrary section $\gamma \in \Gamma_U(\pi)$ the mapping $\bar{\gamma} = \alpha \circ \gamma \circ \alpha_0^{-1}$ is the section of $\bar{\pi}$, i.e. $\bar{\gamma} \in \Gamma_{\bar{U}}$, $\bar{U} = \alpha_0(U)$.

Are the above mentioned assertions valid not only for local isomorphisms, but also for homomorphisms? (Homomorphisms α is not necessarily one-to-one mapping.)

Problem 8.3 Prove that the 1th equivalence of sections is independent of the choice of charts.

Problem 8.4 Prove Proposition 8.3.

Problem 8.5 Construct prolongations of sections of the projection π of the fibred manifold $(Y, \pi, X) = (\mathbf{R}^n \times \mathbf{R}^m, \pi, \mathbf{R}^n)$ to $(J^r Y, \pi_r, X)$. (For $1 \le n \le 3$ the simplified notation of coordinates is used: $x^1 = x$, $x^2 = y$, $x^3 = z$.)

(a) $n = 2, m = 1, r = 2, \gamma = (x, y, x^2 + y^2)$,
(b) $n = 2, m = 2, r = 2, \gamma = (x, y, \ln \sqrt{x^2 + y^2}, \operatorname{arctg} \frac{y}{x})$,
(c) $n = 3, m = 3, r = 1, \gamma = (x, y, z, xy^2z^2, x^2yz^2, x^2y^2z)$,
(d) $n = 2, m = 1, r = 4, \gamma = (x, y, x^5 + y^5)$.

Problem 8.6 Express formally the prolongations $(J^r\alpha, \alpha_0)$ of a local isomorphism (α, α_0) for $r = 1$ and $r = 2$ in coordinates.

Problem 8.7 For a fibred manifold (Y, π, X) with n-dimensional base ($n > 1$) construct tangent bundles to X, Y, J^1Y (and in general J^rY) using the concept of vectors as classes of equivalent parametrized curves.

Problem 8.8 Prove that vector fields on Y are sections of the fibred manifold (TY, Π, Y).

Problem 8.9 Prove part (b) of Proposition 8.5 following considerations in the proof of (a).

Problem 8.10 Let ξ be a π-projectable vector field on Y, (defined on a open set $W \subset Y$)

$$\xi = \xi^i(x^j)\frac{\partial}{\partial x^i} + \Xi^\sigma(x^j, y^\nu)\frac{\partial}{\partial y^\sigma}.$$

Using definition relations (8.10) derive the explicit expression for components Ξ^σ_i of the vector field $J^1\xi$.

Problem 8.11 For the π-projectable vector field ξ on $(\mathbf{R}^2 \times \mathbf{R}^2, p, \mathbf{R}^2)$,

$$\xi = x^2\frac{\partial}{\partial x^1} - x^1\frac{\partial}{\partial x^2} + x^2y^1\frac{\partial}{\partial y^1} + x^1y^2\frac{\partial}{\partial y^2}$$

express components of $J^1\xi$ and $J^2\xi$.

Problem 8.12 Prove part (d) in Example 8.4.

Problem 8.13 Let (Y, π, X) be a fibred manifold with dimension $\dim Y = 3$, $\dim X = 2$, (see Example 8.4). Derive the chart expressions of contact 1-forms, contact 2-forms and contact 3-forms.

Problem 8.14 Solve the same question as in Problem 8.13 for contact k-forms, $1 \le k \le 4$ on the fibred manifold (Y, π, X), $\dim Y = 4$, $\dim X = 2$.

Problem 8.15 Let $(Y, \pi, X) = (\mathbf{R}^2 \times \mathbf{R}, p, \mathbf{R}^2)$ and let η_1 and η_2 be differential 1-forms defined on open sets $W_1 \subset J^1Y$ and $W_2 \subset J^2Y$, respectively, where

$$\eta_1 = A_1\,\mathrm{d}x^1 + A_2\,\mathrm{d}x^2 + B\,\mathrm{d}y + C^1\,\mathrm{d}y_1 + C^2\,\mathrm{d}y_2,$$

$$\begin{aligned}\eta_2 = {} & A_1\,\mathrm{d}x^1 + A_2\,\mathrm{d}x^2 + B\,\mathrm{d}y + C^1\,\mathrm{d}y_1 + C^2\,\mathrm{d}y_2 + D^{11}\,\mathrm{d}y_{11} \\ & + D^{12}\,\mathrm{d}y_{12} + D^{22}\,\mathrm{d}y_{22}.\end{aligned}$$

Write chart expressions of these forms in bases adapted to the contact structure of the underlying fibred manifold.

Problem 8.16 Let (Y, π, X) be a fibred manifold and let f be a function (i.e. 0-form) on J^rY defined on an open set $W_r \subset J^rY$. Express the total derivative of f in the basis (8.17).

Problem 8.17 For $r = 1$ and $r = 2$ express the decomposition (8.19) of an arbitrary

(a) 1-form on $W_r \subset J^rY$,

$$\eta = A_i\,\mathrm{d}x^i + \sum_{|J|=0}^{r} B_\sigma^J\,\mathrm{d}y_J^\sigma,$$

(b) 2-form on $W_r \subset J^rY$,

$$\eta = A_{ij}\,\mathrm{d}x^i \wedge \mathrm{d}x^j + \sum_{|J|=0}^{r} B_{\sigma,i}^J\,\mathrm{d}y_J^\sigma \wedge \mathrm{d}x^i + \sum_{|J|,|L|=0}^{r} C_{\sigma\nu}^{JL}\,\mathrm{d}y_J^\sigma \wedge \mathrm{d}y_L^\nu,$$

$1 \le i, j \le n$, $1 \le \sigma, \nu \le m$, $J = (j_1 \dots j_r)$, $1 \le j_1, \dots, j_r \le n$, $L = (\ell_1 \dots \ell_r)$, $1 \le \ell_1, \dots, \ell_r \le n$.

Problem 8.18 Let $f \in \Lambda^0(TJ^rY)$ be a function (0-form) defined on an open set $W_r \subset J^rY$. Express the decomposition of the exterior derivative $\mathrm{d}f$ into its component of individual degrees of contactness for (a) $r = 1$, (b) $r = 2$, (c) general r.

Problem 8.19 Consider relation (8.22). Prove this expression and the fact that forms μ and $\tilde{\eta}$ do not contain the factor $\mathrm{d}u$.

Problem 8.20 The dynamical form E is defined on the 2nd prolongation of the fibred manifold $(Y, \pi, X) = (\mathbf{R}^2 \times \mathbf{R}^2, p, \mathbf{R}^2)$ with corresponding charts (V, ψ), $(p(V), \varphi)$, $\psi = (x^1, x^2, y^1, y^2)$, $\varphi = (x^1, x^2)$, $(p_{1,0}^{-1}(V), \psi_1)$, $(p_{2,0}^{-1}(V), \psi_2)$,

$$\psi_1 = (x^1, x^2, y^1, y^2, y_1^1, y_2^1, y_1^2, y_2^2),$$
$$\psi_2 = (x^1, x^2, y^1, y^2, y_1^1, y_2^1, y_1^2, y_2^2, y_{11}^1, y_{12}^1 = y_{21}^1, y_{22}^1, y_{11}^2, y_{12}^2 = y_{21}^2, y_{22}^2),$$

as follows:

$$E = E_1\, \mathrm{d}y^1 \wedge \omega_0 + E_2\, \mathrm{d}y^2 \wedge \omega_0, \quad \omega_0 = \mathrm{d}x^1 \wedge \mathrm{d}x^2,$$

$$E_1 = (y_{22}^1 - 2y_1^2), \quad E_2 = (y_{22}^2 + 2y_1^2).$$

Calculate the forms $\mathcal{A}E$ and ϱ, where $\mathcal{A}$ is the operator (8.23) and $\mathrm{d}\varrho = E$.

Problem 8.21 Let (V, ψ), $\psi = (x^i, y^\sigma)$, and $(\bar{V}, \bar{\psi})$, $\bar{\psi} = (\bar{x}^i, \bar{y}^\sigma)$, $V \cap \bar{V} \neq \emptyset$ are fibred charts on Y, $\left(\pi(V), \varphi\right)$, $\varphi = (x^i)$, and $\left(\pi(\bar{V}), \bar{\varphi}\right)$, $\bar{\varphi} = (\bar{x}^i)$, the associated charts on X, and $\left(\pi_{r,0}^{-1}(V), \psi_r\right)$ and $\left(\pi_{r,0}^{-1}(\bar{V}), \bar{\psi}_r\right)$, the corresponding associated fibred charts on J^rY. Derive the transformation relations for an rth order Lagrangian.

Problem 8.22 Execute in detail all steps of the calculation leading to the Poincaré-Cartan form of a first-order Lagrangian (8.25) and to the corresponding Euler-Lagrange form (8.26) (Example 8.8).

Problem 8.23 Calculate the Euler-Lagrange form of Lagrangians presented in Example 8.10 and calculate the trivial Lagrangian as their difference. Prove that this trivial Lagrangian is in agreement with Propositions 8.11 and 8.12

Problem 8.24 Prove items (a) and (e) in Theorem 8.4. (Especially explain why the request of verticality of the vector field ξ can be omitted in item (e).)

Problem 8.25 Perform all calculations leading to the Noether-Bessel-Hagen equation (8.39).

Problem 8.26 Consider the first-order Lagrange structure on (Y, π, X), $\dim X = n > 1$. Suppose that translations along the base X, $\partial/\partial x^i$, and translations along fibres, $\partial/\partial y^\sigma$, are symmetries of the Lagrangian. Express corresponding Noether currents.

Problem 8.27 Express the explicit form of Eq. (8.39) for the wave equation

$$y_{11}^1 - y_{22}^1 - y_{33}^1 - y_{44}^1 = 0, \quad \text{or in physics notation} \quad \frac{\partial^2 y^1}{\partial t^2} - \Delta y^1 = 0,$$

(see Example 8.9).

Problem 8.28 Reformulate the Helmholtz conditions (8.41) for mechanics, i.e. for $\dim X = 1$.

Problem 8.29 Apply the general Helmholtz conditions (8.41) and write them for second-order dynamical form. Write the concrete relation for calculating the Vainberg-Tonti Lagrangian.

$$E = E_\sigma(x^i, y^\nu, y^\nu_j, y^\nu_{ij})\, \omega^\sigma \wedge \omega_0.$$

References

1. D. Krupka, Global variational theory in fibred spaces, in *Handbook of Global Analysis* (Elsevier Sci. B. V., Amsterdam, 2008), pp. 773–836
2. D. Krupka, *Introduction to Global Variational Geometry*. Atlantis Studies in Variational Geometry, vol. 1 (Atlantis Press, Paris, 2015)
3. D. Krupka, J. Musilová, Trivial Lagrangians in field theory. Diff. Geom. Appl. **9**, 293–305 (1998)
4. O. Krupková *The Geometry of Ordinary Variational Equations*. Lecture Notes in Mathematics, vol. 1678 (Springer, Berlin, 1997)
5. O. Rossi, The Lagrangian order reduction in field theories. Preprint 1–22 (2017)

Variational Physics 9

This chapter presents some important applications of the above developed mathematical tools in physics. The concept of fibred manifolds and the corresponding adapted structures (local isomorphisms, vector fields and differential forms) is very natural for the use in physics and in technical applications. We present examples of two types:

- The use of fibred manifolds and their prolongations as underlying geometrical structures as well as the vector fields and differential forms adapted to the fibred structure for formulation of some key variational physical theories, including finding symmetries of Lagrangians and corresponding Noether currents.
- Inverse problem: It appears that important physical theories are variational. On the basis of well-known equations of motion of some of them we prove their variationality, including the constructions of corresponding Vainberg-Tonti Lagrangian as well as the minimal order Lagrangian. See e.g. [1,2].

The interested reader can find the advanced approach to problems of contemporary physics, e.g. in [3–6].

Prerequisites problems exposed in previous chapters: inverse problem in mechanics and field theory on fibred manifolds, symmetries of Lagrangians (Noether theorem and Noether equation) and symmetries of equations of motion (Noether-Bessel-Hagen equation); basic theories in physics (mechanics, fundamentals of special theory of relativity (STR), electrodynamics at least three-dimensional version, quantum mechanics). □

J. Musilová et al., *Calculus of Variations on Fibred Manifolds and Variational Physics*, Lecture Notes in Physics 1033, https://doi.org/10.1007/978-3-031-77408-9_9

9.1 Mechanics and Special Relativity

Classical mechanics and special relativity are typical examples of physical disciplines, whose mathematical tools correspond to variational mechanics. The geometric basis for it, as we know, are fibred spaces with a one-dimensional base. First, for few examples, we will focus on the formulation of the variational problem (Lagrangian and Euler-Lagrange form) in classical Newtonian mechanics and special relativity.

Example 9.1 *Variational mechanics of mass particles and their systems.* Basic knowledge of variational mechanics of particles and their systems is a common part of theoretical mechanics. Now we have in our disposal the mathematical tools to properly deriving, generalizing, and expanding the well-known results of theoretical mechanics. It is enough to "just" apply the general conclusions we have drawn in previous chapters.

Consider a mechanical system with m degrees of freedom. Its mechanical state is determined by the generalized coordinates q^σ and generalized velocities $\dot{q}^\sigma$, $1 \leq \sigma \leq m$. If the system is a single mass particle, the state of which is not limited by constraints, then it has 3 degrees of freedom and (q^1, q^2, q^3) can be Cartesian coordinates. The basic fibred space for such a mass particle is $(Y, \pi, X) = (\mathbf{R} \times \mathbf{R}^3, p, \mathbf{R})$. The coordinate on the base $X = \mathbf{R}$ is the time coordinate. If a studied system contains N mass particles without constraints, its $3N$ degrees of freedom can be again given by Cartesian coordinates. Fibred space is then $(Y, \pi, X) = (\mathbf{R} \times \mathbf{R}^{3N}, p, \mathbf{R})$.

If the system is subjected to some constraints placed only on coordinates (holonomic constraints), its number of degrees of freedom is $m = 3N - k$, where k is the number of constraints. It is often advantageous to choose instead of a reduced number Cartesian coordinates other coordinates that better fit the geometrical symmetry of the problem. For example, for a spherical pendulum (a mass particle on a fiber of the fixed length l and negligible mass with the single constraint $x^2 + y^2 + z^2 = l^2$) it is advisable to choose a spherical angle ϑ and azimuthal angle φ.

Physical reasons lead to the choice of the Lagrangian of the system of mass points as the difference between kinetic and potential energy (see, e.g. the famous classical textbook [7]),

$$\lambda = L\,\mathrm{d}t, \quad L = \frac{1}{2}\sum_{\sigma=1}^{m} m_\sigma (\dot{q}^\sigma)^2 - U(t, q^1, \ldots, q^m),$$

where m_σ are constants (masses, inertia, etc.), or more generally

$$\lambda = L\,\mathrm{d}t, \quad L = \frac{1}{2} M g_{\sigma\nu} \dot{q}^\sigma \dot{q}^\nu - U(t, q^1, \ldots, q^m),$$

where $g_{\sigma\nu}$ are functions of coordinates, $g_{\sigma\nu} = g_{\sigma\nu}(q^\mu)$, symmetric in indices σ a ν (metric tensor). So, we have first-order Lagrangian in a special form—kinetic part is a quadratic form in generalized velocities. Lepage equivalent reads

$$\theta_\lambda = \left(\frac{1}{2} M g_{\sigma\nu}(q^\mu)\dot{q}^\sigma \dot{q}^\nu - U(t, q^\mu)\right) \mathrm{d}t + M g_{\sigma\nu}\dot{q}^\nu \omega^\sigma,$$

and Euler-Lagrange form

$$\begin{aligned} E_\lambda &= \left[\frac{\partial L}{\partial q^\sigma} - \frac{\mathrm{d}}{\mathrm{d}t}\left(\frac{\partial L}{\partial \dot{q}^\sigma}\right)\right] \omega^\sigma \wedge \mathrm{d}t \\ &= \left(-M g_{\sigma\nu}\ddot{q}^\nu + M\Gamma_{\sigma\nu\lambda}\dot{q}^\nu \dot{q}^\lambda - \frac{\partial U}{\partial q^\sigma}\right) \omega^\sigma \wedge \mathrm{d}t, \end{aligned}$$

$$2\Gamma_{\sigma\nu\lambda} = \frac{\partial g_{\nu\lambda}}{\partial q^\sigma} - \frac{\partial g_{\sigma\lambda}}{\partial q^\nu} - \frac{\partial g_{\sigma\nu}}{\partial q^\lambda}$$

are *Christoffel symbols*. (See Example 4.4 as well.) The specific form of equations of motion depends, of course, on the mechanical system in the particular situation, i.e. on the choice of quantities $g_{\sigma\nu}$, $1 \leq \sigma, \nu \leq m$, and functions $U(t, q^1, \dots, q^m)$.

Example 9.2 (*Rigid body*) We say that a body (with a discrete or continuous distribution of mass) is *rigid* if there is a frame of reference in which every point or volume element of the body is at rest. It is then a frame of reference *associated with this body*. When studying the motion of a rigid body, we do not have to follow the motion of each of its elements separately, we just need to deal with two types of motion of a body as a whole—translation and rotation. If the body were not rigid, we would have to use mathematical field theory (fibred manifolds with multidimensional bases—see Chap. 8).

If the motion of a rigid body is not constrained, it is described by six independent generalized coordinates. Three of them, (x, y, z), characterize the position of the center of mass, and three, (α, β, γ), are angles of rotation around three selected independent axes. Usually, these are the so-called principal axis of the tensor of inertia of the body, passing through the center of mass. Given them, the inertia is diagonal, $\operatorname{diag} J = (J_1, J_2, J_3)$. The Lagrangian of this mechanical system is then

$$\begin{aligned} \lambda = L\,\mathrm{d}t &= \frac{1}{m} m(\dot{x}^2 + \dot{y}^2 + \dot{z}^2) + \frac{1}{2}(J_1\dot{\alpha}^2 + J_2\dot{\beta}^2 + J_3\dot{\gamma}^2) \\ &- U(t, x, y, z, \alpha, \beta, \gamma). \end{aligned}$$

However, to solve completely the motion of such a "flywheel" is not easy at all. (In addition, note that forces exerted on a body by its surroundings do not have to be conservative, even when potential energy U does not depend on time. In such a case the potential energy of the body is undefined and the problem is not variational.)

Example 9.3 (*Relativistic particle—three-dimensional approach*) The basic fibred space for a particle within the framework of special relativity can be the same as for classical particle, $(Y, \pi, X) = (\mathbf{R} \times \mathbf{R}^3, p, \mathbf{R})$, the coordinate on the base X is time t, coordinates on fibres are usually Cartesian, $\boldsymbol{r} = (q^1, q^2, q^3)$, velocities are $\boldsymbol{v} = (\dot{q}^1, \dot{q}^2, \dot{q}^3)$. Lagrangian is of the form

$$\lambda = L\,\mathrm{d}t, \quad L = -mc^2\sqrt{1 - \frac{v^2}{c^2}} + (\boldsymbol{v}\boldsymbol{\Phi} - \varphi + mc^2),$$

where $\boldsymbol{\Phi} = \boldsymbol{\Phi}(t, \boldsymbol{r})$ is *vector potential energy* and $\varphi = \varphi(t, \boldsymbol{r})$ is *scalar potential energy*. Components of the corresponding Euler-Lagrange form are

$$\begin{aligned} E_\sigma &= A_\sigma(t, q^\mu, \dot{q}^\mu) + B_{\sigma\nu}(t, q^\mu, \dot{q}^\mu)\,\ddot{q}^\nu \\ &= \frac{\partial L}{\partial q^\sigma} - \frac{\mathrm{d}}{\mathrm{d}t}\left(\frac{\partial L}{\partial \dot{q}^\sigma}\right) = \left(\frac{\partial \Phi_\nu}{\partial q^\sigma} - \frac{\partial \Phi_\sigma}{\partial q^\nu}\right)\dot{q}^\nu - \frac{\partial \Phi_\sigma}{\partial t} - \frac{\partial \varphi}{\partial q^\sigma} \\ &\quad - \frac{mc}{\sqrt{c^2 - \sum_{\mu=1}^3 (\dot{q}^\mu)^2}}\left(\delta_{\sigma\nu} - \frac{\dot{q}^\sigma \dot{q}^\nu}{c^2 - \sum_{\mu=1}^3 (\dot{q}^\mu)^2}\right)\ddot{q}^\nu, \\ B_{\sigma\nu} &= -\frac{mc}{\sqrt{c^2 - \sum_{\mu=1}^3 (\dot{q}^\mu)^2}}\left(\delta_{\sigma\nu} - \frac{\dot{q}^\sigma \dot{q}^\nu}{c^2 - \sum_{\mu=1}^3 (\dot{q}^\mu)^2}\right), \\ A_\sigma &= \left(\frac{\partial \Phi_\nu}{\partial q^\sigma} - \frac{\partial \Phi_\sigma}{\partial q^\nu}\right)\dot{q}^\nu - \frac{\partial \Phi_\sigma}{\partial t} - \frac{\partial \varphi}{\partial q^\sigma}. \end{aligned}$$

Vector form of equations of motion (equations of extremals) reads

$$\frac{\mathrm{d}\boldsymbol{p}}{\mathrm{d}t} = \boldsymbol{F}, \quad \text{where} \quad \boldsymbol{p} = \frac{m\boldsymbol{v}}{\sqrt{1 - \frac{v^2}{c^2}}}$$

is relativistic momentum and

$$\boldsymbol{F} = \boldsymbol{v} \times \operatorname{rot} \boldsymbol{\Phi} - \frac{\partial \boldsymbol{\Phi}}{\partial t} - \operatorname{grad} \varphi$$

is the force.

Example 9.4 (*Relativistic particle as a mechanical system with constraint*) Mechanical system of one relativistic particle can be interpreted also as a particle moving in time-space with (diagonal) Minkowski metric tensor $\operatorname{diag}(g) = (-1, -1, -1, 1)$. Basic fibred space is $(Y, \pi, X) = (\mathbf{R} \times \mathbf{R}^4, p, \mathbf{R})$ with four-dimensional fibres, the coordinate on the base $X = \mathbf{R}$ is a certain parameter s without physical meaning. Coordinates on fibres are (q^1, q^2, q^3, q^4), where q^1, q^2 a q^3 are Cartesian space coordinates and $q^4 = c\tau$, where τ is now time variable. Corresponding velocities $\dot{q}^\sigma$,

$1 \leq \sigma \leq 4$, obey the condition for four-velocity $\sqrt{(\dot{q}^4)^2 - \sum_{p=1}^{3}(\dot{q}^p)^2} = 1$. This constraint is non-holonomic. The corresponding Lagrangian is

$$L = -\frac{1}{2}m\left[(\dot{q}^4)^2 - \sum_{p=1}^{3}(\dot{q}^p)^2\right] + \dot{q}^\sigma \Phi_\sigma - \psi,$$

functions $\Phi_\sigma = \Phi_\sigma(q^1, q^2, q^3, q^4)$, $1 \leq \sigma \leq 4$, and $\psi = \psi(q^1, q^2, q^3, q^4)$ are independent of the mentioned "non-physical" parameter s. The Euler-Lagrange form has following components:

$$E_l = -m\ddot{q}^l + \dot{q}^\sigma\left(\frac{\partial \Phi_\sigma}{\partial q^l} - \frac{\partial \Phi_l}{\partial q^\sigma}\right) - \frac{\partial \psi}{\partial q^l}, \quad 1 \leq l \leq 3,$$

$$E_4 = m\ddot{q}^4 + \dot{q}^\sigma\left(\frac{\partial \Phi_\sigma}{\partial q^4} - \frac{\partial \Phi_4}{\partial q^\sigma}\right) - \frac{\partial \psi}{\partial q^4}, \quad 1 \leq \sigma \leq 4.$$

Four-vector $(\mathbf{\Phi}, \Phi_4)$ has four components. First three are connected with the vector potential, the fourth with the scalar potential, ψ is some additional function. The equations of motion must be solved taking into account the constraint. It is, of course, the problem of non-holonomic mechanics which is not the subject of our text. For details see, e.g. [8].

The following example is devoted to the problem of symmetries and conservation laws in classical mechanics. Our considerations lead to the well-known *Euclidean group of time-space symmetries*.

Example 9.5 (*Time-space symmetries*) Consider a mechanical system of a single mass particle without constraints. Basic fibred space is $(Y, \pi, X) = (\mathbf{R} \times \mathbf{R}^3, p, \mathbf{R})$ and the Lagrangian is of the form

$$\lambda = L\,\mathrm{d}t, \quad L = \frac{1}{2}m\left[(\dot{q}^1)^2 + (\dot{q}^2)^2 + (\dot{q}^3)^2\right] - U(t, q^1, q^2, q^3),$$

Lepage equivalent is then

$$\theta_\lambda = \left[\frac{1}{2}m((\dot{q}^1)^2 + (\dot{q}^2)^2 + (\dot{q}^3)^2) - U(t, q^1, q^2, q^3)\right]\mathrm{d}t$$
$$+ m\dot{q}^1\,\omega^1 + m\dot{q}^2\,\omega^2 + m\dot{q}^3\,\omega^3.$$

Fundamental symmetries of the time-space are:

- time homogeneity (for an observer connected with a free particle all time points are equivalent),
- space homogeneity (for an observer connected with a free particle all space points are equivalent),

- space isotropy (for an observer connected with a free particle all space directions are equivalent).

Corresponding symmetries of the Lagrangian $\lambda \in \Lambda^1_{\mathbf{R}}(TJ^1(\mathbf{R} \times \mathbf{R}^3))$,

$$L = \frac{1}{2}m\big((\dot{q}^1)^2 + (\dot{q}^2)^2 + (\dot{q}^3)^2\big),$$

thus ought to be time and space translations and space rotations. Let us express them as local isomorphisms of the basic fibred space.

- Time translations are given by local isomorphisms

$$\begin{aligned}\alpha_u : \mathbf{R} \times \mathbf{R}^3 \ni (t, q^1, q^2, q^3) &\longrightarrow \alpha_u(t, q^1, q^2, q^3)\\ &= (t + u, q^1, q^2, q^3) \in \mathbf{R} \times \mathbf{R}^3,\\ \alpha_{0u} : \mathbf{R} \ni t &\longrightarrow \alpha_{0u}(t) = t + u \in \mathbf{R},\end{aligned}$$

  forming the one-parameter group. The corresponding vector field with components

$$\xi^0 = \frac{dt\alpha_u(t, q^\nu)}{du}\bigg|_{u=0} = 1, \quad \Xi^\sigma = \frac{dq^\sigma\alpha_u(t, q^\nu)}{du}\bigg|_{u=0} = 0 \Rightarrow \xi = \frac{\partial}{\partial t},$$

  is the *generator of time translation*.
- Space translations are given by local isomorphisms $(\alpha_u^\sigma, \alpha_{0u})$, $1 \le \sigma \le 3$,

$$\begin{aligned}\alpha_u^1 : \mathbf{R} \times \mathbf{R}^3 \ni (t, q^1, q^2, q^3) &\longrightarrow \alpha_u^1(t, q^1, q^2, q^3)\\ &= (t, q^1 + u, q^2, q^3) \in \mathbf{R} \times \mathbf{R}^3,\\ \alpha_u^2 : \mathbf{R} \times \mathbf{R}^3 \ni (t, q^1, q^2, q^3) &\longrightarrow \alpha_u^2(t, q^1, q^2, q^3)\\ &= (t, q^1, q^2 + u, q^3) \in \mathbf{R} \times \mathbf{R}^3,\\ \alpha_u^3 : \mathbf{R} \times \mathbf{R}^3 \ni (t, q^1, q^2, q^3) &\longrightarrow \alpha_u^3(t, q^1, q^2, q^3)\\ &= (t, q^1, q^2, q^3 + u) \in \mathbf{R} \times \mathbf{R}^3.\end{aligned}$$

  Projection of every of these mappings onto the base is the identity.

$$\alpha_{0u} : \mathbf{R} \ni t \longrightarrow \alpha_{0u}(t) = t \in \mathbf{R}.$$

  Vector fields corresponding to those one-parameter groups are *generators of space translations*

$$\frac{\partial}{\partial q^1}, \quad \frac{\partial}{\partial q^2}, \quad \frac{\partial}{\partial q^3}.$$

For example, for one-parameter group $(\alpha^1_u, \alpha^1_{0u})$ we obtain

$$\xi^0 = \left.\frac{dt\alpha^1_u}{du}\right|_{u=0} = 0, \quad \Xi^1 = \left.\frac{dq^1\alpha^1_u}{du}\right|_{u=0} = 1, \quad \Xi^2 = \left.\frac{dq^2\alpha^1_u}{du}\right|_{u=0} = 0,$$
$$\Xi^3 = \left.\frac{dq^3\alpha^1_u}{du}\right|_{u=0} = 0.$$

For the other groups the calculation is analogous.

- Space rotations are generated by rotations around coordinate axes q^1, q^2 a q^3. For q^3 concretely

$$\alpha^{(3)}_u : \mathbf{R} \times \mathbf{R}^3 \ni (t, q^1, q^2, q^3) \longrightarrow \alpha^{(3)}_u(t, q^1, q^2, q^3)$$
$$= (t, q^1 \cos u + q^2 \sin u, -q^1 \sin u + q^2 \cos u, q^3) \in \mathbf{R} \times \mathbf{R}^3.$$

The projection of $\alpha^{(3)}_u$ is again the identity on $\mathbf{R}$. The corresponding *generator of space rotation* is the vector field

$$q^2 \frac{\partial}{\partial q^1} - q^1 \frac{\partial}{\partial q^2}.$$

Other generators of space rotations we obtain by cyclic permutation of indices,

$$q^1 \frac{\partial}{\partial q^3} - q^3 \frac{\partial}{\partial q^1}, \quad q^3 \frac{\partial}{\partial q^2} - q^2 \frac{\partial}{\partial q^3}.$$

We now construct the Noether equation for individual generators. This enables us to find out the requirements for the Lagrangian

$$L = \frac{1}{2} m \left[(\dot{q}^1)^2 + (\dot{q}^2)^2 + (\dot{q}^3)^2 \right] - U(t, q^1, q^2, q^3)$$

of a mass particle ensuring that individual time-space transformations will be invariance transformations. (Of course, it is to be expected that these requirements will be placed on the potential energy $U(t, q^1, q^2, q^3)$, not on the kinetic part of the Lagrangian.)

- For the time translation generator it holds

$$\xi = \frac{\partial}{\partial t}, \quad J^1\xi = \frac{\partial}{\partial t}$$

and the Noether equation is very simple, $\frac{\partial L}{\partial t} = 0$. This means that the time translation generator is the symmetry of a particle if its Lagrangian is independent of time. Thus the potential energy U is independent of time too. The corresponding Noether current is

$$\Phi\left(\frac{\partial}{\partial t}\right) = -L + \frac{\partial L}{\partial \dot{q}^\sigma}\dot{q}^\sigma = -\left[\frac{1}{2}m\left((\dot{q}^1)^2 + (\dot{q}^2)^2 + (\dot{q}^3)^2\right) + U(q^1, q^2, q^3)\right].$$

This is the oppositely taken mechanical energy. We have the *conservation law of mechanical energy*.

- Noether equation of the space translation generator, e.g. along q^1, is again very simple, $\frac{\partial L}{\partial q^1} = 0$, because

$$\xi = \frac{\partial}{\partial q^1}, \quad J^1\xi = \frac{\partial}{\partial q^1}.$$

If the space translation generator along q^1 is the symmetry of the Lagrangian, then the Lagrange function is independent of the coordinate q^1. (Coordinate q^1 is *cyclic*.) Conserved quantity is the first component of momentum

$$\Phi\left(\frac{\partial}{\partial q^1}\right) = \frac{\partial L}{\partial \dot{q}^1} = m\dot{q}^1.$$

The analogous situation is as for other generators of space translations. (We obtain the same conclusion as in classical Newtonian mechanics based on Newton laws of motion.)

- For the generator of space rotation, e.g. around the coordinate axis q^3 it holds

$$\xi = q^2\frac{\partial}{\partial q^1} - q^1\frac{\partial}{\partial q^2}, \quad J^1\xi = q^2\frac{\partial}{\partial q^1} - q^1\frac{\partial}{\partial q^2} + \dot{q}^2\frac{\partial}{\partial \dot{q}^1} - \dot{q}^1\frac{\partial}{\partial \dot{q}^2}.$$

Noether equation reads

$$\frac{\partial L}{\partial q^1}q^2 - \frac{\partial L}{\partial q^2}q^1 + \frac{\partial L}{\partial \dot{q}^1}\dot{q}^2 - \frac{\partial L}{\partial \dot{q}^2}\dot{q}^1 = \frac{\partial L}{\partial q^1}q^2 - \frac{\partial L}{\partial q^2}q^1$$

because $\frac{\partial L}{\partial \dot{q}^1}\dot{q}^2 - \frac{\partial L}{\partial \dot{q}^2}\dot{q}^1 = m\dot{q}^1\dot{q}^2 - m\dot{q}^2\dot{q}^1 = 0$. Then the Noether equation gives

$$\frac{\partial L}{\partial q^1}q^2 - \frac{\partial L}{\partial q^2}q^1 = -\frac{\partial U}{\partial q^1}q^2 + \frac{\partial U}{\partial q^2}q^1 = 0.$$

The corresponding Noether current is

$$\phi = q^2\frac{\partial L}{\partial \dot{q}^1} - q^1\frac{\partial L}{\partial \dot{q}^2} = m(q^2\dot{q}^1 - q^1\dot{q}^2),$$

which is the (negatively taken) third component of the particle angular momentum. The quantity

$$\begin{aligned} \boldsymbol{M} &= -\left(q^3\frac{\partial U}{\partial q^2} - q^2\frac{\partial U}{\partial q^3}, q^1\frac{\partial U}{\partial q^3} - q^3\frac{\partial U}{\partial q^1}, q^2\frac{\partial U}{\partial q^1} - q^1\frac{\partial U}{\partial q^2}\right) \\ &= -\boldsymbol{r} \times \operatorname{grad} U = \boldsymbol{r} \times \boldsymbol{F} \end{aligned}$$

has the meaning of the torque of the resulting (conservative) force $\boldsymbol{F} = -\operatorname{grad} U$ acting on the particle with respect to the coordinate system origin. Noether currents corresponding to individual generators of space rotations around coordinate axes q^1, q^2 a q^3 are

$$\begin{aligned} \xi^{(1)} &= q^3\frac{\partial}{\partial q^2} - q^2\frac{\partial}{\partial q^3}, & \phi(\xi^{(1)}) &= q^3\frac{\partial L}{\partial \dot{q}^2} - q^2\frac{\partial L}{\partial \dot{q}^3}, \\ \xi^{(2)} &= -q^3\frac{\partial}{\partial q^1} + q^1\frac{\partial}{\partial q^3}, & \phi(\xi^{(2)}) &= q^3\frac{\partial L}{\partial \dot{q}^1} - q^1\frac{\partial L}{\partial \dot{q}^3}, \\ \xi^{(3)} &= q^2\frac{\partial}{\partial q^1} - q^1\frac{\partial}{\partial q^2}, & \phi(\xi^{(3)}) &= q^2\frac{\partial L}{\partial \dot{q}^1} - q^1\frac{\partial L}{\partial \dot{q}^2}. \end{aligned}$$

These currents are components of the angular momentum with respect to the coordinate system origin,

$$\boldsymbol{\ell} = \boldsymbol{r} \times \boldsymbol{p} = m\left(q^2\dot{q}^3 - q^3\dot{q}^2, q^3\dot{q}^1 - q^1\dot{q}^3, q^1\dot{q}^2 - q^2\dot{q}^1\right),$$

which again represents the well-known result of the Newtonian mechanics.

Let $(Y, \pi, X) = (\mathbf{R} \times \mathbf{R}^3, p, \mathbf{R})$ be the fibred space. Time translations $t \to t + u$, space translations $q^\sigma \to q^\sigma + u$, $1 \le \sigma \le 3$, and space rotations $(q^1, q^2, q^3) \to (q^1 \cos u + q^2 \sin u, -q^1 \sin u + q^2 \cos u, q^3)$ and cyclic permutations, form the group called the *Euclidean group*. This group is generated by 7 independent vector fields corresponding to 7 local one-parameter groups (see Table 9.1).

Euclidean group is the symmetry group of the Lagrangian of a free particle. Conserved quantities along extremals (Noether currents) are mechanical energy $\frac{1}{2}m\left[(\dot{q}^1)^2 + (\dot{q}^2)^2 + (\dot{q}^3)^2\right] + U(q^1, q^2, q^3)$, momentum $\boldsymbol{p} = (m\dot{q}^1, m\dot{q}^2, m\dot{q}^3)$ and angular momentum $\boldsymbol{r} \times \boldsymbol{p} = m(q^2\dot{q}^3 - q^3\dot{q}^2, q^3\dot{q}^1 - q^1\dot{q}^3, q^1\dot{q}^2 - q^2\dot{q}^1)$.

The "analogy" of the Euclidean group in the theory of relativity is the *Poincaré group*. This group has 10 generators represented by vector fields

$$\begin{aligned} &\frac{\partial}{\partial q^1}, \quad \frac{\partial}{\partial q^2}, \quad \frac{\partial}{\partial q^3}, \quad \frac{\partial}{\partial q^4}, \\ &q^1\frac{\partial}{\partial q^4} + q^4\frac{\partial}{\partial q^1}, \quad q^2\frac{\partial}{\partial q^4} + q^4\frac{\partial}{\partial q^2}, \quad q^3\frac{\partial}{\partial q^4} + q^4\frac{\partial}{\partial q^3}, \\ &q^3\frac{\partial}{\partial q^2} - q^2\frac{\partial}{\partial q^3}, \quad q^1\frac{\partial}{\partial q^3} - q^3\frac{\partial}{\partial q^1}, \quad q^2\frac{\partial}{\partial q^1} - q^1\frac{\partial}{\partial q^2}. \end{aligned} \tag{9.1}$$

Table 9.1 Euclidean group

Transformation	α_{0u}, α_u	Generators
Time translations	$t \longrightarrow t + u$	$\frac{\partial}{\partial t}$
	$(t, q^\sigma) \longrightarrow (t + u, q^\sigma)$	
Space translations	$t \longrightarrow t,\ (t, q^1, q^2, q^3) \longrightarrow (t, q^1 + u, q^2, q^3)$	$\frac{\partial}{\partial q^1}, \frac{\partial}{\partial q^2}, \frac{\partial}{\partial q^3}$
	$t \longrightarrow t,\ (t, q^1, q^2, q^3) \longrightarrow (t, q^1, q^2 + u, q^3)$	
	$t \longrightarrow t,\ (t, q^1, q^2, q^3) \longrightarrow (t, q^1, q^2, q^3 + u)$	
Space rotations	$t \longrightarrow t,\ (t, q^1, q^2, q^3) \longrightarrow \longrightarrow (t, q^1, q^2 \cos u + q^3 \sin u, -q^2 \sin u + q^3 \cos u)$	$q^3 \frac{\partial}{\partial q^2} - q^2 \frac{\partial}{\partial q^3}$
	$t \longrightarrow t,\ (t, q^1, q^2, q^3) \longrightarrow (-q^3 \sin u + q^1 \cos u, q^2, q^3 \cos u + q^1 \sin u)$	$q^1 \frac{\partial}{\partial q^3} - q^3 \frac{\partial}{\partial q^1}$
	$t \longrightarrow t,\ (t, q^1, q^2, q^3) \longrightarrow (q^1 \cos u + q^2 \sin u, -q^1 \sin u + q^2 \cos u, q^3)$	$q^2 \frac{\partial}{\partial q^1} - q^1 \frac{\partial}{\partial q^2}$

In the following part we solve the inverse problem in classical mechanics. So, we will assess the variationality of the second-order dynamical form $E = E_\sigma\, \omega^\sigma \wedge dt$, $E_\sigma = A(t, q^\mu, \dot{q}^\mu) + B_{\sigma\nu}(t, q^\mu, \dot{q}^\mu)\ddot{q}^\nu$, $1 \leq \sigma \leq m$. It is well-known that such a dynamical form is variational if and only if it obeys Helmholtz conditions

$$B_{\sigma\nu} = B_{\nu\sigma}, \qquad \frac{\partial B_{\sigma\nu}}{\partial \dot{q}^\mu} = \frac{\partial B_{\sigma\mu}}{\partial \dot{q}^\nu}$$

and function A_σ are of the form

$$A_\sigma = \dot{q}^\nu \left[\int_0^1 \left(\frac{\partial B_{\nu\lambda}}{\partial q^\sigma} \dot{q}^\lambda + \frac{\partial B_{\sigma\nu}}{\partial t} \right) \circ \chi \, du + \dot{q}^\lambda \int_0^1 \left(\frac{\partial B_{\sigma\lambda}}{\partial q^\nu} - \frac{\partial B_{\nu\lambda}}{\partial q^\sigma} \right) \circ \chi \, du \right]$$
$$+ \dot{q}^\nu \left(\frac{\partial \Phi_\nu}{\partial q^\sigma} - \frac{\partial \Phi_\sigma}{\partial q^\nu} \right) - \left(\frac{\partial \varphi}{\partial q^\sigma} + \frac{\partial \Phi_\sigma}{\partial t} \right). \tag{9.2}$$

Example 9.6 (*Inverse problem for a non-relativistic particle*) Basic fibred space for a non-relativistic particle without constraints is $(\mathbf{R} \times \mathbf{R}^3, p, \mathbf{R})$, where (t, q^1, q^2, q^3) means time-space coordinates (time and Cartesian space coordinates). The standard form of equations of motion is $F_\sigma - m g_{\sigma\nu} \ddot{q}^\nu = 0$, where $\boldsymbol{F} = (F_1, F_2, F_3)$ is the resulting force acting on a particle of mass m, $g = (g_{\mu\nu})$ is the metric tensor. Its components obey the symmetry relations $g_{\sigma\nu} = g_{\nu\sigma}$ a they depend only on space coordinates $g_{\sigma\nu} = g_{\sigma\nu}(q^1, q^2, q^3)$. Under those assumptions the conditions for $B_{\sigma\nu} = -mg_{\sigma\nu}$ are automatically fulfilled. Coefficients A_σ represent the components of the force $\boldsymbol{F}$. In order this force to be variational, it must hold relations (9.2). We now put into (9.2) functions $B_{\sigma\nu}$. We will remind that the mapping χ assigns to the point $(t, q^\mu, \dot{q}^\mu)$ in the phase space the point $(t, q^\mu, u\dot{q}^\mu)$. Here $u \in [0, 1]$ is integration variable. Then

$$F_\sigma = -m\dot{q}^\nu \left[\int_0^1 \left(\frac{\partial g_{\nu\lambda}}{\partial q^\sigma} \right) u\dot{q}^\lambda \, du + \dot{q}^\lambda \int_0^1 \left(\frac{\partial g_{\sigma\lambda}}{\partial q^\nu} - \frac{\partial g_{\nu\lambda}}{\partial q^\sigma} \right) du \right] + \dot{q}^\nu \left(\frac{\partial \Phi_\nu}{\partial q^\sigma} - \frac{\partial \Phi_\sigma}{\partial q^\nu} \right) - \left(\frac{\partial \varphi}{\partial q^\sigma} + \frac{\partial \Phi_\sigma}{\partial t} \right),$$

where Φ_σ, $1 \le \sigma \le 3$, and φ are arbitrary functions on $Y = \mathbf{R} \times \mathbf{R}^3$, i.e. function of time and coordinates. Calculating integrals we obtain (see also Example 9.1)

$$F_\sigma = \dot{q}^\nu \dot{q}^\lambda \Gamma_{\sigma\nu\lambda} + \dot{q}^\nu \left(\frac{\partial \Phi_\nu}{\partial q^\sigma} - \frac{\partial \Phi_\sigma}{\partial q^\nu} \right) - \left(\frac{\partial \varphi}{\partial q^\sigma} + \frac{\partial \Phi_\sigma}{\partial t} \right),$$

$$\Gamma_{\sigma\nu\lambda} = -\frac{1}{2} m \left(\frac{\partial g_{\sigma\nu}}{\partial q^\lambda} + \frac{\partial g_{\sigma\lambda}}{\partial q^\nu} - \frac{\partial g_{\nu\lambda}}{\partial q^\sigma} \right).$$

The underlying space for classical Newtonian mechanics is Euclidean, the metric g is thus Euclidean, $g = (g_{\sigma\nu}) = (\delta_{\sigma\nu})$. In such a case functions $\Gamma_{\sigma\nu\lambda}$ are zero and the components of the variational force are given only by functions $\boldsymbol{\Phi} = (\Phi_1, \Phi_2, \Phi_3)$ and φ,

$$F_\sigma = \dot{q}^\nu \left(\frac{\partial \Phi_\nu}{\partial q^\sigma} - \frac{\partial \Phi_\sigma}{\partial q^\nu} \right) - \left(\frac{\partial \varphi}{\partial q^\sigma} + \frac{\partial \Phi_\sigma}{\partial t} \right).$$

In the vector notation $\boldsymbol{r} = (q^1, q^2, q^3)$ and $\boldsymbol{v} = (\dot{q}^1, \dot{q}^2, \dot{q}^3)$, $\boldsymbol{F} = (F_3, F_2, F_3)$ we have

$$\boldsymbol{F} = \boldsymbol{v} \times \operatorname{rot} \boldsymbol{\Phi} - \frac{\partial \boldsymbol{\Phi}}{\partial t} - \operatorname{grad} \varphi.$$

If we denote $\operatorname{rot} \boldsymbol{\Phi} = Q\boldsymbol{B}$ and $Q\boldsymbol{E} = -\frac{\partial \boldsymbol{\Phi}}{\partial t} - \operatorname{grad} \varphi$, then

$$\boldsymbol{F} = Q\boldsymbol{v} \times \boldsymbol{B} + Q\boldsymbol{E}, \quad \operatorname{div} \boldsymbol{B} = 0, \quad \operatorname{rot} \boldsymbol{E} = -\frac{\partial \boldsymbol{B}}{\partial t}.$$

The character of every variational force is Lorentzian, $\boldsymbol{F} = Q\boldsymbol{v} \times \boldsymbol{B} + Q\boldsymbol{E}$. In addition, relations $\operatorname{div} \boldsymbol{B} = 0$ a $\operatorname{rot} \boldsymbol{E} = -\frac{\partial \boldsymbol{B}}{\partial t}$ represent the formal expression of two well-known Maxwell equations. Note, that this result is solely a consequence of the requirement of variationality of equations of motion.

Example 9.7 (*Variationality and inertial forces*)The second Newton law for a particle in non-inertial reference frame contains also *inertial* or *fictional* forces connected with the acceleration of the non-inertial reference frame with respect inertial ones.

This acceleration is the sum of the translational term $\boldsymbol{A}(t)$ and rotational terms given by the angular velocity $\boldsymbol{\omega}(t)$ and angular acceleration $\boldsymbol{\varepsilon}(t)$. The resulting relation has the well-known form

$$\tilde{\boldsymbol{a}} = \boldsymbol{A} + 2\boldsymbol{\omega} \times \boldsymbol{v} + \boldsymbol{\varepsilon} \times \boldsymbol{r} + \boldsymbol{\omega} \times (\boldsymbol{\omega} \times \boldsymbol{r}),$$

where $\boldsymbol{r}$ and $\boldsymbol{v}$ are the position vector and velocity of the studied particle with respect to the non-inertial reference frame. Inertial forces are then *translational* $\boldsymbol{F}_T^* = -m\boldsymbol{A}$, *Coriolis* $\boldsymbol{F}_C^* = -2m\boldsymbol{\omega} \times \boldsymbol{v}$, *Euler* $\boldsymbol{F}_E^* = -m\boldsymbol{\varepsilon} \times \boldsymbol{r}$ and *centrifugal* $\boldsymbol{F}_o^* = -m\boldsymbol{\omega} \times (\boldsymbol{\omega} \times \boldsymbol{r})$, i.e.

$$\boldsymbol{F}^* = -m\boldsymbol{A} + (-2\,m\boldsymbol{\omega} \times \boldsymbol{v}) + (-m\boldsymbol{\varepsilon} \times \boldsymbol{r}) + \big(-m\boldsymbol{\omega} \times (\boldsymbol{\omega} \times \boldsymbol{r})\big).$$

The question is whether these forces are variational. This means that we search a vector function $\boldsymbol{\Phi}$ and scalar function φ for which

$$\boldsymbol{F}^* = \boldsymbol{v} \times \operatorname{rot} \boldsymbol{\Phi} + \left(-\frac{\partial \boldsymbol{\Phi}}{\partial t}\right) + (-\operatorname{grad} \varphi).$$

Comparing these two expressions, we can see that the "candidate" for the role of $\boldsymbol{\Phi}$ is the vector product $m\boldsymbol{\omega} \times \boldsymbol{r}$. (The angular velocity $\boldsymbol{\omega}$ depends only on time, $\boldsymbol{\omega} = \boldsymbol{\omega}(t)$.) Euler force $\boldsymbol{F}_E^* = -m\boldsymbol{\varepsilon} \times \boldsymbol{r}$ could be connected with the term $-\frac{\partial \boldsymbol{\Phi}}{\partial t}$. However, for this to be the case, it must be compatible with the expression for the Coriolis force $\boldsymbol{F}_C^* = -2\,m\boldsymbol{\omega} \times \boldsymbol{v} = \boldsymbol{v} \times (2\,m\boldsymbol{\omega})$. Let us calculate the components of rotation of the vector field $\boldsymbol{\Phi} = m\boldsymbol{\omega} \times \boldsymbol{r}$. For example, for the fist component it holds

$$\operatorname{rot}_1 \boldsymbol{\Phi} = \frac{\partial \Phi_3}{\partial q^2} - \frac{\partial \Phi_2}{\partial q^3} = m\frac{\partial}{\partial q^2}(\omega_1 q^2 - \omega_2 q^1) - m\frac{\partial}{\partial q^3}(\omega_3 q^1 - \omega_1 q^3) = 2\,m\omega_1.$$

Completely analogous calculations for remaining components show that we finally have $\operatorname{rot}(m\boldsymbol{\omega} \times \boldsymbol{r}) = 2\,m\boldsymbol{\omega}$. So, the choice of the vector field $\boldsymbol{\Phi}$ in the form $\boldsymbol{\Phi} = m\boldsymbol{\omega} \times \boldsymbol{r}$ is suitable. In the following step we will search the scalar function φ, for which it holds

$$\operatorname{grad} \varphi = m\boldsymbol{A} + m\boldsymbol{\omega} \times (\boldsymbol{\omega} \times \boldsymbol{r}) = m\boldsymbol{A} + m(\boldsymbol{\omega}\boldsymbol{r})\boldsymbol{\omega} - m(\boldsymbol{\omega}^2\boldsymbol{r}).$$

Step-by-step integration of corresponding equations leads to following results:

$$\begin{aligned}
\frac{\partial \varphi}{\partial q^1} &= mA_1 - m\left[q^1\left(\omega_2^2 + \omega_3^2\right) - q^2\omega_1\omega_2 - q^3\omega_1\omega_3\right],\\
\frac{\partial \varphi}{\partial q^2} &= mA_2 - m\left[-q^1\omega_1\omega_2 + q^2\left(\omega_1^2 + \omega_3^2\right) - q^3\omega_2\omega_3\right],\\
\frac{\partial \varphi}{\partial q^3} &= mA_3 - m\left[-q^1\omega_1\omega_3 - q^2\omega_2\omega_3 + q^3\left(\omega_1^2 + \omega_2^2\right)\right],
\end{aligned}$$

and moreover, taking into account that $\boldsymbol{\omega}$ a $\boldsymbol{A}$ depend only on time:

$$\varphi = m\boldsymbol{A}\boldsymbol{r} - \frac{1}{2}m\left(q^1\ q^2\ q^3\right)\begin{pmatrix} \omega_2^2+\omega_3^2 & -\omega_1\omega_2 & -\omega_1\omega_3 \\ -\omega_1\omega_2 & \omega_1^2+\omega_3^2 & -\omega_2\omega_3 \\ -\omega_1\omega_3 & -\omega_2\omega_3 & \omega_1^2+\omega_2^2 \end{pmatrix}\begin{pmatrix} q^1 \\ q^2 \\ q^3 \end{pmatrix}.$$

Here $\boldsymbol{A}(t)$ the translational acceleration of the non-inertial reference frame with respect to the inertial one. We can conclude that inertial forces are variational.

9.2 Mechanics of Continuous Media and Waves

In general, mechanics of continuous media requires as the underlying geometrical structure a fibred manifold with four-dimensional base X with coordinates, e.g. (t, x, y, z) (time and Cartesian coordinates). The coordinates on fibres can be, e.g. components of (symmetric) deformation tensor, or alternatively stress tensor. We show here only the mathematical description of a very simple model—oscillations of an elastic membrane. The formulation of the mentioned variational problem leads, as we expect, to the wave equation as equation of motion.

On the other hand, if we take the wave equation as a starting point, we can prove, that it is variational. We then construct the Vainberg-Tonti Lagrangian as well as the minimal order Lagrangian.

Example 9.8 (*Oscillating membrane*) Consider transversal oscillations of a membrane which in equilibrium (at rest) occupies an open connected set $M \subset \mathbf{R}^2$. Denote the displacement of the point $(x, y) \in M$ of the membrane at the time "point" t as $u(t, x, y)$. The corresponding fibred space is $(Y, \pi, X) = (\mathbf{R}^3 \times \mathbf{R}, p, \mathbf{R}^3)$, the domain of the Lagrangian is such a subset of J^1Y, for which $(x, y) \in M$. The Lagrangian reads

$$\lambda = L\,\mathrm{d}t \wedge \mathrm{d}x \wedge \mathrm{d}y, \quad L(t, x, y, u, u_x, u_y) = \frac{1}{2}\sigma u_t^2 - \frac{1}{2}\tau(u_x^2 + u_y^2),$$

i.e. it is the difference of the kinetic energy and the elastic potential energy of the deformed membrane, τ is the quantity characterizing the stress in the membrane. The Euler-Lagrange equation has the form

$$-\mathrm{d}_t\frac{\partial L}{\partial u_t} - \mathrm{d}_x\frac{\partial L}{\partial u_x} - \mathrm{d}_y\frac{\partial L}{\partial u_y} = 0 \implies \frac{\partial^2 u}{\partial x^2} + \frac{\partial^2 u}{\partial y^2} - \frac{\sigma}{\tau}\frac{\partial^2 u}{\partial t^2} = 0.$$

This is the wave equation in two space dimensions. The phase velocity of oscillations is $v_f = \sqrt{\tau/\sigma}$.

Now we determine symmetries of the Lagrangian and corresponding Noether currents. Noether equation for the vector field

$$\xi = \xi^t \frac{\partial}{\partial t} + \xi^x \frac{\partial}{\partial x} + \xi^y \frac{\partial}{\partial y} + \Xi^u \frac{\partial}{\partial u}$$

has the form

$$\begin{aligned}
&\sigma u_t (\mathrm{d}_t \Xi^u - u_t \, \mathrm{d}_t \xi^t - u_x \, \mathrm{d}_t \xi^x - u_y \, \mathrm{d}_t \xi^y) \\
&- \tau u_x (\mathrm{d}_x \Xi^u - u_t \, \mathrm{d}_x \xi^t - u_x \, \mathrm{d}_x \xi^x - u_y \, \mathrm{d}_x \xi^y) \\
&- \tau u_y (\mathrm{d}_y \Xi^u - u_t \, \mathrm{d}_y \xi^t - u_x \, \mathrm{d}_y \xi^x - u_y \, \mathrm{d}_y \xi^y) \\
&+ \frac{1}{2}\left[\sigma u_t^2 - \tau (u_x^2 + u_y^2)\right](\mathrm{d}_t \xi^t + \mathrm{d}_x \xi^x + \mathrm{d}_y \xi^y) = 0.
\end{aligned}$$

It can be immediately seen that the base vector fields on the base X, i.e. $\frac{\partial}{\partial t}$, $\frac{\partial}{\partial x}$ and $\frac{\partial}{\partial y}$, as well as the vector field $\frac{\partial}{\partial u}$ (the generator of translations along fibres) belong to the set of solutions of this equation, and thus they are symmetries of the Lagrangian. Corresponding currents are

$$\begin{aligned}
\Phi\left(\frac{\partial}{\partial t}\right) &= -\left(\frac{1}{2}\sigma u_t^2 + \frac{1}{2}\tau u_x^2 + \frac{1}{2}\tau u_y^2\right) \mathrm{d}x \wedge \mathrm{d}y - \tau u_t u_x, \mathrm{d}t \wedge \mathrm{d}y \\
&\quad + \tau u_t u_y \, \mathrm{d}t \wedge \mathrm{d}x, \\
\Phi\left(\frac{\partial}{\partial x}\right) &= \left(\frac{1}{2}\sigma u_t^2 + \frac{1}{2}\tau u_x^2 - \frac{1}{2}\tau u_y^2\right) \mathrm{d}y \wedge \mathrm{d}t - \sigma u_t u_x \, \mathrm{d}x \wedge \mathrm{d}y \\
&\quad + \tau u_x u_y \, \mathrm{d}t \wedge \mathrm{d}x, \\
\Phi\left(\frac{\partial}{\partial y}\right) &= \left(\frac{1}{2}\sigma u_t^2 - \frac{1}{2}\tau u_x^2 + \frac{1}{2}\tau u_y^2\right) \mathrm{d}t \wedge \mathrm{d}x - \sigma u_t u_y \, \mathrm{d}x \wedge \mathrm{d}y \\
&\quad + \tau u_x u_y \, \mathrm{d}t \wedge \mathrm{d}y, \\
\Phi\left(\frac{\partial}{\partial u}\right) &= \sigma u_t \, \mathrm{d}x \wedge \mathrm{d}y - \tau u_x \, \mathrm{d}y \wedge \mathrm{d}t - \tau u_y \, \mathrm{d}t \wedge \mathrm{d}x.
\end{aligned}$$

The quantity $\frac{1}{2}\sigma u_t^2 + \frac{1}{2}\tau (u_x^2 + u_y^2)$ is the density of total mechanical energy (kinetic energy plus elastic potential energy).

Example 9.9 (*Variationality of the wave equation*) Let us consider the wave equation for the unknown function $u = u(t, x)$ of time and Cartesian coordinate (a longitudinal or transversal wave progresses along the x-axis with the phase speed v),

$$\frac{\partial^2 u(t, x)}{\partial x^2} - \frac{1}{v^2} \frac{\partial^2 u(t, x)}{\partial t^2} = 0. \tag{9.3}$$

(The quantity v depends on a concrete situation. For example, for electromagnetic waves in vacuum it holds $v = c = \frac{1}{\sqrt{\varepsilon_0 \mu_0}}$, for mechanical waves we have $v = \sqrt{E/\varrho}$, where E is the Young module and ϱ is the density of a medium, etc.). In the following let us consider $v = 1$ for simplicity.

The basic fibred space is $(Y, \pi, X) = (\mathbf{R}^2 \times \mathbf{R}, p, \mathbf{R}^2)$, i.e. $n = 2$, $m = 1$, with charts (V, ψ), $\psi = (t, x, u)$, on Y and $(\pi(V), \varphi)$, $\varphi = (t, x)$, on X. Wave equation is the second-order partial differential equation, i.e. the corresponding dynamical form $E = E_1 \omega^1 \wedge \omega_0 = E_1 \omega^1 \wedge \mathrm{d}t \wedge \mathrm{d}x$ is defined on (J^2Y, π_2, X). Associated fibred charts on J^1Y and J^2Y are $(\pi_{1,0}^{-1}(V), \psi_1)$ and $(\pi_{2,0}^{-2}(V), \psi_2)$, where $\psi_1 = (t, x, u, u_t, u_x)$ and $\psi_2 = (t, x, u, u_t, u_x, u_{tt}, u_{tx}, u_{xx})$, respectively. Equation (9.3) gives

$$E = E_1 \omega^1 \wedge \omega_0 = (u_{xx} - u_{tt})\, \omega^1 \wedge \mathrm{d}t \wedge \mathrm{d}x, \quad \omega^1 = \mathrm{d}u - u_t\, \mathrm{d}t - u_x \mathrm{d}x.$$

General Helmholtz conditions of variationality for second-order dynamical form are (see Chap. 8)

$$\begin{aligned} \frac{\partial E_\sigma}{\partial y^\nu} - \frac{\partial E_\nu}{\partial y^\sigma} + \mathrm{d}_i \frac{\partial E_\nu}{\partial y_i^\sigma} - \mathrm{d}_i\, \mathrm{d}_j \frac{\partial E_\nu}{\partial y_{ij}^\sigma} &= 0, \\ \frac{\partial E_\sigma}{\partial y_i^\nu} + \frac{\partial E_\nu}{\partial y_i^\sigma} - 2\, \mathrm{d}_j \frac{\partial E_\nu}{\partial y_{ij}^\sigma} &= 0, \\ \frac{\partial E_\sigma}{\partial y_{ij}^\nu} - \frac{\partial E_\nu}{\partial y_{ij}^\sigma} &= 0, \end{aligned} \tag{9.4}$$

$1 \le \sigma, \nu \le m$, $1 \le i, j \le n$. In the case of the form E corresponding to the wave equation we have $\sigma = \nu = 1$, $1 \le i, j \le 2$, and we can immediately see that all conditions of (9.4) are trivially fulfilled. The mapping χ for calculation of Vainberg-Tonti Lagrangian $\Lambda = \mathcal{L}\, \mathrm{d}t \wedge \mathrm{d}x$ is

$$\begin{aligned} &\chi : [0, 1] \times W_2 \ni (s, t, x, u, u_t, u_x, u_{tt}, u_{tx}, u_{xx}) \\ &\quad \longrightarrow (t, x, su, su_t, su_x, su_{tt}, su_{tx}, su_{xx}) \in W_2, \quad W_2 \subset \pi_{2,0}^{-1}(V), \end{aligned}$$

$$\Lambda = \left[u \int_0^1 (su_{xx} - su_{tt})\, \mathrm{d}s \right] \mathrm{d}t \wedge \mathrm{d}x = \frac{1}{2}(uu_{xx} - uu_{tt})\, \mathrm{d}t \wedge \mathrm{d}x.$$

The minimal Lagrangian λ we obtain as $\Lambda - \frac{1}{2}\big[\mathrm{d}_x(uu_x) - \mathrm{d}_t(uu_t)\big]\mathrm{d}t \wedge \mathrm{d}x$, i.e. subtracting from Λ a trivial Lagrangian. The minimal Lagrangian is of the first order:

$$\lambda = \frac{1}{2}\left(u_t^2 - u_x^2\right) \mathrm{d}t \wedge \mathrm{d}x.$$

Now we find Noether symmetries and corresponding currents for completeness. Let

$$\xi = \xi^t(t, x) \frac{\partial}{\partial t} + \xi^x(t, x) \frac{\partial}{\partial x} + \Xi(t, x, u) \frac{\partial}{\partial u}$$

be a π-projectable vector field on Y. Its first prolongation is

$$J^1\xi = \xi^t \frac{\partial}{\partial t} + \xi^x \frac{\partial}{\partial x} + \Xi \frac{\partial}{\partial u} + \Xi_t \frac{\partial}{\partial u_t} + \Xi_x \frac{\partial}{\partial u_x}, \quad \text{where}$$
$$\Xi_t = \frac{\partial \Xi}{\partial t} + \frac{\partial \Xi}{\partial u} u_t - u_t \frac{\mathrm{d}\xi^t}{\mathrm{d}t} - u_x \frac{\mathrm{d}\xi^x}{\mathrm{d}t},$$
$$\Xi_x = \frac{\partial \Xi}{\partial x} + \frac{\partial \Xi}{\partial u} u_x - u_t \frac{\mathrm{d}\xi^t}{\mathrm{d}x} - u_x \frac{\mathrm{d}\xi^x}{\mathrm{d}x}.$$

The Noether equation (see Eq. 8.38)

$$\partial_{J^1\xi}\lambda = i_{J^1\xi}\,\mathrm{d}\lambda + \mathrm{d}i_{J^1\xi}\lambda = 0 \implies i_{J^1\xi}\,\mathrm{d}L + L\left(\frac{\mathrm{d}\xi^t}{\mathrm{d}t} + \frac{\mathrm{d}\xi^x}{\mathrm{d}x}\right) = 0$$

reads

$$u_t\left(\frac{\partial \Xi}{\partial t} + \frac{\partial \Xi}{\partial u} u_t - u_t \frac{\mathrm{d}\xi^t}{\mathrm{d}t} - u_x \frac{\mathrm{d}\xi^x}{\mathrm{d}t}\right) - u_x\left(\frac{\partial \Xi}{\partial x} + \frac{\partial \Xi}{\partial u} u_x - u_t \frac{\mathrm{d}\xi^t}{\mathrm{d}x} - u_x \frac{\mathrm{d}\xi^x}{\mathrm{d}x}\right)$$
$$+ \frac{1}{2}\left(u_t^2 - u_x^2\right)\left(\frac{\mathrm{d}\xi^t}{\mathrm{d}t} + \frac{\mathrm{d}\xi^x}{\mathrm{d}x}\right) = 0$$
$$\implies u_t^2\left(\frac{\partial \Xi}{\partial u} - \frac{1}{2}\frac{\mathrm{d}\xi^t}{\mathrm{d}t} + \frac{1}{2}\frac{\mathrm{d}\xi^x}{\mathrm{d}x}\right) + u_x^2\left(-\frac{\partial \Xi}{\partial u} + \frac{1}{2}\frac{\mathrm{d}\xi^x}{\mathrm{d}x} - \frac{1}{2}\frac{\mathrm{d}\xi^t}{\mathrm{d}t}\right)$$
$$+ u_t u_x\left(\frac{\mathrm{d}\xi^t}{\mathrm{d}x} - \frac{\mathrm{d}\xi^x}{\mathrm{d}t}\right) + u_t \frac{\partial \Xi}{\partial t} - u_x \frac{\partial \Xi}{\partial x} = 0.$$

For components of Noether symmetries we obtain the following equations:

$$\begin{aligned}
\frac{\partial \Xi}{\partial u} - \frac{1}{2}\frac{\mathrm{d}\xi^t}{\mathrm{d}t} + \frac{1}{2}\frac{\mathrm{d}\xi^x}{\mathrm{d}x} &= 0,\\
-\frac{\partial \Xi}{\partial u} - \frac{1}{2}\frac{\mathrm{d}\xi^t}{\mathrm{d}t} + \frac{1}{2}\frac{\mathrm{d}\xi^x}{\mathrm{d}x} &= 0,\\
\frac{\mathrm{d}\xi^t}{\mathrm{d}x} - \frac{\mathrm{d}\xi^x}{\mathrm{d}t} &= 0,\\
\frac{\partial \Xi}{\partial t} = \frac{\partial \Xi}{\partial x} &= 0.
\end{aligned} \tag{9.5}$$

The following vector fields are solutions of these equations, and thus they are the Noether symmetries:

$$\xi_1 = \frac{\partial}{\partial t}, \quad \xi_2 = \frac{\partial}{\partial x}, \quad \xi_3 = \frac{\partial}{\partial u},$$

$$\xi_4 = \big(f(x+t) + g(x-t)\big)\frac{\partial}{\partial t} + \big(f(x+t) - g(x-t)\big)\frac{\partial}{\partial x} + \Xi\frac{\partial}{\partial u}, \quad \Xi = \text{const.}$$

The form of the symmetry ξ_4 follows from the fact that components ξ^t and ξ^x are solutions of the wave equation. (This is evident from Eq. (9.5): adding first two equations, taking into account the third one and calculating second derivatives.) If we choose for example $f(x+t) = \frac{1}{2}(x+t)$, $g(x-t) = \frac{1}{2}(x-t)$ and $\Xi = 0$ we can see that the vector field

$$\xi = x\frac{\partial}{\partial t} + t\frac{\partial}{\partial x}$$

is one of the symmetries of the Lagrange structure corresponding to the wave equation.

Now let us calculate the Noether currents corresponding to obtained symmetries. For Noether currents connected with symmetries of first-order Lagrangian it holds (see general relations in Chap. 8)

$$\left[L\xi^i + \frac{\partial L}{\partial y_i^\sigma}\left(\Xi^\sigma - y_j^\sigma \xi^j\right)\right]\omega_i.$$

Noether currents corresponding to symmetries ξ_1, ξ_2, ξ_3 and ξ_4 are

$$\begin{aligned}
\Phi(\xi_1) &= -u_t u_x \,\mathrm{d}t - \frac{1}{2}\left(u_t^2 + u_x^2\right)\mathrm{d}x,\\
\Phi(\xi_2) &= -\frac{1}{2}\left(u_t^2 + u_x^2\right)\mathrm{d}t - u_t u_x \,\mathrm{d}x,\\
\Phi(\xi_3) &= u_x \,\mathrm{d}t + u_t \,\mathrm{d}x,\\
\Phi(\xi_4) &= \big(f(x+t) + g(x-t)\big)\Phi(\xi_1) + \big(f(x+t) - g(x-t)\big)\Phi(\xi_2),
\end{aligned}$$

where f and g are arbitrary functions. (The current $\Phi(\xi_4)$ is a linear combination of $\Phi(\xi_1)$ and $\Phi(\xi_2)$.) More general considerations concerning Noether symmetries and currents of the wave equation can be found in [9]. In the cited work the case of wave equation for n-dimensional base X, $n \geq 2$, with Lagrangian and its Euler-Lagrange form

$$\lambda = \frac{1}{2}g^{ij}u_i u_j \,\omega_0, \quad E_\lambda = g^{ij}u_{ij}\,\omega^1 \wedge \omega_0,$$

is solved in detail. Here $g = g_{ij}\,\mathrm{d}x^i \otimes \mathrm{d}x^j$, $g_{ij} = g_{ji}$, $1 \leq i, j \leq n$, is a general covariant metrics on X, $g^{-1} = (g^{ij})$ is the corresponding contravariant metrics. Fibres of the considered fibred manifold are one-dimensional, $m = 1$. In [9] the NBH equations for wave equation is formulated and solved.

9.3 Classical Electrodynamics

Classical electrodynamics is a typical field theory. Its equations of motion (in three-dimensional case) are Maxwell equations

$$\varepsilon_0 \operatorname{div} \boldsymbol{E} = \varrho, \quad \operatorname{div} \boldsymbol{B} = 0, \quad \operatorname{rot} \boldsymbol{E} = -\frac{\partial \boldsymbol{B}}{\partial t}, \quad \operatorname{rot} \boldsymbol{B} = \varepsilon_0\mu_0\frac{\partial \boldsymbol{E}}{\partial t} + \mu_0 \boldsymbol{j}$$

with the usual meaning of symbols. The question is: Is classical electrodynamics variational theory? The answer is positive—we can find the Lagrangian in classical textbooks, see, e.g. [10]. Nevertheless, in this chapter we show the variationality of classical electrodynamics by solving the inverse problem for Maxwell equations.

In Example 9.3 we considered a relativistic particle with the Lagrangian and the equation of motion in the vector form

$$\lambda = L\,\mathrm{d}t, \quad L = -mc^2\sqrt{1 - \frac{v^2}{c^2}} + (\boldsymbol{v}\boldsymbol{\Phi} - \varphi + mc^2),$$

$$\frac{\mathrm{d}}{\mathrm{d}t}\left(\frac{m\boldsymbol{v}}{\sqrt{1 - \frac{v^2}{c^2}}}\right) = \boldsymbol{v} \times \operatorname{rot}\boldsymbol{\Phi} - \frac{\partial \boldsymbol{\Phi}}{\partial t} - \operatorname{grad}\varphi.$$

If a charged particle moves in the electromagnetic field, then

$$\varphi = eV, \quad \boldsymbol{\Phi} = e\boldsymbol{A},$$

where V is the *electric scalar potential* and $\boldsymbol{A}$ is the *magnetic vector potential* of the field, e is the particle charge. Relations between vector fields $\boldsymbol{B}$, $\boldsymbol{E}$ and potentials V, $\boldsymbol{A}$ are

$$\boldsymbol{B} = \operatorname{rot}\boldsymbol{A}, \quad \boldsymbol{E} = -\operatorname{grad}V - \frac{\partial \boldsymbol{A}}{\partial t}.$$

We know (see Example 9.3) that equations $\operatorname{div}\boldsymbol{B} = 0$ and $\operatorname{rot}\boldsymbol{E} = -\frac{\partial \boldsymbol{B}}{\partial t}$ are purely the consequence of the variationality of equations of motion of a relativistic particle. The variationality of the second pair of Maxwell equations we now derive by solving the inverse problem. Let us move on to the four-dimensional version of Maxwell equations in standard four-dimensional notation [10]. The underlying fibred manifold is $(Y, \pi, X) = (\mathbf{R}^4 \times \mathbf{R}^4, p, \mathbf{R}^4)$ with the Minkowski metric tensor on the base X (four-dimensional time-space). Coordinates on $X = \mathbf{R}^4$ are (x^0, x^1, x^2, x^3), where $x^0 = ct$ and x^1, x^2, x^3 are Cartesian coordinates, (covariant) metric tensor is $g = g_{ij}\,\mathrm{d}x^i \otimes \mathrm{d}x^j, 0 \le i, j \le 3, \operatorname{diag} g = (1, -1, -1, -1)$. Fibres are four-dimensional spaces again with Minkowski metric tensor g, coordinates on a fibre are components of four-co-vectors

$$(\varphi_A) = (A_0, A_1, A_2, A_3),$$

meanwhile the four-vectors are o the form

$$(A) = (A^0, A^1, A^2, A^3) = (A^0, \boldsymbol{A}), \quad A^0 = A_0,$$
$$A^i = -A_i, \quad i = 1, 2, 3, \quad (\varphi_A) = (A_0, -\boldsymbol{A}),$$

with $A^0 = c^{-1}V$, $\boldsymbol{A} = (A^1, A^2, A^3)$,

$$A^i = g^{ij}A_j, \quad A_i = g_{ij}A^j, \quad g^{-1} = g^{ij}\frac{\partial}{\partial x^i} \otimes \frac{\partial}{\partial x^j}.$$

Coordinates on (J^1Y, π_1, X) and (J^2Y, π_2, X) are then

$$(x^i, A_j, A_{j,i}), \quad (x^i, A_j, A_{j,i}, A_{j,ik}), \quad 0 \leq i, j, k \leq 3.$$

For four-dimensional formulation of Maxwell equations introduce the (covariant) electromagnetic field tensor $F = (F_{ij})$

$$F_{ij} = A_{j,i} - A_{i,j} = \frac{\partial A_j}{\partial x^i} - \frac{\partial A_i}{\partial x^j}.$$

Note, that it holds

$$F_{ij} = g_{ir} g_{js} F^{rs}, \quad F^{ij} = g^{ir} g^{js} F_{rs}, \quad F_i^j = g_{ir} F^{jr} = g^{jr} F_{ir}.$$

In matrix notation we have

$$(F_{ij}) = \begin{pmatrix} 0 & c^{-1}E^1 & c^{-1}E^2 & c^{-1}E^3 \\ -c^{-1}E^1 & 0 & -B^3 & B^2 \\ -c^{-1}E^2 & B^3 & 0 & -B^1 \\ -c^{-1}E^3 & -B^2 & B^1 & 0 \end{pmatrix},$$

$$(F^{ij}) = \begin{pmatrix} 0 & -c^{-1}E^1 & -c^{-1}E^2 & -c^{-1}E^3 \\ c^{-1}E^1 & 0 & B^3 & -B^2 \\ c^{-1}E^2 & -B^3 & 0 & B^1 \\ c^{-1}E^3 & B^2 & -B^1 & 0 \end{pmatrix}.$$

Denoting $(j) = (c\varrho, \boldsymbol{j})$ (the four-vector of electrical current, ϱ being the charge density) we obtain the four-dimensional formulation of first and last Maxwell equations (in the contravariant form), $\varepsilon_0 \operatorname{div} \boldsymbol{E} = \varrho$ and $\operatorname{rot} \boldsymbol{B} = \varepsilon_0 \mu_0 \frac{\partial \boldsymbol{E}}{\partial t} + \mu_0 \boldsymbol{j}$

$$E^i = -\frac{1}{c}\left(j^i + \frac{1}{\mu_0}\frac{\partial F^{ij}}{\partial x^j}\right) = 0, \quad 0 \leq i, j \leq 3, \tag{9.6}$$

which can be, of course, also found in textbooks of classical electrodynamics (see, e.g. [10]). (Since the vectors $\boldsymbol{E}$ and $\boldsymbol{B}$ no longer appear directly in four-dimensional notation, there can be no undesirable confusion if we use the standard notation E^i, or E_i for the components of the Euler-Lagrange form $E_i \omega^i \wedge \omega_0$.) The meaning of quantities c, ε_0 and μ_0 is of course standard and their values are

$$c = 299{,}792{,}458\,\mathrm{m\,s^{-1}} \text{ exact}, \mu_0 = 4\pi \cdot 10^7\,\mathrm{H\,m^{-1}},$$
$$\varepsilon_0 \doteq 8.85 \cdot 10^{-12}\,\mathrm{F\,m^{-1}}, \varepsilon_0\mu_0 = c^{-2}.$$

(The four-vector of electrical current obeys the continuity equation

$$\frac{\partial j^i}{\partial x^i} = 0, \quad \text{in three-dimensional version} \quad c\frac{\partial \varrho}{\partial t} + \operatorname{div} \boldsymbol{j} = 0,$$

representing the conservation of charge.) We need to rewrite the equations of motion (9.6) into the form corresponding to our choice of the basic fibred manifold and their first and second prolongations. We obtain

$$E^i \circ J^2\gamma = -\frac{1}{c}\left[g^{ik} j_k + \frac{1}{\mu_0} g^{ij} g^{k\ell} \left(A_{\ell,jk} - A_{j,\ell k}\right)\right] \circ J^2\gamma = 0, \quad 0 \le i, j, k \le 3.$$

The Helmholtz conditions of variationality of a second-order dynamical form are given by (9.4). For our situation, the meaning of the quantities contained in them is as follows:

$$\sigma \sim i, \quad y^\sigma \sim A_i, \quad y^\sigma_k \sim A_{i,k}, \quad y^\sigma_{k\ell} \sim A_{i,k\ell}, \quad E_\sigma \sim E^i.$$

Verifications of first two conditions (9.4) is trivial, let us verify the last one:

$$\frac{\partial E^i}{\partial A_{p,rs}} = -\frac{1}{c\mu_0} g^{ij} g^{k\ell} \left(\delta_{\ell p}\delta_{rj}\delta_{sk} - \delta_{jp}\delta_{\ell r}\delta_{ks}\right) = -\frac{1}{c\mu_0}\left(g^{ir} g^{sp} - g^{ip} g^{sr}\right),$$

$$\frac{\partial E^p}{\partial A_{i,rs}} = -\frac{1}{c\mu_0}\left(g^{pr} g^{si} - g^{pi} g^{sr}\right) = \frac{\partial E^i}{\partial A_{p,rs}}.$$

The equality $\frac{\partial E^i}{\partial A_{p,rs}} = \frac{\partial E^p}{\partial A_{i,rs}}$ holds due to symmetry of starting expressions in indices r and s. We conclude that the dynamical form for classical electrodynamics is variational. The mapping χ for the construction of the corresponding Vainberg-Tonti Lagrangian Λ is

$$\chi : \left(u, (x^i, A_j, A_{j,k}, A_{j,k\ell})\right) \longrightarrow (x^i, uA_j, uA_{j,k}, uA_{j,k\ell}),$$

$$\Lambda = A_i \int_0^1 (E^i \circ \chi)\, \mathrm{d}u = -\frac{1}{c} A_i \left[g^{i\ell} j_\ell + \frac{1}{\mu_0} g^{ij} g^{k\ell} \left(A_{\ell,jk} - A_{j,\ell k}\right) u\, \mathrm{d}u\right] \omega_0$$

$$= -\frac{1}{c}\left[A_i j^i + \frac{1}{2\mu_0} g^{ij} g^{k\ell} A_i \left(A_{\ell,jk} - A_{j,\ell k}\right)\right] \omega_0.$$

This is a second-order Lagrangian. The well-known minimal Lagrangian (see [10]) we obtain by subtracting trivial Lagrangian, taking into account the relation

$$\mathrm{d}_k \left[A_i \left(A_{\ell,j} - A_{j,\ell}\right)\right] = A_{i,k}(A_{\ell,j} - A_{j,\ell}) + A_i(A_{\ell,jk} - A_{j,\ell k})$$

and taking into account the relation $A_{i,k}F^{ik} = -A_{k,i}F^{ik}$ due to antisymmetry of the tensor $F = (F^{ij})$. The minimal Lagrangian reads

$$\lambda = L\omega_0 = -\frac{1}{c}\left(j^i A_i + \frac{1}{4\mu_0}F_{ik}F^{ik}\right)\omega_0. \tag{9.7}$$

The problem of Noether symmetries and currents will be solved in Sect. 9.6.

9.4 Quantum Mechanics

The equation of motion for a quantum mechanical system is the Schrödinger equation

$$\mathrm{i}\hbar\frac{\partial|\psi\rangle}{\partial t} = \hat{H}|\psi\rangle$$

where $|\psi\rangle$ represents the state of the system (an element of the corresponding Hilbert space) and $\hat{H}$ is the Hamilton operator (operator of energy), see e.g. [11]. For simplicity we consider a free quantum mechanical particle of mass m executing the one-dimensional motion along the x-axis and express its Schrödinger equation in the coordinate representation. Then

$$\psi = \psi(t,x) = v(t,x) + \mathrm{i}w(t,x), \quad v = \mathrm{Re}\,\psi, \quad w = \mathrm{Im}\,\psi,$$

is the wave function (complex function of time t and coordinate x). Schrödinger equation obtains the form

$$\mathrm{i}\hbar\frac{\partial\psi}{\partial t} = -\frac{\hbar^2}{2m}\frac{\partial^2\psi}{\partial x^2}, \quad \psi = \psi(t,x),$$

and separating real and imaginary part

$$\left(\frac{h^2}{2m}\frac{\partial^2 v}{\partial x^2} - \hbar\frac{\partial w}{\partial t}\right) = 0, \quad \left(\frac{h^2}{2m}\frac{\partial^2 w}{\partial x^2} + \hbar\frac{\partial v}{\partial t}\right) = 0. \tag{9.8}$$

The underlying fibred manifold is $(Y, \pi, X) = (\mathbf{R}^2 \times \mathbf{R}^2, p, \mathbf{R}^2)$ with coordinates on the base X $(x^1, x^2) = (t, x)$, $\omega_0 = \mathrm{d}t \wedge \mathrm{d}x$, and on the fibre $\mathbf{R}^2$ $(y^1, y^2) = (v, w)$. The corresponding dynamical form is defined on (J^2Y, π_2, X), coordinates on J^2Y are

$$\begin{aligned}
&(t, x, v, w, v_t, v_x, w_t, w_x, v_{tt}, v_{tx}, v_{xx}, w_{tt}, w_{tx}, w_{xx}),\\
E &= E_1\omega^1 \wedge \omega_0 + E_2\omega^2 \wedge \omega_0\\
&= \left(\frac{h^2}{2m}\frac{\partial^2 v}{\partial x^2} - \hbar\frac{\partial w}{\partial t}\right)\mathrm{d}v \wedge \mathrm{d}t \wedge \mathrm{d}x + \left(\frac{h^2}{2m}\frac{\partial^2 w}{\partial x^2} + \hbar\frac{\partial v}{\partial t}\right)\mathrm{d}w \wedge \mathrm{d}t \wedge \mathrm{d}x\,.
\end{aligned}$$

We will prove that this form is variational and find the Vainberg-Tonti Lagrangian as well as the minimal Lagrangian. Rewriting the Helmholtz conditions (9.4) into the notation adopted for Schrödinger equation (9.8) we can immediately see that they are trivially fulfilled. The mapping χ has the form

$$\begin{aligned}\chi : &(u, t, x, v, w, v_t, v_x, w_t, w_x, v_{tt}, v_{tx}, v_{xx}, w_{tt}, w_{tx}, w_{xx})\\ &\longrightarrow (t, x, uv, uw, uv_t, uv_x, uw_t, uw_x, uv_{tt}, uv_{tx}, uv_{xx}, uw_{tt}, uw_{tx}, uw_{xx}),\end{aligned}$$

Vainberg-Tonti Lagrangian is

$$\begin{aligned}\Lambda &= \left[v \int_0^1 u \left(\frac{\hbar^2}{2m} v_{xx} - \hbar w_t \right) \mathrm{d}u + w \int_0^1 u \left(\frac{\hbar^2}{2m} w_{xx} + \hbar v_t \right) \right] \mathrm{d}t \wedge \mathrm{d}x\\ &= \left[\frac{\hbar^2}{4m} (v v_{xx} + w w_{xx}) - \frac{\hbar}{2} (v w_t - w v_t) \right] \mathrm{d}t \wedge \mathrm{d}x.\end{aligned}$$

We obtain the minimal Lagrangian by subtracting a trivial Lagrangian from Λ. The Lagrange function of a trivial Lagrangian has the form $\mathrm{d}_t f^t + \mathrm{d}_x f^x$ for some functions f^t and f^x of the type

$$f = f(t, x, v, w, v_t, w_t, v_x, w_x).$$

$$\begin{aligned}\mathrm{d}_t f &= \frac{\partial f}{\partial t} + \frac{\partial f}{\partial v} v_t + \frac{\partial f}{\partial w} w_t + \frac{\partial f}{\partial v_t} v_{tt} + \frac{\partial f}{\partial v_x} v_{xt} + \frac{\partial f}{\partial w_t} w_{tt} + \frac{\partial f}{\partial w_x} w_{xt},\\ \mathrm{d}_x f &= \frac{\partial f}{\partial x} + \frac{\partial f}{\partial v} v_x + \frac{\partial f}{\partial w} w_x + \frac{\partial f}{\partial v_t} v_{tx} + \frac{\partial f}{\partial v_x} v_{xx} + \frac{\partial f}{\partial w_t} w_{tx} + \frac{\partial f}{\partial w_x} w_{xx}.\end{aligned}$$

Denoting $f = \frac{\hbar^2}{4m}(v v_x + w w_x)$ we obtain

$$\mathrm{d}_x f = \frac{\hbar^2}{4m}(v_x^2 + w_x^2) + \frac{\hbar^2}{4m}(v v_{xx} + w w_{xx}), \quad \text{and then}$$

$$\frac{\hbar^2}{4m}(v v_{xx} + w w_{xx}) = -\frac{\hbar^2}{4m}(v_x^2 + w_x^2) + \frac{\hbar^2}{4m} \mathrm{d}_x (v v_x + w w_x).$$

The form $\frac{\hbar^2}{4m} \mathrm{d}_x (v v_x + w w_x)\, \mathrm{d}t \wedge \mathrm{d}x$ is the trivial Lagrangian. The minimal Lagrangian is then

$$\lambda = \left[-\frac{\hbar^2}{4m}(v_x^2 + w_x^2) - \frac{\hbar}{2}(v w_t - w v_t) \right] \mathrm{d}t \wedge \mathrm{d}x. \tag{9.9}$$

We express it with help of complex functions ψ a ψ^*. It holds

$$v = \frac{1}{2}(\psi + \psi^*), \quad w = \frac{1}{2\mathrm{i}}(\psi - \psi^*).$$

After some calculations we obtain

$$\lambda = \left[-\frac{\hbar^2}{4m} (\psi_x \psi_x^*) + \frac{i\hbar}{4} (\psi^* \psi_t - \psi \psi_t^*) \right] dt \wedge dx.$$

If the particle is not free and it moves in a field with potential energy $V = V(t, x)$, the Schrödinger equation is variational as well and Lagrangian contains the additional term $-\frac{1}{2} V(v^2 + w^2) = -\frac{1}{2} V \psi \psi^*$. (See Sect. 9.6.)

For finding Noether symmetries and currents of the Lagrangian (9.9) we formulate the Noether equation: In general

$$\partial_{J^1\xi} \lambda = 0 \implies i_{J^1\xi} \, dL + L \frac{d\xi^i}{dx^i} = 0,$$

for a π-projectable vector field ξ on Y,

$$\xi = \xi^i(x^j) \frac{\partial}{\partial x^i} + \Xi^\sigma(x^j, y^\sigma) \frac{\partial}{\partial y^\sigma},$$

and for our case

$$i_{J^1\xi} \, d \left[-\frac{\hbar^2}{4m} (v_x^2 + w_x^2) - \frac{\hbar}{2} (v w_t - w v_t) \right]$$
$$+ \left[-\frac{\hbar^2}{4m} (v_x^2 + w_x^2) - \frac{\hbar}{2} (v w_t - w v_t) \right] \left(\frac{d\xi^t}{dt} + \frac{d\xi^x}{dx} \right) = 0,$$

$$\xi = \xi^t(t, x) \frac{\partial}{\partial t} + \xi^x(t, x) \frac{\partial}{\partial x} + \Xi^v(t, x, v, w) \frac{\partial}{\partial v} + \Xi^w(t, x, v, w) \frac{\partial}{\partial w},$$

$$J^1\xi = \xi^t \frac{\partial}{\partial t} + \xi^x \frac{\partial}{\partial x} + \Xi^v \frac{\partial}{\partial v} + \Xi^w \frac{\partial}{\partial w} + \Xi_t^v \frac{\partial}{\partial v_t} + \Xi_x^v \frac{\partial}{\partial v_x} + \Xi_t^w \frac{\partial}{\partial w_t} + \Xi_x^w \frac{\partial}{\partial w_x},$$

$$\Xi_t^v = \frac{d\Xi^v}{dt} - v_t \frac{d\xi^t}{dt} - v_x \frac{d\xi^x}{dt}, \quad \Xi_x^v = \frac{d\Xi^v}{dx} - v_t \frac{d\xi^t}{dx} - v_x \frac{d\xi^x}{dx},$$

$$\Xi_t^w = \frac{d\Xi^w}{dt} - w_t \frac{d\xi^t}{dt} - w_x \frac{d\xi^x}{dt}, \quad \Xi_x^w = \frac{d\Xi^w}{dx} - w_t \frac{d\xi^t}{dx} - w_x \frac{d\xi^x}{dx},$$

$$\frac{d\Xi^v}{dt} = \frac{\partial \Xi^v}{\partial t} + \frac{\partial \Xi^v}{dv} v_t + \frac{\partial \Xi^v}{\partial w} w_t,$$

and similarly other total derivatives, i.e. Ξ^v with respect to x and Ξ^w with respect to t and x. After some calculations and separating the terms standing by v_x^2, w_x^2, $v_x w_x$, $w_t w_x$, $v_t v_x$, v_t, v_x, w_t, w_x and the term without derivatives we obtain the set of 9 independent Noether equations (equations corresponding to coefficients standing by $v_x v_t$ and $w_x w_t$ are identical):

$$
\begin{aligned}
\frac{1}{2}\left(\frac{\partial \xi^x}{\partial x} - \frac{\partial \xi^t}{\partial t}\right) - \frac{\partial \Xi^w}{\partial w} &= 0,\\
\frac{1}{2}\left(\frac{\partial \xi^x}{\partial x} - \frac{\partial \xi^t}{\partial t}\right) - \frac{\partial \Xi^v}{\partial v} &= 0,\\
\frac{\partial \Xi^w}{\partial v} + \frac{\partial \Xi^v}{\partial w} &= 0,\\
-v\frac{\partial \xi^x}{\partial x} - \Xi^v + w\frac{\partial \Xi^v}{\partial w} - v\frac{\partial \Xi^w}{\partial w} &= 0,\\
w\frac{\partial \xi^x}{\partial x} + \Xi^w + w\frac{\partial \Xi^v}{\partial v} - v\frac{\partial \Xi^w}{\partial v} &= 0,\\
v\frac{\partial \xi^x}{\partial t} - \frac{\hbar}{m}\frac{\partial \Xi^w}{\partial x} &= 0,\\
-w\frac{\partial \xi^x}{\partial t} - \frac{\hbar}{m}\frac{\partial \Xi^v}{\partial x} &= 0,\\
w\frac{\partial \Xi^v}{\partial t} - v\frac{\partial \Xi^w}{\partial t} &= 0,\\
\frac{\mathrm{d}\xi^t}{\mathrm{d}x} &= 0.
\end{aligned}
$$

Solving these equations (Problem 9.12) we obtain the general expressions for components of Noether symmetries:

$$
\begin{aligned}
\xi^t &= -2at^2 - 4bt + P,\\
\xi^x &= -2x(at+b) + et + Q,\\
\Xi^v &= (at+b)v - \left(-\frac{ma}{\hbar}x^2 + e\frac{m}{\hbar}x + c\right)w,\\
\Xi^w &= (at+b)w + \left(-\frac{ma}{\hbar}x^2 + e\frac{m}{\hbar}x + c\right)v,
\end{aligned}
$$

where a, b, c, e, P, Q are constants. With help of six basic choices of (a, b, c, e, P, Q) as $(1, 0, 0, 0, 0, 0)$, $(0, 1, 0, 0, 0, 0)$, …, $(0, 0, 0, 0, 0, 1)$ we obtain six independent Noether symmetries.

Example 9.10 (*Klein-Gordon equation*) Klein-Gordon equation describes a scalar field $u = u(x^0, x^1, x^2, x^3)$ of an uncharged particle of mass m with zero spin (e.g. meson π^0). The corresponding fibred space is $(Y, \pi, X) = (\mathbf{R}^4 \times \mathbf{R}, p, \mathbf{R}^4$, equation of motion is

$$
-m^2 u - u_{00} + u_{11} + u_{22} + u_{33} = 0, \quad \text{i.e.} \quad \left(\frac{\partial^2}{\partial (x^0)^2} - \Delta - m^2\right) u = 0,
$$

where $\frac{\partial^2}{\partial (x^0)^2} - \Delta$ is D'Alembert operator. This equation is also variational (see Sect. 9.6).

9.5 A Note to String Theories

String theory is a separate and difficult discipline of theoretical physics. It is an alternative approach to the *standard model* to explain the origin of the universe. Here we present only one simple example of string description using variational physics.

One can study the string within field theory as a typical first-order problem. Basic fibred space (Y, π, X) has two-dimensional base with local coordinates x^0 (time variable) and x^1 (position on the string). Every fibre is four-dimensional time-space with coordinates (q^0, q^1, q^2, q^3) (standard relativistic notation). So, $m = 4$, $\dim Y = 6$. Fibred chart on Y is thus (V, ψ), $\psi = (x^i, q^\mu)$, $i = 0, 1$, $\mu = 0, 1, 2, 3$, associated chart on X is (U, φ), $U = \pi(V)$, $\varphi = (x^i)$, associated fibred chart on J^1Y is (V_1, ψ_1), $V_1 = \pi^{-1}(V)$, $\psi_1 = (x^i, q^\mu, q_i^\mu)$. Base and fibres are spaces with metric tensors. Covariant metric tensor on the base denote $h = h_{ij}\, \mathrm{d}x^i \otimes \mathrm{d}x^j$, covariant metric tensor on fibres is $g = g_{\mu\nu}\, \mathrm{d}q^\mu \otimes \mathrm{d}q^\nu$. (Both metric tensors are of course symmetric.)

Lagrangian of the string λ is in specific situations constructed on the basis of physical arguments, and it has the form

$$\lambda = L(x^i, q^\mu, q_i^\mu)\, \mathrm{d}x^0 \wedge \mathrm{d}x^1, \quad i = 0, 1, \quad \mu = 0, 1, 2, 3.$$

Principal Lepage equivalent (Poincaré-Cartan form) of this Lagrangian is

$$\begin{aligned} \theta_\lambda &= L\, \mathrm{d}x^0 \wedge \mathrm{d}x^1 + \frac{\partial L}{\partial q_i^\mu}\, \omega^\mu \wedge \omega_i, \quad \omega_i = i_{\frac{\partial}{\partial x^i}} (\mathrm{d}x^0 \wedge \mathrm{d}x^1), \\ \theta_\lambda &= \left(L - \frac{\partial L}{\partial q_0^\mu}\, q_0^\mu - \frac{\partial L}{\partial q_1^\mu}\, q_1^\mu \right) \mathrm{d}x^0 \wedge \mathrm{d}x^1 \\ &\quad + \left(\frac{\partial L}{\partial q_0^\mu}\, \mathrm{d}q^\mu \wedge \mathrm{d}x^1 - \frac{\partial L}{\partial q_1^\mu}\, \mathrm{d}q^\mu \wedge \mathrm{d}x^0 \right). \end{aligned} \tag{9.10}$$

For so called *boson string* is used *Nambu-Goto* functional (see [12])

$$S = -T \int_\Sigma d\Sigma, \quad d\Sigma = \sqrt{-\det h}\, \mathrm{d}x^0 \wedge \mathrm{d}x^1,$$

where constant T characterizes the stress of the string. In our considerations, we use units in which the light speed in vacuum is $c = 1$. The form $\mathrm{d}\Sigma$ is the surface element on the base. The metric tensor h is supposed to be induced by the time-space metric tensors g, i.e.

$$h = \gamma^* g = \left[(g_{\mu\nu} \circ \gamma)\, \frac{\partial q^\mu}{\partial x^i} \frac{\partial q^\nu}{\partial x^j} \right] \mathrm{d}x^i \otimes x^j,$$

$$\gamma : (x^0, x^1) \longrightarrow (x^0, x^1, q^0\gamma(x^0, x^1), q^1\gamma(x^0, x^1), q^2\gamma(x^0, x^1), q^3\gamma(x^0, x^1))$$

is a section of the projection π. The metric tensor on J^1Y $\tilde{h} = \tilde{h}_{ij}(x^k, q^\lambda, q^\lambda_k)\,\mathrm{d}x^i \otimes \mathrm{d}x^j$, is defined by the condition $h = J^1\gamma^*\tilde{h}$, i.e.

$$\tilde{h} = \tilde{h}_{ij}\,\mathrm{d}x^i \otimes \mathrm{d}x^j = g_{\mu\nu}\, q^\mu_i\, q^\nu_j\,\mathrm{d}x^i \otimes \mathrm{d}x^j.$$

Denote $D = \det \tilde{h}$. Lagrangian $\lambda = L\,\mathrm{d}x^0 \wedge \mathrm{d}x^1$ corresponding to Nambu-Goto functional is

$$\begin{aligned}\lambda &= -T\sqrt{-D}\,\mathrm{d}x^0 \wedge \mathrm{d}x^1\\ &= -T\sqrt{(g_{\alpha\beta}\, g_{\mu\nu} - g_{\alpha\mu}\, g_{\beta\nu})\, q^\alpha_0 q^\mu_0 q^\beta_1 q^\nu_1}\,\mathrm{d}x^0 \wedge \mathrm{d}x^1.\end{aligned} \tag{9.11}$$

Putting this Lagrangian (9.11) into (9.10), we obtain the principal Lepage equivalent for the boson string,

$$\begin{aligned}\theta_\lambda &= T\sqrt{-D}\,\mathrm{d}x^0 \wedge \mathrm{d}x^1\\ &+ \frac{T}{\sqrt{-D}}(g_{\alpha\mu}\, g_{\beta\nu} - g_{\alpha\beta}\, g_{\mu\nu})\,\mathrm{d}q^\mu \wedge (q^\alpha_0\, q^\beta_0\, q^\nu_1\,\mathrm{d}x^0 - q^\alpha_1\, q^\beta_1\, q^\nu_0\,\mathrm{d}x^1).\end{aligned}$$

(Note that the functional for the boson string is a special case of so called *Polyakov functional*, which considers metric tensors h and g as independent.)

We focus on Noether symmetries and currents of the string Lagrangian (see also [13]). We express the Noether equation $\partial_{J^1\xi}\lambda = 0$ for a vector field

$$\xi = \xi^i \frac{\partial}{\partial x^i} + \Xi^\mu \frac{\partial}{\partial q^\mu}, \quad J^1\Xi = \xi^i \frac{\partial}{\partial x^i} + \Xi^\mu \frac{\partial}{\partial q^\mu} + \left(d_i\,\Xi^\mu - q^\mu_j \frac{\partial \xi^j}{\partial x^i}\right)\frac{\partial}{\partial q^\mu_i},$$

as well as Noether currents:

$$\begin{aligned}\Phi = h\left(i_{J^1\xi}\theta_\lambda\right) \circ J^1\gamma &= -\frac{T}{\sqrt{-D}}\left(g_{\alpha\beta}g_{\mu\nu} - g_{\alpha\mu}g_{\beta\nu}\right)\\ &\times\left[q^\alpha_0 q^\beta_0 q^\nu_1 (\Xi^\mu - q^\mu_0\xi^0)\,\mathrm{d}x^0 - q^\alpha_1 q^\beta_1 q^\nu_0 (\Xi^\mu - q^\mu_1\xi^1)\,\mathrm{d}x^1\right].\end{aligned}$$

A clearer form of the Noether equation and currents is obtained by using generalized momenta. It holds (using admissible interchanges of indices)

$$\begin{aligned}p^0_\mu &= \frac{\partial L}{\partial q^\mu_0} = -T\frac{\partial\sqrt{-D}}{\partial q^\mu_0}\\ &= -\frac{T}{\sqrt{-D}}(g_{\alpha\beta}g_{\mu\nu} - g_{\alpha\mu}g_{\beta\nu})q^\alpha_0 q^\beta_1 q^\nu_1,\\ p^1_\mu &= \frac{\partial L}{\partial q^\mu_1} = -T\frac{\partial\sqrt{-D}}{\partial q^\mu_1}\\ &= -\frac{T}{\sqrt{-D}}(g_{\alpha\beta}g_{\mu\nu} - g_{\alpha\mu}g_{\beta\nu})q^\alpha_1 q^\beta_0 q^\nu_0.\end{aligned}$$

We express $p^0_\mu q^\mu_0$, $p^1_\mu q^\mu_0$, $p^0_\mu q^\mu_1$, $p^1_\mu q^\mu_1$. For the first of them (the others are given by analogy) it holds (using admissible interchanges of indices)

$$\begin{aligned} p^0_\mu q^\mu_1 &= -\frac{T}{\sqrt{-D}}(g_{\alpha\beta}g_{\mu\nu} - g_{\alpha\mu}g_{\beta\nu})q^\alpha_0 q^\beta_1 q^\nu_1 q^\mu_1 \\ &= -\frac{T}{\sqrt{-D}}\left[g_{\alpha\beta}g_{\mu\nu}q^\alpha_0 q^\beta_1 q^\nu_1 q^\mu_1 - g_{\alpha\mu}g_{\beta\nu}q^\alpha_0 q^\beta_1 q^\nu_1 q^\mu_1\right] \\ &= -\frac{T}{\sqrt{-D}}\left[g_{\alpha\beta}g_{\mu\nu}q^\alpha_0 q^\beta_1 q^\nu_1 q^\mu_1 - g_{\alpha\beta}g_{\mu\nu}q^\alpha_0 q^\mu_1 q^\nu_1 q^\beta_1\right] = 0. \end{aligned}$$

Finally we obtain

$$p^0_\mu q^\mu_0 = -T\sqrt{-D}, \quad p^1_\mu x^\mu_1 = -T\sqrt{-D}, \quad p^0_\mu q^\mu_1 = 0, \quad p^1_\mu q^\mu_0 = 0. \tag{9.12}$$

Because D is independent of variables on the base, x^0 a x^1, we obtain the Noether equation as follows:

$$-T\left[\frac{\partial\sqrt{-D}}{\partial q^\mu}\Xi^\mu + \frac{\partial\sqrt{-D}}{\partial q^\mu_i}\Xi^\mu_i + \sqrt{-D}\left(\frac{\partial\xi^0}{\partial x^0} + \frac{\partial\xi^1}{\partial x^1}\right)\right] \mathrm{d}x^0 \wedge \mathrm{d}x^1 = 0.$$

Considering that $-T\frac{\partial\sqrt{-D}}{\partial q^\mu_i} = p^i_\mu$, using (9.12) and putting $\Xi^\mu_i = \mathrm{d}_i\,\Xi^\mu - q^\mu_j\frac{\partial\xi^j}{\partial x^i}$, the formal expression of the Noether equation will be further simplified

$$-T\frac{\partial\sqrt{-D}}{\partial q^\mu}\Xi^\mu + p^i_\mu\,\mathrm{d}_i\,\Xi^\mu = 0. \tag{9.13}$$

Noether current can be also expressed with help of generalized momenta. For this purpose we use the principal Lepage equivalent (9.10), where we put $\frac{\partial L}{\partial q^\mu_i} = p^i_\mu$. Then

$$\theta_\lambda = \left(L - p^0_\mu q^\mu_0 - p^1_\mu q^\mu_1\right)\mathrm{d}x^0 \wedge \mathrm{d}x^1 + \left(p^0_\mu\,\mathrm{d}q^\mu \wedge \mathrm{d}x^1 - p^1_\mu\,\mathrm{d}q^\mu \wedge \mathrm{d}x^0\right).$$

Putting $L = -T\sqrt{-D}$ and taking into account relations (9.12) we have

$$\theta_\lambda = T\sqrt{-D}\,\mathrm{d}x^0 \wedge \mathrm{d}x^1 + \left(p^0_\mu\,\mathrm{d}q^\mu \wedge \mathrm{d}x^1 - p^1_\mu\,\mathrm{d}q^\mu \wedge \mathrm{d}x^0\right)$$

and for Noether current

$$\begin{aligned} \Phi(\xi) = hi_{J^1\xi}\theta_\lambda &= h\Big[-\left(T\sqrt{-D}\xi^1 + p^1_\mu\Xi^\mu\right)\mathrm{d}x^0 + \\ &\quad + \left(T\sqrt{-D}\xi^0 + p^0_\mu\Xi^\mu\right)\mathrm{d}x^1 + \left(p^1_\mu\xi^0 - p^0_\mu\xi^1\right)\mathrm{d}q^\mu\Big] \\ \Longrightarrow \Phi(\xi) &= \Xi^\mu\left(-p^1_\mu\,\mathrm{d}x^0 + p^0_\mu\,\mathrm{d}x^1\right). \end{aligned} \tag{9.14}$$

Table 9.2 Poincaré group

Symmetry ξ	Noether current $\Phi(\xi)$
$\xi = \frac{\partial}{\partial q^{\mu}}$	$\Phi = -p^1_{\mu}\,\mathrm{d}x^0 + p^0_{\mu}\,\mathrm{d}x^1$
$\xi = q^s \frac{\partial}{\partial q^0} + q^0 \frac{\partial}{\partial q^s}$	$\Phi = (-p^1_0 q^s - p^1_s q^0)\,\mathrm{d}x^0 + (p^0_0 q^s + p^0_s q^0)\,\mathrm{d}x^1$
$\xi = q^2 \frac{\partial}{\partial q^1} - q^1 \frac{\partial}{\partial q^2}$ and cycl.	$\Phi = (-p^1_1 q^2 + p^1_2 q^1)\,\mathrm{d}x^0 + (p^0_1 q^2 - p^0_2 q^1)\,\mathrm{d}x^1$

Note that the Noether equation (9.13) does not depend on the projection of the vector field ξ on the base. This means that every vector field $\xi_0 = \xi^0 \frac{\partial}{\partial x^0} + \xi^1 \frac{\partial}{\partial x^1}$ is a symmetry of the string Lagrangian. The corresponding Noether current is if course zero. This follows from relation (9.14).

For Minkowski metric tensor diag $g = (1, -1, -1, -1)$ (as the special case) D does not depend on time-space coordinates and the Noether equations reads

$$-\frac{T}{\sqrt{-D}}(g_{\alpha\beta}g_{\mu\nu} - g_{\alpha\mu}g_{\beta\nu})$$
$$\times \left[q^{\alpha}_0 q^{\beta}_1 q^{\nu}_1 \Xi^{\mu}_0 + q^{\alpha}_1 q^{\beta}_0 q^{\nu}_0 \Xi^{\mu}_1 + q^{\alpha}_0 q^{\mu}_0 q^{\beta}_1 q^{\nu}_1 \left(\frac{\partial \xi^0}{\partial x^0} + \frac{\partial \xi^1}{\partial x^1} \right) \right] = 0,$$

and using momenta

$$-p^i_{\mu}\,\mathrm{d}_i \Xi^{\mu} = 0. \tag{9.15}$$

This equation is fulfilled for all elements of Poincaré group. Symmetries and currents are summarized in Table 9.2 ($s = 1, 2, 3$, $\mu = 0, 1, 2, 3$).

9.6 Problems

Problem 9.1 The equation of motion of a linear harmonic oscillator is $-(m\ddot{x} + kx) = 0$, where m and k are positive constants. By the direct calculation verify the variationality of this equation. Construct the Vainberg-Tonti Lagrangian and minimal Lagrangian.

Problem 9.2 Find the Noether symmetrics and currents of the linear harmonic oscillator with equation of motion given in Problem 9.1.

Problem 9.3 For a linear harmonic oscillator formulate and solve the Noether-Bessel-Hagen equation $\partial_{J^2\xi}\mathcal{E}$, where

$$\mathcal{E} = E\,\mathrm{d}x \wedge \mathrm{d}t = -(m\ddot{x} + kx)\,\mathrm{d}x \wedge \mathrm{d}t$$

is the dynamical form. (Recall that the NBH equation gives us symmetries and corresponding invariant transformations of equations of motion.)

Problem 9.4 Consider again the equation of motion of a linear harmonic oscillator and its dynamical form (Problem 9.2). We know that a symmetry of a second-order dynamical form E, i.e. a π-projectable vector field ξ on Y for which $\partial_{J^2\xi} E = 0$, is not in general the symmetry of corresponding (first order) Lagrangian λ, but for such a vector field the form $\partial_{J^1\xi}\lambda$ is trivial Lagrangian (see Sect. 4.5). Verify this assertion for vector fields ξ_j, $j = 1, 2, 3, 4, 5$ obtained solving problem 9.3.

Problem 9.5 (*A generalization of the wave equation*) In Sect. 9.2 we have considered the plane wave equation for a wave propagating along x-axis with Euclidean metric on the time-space base $X = \mathbf{R}^2$ of the underlying fibred manifold. Now we generalize the situation for a n dimensional base with (covariant) metric tensor $g = g_{ij}\, dx^i \otimes dx^j$, $g^{-1} = (g^{ij})$, $g^{ij} = g^{ji}$, $0 \le i, j \le n-1$. The base X is n-dimensional time space, the metric tensor is diagonal, $\operatorname{diag} g = (1, -1, \ldots, -1)$. Prove the variationality of this generalized wave equation, find minimal Lagrangian and equations for its Noether symmetries (the complete solution see in [9]).

Problem 9.6 Find equations for Noether symmetries and currents of the first-order Lagrangian corresponding to four-dimensional Maxwell equations (see Sect. 9.3) and prove that the vector fields $\frac{\partial}{\partial x^i}$ and $\frac{\partial}{\partial A_i}$, $0 \le i \le 3$, are Noether symmetries. Express the Noether current for $\xi = \xi^i \frac{\partial}{\partial x^i}$.

Problem 9.7 Complete the calculations in Sect. 9.4 leading to the Noether equations, solve them, and calculate the corresponding currents. By an appropriate choice of constants in the general expression for Noether symmetries (always one 1 and the other zeros), express six independent generators of symmetries.

Problem 9.8 Calculate Noether currents corresponding to generators ξ_1, ξ_2, ξ_3, ξ_4, ξ_5 obtained in Sect. 9.4.

Problem 9.9 Prove that the Noether current $\Phi(\xi_5)$ obtained in Problem 9.8 is closely related to the continuity equation for probability density.

Problem 9.10 In Sect. 9.5, make omitted formal calculations.

Problem 9.11 Calculate the additional term in the Lagrangian of a quantum mechanical mass particles moving in the field with potential energy $V = V(t, x)$.

Problem 9.12 Verify the variationality of Klein-Gordon equation (see Example 9.10).

Problem 9.13 By direct calculations prove that all elements of the Poincaré group obey the Noether equation (9.15).

References

1. J. Musilová, S. Hronek, The calculus of variations on jet bundles as a universal approach for a variational formulation of fundamental physical theories. Commun. Math. **24**(2), 173–193 (2016)
2. S. Hronek, Differential forms and variational theories. Thesis. Masaryk University, Brno, 2017. 71 pp. (Supervisor J. Musilová) (In Czech)
3. M. Palese, E. Winterroth, A variational perspective on classical Higgs fields in gauge-natural theories. Theor. Math. Phys. **168**, 1002–1008 (2011). arXiv:1110.5426
4. M. Palese, E. Winterroth, Symmetries of Helmholtz forms and globally variational dynamical forms. J. Phys. Conf. Ser. **343**, 012129, 4 p (2012), arXiv:1110.5764
5. M. Palese, E. Winterroth, Higgs fields on spinor gauge-natural bundles. J. Phys. Conf. Ser. **411**, 012025, 5 p (2013)
6. M. Palese, E. Winterroth, Generalized symmetries generating Noether currents and canonical conserved quantities. J. Phys. Conf. Ser. **563**, 012023, 4 p (2014)
7. L.D. Landau, E.M. Lifshitz, *Mechanics*, 3rd edn. (Pergamon Press, 1975)
8. O. Rossi, J. Musilová, The relativistic mechanics in nonholonomic setting: a unified approach to particles with non-zero mass and massless particles. J. Phys. A: Math. Theor. **45**, 255202, 27 pp (2012)
9. E. Búš, Symmetries and conservation laws in variational theories. Thesis, Masaryk University, Brno, 2023, 86 pp. (Supervisor J. Musilová) (In Slovak)
10. L.D. Landau, E.M. Lifshitz, *The Classical Theory of Fields*, 3rd edn. (Pergamon Press, 1975)
11. L.D. Landau, E.M. Lifshitz, *Quantum Mechanics: Non-relativistic Theory*, 3rd edn. (Pergamon Press, 1977)
12. T. Goto, Prog. Theor. Phys. **46**, 1560 (1971)
13. J. Musilová, M. Lenc, Lepage forms in variational theories: from Lepage's idea to the variational sequence, in *Geometry and Physics*, ed. by O. Krupková, D. Saunders (Nova Science Publishers Inc., 2009), pp. 3–26

Solutions and Hints

Problems of Chapter 2

Problem 2.1

Hint Write the parametric equations of the surface Y with help of parameters α and θ shown in Fig. 2.17 and express the mapping π also with help of these parameters. Calculate the basic tangent vectors

$$\xi_1 = \left(\frac{\partial \bar{x}}{\partial \theta}, \frac{\partial \bar{y}}{\partial \theta}, \frac{\partial \bar{z}}{\partial \theta} \right), \quad \xi_2 = \left(\frac{\partial \bar{x}}{\partial \alpha}, \frac{\partial \bar{y}}{\partial \alpha}, \frac{\partial \bar{z}}{\partial \alpha} \right)$$

to Y. Write the parametric equations of the circle S^1 with help of θ and calculate the basic tangent vector

$$\zeta = \left(\frac{\partial x}{\partial \theta}, \frac{\partial y}{\partial \theta}, \frac{\partial z}{\partial \theta} \right).$$

Express the Jacobi matrix $D\pi$ of π representing the tangent mapping $T\pi$, determine its rank and estimate if $T\pi(\xi_1)$ and $T\pi(\xi_2)$ are collinear with ζ. □

Problem 2.2

Solution The mentioned fibred charts adapted to the Cartesian product arise as Cartesian products of charts on the base X and on the fibre M. Let $\mathcal{A}_X = \{(U_\iota, \varphi_\iota)\}$ and $\mathcal{A}_M = \{(W_\kappa, \chi_\kappa)\}$ be atlases on X and on M, respectively, where

$$\varphi_\iota : U_\iota \ni x \longrightarrow \varphi_\iota(x) = \big(t\varphi_\iota(x)\big) \in \varphi_\iota(U_\iota) \subset \mathbf{R},$$
$$\chi_\kappa : W_\kappa \ni z \longrightarrow \chi_\kappa(z) = \big(q^1\chi_\kappa(z), \ldots, q^m\chi_\kappa(z)\big) \in \chi_\kappa(W_\kappa) \subset \mathbf{R}^m.$$

J. Musilová et al., *Calculus of Variations on Fibred Manifolds and Variational Physics*, Lecture Notes in Physics 1033, https://doi.org/10.1007/978-3-031-77408-9

Denote $V_{\iota,\kappa} = U_\iota \times W_\kappa$, $y = (x, z) \in X \times M$, $x \in X$, $z \in M$, and $\psi_{\iota,\kappa} = \varphi_\iota \times \chi_\kappa$, i.e.

$$\begin{aligned}\psi_{\iota,\kappa} : V_{\iota,\kappa} \ni y &\longrightarrow \psi_{\iota,\kappa}(y)\big(t\psi_{\iota,\kappa}(y), q^1\psi_{\iota,\kappa}(y), \ldots, q^m\psi_{\iota,\kappa}(y)\big)\\ &= \big(t\varphi_\iota(x), q^1\chi_\kappa(z), \ldots, q^m\chi_\kappa(z)\big) \in \mathbf{R} \times \mathbf{R}^m.\end{aligned}$$

□

Problem 2.3

Solution

(a) Let $(U, \varphi), (\bar{U}, \bar{\varphi}) \in \mathcal{A}$, $\varphi = (t)$, $\bar{\varphi} = (\bar{t})$ be two coordinate systems such that $\bar{U} \cap U \neq \emptyset$. Suppose that $C_{x,1}$ and $C_{x,2}$ are two curves going through a point $x \in \bar{U} \cap U$. Let

$$\left.\frac{\mathrm{d}t\varphi(C_{x,1})(s)}{\mathrm{d}s}\right|_{s=0} = \left.\frac{\mathrm{d}t\varphi(C_{x,2})(s)}{\mathrm{d}s}\right|_{s=0}.$$

Then

$$\begin{aligned}\left.\frac{\mathrm{d}\bar{t}\bar{\varphi}(C_{x,1})(s)}{\mathrm{d}s}\right|_{s=0} &= \left.\frac{\mathrm{d}\bar{t}(\bar{\varphi}\circ\varphi^{-1})(t)}{\mathrm{d}t}\cdot\frac{\mathrm{d}t\varphi(C_{x,1})(s)}{\mathrm{d}s}\right|_{s=0}\\ &= \left.\frac{\mathrm{d}\bar{t}(\bar{\varphi}\circ\varphi^{-1})(t)}{\mathrm{d}t}\cdot\frac{\mathrm{d}t\varphi(C_{x,2})(s)}{\mathrm{d}s}\right|_{s=0} = \left.\frac{\mathrm{d}\bar{t}\bar{\varphi}(C_{x,2})(s)}{\mathrm{d}s}\right|_{s=0}.\end{aligned}$$

(b) Let $\xi(x), \zeta(x) \in T_xX$ be two tangent vectors to the manifold X at the point x. Denote $C_x^{(\xi)}$ and $C_x^{(\zeta)}$ the arbitrarily chosen representatives of vectors $\xi(x)$ and $\zeta(x)$, respectively. The mappings

$$\begin{aligned}&\text{“}+\text{”} : T_xX \times T_xX \ni [\xi(x), \zeta(x)] \longrightarrow \xi(x) + \zeta(x) = [C_x^{(\xi)} + C_x^{(\zeta)}] \in T_xX,\\ &\text{“}\cdot\text{”} : \mathbf{R} \times T_xX \ni [\alpha, \xi(x)] \longrightarrow \alpha\xi(x) = [\alpha C_x^{(\xi)}] \in T_xX.\end{aligned}$$

These mappings obey the rules for the addition of vectors and multiplication of a vector by a scalar prescribed for vector spaces. Thus, T_xX is the vector space. Its dimension is 1.

(c) Denote $V_\iota = \{\xi(x) \in T_xX \mid x \in U_\iota\} \subset TX$. Define the mapping

$$\begin{aligned}\psi_\iota : V_\iota \ni \xi(x) &\longrightarrow \psi_\iota(\xi(x)) = \big(t\psi_\iota(\xi(x)), \xi^1\psi_\iota(\xi(x))\big)\\ &= \left(t\varphi_\iota(x), \left.\frac{\mathrm{d}t(\varphi_\iota C_x)(s)}{\mathrm{d}s}\right|_{s=0}\right) \in \mathbf{R}^2.\end{aligned}$$

Let us denote $\varphi_\iota = \varphi$, $\varphi_\kappa = \bar{\varphi}$, $\psi_\iota = \psi$, $\psi_\kappa = \bar{\psi}$, $V_\kappa = \bar{V}$, $V_\iota = V$, $U_\kappa = \bar{U}$, $U_\iota = U$ for simplicity.

$$\begin{aligned}\bar{\psi}\circ\psi^{-1} : \psi(\bar{V} \cap V) \ni (t, \xi^1) &\longrightarrow \bar{\psi} \circ \psi^{-1}(t, \xi^1) \\ &= \left(\bar{t}(\bar{\varphi} \circ \varphi^{-1})(t), \bar{\xi}^1(\bar{\psi} \circ \psi^{-1})(t, \xi^1)\right) \in \bar{\psi}(\bar{V} \cap V),\end{aligned}$$

where

$$\begin{aligned}\bar{\xi}^1\bar{\psi}\left(\xi(x)\right) &= \bar{\xi}^1(\bar{\psi} \circ \psi^{-1})(t, \xi^1) \\ &= \left.\frac{\mathrm{d}\bar{t}\left(\bar{\varphi}C_x(s)\right)}{\mathrm{d}s}\right|_{s=0} = \left.\frac{\mathrm{d}\left(\bar{\varphi} \circ \varphi^{-1}(t)\right)}{\mathrm{d}t} \cdot \frac{\mathrm{d}t(\varphi C_x)(s)}{\mathrm{d}s}\right|_{s=0} \\ &= \frac{\mathrm{d}\left(\bar{\varphi} \circ \varphi^{-1}(t)\right)}{\mathrm{d}t} \cdot \xi^1\psi\left(\xi(x)\right), \quad \text{shortly } \bar{\xi}^1 = \frac{\mathrm{d}\bar{t}(\bar{\varphi} \circ \varphi^{-1})(t)}{\mathrm{d}t}\,\xi^1.\end{aligned}$$

Since the mappings $\bar{\varphi} \circ \varphi^{-1}$ (and, of course, $\varphi \circ \bar{\varphi}^{-1}$ as well) are diffeomorphisms and mappings C_x are smooth, it is evident that mappings $\bar{\psi} \circ \psi^{-1}$ and $\psi \circ \bar{\psi}^{-1}$ are diffeomorphisms. So, the system $\mathcal{A}_{TX} = \left\{(V_\iota, \psi_\iota)\right\}$, $\iota \in I$, is the atlas on TX and the set TX has the structure of 2-dimensional smooth manifold.

(d) Following above derived results we can see that the chart representations of the mapping π, i.e.

$$\varphi_\iota \circ \pi \circ \psi_\iota^{-1} : \psi_\iota(V_\iota)\,(t, \xi^1) \longrightarrow (\varphi_\iota \circ \pi \circ \psi_\iota^{-1})(t, \xi^1) = t,$$

where $\iota \in I$, are the Cartesian projections. This proves that π is the surjective submersion. The bundle (TX, π, X) is the fibred manifold with fibres (globally) isomorphic with $\mathbf{R}$. It holds $\psi_\iota(V_\iota) = \varphi_\iota(U_\iota) \times \mathbf{R}$. If X is connected then the tangent bundle (TX, π, X) is the fibration with the type fibre $\mathbf{R}$. As the example of such a situation is

$$\pi : S^1 \times \mathbf{R} \ni (t, z) \longrightarrow \pi(t, z) = t \in S^1, \quad t \in [0, 2\pi],$$

discussed in Example 2.3 and represented in Fig. 2.3 (left). If the base X is one-dimensional, then the tangent bundle (TX, π, X) has always the structure of the Cartesian product $X \times \mathbf{R}$. (In a general situation of $n = \dim X > 1$ the tangent bundle is defined by quite analogous way, i.e. with help of equivalent parametrized curves, and $\dim TX = 2n$. However, it need not have the structure of the Cartesian product $X \times \mathbf{R}^n$.)

(e) Transformation rules for components of tangent vectors (ξ^1) and $(\bar{\xi}^1)$ in various charts were derived in the item (c).

Note that the projection π assigns to every tangent vector $\xi(x)$ the point x at which the tangent vector is bounded. □

Problem 2.4

Solution In the proof of Proposition 2.2 we showed that the equivalence conditions for sections γ_1 and γ_2 1-equivalent at a point $x \in X$ are independent of coordinates, i.e. they are fulfilled in all fibred charts. We will prove the same for two 2-equivalent sections γ_1 and γ_2. Let (V, ψ), $\psi = (t, q^\sigma)$ and $(\bar{V}, \bar{\psi})$, $\psi = (\bar{t}, \bar{q}^\sigma)$ are two fibred charts on Y, $(U, \varphi), \varphi = (t)$, and $(\bar{U}, \bar{\varphi}), \bar{\varphi} = (\bar{t})$, being the corresponding associated charts on X. Let $x \in \bar{U} \cap U$. Suppose that for sections γ_1 and γ_2 it holds $\gamma_1(x) = \gamma_2(x)$ and relations of the 2-equivalence (2.6) are fulfilled in coordinates (V, ψ), (U, φ). Then following (2.7) we obtain

$$\begin{aligned}
\ddot{\bar{q}}^\sigma J^2_x\gamma(t) &= \frac{\mathrm{d}}{\mathrm{d}\bar{t}}\left(\frac{\mathrm{d}t}{\mathrm{d}\bar{t}}\left(\frac{\partial \bar{q}^\sigma(t, q^\lambda)}{\partial t} + \frac{\partial \bar{q}^\sigma(t, q^\lambda)}{\partial q^\nu}\dot{q}^\nu\right)(J^2_x\gamma)\right) \\
&= \frac{\mathrm{d}^2 t}{\mathrm{d}\bar{t}^2}\left(\frac{\partial \bar{q}^\sigma(t, q^\lambda)}{\partial t} + \frac{\partial \bar{q}^\sigma(t, q^\lambda)}{\partial q^\nu}\dot{q}^\nu\right)(J^2_x\gamma) \\
&\quad + \left(\frac{\mathrm{d}t}{\mathrm{d}\bar{t}}\right)^2\left(\frac{\partial^2 \bar{q}^\sigma(t, q^\lambda)}{\partial t^2} + 2\frac{\partial^2 \bar{q}^\sigma(t, q^\lambda)}{\partial t \partial q^\nu}\dot{q}^\nu + \frac{\partial^2 \bar{q}^\sigma(t, q^\lambda)}{\partial q^\nu \partial q^\mu}\dot{q}^\nu\dot{q}^\mu\right. \\
&\quad \left. + \frac{\partial \bar{q}^\sigma(t, q^\lambda)}{\partial q^\nu}\ddot{q}^\nu\right)(J^2_x\gamma).
\end{aligned}$$

As we suppose that $\gamma_1(x) = \gamma_2(x)$ and in the fibred chart (V, ψ), $\psi = (t, q^\sigma)$, $1 \le \sigma \le m$, it holds

$$\dot{q}^\sigma J^2_x\gamma_1 = \dot{q}^\sigma J^2_x\gamma_1, \quad \ddot{q}^\sigma J^2_x\gamma_1 = \ddot{q}^\sigma J^2_x\gamma_1,$$

we can conclude from just above derived transformation rule that the same relations hold also in the fibred chart $(\bar{V}, \bar{\psi})$. □

Problem 2.5

Hint Show that $\pi \circ \bar{\gamma} = \mathrm{id}_{\alpha_0 \circ \pi(W)}$ using the relations $\pi \circ \alpha = \alpha_0 \circ \pi$ and $\pi \circ \gamma = \mathrm{id}_{\pi(W)}$. □

Problem 2.6

Hint The fact that $J^1\alpha$ is a (local) diffeomorphism was proved in Sect. 2.4. It only remains to prove the property $\pi_1 \circ J^1\alpha = \alpha_0 \circ \pi_1$. This means that we need prove that for an arbitrary 1-jet $J^1_x\gamma$, $x \in \pi(W)$ it holds

$$\pi_1 \circ J^1\alpha(J^1_x\gamma) = \alpha_0 \circ \pi_1(J^1_x\gamma).$$

Finish the proof putting

$$J^1\alpha(J^1_x\gamma) = J^1_{\alpha_0(x)}(\alpha \circ \gamma \circ \alpha_0^{-1})$$

and taking into account the definition of jets of sections. □

Problems of Chapter 3

Problem 3.1

Solution The Jacobi matrix of the mapping α_u at a point $y \in V$ in the chosen charts is

$$D\alpha_u = \begin{pmatrix} \frac{\partial t(\psi\circ\alpha_u\circ\psi^{-1})(t,q^\lambda)}{\partial t} & \frac{\partial q^1(\psi\circ\alpha_u\circ\psi^{-1})(t,q^\lambda)}{\partial t} & \dots & \frac{\partial q^m(\psi\circ\alpha_u\circ\psi^{-1})(t,q^\lambda)}{\partial t} \\ \frac{\partial t(\psi\circ\alpha_u\circ\psi^{-1})(t,q^\lambda)}{\partial q^1} & \frac{\partial q^1(\psi\circ\alpha_u\circ\psi^{-1})(t,q^\lambda)}{\partial q^1} & \dots & \frac{\partial q^m(\psi\circ\alpha_u\circ\psi^{-1})(t,q^\lambda)}{\partial q^1} \\ \dots & \dots & \dots & \dots \\ \frac{\partial t(\psi\circ\alpha_u\circ\psi^{-1})(t,q^\lambda)}{\partial q^m} & \frac{\partial q^1(\psi\circ\alpha_u\circ\psi^{-1})(t,q^\lambda)}{\partial q^m} & \dots & \frac{\partial q^m(\psi\circ\alpha_u\circ\psi^{-1})(t,q^\lambda)}{\partial q^m} \end{pmatrix}.$$

Denote $\zeta\big(\alpha_u(y)\big) = T\alpha_u\big(\xi(y)\big)$. Then we have for the vector $\zeta\big(\alpha_u(y)\big)$

$$\zeta\big(\alpha_u(y)\big) = \left[\zeta^0\frac{\partial}{\partial t} + \zeta^\sigma\frac{\partial}{\partial q^\sigma}\right]_{\alpha_u(y)},$$

$$\zeta^0\big(\alpha_u(y)\big) = \xi^0(y)\frac{\partial t(\psi\circ\alpha_u\circ\psi^{-1})(t,q^\lambda)}{\partial t} + \xi^\nu(y)\frac{\partial t(\psi\circ\alpha_u\circ\psi^{-1})(t,q^\lambda)}{\partial q^\nu},$$

$$\zeta^\sigma\big(\alpha_u(y)\big) = \xi^0(y)\frac{\partial q^\sigma(\psi\circ\alpha_u\circ\psi^{-1})(t,q^\lambda)}{\partial t} + \xi^\nu(y)\frac{\partial q^\sigma(\psi\circ\alpha_u\circ\psi^{-1})(t,q^\lambda)}{\partial q^\nu}.$$

Now we calculate the components of the vector $\xi\big(\alpha_u(y)\big)$ defined by relations (3.4). In this calculation we take into account the group operations in the local one-parameter group, i.e.

$$\alpha_{u+s} = \alpha_s\circ\alpha_u = \alpha_u\circ\alpha_s, \quad \alpha_{u=0} = \mathrm{id}.$$

First, we calculate the component $\xi^0\big(\alpha_u(y)\big)$, using the chain rule for derivatives of functions:

$$\begin{aligned} \xi^0\big(\alpha_u(y)\big) &= \left.\frac{\mathrm{d}t(\psi\circ\alpha_{u+s})(y)}{\mathrm{d}s}\right|_{s=0} = \left[\frac{\mathrm{d}}{\mathrm{d}s}\Big(t\big(\psi\circ\alpha_s\circ\alpha_u(y)\big)\Big)\right]_{s=0} \\ &= \left[\frac{\partial t(\psi\circ\alpha_u\circ\psi^{-1})\big(t(\psi\circ\alpha_s)(y), q^\lambda(\psi\circ\alpha_s)(y)\big)}{\partial t}\right]_{s=0}\left.\frac{\mathrm{d}t(\psi\circ\alpha_s)(y)}{\mathrm{d}s}\right|_{s=0} \\ &+ \left[\frac{\partial t(\psi\circ\alpha_u\circ\psi^{-1})\big(t(\psi\circ\alpha_s)(y), q^\lambda(\psi\circ\alpha_s)(y)\big)}{\partial q^\nu}\right]_{s=0}\left.\frac{\mathrm{d}q^\nu(\psi\circ\alpha_s)(y)}{\mathrm{d}s}\right|_{s=0}. \end{aligned}$$

However, Eq. (3.4) gives

$$\left.\frac{\mathrm{d}t(\psi\circ\alpha_s)(y)}{\mathrm{d}s}\right|_{s=0} = \xi^0(y) \quad \text{and} \quad \left.\frac{\mathrm{d}q^\nu(\psi\circ\alpha_s)(y)}{\mathrm{d}s}\right|_{s=0} = \xi^\nu(y)$$

and thus $\xi(\alpha_u(y)) = \zeta(\alpha_u(y))$, i.e. $\xi(\alpha_u(y)) = T\alpha_u(\xi(y))$.

Similarly, for a vector field ξ_0 on an open subset of X and its local one-parameter group $\{\alpha_{0u}\}$ prove that $T\alpha_{0u} \circ \xi_0 = \xi_0 \circ \alpha_{0u}$. It is easy to obtain the relation for components

$$\begin{aligned}
\xi_0^0(\alpha_{0u}(x)) &= \left.\frac{dt(\varphi \circ \alpha_{0,u+s})(x)}{ds}\right|_{s=0} = \left[\frac{d}{ds}(t(\varphi \circ \alpha_{0\,s} \circ \alpha_{0u}))\right]_{s=0} \\
&= \left[\frac{\partial t(\varphi \circ \alpha_{0u} \circ \varphi^{-1})(t(\varphi \circ \alpha_{0\,s})(x))}{\partial t}\right]_{s=0} \left.\frac{dt(\varphi \circ \alpha_{0\,s})(x)}{ds}\right|_{s=0} \\
&= \left[\frac{\partial t(\varphi \circ \alpha_{0u} \circ \varphi^{-1})(t(\varphi \circ \alpha_{0\,s})(x))}{\partial t}\right]_{s=0} \xi_0^0(x).
\end{aligned}$$

□

Problem 3.2

Hint Express the components ξ^0 and ξ^σ of the vector field ξ and the component ξ_0^0 of its projection in the canonical (global) fibred chart $(\mathbf{R} \times \mathbf{R}^m, \mathrm{id}_{\mathbf{R}\times\mathbf{R}^m})$, the associated chart being $(\mathbf{R}, \mathrm{id}_{\mathbf{R}})$. You obtain a the set of ordinary differential equations for local one-parameter groups $\{\alpha_u\}$, $\{\alpha_{0u}\}$, $u \in (-\varepsilon, \varepsilon)$, of vector fields ξ and ξ_0, respectively. (Write down these equations and use the fact that ξ_0 is the projection of ξ and the initial conditions for $u = 0$.) □

Problem 3.3

Hint Taking into account the definition of the jet prolongation of isomorphisms prove the group properties of $\{J^1\alpha_u\}$, i.e. $J^1\alpha_{u+s} = J^1\alpha_u \circ J^1\alpha_s = J^1\alpha_u \circ J^1\alpha_s$, and $J^1\alpha_0 = \mathrm{id}$. And analogously for $\{J^2\alpha_u\}$. □

Problem 3.4

Hint First, prove the assertions of the proposition for 1-forms by calculating $i_\xi\, dt$, $i_\xi\, dq^\sigma$, $i_\xi\, d\dot{q}^\sigma$, $i_\xi\, d\ddot{q}^\sigma$ for a π_r-vertical vector field ξ. Then use the properties of the exterior (wedge) product of forms. □

Problem 3.5

Hint Taking into account Proposition 3.3 (see also Problem 3.4) it satisfies to prove the assertion for 1-forms. For the proof use the direct calculation of $J^r\gamma^*\omega$ in fibred charts. □

Problem 3.6

Hint Follow step by step the proof made for $r = 0$: First it is necessary to prove the assertion for 1-forms on J^1Y. Write down the chart expression of a 1-form ω on J^1Y and the chart expression if $J^1\gamma$ for a section γ. Then calculate the pullback $J^1\gamma^*\omega$ and compare the result with the pullback $J^2\gamma^*\eta$ for a horizontal 1-form η on J^2Y. □

Problem 3.7

Hints

(a) For an arbitrary section γ calculate the form $J^{r+1}\gamma^*\pi^*_{r+1,r}\omega$. The **R**-linearity is a direct consequence of the definition and the **R**-linearity of the horizontalization.
(b) Suppose that $\omega = \eta \wedge \chi$, where η is a contact form. Using the properties of the pullback mapping calculate $J^r\gamma^*\omega$.
(c) For a 2-form use the direct calculation of $J^r\gamma^*\omega$ for a section γ. The conclusion for $k > 2$ then results from (b).
(d) Write the chart expression of the form η and transform it into the base adapted for the contact structure.
(e) Use the direct calculation in coordinates. □

Problem 3.8

Hint Using the direct calculation express the difference $p(\omega \wedge \eta) - p\omega \wedge p\eta$. Take into account that the image of a k-form by the horizontalization mapping is zero for $k > 1$. □

Problem 3.9

Hint One possibility is the direct use of Definition 3.8. Prove that the exterior product $\omega \wedge \eta$ of a 1-contact 1-form ω and a $(k-1)$-contact form η, $k > 1$, is a k-contact form. Remember that the definition of a 1-contact 1-form is different from the definition of a 1-contact q-form for $q > 1$. □

Problem 3.10

Hint For a k-form η on Y express the 1-forms $\pi^*_{1,0}dq^\sigma$ in the chart expression of η in the basis adapted to the contact structure and execute the exterior products. For a k-form η on J^1Y express the 1-forms $\pi^*_{2,1}dq^\sigma$ and 1-forms $\pi^*_{2,1}d\dot{q}^\sigma$ in the adapted basis and execute the exterior products. □

Problem 3.11

Hint Following the considerations in the proof of Proposition 3.6 for $r = 1$ calculate $\partial_\xi \dot{\omega}^\sigma$. □

Problem 3.12

Solution We can use the definition of the Lie derivative

$$\partial_\zeta \eta = \lim_{u \to 0} \frac{\phi_u^* \eta - \eta}{u},$$

where η is a form on J^rY and $\{\phi_u\}$ is the local one-parameter group of the given vector field ζ. (The mappings of the one-parameter group of a generally defined vector field on J^rY are one to one and differentiable, even though they need not be local isomorphisms of the given fibred manifold.) Another possibility for solving the problem is the use of relation (3.19). Let η be a π_r-horizontal form on J^rY, $r = 0, 1, 2$. This means that it is a 1-form and its contraction by every π_r-vertical vector field ξ on J^rY vanishes. Alternatively, the chart expression of η is $\eta = A\,\mathrm{d}t$, where A is a function on J^rY. First, use the definition of the Lie derivative.

$$\partial_\zeta \eta = \lim_{u \to 0} \frac{\phi_u^* \eta - \eta}{u},$$

$$i_\xi \partial_\zeta \eta = \lim_{u \to 0} \frac{i_\xi \phi_u^* \eta - i_\xi \eta}{u} = \lim_{u \to 0} \frac{i_\xi \phi_u^* \eta}{u}$$

because $i_\xi \eta = 0$. Continuing the calculation we get

$$i_\xi \partial_\zeta \eta = \lim_{u \to 0} \frac{i_\xi \phi_u^* \eta}{u} = \lim_{u \to 0} \frac{(\phi_u^* \eta)(\xi)}{u} = \lim_{u \to 0} \frac{\eta\big(T\phi_u(\xi)\big)}{u}.$$

Because of the fact that the vector field $T\phi_u(\xi)$ need not be π_r-vertical in general, we can conclude that the Lie derivative do not preserve the horizontality of forms. On the other hand, it is clear (express the Jacobi matrix of ϕ_u) that the vector field $T\phi_u(\xi)$ is π_r-vertical iff the first column of the Jacobi matrix of ϕ_u is zero except for the first element. This is valid just for π_r-projectable vector field ζ. The same result we obtain using relation (3.19), e.g. for $r = 2$:

$$\partial_\zeta (A\,\mathrm{d}t) = i_\zeta \left(\frac{\partial A}{\partial q^\sigma} \mathrm{d}q^\sigma \wedge \mathrm{d}t + \frac{\partial A}{\partial \dot{q}^\sigma} \mathrm{d}\dot{q}^\sigma \wedge \mathrm{d}t + \frac{\partial A}{\partial \ddot{q}^\sigma} \mathrm{d}\ddot{q}^\sigma \wedge \mathrm{d}t \right) + \mathrm{d}(A\zeta^0).$$

Here ζ is a vector field on J^rY, i.e. for $r = 2$

$$\zeta = \zeta^0 \frac{\partial}{\partial t} + \zeta^\sigma \frac{\partial}{\partial q^\sigma} + \tilde{\zeta}^\sigma \frac{\partial}{\partial \dot{q}^\sigma} + \hat{\zeta}^\sigma \frac{\partial}{\partial \ddot{q}^\sigma},$$

where components are functions on J^2Y. For $r = 1$ and $r = 0$ the "redundant" components are zero and remaining are function on J^rY. For example, for $r = 1$ we have after some calculations:

$$\partial_\zeta (A\,\mathrm{d}t) = \left(\frac{\partial A}{\partial t} \zeta^0 + \frac{\partial A}{\partial q^\sigma} \zeta^\sigma + \frac{\partial A}{\partial \dot{q}^\sigma} \tilde{\zeta}^\sigma + A \frac{\partial \zeta^0}{\partial t} \right) \mathrm{d}t$$
$$+ A \frac{\partial \zeta^0}{\partial q^\sigma} \mathrm{d}q^\sigma + A \frac{\partial \zeta^0}{\partial \dot{q}^\sigma} \mathrm{d}\dot{q}^\sigma.$$

This form is π_1-horizontal iff ζ^0 depends only on the variable t. This means that the vector field ζ is π_1-projectable. The situation is the same for J^2Y. We can conclude that the Lie derivative of a form on J^rY with respect to a π_r-projectable vector field defined on J^rY preserves the horizontality of forms. The solution of the problem for contact forms is very easy because it is fully contained in Proposition 3.6 stating that the Lie derivative of a contact form on J^rY, $r = 1, 2$, with respect to a vector field ζ

on J^rY is contact iff ζ is a contact symmetry, i.e. there exists a π-projectable vector field ξ on Y such that $\zeta = J^r\xi$. So, the importance of prolongations of π-projectable vector fields on Y lies, among other, in the fact that they preserve the properties of horizontality and contactness of forms via the Lie derivative. □

Problem 3.13

Solution We use the definition relations (2.8) and (2.9) of the first and second prolongation of a local isomorphism α:

$$J^r\alpha \circ J^r\gamma(x) = J^r(\alpha \circ \gamma \circ \alpha_0^{-1}) \circ \alpha_0(x), \quad x \in X, \quad r = 1, 2.$$

Let η be a contact 1-form on $J^rY, r = 1, 2$. Then $J^r\gamma^*\eta = 0$ for an arbitrary section $\gamma \in \Gamma(\pi)$. Let us estimate whether the form $J^r\alpha^*\eta$ is contact as well, i.e. whether $J^r\gamma^*(J^r\alpha^*\eta) = 0$.

$$\begin{aligned} J^1\gamma^*(J^1\alpha^*\eta) &= \left(J^1\alpha \circ J^1\gamma\right)^* \eta = \left(J^1(\alpha \circ \gamma \circ \alpha_0^{-1}) \circ \alpha_0\right)^* \eta \\ &= \alpha_0^* \circ J^1(\alpha \circ \gamma \circ \alpha_0^{-1})^*\eta = 0. \end{aligned}$$

The last equality holds because the mapping $\alpha_0 \circ \gamma \circ \alpha_0^{-1}$ is the section of π. So, we can conclude that the pullback of a contact form η by the mapping $J^1\alpha$ is the contact form, especially, if η is a contact 1-form, then $J^r\alpha^*\eta$ is the contact 1-form. Now let ω be a horizontal 1-form on J^rY, $\omega = A\,\mathrm{d}t$, where A is a function on J^rY. Then

$$J^r\alpha^*\omega = (A \circ J^r\alpha)\, J^r\alpha^*\mathrm{d}t = (A \circ J^r\alpha)\,\mathrm{d}t,$$

which proves that the form $J^r\alpha^*\omega$ is horizontal. Using the property of pullback of forms by a mapping f, $f^*(\chi \wedge \varrho) = f^*\chi \wedge f^*\varrho$ for arbitrary forms χ and ϱ, we can easily finish the proof. □

Problem 3.14

Hint Formulate the Poincaré lemma for general 1-forms on Y and on J^1Y, without specifying their degree of contactness, i.e.

$$\eta = A_0\,\mathrm{d}t + B_\sigma\,\mathrm{d}q^\sigma, \quad \text{or} \quad \eta = A_0\,\mathrm{d}t + B_\sigma\,\mathrm{d}q^\sigma + \tilde{B}_\sigma\,\mathrm{d}\dot{q}^\sigma,$$

where A_0, B_σ, $\tilde{B}_\sigma$ are functions on J^rY, $r = 0, 1$. For a 1-form η on Y express $\pi_{1,0}^*\eta$, i.e. $\pi_{1,0}^*\eta = A_0\,\mathrm{d}t + B_\sigma\,\omega^\sigma$, and reproduce the procedure presented in the proof of Theorem 3.2 for k-forms with $k > 1$. For a 1-form η on J^1Y it holds $\pi_{2,1}^*\eta = A_0\,\mathrm{d}t + B_\sigma\,\omega^\sigma + \tilde{B}_\sigma\,\dot{\omega}^\sigma$. Apply the same procedure as in the proof of Theorem 3.2, using the mapping χ introduced for J^2Y. (Remember that the form ϱ is a 0-form, i.e. a function.) □

Problem 3.15

Hint for (a) Solve this problem in charts. For example, for a 1-contact 2-form $\eta = A_\sigma(t, q^\nu, \dot{q}^\nu)\, \omega^\sigma \wedge \mathrm{d}t$ we have

$$\mathrm{d}\eta = \frac{\partial A_\sigma}{\partial q^\nu} \omega^\nu \wedge \omega^\sigma \wedge \mathrm{d}t + \frac{\partial A_\sigma}{\partial \dot{q}^\nu} \mathrm{d}\dot{q}^\nu \wedge \omega^\sigma \wedge \mathrm{d}t.$$

The requirement $\mathrm{d}\eta = 0$ leads to the following conditions for coefficients A_σ:

$$\frac{\partial A_\sigma}{\partial q^\nu} + \frac{\partial A_\nu}{\partial q^\sigma} = 0, \quad \frac{\partial A_\sigma}{\partial \dot{q}^\nu} = 0, \quad 1 \le \sigma, \nu \le m,$$

while there are no conditions for coefficients A_σ as functions of the coordinate t. (An example of such coefficients: $A_\sigma = f(t)\varepsilon_{\sigma\alpha\beta} q^\alpha q^\beta$, where $(\varepsilon_{\sigma\alpha\beta})$ is Levi-Civita tensor.) □

Solution of (b) We show the complete solution with all calculations in detail for 1-contact 2-forms on J^1Y. For 2-contact 2-forms and for J^2Y the procedure is the same, although technically more complicated. Let η be a 1-contact 2-form on J^1Y. Then the chart expression of η is following (3.17)

$$\eta = A_\sigma \omega^\sigma \wedge \mathrm{d}t, \quad A = A(t, q^\nu, \dot{q}^\nu).$$

The mapping χ has the form

$$\chi : [0, 1] \times J^1Y \ni (u, t, q^\sigma, \dot{q}^\sigma) \longrightarrow (t, uq^\sigma, u\dot{q}^\sigma) \in J^1Y,$$

$$\chi^*\eta = (A_\sigma \circ \chi)(q^\sigma\, \mathrm{d}u + u\omega^\sigma) \wedge \mathrm{d}t = \mathrm{d}u \wedge q^\sigma (A_\sigma \circ \chi)\, \mathrm{d}t + u(A_\sigma \circ \chi)\omega^\sigma \wedge \mathrm{d}t.$$

$$\mathcal{A}\eta = \left(q^\sigma \int_0^1 (A_\sigma \circ \chi)\, \mathrm{d}u \right) dt$$

$$\Longrightarrow \mathrm{d}\mathcal{A}\eta = \left[\int_0^1 (A_\sigma \circ \chi)\, \mathrm{d}u + q^\nu \int_0^1 \frac{\partial (A_\nu \circ \chi)}{\partial q^\sigma} \mathrm{d}u \right] \omega^\sigma \wedge \mathrm{d}t$$
$$+ \left[q^\sigma \int_0^1 \frac{\partial (A_\sigma \circ \chi)}{\partial \dot{q}^\nu} \mathrm{d}u \right] \mathrm{d}\dot{q}^\nu \wedge \mathrm{d}t.$$

Taking into account the following relation resulting form the chain rule for derivatives of composed functions, we can express $\mathrm{d}\mathcal{A}\eta$ by an alternative way (step by step calculations):

$$\frac{\partial (A_\sigma \circ \chi)}{\partial q^\nu} = \frac{\partial (A_\sigma \circ \chi)}{\partial (uq^\nu)} \frac{\partial (uq^\nu)}{\partial q^\nu} = u \left(\frac{\partial A_\sigma}{\partial q^\nu} \circ \chi \right),$$

and analogously

$$\frac{\partial(A_\sigma \circ \chi)}{\partial \dot{q}^\nu} = \frac{\partial(A_\sigma \circ \chi)}{\partial(u\dot{q}^\nu)}\frac{\partial(u\dot{q}^\nu)}{\partial \dot{q}^\nu} = u\left(\frac{\partial A_\sigma}{\partial \dot{q}^\nu} \circ \chi\right),$$

$$\mathrm{d}\mathcal{A}\eta = \left[\int_0^1 (A_\sigma \circ \chi)\,\mathrm{d}u + q^\nu \int_0^1 u\left(\frac{\partial A_\nu}{\partial q^\sigma} \circ \chi\right)\mathrm{d}u\right]\omega^\sigma \wedge \mathrm{d}t + \left[q^\sigma \int_0^1 u\left(\frac{\partial A_\sigma}{\partial \dot{q}^\nu} \circ \chi\right)\mathrm{d}u\right]\mathrm{d}\dot{q}^\nu \wedge \mathrm{d}t.$$

On the other hand, for $\mathrm{d}\eta$ and $\chi^*\,\mathrm{d}\eta$ we get

$$\begin{aligned}
\mathrm{d}\eta &= \frac{\partial A_\sigma}{\partial q^\nu}\omega^\nu \wedge \omega^\sigma \wedge \mathrm{d}t + \frac{\partial A_\sigma}{\partial \dot{q}^\nu}\,\mathrm{d}\dot{q}^\nu \wedge \omega^\sigma \wedge \mathrm{d}t,\\
\chi^*\,\mathrm{d}\eta &= \left(\frac{\partial A_\sigma}{\partial q^\nu} \circ \chi\right)(q^\nu\,\mathrm{d}u + u\omega^\nu) \wedge (q^\sigma\,\mathrm{d}u + u\omega^\sigma) \wedge \mathrm{d}t\\
&\quad + \left(\frac{\partial A_\sigma}{\partial \dot{q}^\nu} \circ \chi\right)(\dot{q}^\nu\,\mathrm{d}u + u\,\mathrm{d}\dot{q}^\nu) \wedge (q^\sigma\,\mathrm{d}u + u\omega^\sigma) \wedge \mathrm{d}t\\
&= \mathrm{d}u \wedge \left[uq^\nu\left(\frac{\partial A_\sigma}{\partial q^\nu} - \frac{\partial A_\nu}{\partial q^\sigma}\right) \circ \chi + u\dot{q}^\nu\left(\frac{\partial A_\sigma}{\partial \dot{q}^\nu} \circ \chi\right)\right]\omega^\sigma \wedge \mathrm{d}t\\
&\quad - \left[uq^\sigma\left(\frac{\partial A_\sigma}{\partial \dot{q}^\nu} \circ \chi\right)\right]\mathrm{d}\dot{q}^\nu \wedge \mathrm{d}t + \bar{\eta},
\end{aligned}$$

where $\bar{\eta}$ is not meaningful for the construction of the operator $\mathcal{A}$. For $\mathcal{A}\,\mathrm{d}\eta$ we have

$$\mathcal{A}\,\mathrm{d}\eta = \left[q^\nu \int_0^1 u\left(\frac{\partial A_\sigma}{\partial q^\nu} - \frac{\partial A_\nu}{\partial q^\sigma}\right) \circ \chi\,\mathrm{d}u + \dot{q}^\nu \int_0^1 u\left(\frac{\partial A_\sigma}{\partial \dot{q}^\nu} \circ \chi\right)\mathrm{d}u\right]\omega^\sigma \wedge \mathrm{d}t - \left[q^\sigma \int_0^1 u\left(\frac{\partial A_\sigma}{\partial \dot{q}^\nu} \circ \chi\right)\mathrm{d}u\right]\mathrm{d}\dot{q}^\nu \wedge \mathrm{d}t.$$

The sum $\mathrm{d}\mathcal{A}\eta + \mathcal{A}\,\mathrm{d}\eta$ is

$$\mathrm{d}\mathcal{A}\eta + \mathcal{A}\,\mathrm{d}\eta = \left\{\int_0^1 (A_\sigma \circ \chi)\,\mathrm{d}u + \int_0^1 u\left[q^\nu\left(\frac{\partial A_\sigma}{\partial q^\nu} \circ \chi\right) + \dot{q}^\nu\left(\frac{\partial A_\sigma}{\partial \dot{q}^\nu} \circ \chi\right)\right]\mathrm{d}u\right\}\omega^\sigma \wedge \mathrm{d}t.$$

Rewriting the first integral per partes we obtain

$$\int_0^1 (A_\sigma \circ \chi)\,\mathrm{d}u = u(A_\sigma \circ \chi)\Big|_0^1 - \int_0^1 u \frac{\mathrm{d}(A_\sigma \circ \chi)}{\mathrm{d}u}\,\mathrm{d}u$$

$$= A_\sigma - \int_0^1 u \left(\frac{\partial (A_\sigma \circ \chi)}{\partial (uq^\nu)} \frac{\mathrm{d}(uq^\nu)}{\mathrm{d}u} + \frac{\partial (A_\sigma \circ \chi)}{\partial (u\dot{q}^\nu)} \frac{\mathrm{d}(u\dot{q}^\nu)}{\mathrm{d}u} \right) \mathrm{d}u$$

$$= A_\sigma - \int_0^1 u \left(q^\nu \frac{\partial A_\sigma}{\partial q^\nu} \circ \chi + \dot{q}^\nu \frac{\partial A_\sigma}{\partial \dot{q}^\nu} \circ \chi \right) \mathrm{d}u.$$

Now we can immediately see that $\mathrm{d}\mathcal{A}\eta + \mathcal{A}\,\mathrm{d}\eta = \eta$. Thus, for a closed 1-contact 2-form $\eta = A_\sigma\, \omega^\sigma \wedge \mathrm{d}t$ on $J^1 Y$ holds

$$\eta = \mathrm{d}\varrho, \quad \text{where} \quad \varrho = \mathcal{A}\eta = \left[q^\sigma \int_0^1 (A_\sigma \circ \chi)\,\mathrm{d}u \right] \mathrm{d}t.$$

Construct the form ϱ via the operator $\mathcal{A}$ for a closed form $\eta = f(t)\varepsilon_{\sigma\alpha\beta} q^\alpha q^\beta\, \omega^\sigma \wedge \mathrm{d}t$ presented in the part (a) of the problem. Reproduce all steps of the proof for 2-contact 2-forms on $J^2 Y$ and for 1-contact and 2-contact 2-forms on $J^2 Y$. □

Solution of (c) The decomposition of the form η defined on $J^1 Y$ into its 1-contact and 2-contact components reads

$$\pi_{2,1}^* \eta = E_\sigma \omega^\sigma \wedge \mathrm{d}t + G_\sigma \dot{\omega}^\sigma \wedge \mathrm{d}t + F_{\sigma\nu} \omega^\sigma \wedge \omega^\nu + B_{\sigma\nu} \omega^\sigma \wedge \dot{\omega}^\nu + Q_{\sigma\nu} \dot{\omega}^\sigma \wedge \dot{\omega}^\nu,$$

$E_\sigma = A_\sigma + B_{\sigma\nu} \ddot{q}^\nu, G_\sigma = P_\sigma + 2Q_{\sigma\nu}\ddot{q}^\nu$. We do not differentiate between functions on $J^1 Y$ and their lifts on $J^2 Y$, i.e. we denote $\pi_{2,1}^* f = f \circ \pi_{2,1} \equiv f$ for a function f on $J^1 Y$. Using the mapping χ (3.25) and relations (3.26) we obtain

$$\begin{aligned}\pi_{2,1}^* \eta = {} & (E_\sigma \circ \chi)(q^\sigma\,\mathrm{d}u + u\omega^\sigma) \wedge \mathrm{d}t + (G_\sigma \circ \chi)(\dot{q}^\sigma\,\mathrm{d}u + u\dot{\omega}^\sigma) \wedge \mathrm{d}t \\ & + (F_{\sigma\nu} \circ \chi)(q^\sigma\,\mathrm{d}u + u\omega^\sigma) \wedge (q^\nu\,\mathrm{d}u + u\omega^\nu) \\ & + (B_{\sigma\nu} \circ \chi)(q^\sigma\,\mathrm{d}u + u\omega^\sigma) \wedge (\dot{q}^\nu\,\mathrm{d}u + u\dot{\omega}^\nu) \\ & + (Q_{\sigma\nu} \circ \chi)(\dot{q}^\sigma\,\mathrm{d}u + u\dot{\omega}^\sigma) \wedge (\dot{q}^\nu\,\mathrm{d}u + u\dot{\omega}^\nu) = \mathrm{d}u \wedge \mu + \bar{\eta},\end{aligned}$$

where $(F_{\sigma\nu} + F_{\nu\sigma} = 0,\ Q_{\sigma\nu} + Q_{\nu\sigma} = 0)$

$$\begin{aligned}\mu = {} & \big[(E_\sigma \circ \chi) q^\sigma + (G_\sigma \circ \chi)\dot{q}^\sigma\big]\mathrm{d}t + u\big[(2F_{\nu\sigma} \circ \chi) q^\nu - B_{\sigma\nu}\dot{q}^\nu\big]\omega^\sigma \\ & + u\big[(2Q_{\nu\sigma} \circ \chi)\dot{q}^\nu + (B_{\nu\sigma} \circ \chi) q^\nu\big]\dot{\omega}^\sigma, \\ \bar{\eta} = {} & u\big[(E_\sigma \circ \chi)\omega^\sigma \wedge \mathrm{d}t + (G_\sigma \circ \chi)\dot{\omega}^\sigma \wedge \mathrm{d}t\big] \\ & + u^2\big[(F_{\sigma\nu} \circ \chi)\omega^\sigma \wedge \omega^\nu + (B_{\sigma\nu} \circ \chi)\omega^\sigma \wedge \dot{\omega}^\nu + (Q_{\sigma\nu} \circ \chi)\dot{\omega}^\sigma \wedge \dot{\omega}^\nu\big].\end{aligned}$$

The form $\bar{\eta}$ is not relevant for the construction of the operator $\mathcal{A}$. We apply the operator $\mathcal{A}$ to the form $\pi_{2,1}^* \eta$ as follows:

$$\begin{aligned}\mathcal{A}(\pi_{2,1}^* \eta) = &\left[q^\sigma \int_0^1 (E_\sigma \circ \chi)\, \mathrm{d}u + \dot{q}^\sigma \int_0^1 (G_\sigma \circ \chi) \mathrm{d}u \right] \mathrm{d}t \\ &+ \left[2q^\nu \int_0^1 u (F_{\nu\sigma} \circ \chi)\, \mathrm{d}u - \dot{q}^\nu \int_0^1 u (B_{\sigma\nu} \circ \chi)\, \mathrm{d}u \right] \omega^\sigma \\ &+ \left[2\dot{q}^\nu \int_0^1 u (Q_{\nu\sigma} \circ \chi)\, \mathrm{d}u + q^\nu \int_0^1 u (B_{\nu\sigma} \circ \chi)\, \mathrm{d}u \right] \dot{\omega}^\sigma .\end{aligned} \tag{9.16}$$

(Prove that it holds $\pi_{2,1}^* \eta = \mathrm{d}\mathcal{A}(\pi_{2,1}^* \eta) + \mathcal{A}(\mathrm{d}\pi_{2,1}^* \eta)$.) We have obtained directly the decomposition of the 1-form $\mathcal{A}(\pi_{2,1}^* \eta)$ into its horizontal and contact components. This decomposition is formally given on $J^2 Y$. However, it is $\pi_{2,1}$-projectable. Indeed, putting into (9.16) $\dot{\omega}^\sigma = \mathrm{d}\dot{q}^\sigma - \ddot{q}^\sigma \, \mathrm{d}t$, and $E_\sigma = A_\sigma + B_{\sigma\nu} \ddot{q}^\nu$, $G_\sigma = P_\sigma + 2Q_{\sigma\nu} \ddot{q}^\nu$ we obtain

$$\begin{aligned}\mathcal{A}\eta = &\left[q^\sigma \int_0^1 (A_\sigma \circ \chi)\, \mathrm{d}u + \dot{q}^\sigma \int_0^1 (P_\sigma \circ \chi) \mathrm{d}u \right] \mathrm{d}t \\ &+ \left[2q^\nu \int_0^1 u (F_{\nu\sigma} \circ \chi)\, \mathrm{d}u - \dot{q}^\nu \int_0^1 u (B_{\sigma\nu} \circ \chi)\, \mathrm{d}u \right] \omega^\sigma \\ &+ \left[2\dot{q}^\nu \int_0^1 u (Q_{\nu\sigma} \circ \chi)\, \mathrm{d}u + q^\nu \int_0^1 u (B_{\nu\sigma} \circ \chi)\, \mathrm{d}u \right] \mathrm{d}\dot{q}^\sigma .\end{aligned} \tag{9.17}$$

Especially for the horizontal component of the form $\mathcal{A}\eta$ we have

$$h\mathcal{A}\eta = \left[q^\sigma \int_0^1 (E_\sigma \circ \chi)\, \mathrm{d}u + \dot{q}^\sigma \int_0^1 (G_\sigma \circ \chi) \mathrm{d}u \right] \mathrm{d}t .$$

If the form η is closed, then of course $\eta = \mathrm{d}\mathcal{A}\eta$. □

Problem 3.16

Solution Let α and $\bar{\alpha}$ be the arbitrary chosen Lepage 2-forms associated with E and Δ_α and $\Delta_{\bar{\alpha}}$ the corresponding dynamical distributions. Suppose that $\delta = J^1 \gamma$ is a holonomic integral section of the distribution Δ_α, i.e. using (3.31) and the definition property of contact forms, $J^1\gamma^*\omega^\sigma = 0$ we obtain the condition under what the section $J^1\gamma$ is an integral section of Δ_α,

$$J^1\gamma^*(A_\sigma\, \mathrm{d}t + B_{\sigma\nu}\, \mathrm{d}\dot{q}^\nu) = 0, \quad \text{for every} \quad 1 \le \sigma \le m.$$

This condition is the same for $\Delta_{\tilde{\alpha}}$. This finishes the proof of Proposition 3.7(a). Now suppose that $\delta = J^1\gamma$ is a holonomic integral section of Δ_α, i.e. $J^1\gamma^* i_{\tilde{\xi}}\alpha = 0$ for every π_1-vertical vector field $\tilde{\xi}$ on J^1Y. Denote $\hat{\xi}$ the lift of the vector field $\tilde{\xi}$ (recall that the vector fields $\hat{\xi}$ and $\tilde{\xi}$ have identical chart expressions). Then

$$J^2\gamma^*\pi_{2,1}^* i_{\tilde{\xi}}\alpha = J^1\gamma^* i_{\tilde{\xi}}\alpha = 0,$$

because of the fact that $J^1\gamma$ is an integral section of Δ_α. On the other hand,

$$J^2\gamma^*\pi_{2,1}^* i_{\tilde{\xi}}\alpha = J^2\gamma^* i_{\hat{\xi}}(p_1\alpha + p_2\alpha) = J^2\gamma^* i_{\hat{\xi}} E = (E_\sigma \xi^\sigma) \circ J^2\gamma\, \mathrm{d}t$$

for arbitrary choice of ξ^σ. This implies that $E_\sigma \circ J^2\gamma = 0$. □

Problem 3.29

Results

$$\frac{\partial A_1}{\partial \dot{q}^1} = \frac{\mathrm{d}' B_{11}}{\mathrm{d}t},$$

$$\varrho = A\eta = \left[q^1 \int_0^1 (A_1 \circ \chi)\, \mathrm{d}u \right] \mathrm{d}t - \left[\dot{q}^1 \int_0^1 u(B_{11} \circ \chi)\, \mathrm{d}u \right] \omega^1$$
$$+ \left[q^1 \int_0^1 u(B_{11} \circ \chi)\, \mathrm{d}u \right] \mathrm{d}\dot{q}^1,$$
$$\varrho = \frac{1}{2}k(q^1)^2\, \mathrm{d}t - \frac{1}{2}m\dot{q}^1\omega^1 + \frac{1}{2}mq^1\, \mathrm{d}\dot{q}^1,$$
$$\varrho = \frac{g}{l}(1 - \cos q^1)\, \mathrm{d}t - \frac{1}{2}\dot{q}^1\, \omega^1 + \frac{1}{2}q^1\, \mathrm{d}\dot{q}^1.$$

□

Problems of Chapter 4

Problem 4.1

Hint Take into account that the form $\pi_{2,1}^*\varrho$ defined on J^2Y is projectable on J^1Y.□

Problem 4.2

Solution Suppose that the opposite holds, i.e. there exists a point $t_0 \in (a, b)$ such that $f(t) \neq 0$. Without loss of generality we can suppose that $f(t_0) > 0$. Function

f is continuous at the point t_0, then there exists an interval (t_1, t_2) containing t_0 such that $f(t) > 0$ for $t \in (t_1, t_2)$. Choose the function $u(t)$ as follows:

$$u(t) = (t - t_1)(t_2 - t), \quad \text{for} \quad (t_1, t_2)$$

and zero on $[a, b] \backslash [t_1, t_2]$. This function is non-negative on $[a, b]$. Then taking into account the assumptions of the proposition we have

$$\int_{\Omega} f(t)u(t)\,\mathrm{d}t = \int_{t_1}^{t_2} f(t)(t - t_1)(t_2 - t)\,\mathrm{d}t = 0.$$

On the other hand, the integrand is a non-negative function on $[t_1, t_2]$. Moreover, it is nonzero on (t_1, t_2). Thus its integral in limits t_1 and t_2 is positive. This result is in contradiction with the starting assumptions. □

Problem 4.3

Hint Let γ be an extremal of the given Lagrange structure. Condition (c) of Theorem 4.2 holds. Lifting this condition on J^2Y and using the fact that $p_1\,\mathrm{d}\theta_\lambda = p_1\alpha$ we obtain relation (4.17). □

Problem 4.4

Hint The condition $E_\lambda = 0$ means that

$$\frac{\partial^2 L}{\partial \dot{q}^\sigma \partial \dot{q}^\nu} = 0, \qquad \frac{\partial L}{\partial q^\sigma} - \frac{\mathrm{d}'}{\mathrm{d}t}\frac{\partial L}{\partial \dot{q}^\sigma} = 0, \quad 1 \le \sigma, \nu \le m.$$

From the first set of these relations it follows that the Lagrange function is affine in variables $\dot{q}^\sigma$, i.e. $L = P(t, q^\nu) + Q_\sigma(t, q^\nu)\dot{q}^\sigma$. Using the second set of previous relations find the functions $P(t, q^\nu)$ and $Q_\sigma(t, q^\nu)$. □

Problem 4.5

Hint The element of the rotational surface can be expressed a $\mathrm{d}S = 2\pi q\,\mathrm{d}l = 2\pi q\sqrt{1 + (\dot{q}(t))^2}\,\mathrm{d}t$. Explain this fact (the situation is depicted in Fig. A.1). What is the underlying fibred manifold? Formulate the variational integral. What is the Lagrangian? Write its Lepage equivalent and Euler-Lagrange form. Solve the Euler-Lagrange equations for extremals. □

Problem 4.6

Solution Consider the fibred manifold $(\mathbf{R} \times \mathbf{R}^3, p, \mathbf{R})$ and its submanifold $(\mathbf{R} \times S^2, p, \mathbf{R})$. Let $(\mathbf{R} \times \mathbf{R}^3, \psi)$, $\psi = (t, x, y, z)$, $(\mathbf{R}, \varphi)$, $\phi = (t)$, $(J^1(\mathbf{R} \times \mathbf{R}^3), \psi_1)$, $\psi_1 = (t, x, y, z, \dot{x}, \dot{y}, \dot{z})$ be a fibred chart on $\mathbf{R} \times \mathbf{R}^3$, the associated chart on $\mathbf{R}$ and the associated fibred chart on $J^1(\mathbf{R} \times \mathbf{R}^3)$, respectively. Consider adapted fibred

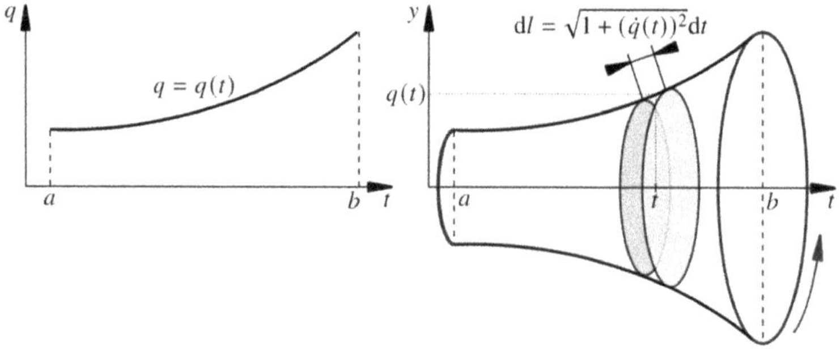

Fig. A.1 Problem of the minimal rotational surface

charts on S^2, i.e. use the spherical coordinates $q^1 = \vartheta$ and $q^2 = \varphi$ instead of the Cartesian ones, i.e. introduce the mapping

$$\mathcal{S} : \mathbf{R} \times \left[-\frac{\pi}{2}, \frac{\pi}{2}\right] \times [0, 2\pi] \ni (t, \vartheta, \varphi) \longrightarrow \mathcal{S}(t, \vartheta, \varphi) = (t, x, y, z) \in \mathbf{R}^3.$$

$$x = \cos\vartheta\cos\varphi, \quad x = \cos\vartheta\sin\varphi, \quad z = \sin\vartheta.$$

Every fibre $\mathbf{R}^3$ is the Euclidean space with metric $g = \mathrm{d}x \otimes \mathrm{d}x + \mathrm{d}y \otimes \mathrm{d}y + \mathrm{d}z \otimes \mathrm{d}z$. The induced metric on the sphere S^2 is (execute the needed calculations)

$$\tilde{g} = \mathcal{S}^* g = \mathrm{d}\vartheta \otimes \mathrm{d}\vartheta + \cos^2\vartheta\, \mathrm{d}\varphi \otimes \mathrm{d}\varphi.$$

The corresponding Christoffel symbols $\Gamma_{\lambda\nu\sigma}$, $1 \le \lambda, \nu, \sigma \le 2$, are (execute the calculations)

$$\Gamma_{122} = \Gamma_{212} = -\Gamma_{221} = \cos\vartheta\sin\vartheta,$$

the remaining are zero. The equations of geodesics are

$$\ddot{\vartheta} + \cos\vartheta\sin\vartheta\,\dot{\varphi}^2 = 0, \quad \ddot{\varphi} = 0.$$

Solve these equations. Consider φ as a function of ϑ (Example 4.2). Denote

$$\varphi' = \frac{\mathrm{d}\varphi}{\mathrm{d}\vartheta}, \quad \varphi'' = \frac{\mathrm{d}^2\varphi}{\mathrm{d}\vartheta^2}.$$

Then

$$\dot{\varphi} = \varphi'\dot{\vartheta}, \quad \ddot{\varphi} = \varphi''\dot{\vartheta}^2 + \varphi'\ddot{\vartheta}.$$

Prove that the equation

$$\frac{\mathrm{d}\varphi}{\mathrm{d}\vartheta} = \frac{K}{\cos\vartheta\sqrt{\cos^2\vartheta - K^2}}$$

derived in Example 4.2 is satisfied. □

Problem 4.7

Hint Construct the Euler-Lagrange forms of both Lagrangians. □

Problem 4.8

Hint Put the Lagrange function into the Noether equation (4.19) and express the total derivatives of components ξ^0 and ξ^i, $i = 1, 2, 3$, occurring in this equation through their partial derivatives with respect to variables t, x^1, x^2, x^3. You obtain the second degree polynomial in variables $\dot{x}^i$, $i = 1, 2, 3$, its coefficients being functions on Y. This polynomial is zero. Putting the coefficients equal to zero you obtain relatively simple equations for components of symmetries. □

Problem 4.9

Hint For example, from the relation $\frac{\partial \zeta^1}{\partial x^2} = -\frac{\partial \zeta^2}{\partial x^1}$ that both sides of this relation are independent of the variables x^1 and x^2 (the left-hand side does not depend on x^1 and the right-hand side does not depend on x^2). Thus

$$\frac{\partial \zeta^1}{\partial x^2} = -\frac{\partial \zeta^2}{\partial x^1} = A^1_2(x^3) \Longrightarrow \zeta^1 = A^1_2(x^3)x^2 + B^1(x^3),$$
$$\zeta^2 = -A^1_2(x^3)x^1 + B^2(x^3),$$

where $A^1_2(x^3)$, $B^1(x^3)$ and $B^2(x^3)$ are some (yet unknown) functions of the variable x^3. Apply the same procedure on other two antisymmetry relations. You obtain two relations for each of functions ζ^i. Comparing them you obtain the functions A^i_j and B^k. □

Problems of Chapter 5

Problem 5.1

Hint Use the properties of such dynamical forms and their dynamical systems presented at the end of Sect. 3.4, and follow the procedures proving the variationality of second-order dynamical forms. □

Problem 5.2

Hint Write down a general chart expression of a 2-form α on J^2Y using the basis of 1-forms adapted to the contact structure $(dt, \omega^\sigma, \dot{\omega}^\sigma, d\ddot{q}^\sigma)$, $1 \le \sigma \le m$, and decompose the form $\pi^*_{3,2}\alpha$ into its 1-contact and 2-contact components. The condition $p_1\alpha = \pi^*_{3,2}E$ and the requirement of projectability of this decomposition on J^2Y prove the assertion. □

Problem 5.3

Solution Taking into account expression (5.28) for functions F_σ we obtain

$$\frac{\partial F_\sigma}{\partial \dot{q}^\nu} = \Gamma_{\sigma\mu\beta}\left(\delta^\mu_\nu \dot{q}^\beta + \dot{q}^\mu \delta^\beta_\nu\right) + \phi_{\sigma\nu} = (\Gamma_{\sigma\nu\mu} + \Gamma_{\sigma\mu\nu})\dot{q}^\mu + \phi_{\sigma\nu} = 2\Gamma_{\sigma\nu\mu}\dot{q}^\mu + \phi_{\sigma\nu},$$
$$\frac{\partial F_\sigma}{\partial q^\nu} = \frac{\partial \Gamma_{\sigma\mu\beta}}{\partial q^\nu}\dot{q}^\mu \dot{q}^\beta + \frac{\partial \phi_{\sigma\mu}}{\partial q^\nu}\dot{q}^\mu + \frac{\partial \psi_\sigma}{\partial q^\nu}.$$

The first of conditions (5.27) then gives

$$\phi_{\sigma\nu} + \phi_{\nu\sigma} = 0, \tag{9.18}$$

and

$$(2\Gamma_{\sigma\nu\mu} + 2\Gamma_{\nu\sigma\mu})\dot{q}^\mu = -2m_0 \frac{\partial g_{\sigma\nu}}{\partial q^\mu}\dot{q}^\mu \implies$$

$$\Gamma_{\sigma\nu\mu} + \Gamma_{\nu\sigma\mu} = -m_0 \frac{\partial g_{\sigma\nu}}{\partial q^\mu}. \tag{9.19}$$

From the second condition in (5.27) we obtain

$$\left(\frac{\partial \Gamma_{\sigma\mu\beta}}{\partial q^\nu} - \frac{\partial \Gamma_{\nu\mu\beta}}{\partial q^\sigma}\right)\dot{q}^\mu \dot{q}^\beta + \left(\frac{\partial \phi_{\sigma\mu}}{\partial q^\nu} - \frac{\partial \phi_{\nu\mu}}{\partial q^\sigma}\right)\dot{q}^\mu + \left(\frac{\partial \psi_\sigma}{\partial q^\nu} - \frac{\partial \psi_\nu}{\partial q^\sigma}\right)$$
$$= \frac{1}{2}\frac{\mathrm{d}'}{\mathrm{d}t}\left[\left(2\Gamma_{\sigma\nu\mu} - 2\Gamma_{\nu\sigma\mu}\right)\dot{q}^\mu + 2\phi_{\sigma\nu}\right]$$
$$= \left[\frac{\partial \Gamma_{\sigma\nu\mu}}{\partial q^\beta} - \frac{\partial \Gamma_{\nu\sigma\mu}}{\partial q^\beta}\right]_{\mathrm{sym}\,(\beta,\mu)} \dot{q}^\beta \dot{q}^\mu$$
$$+ \left(\frac{\partial \Gamma_{\sigma\nu\mu}}{\partial t} - \frac{\partial \Gamma_{\nu\sigma\mu}}{\partial t} + \frac{\partial \phi_{\sigma\nu}}{\partial q^\mu}\right)\dot{q}^\mu + \frac{\partial \phi_{\sigma\nu}}{\partial t},$$

i.e.

$$\frac{\partial \Gamma_{\sigma\mu\beta}}{\partial q^\nu} - \frac{\partial \Gamma_{\nu\mu\beta}}{\partial q^\sigma} = \left[\frac{\partial \Gamma_{\sigma\nu\mu}}{\partial q^\beta} - \frac{\partial \Gamma_{\nu\sigma\mu}}{\partial q^\beta}\right]_{\mathrm{sym}\,(\beta,\mu)}, \tag{9.20}$$

$$\frac{\partial \phi_{\sigma\mu}}{\partial q^\nu} - \frac{\partial \phi_{\nu\mu}}{\partial q^\sigma} - \frac{\partial \phi_{\sigma\nu}}{\partial q^\mu} = \frac{\Gamma_{\sigma\nu\mu}}{\partial t} - \frac{\partial \Gamma_{\nu\sigma\mu}}{\partial t}, \tag{9.21}$$

$$\frac{\partial \psi_\sigma}{\partial q^\nu} - \frac{\partial \psi_\nu}{\partial q^\sigma} = \frac{\partial \phi_{\sigma\nu}}{\partial t}. \tag{9.22}$$

Using relation (5.28) for triples of indices (σ, μ, ν), (σ, ν, μ) and (ν, μ, σ) we obtain for the quantities $\Gamma_{\sigma\nu\mu}$, $1 \le \sigma, \nu, \mu \le m$, unique expressions

$$\Gamma_{\sigma\nu\mu} = -\frac{m_0}{2}\left(\frac{\partial g_{\sigma\mu}}{\partial q^\nu} + \frac{\partial g_{\sigma\nu}}{\partial q^\mu} - \frac{\partial g_{\nu\mu}}{\partial q^\sigma}\right).$$

We have obtained relations (5.29). Relations (9.20) are then satisfied. Moreover, it is evident from (9.19) that the quantities $\Gamma_{\sigma\nu\mu}$, are independent of the variable t. Then conditions (9.18), (9.21) and (9.22) for remaining yet uncertain functions $\phi_{\sigma\nu}$ and ψ_σ read

$$0 = \phi_{\sigma\nu} + \phi_{\nu\sigma},$$
$$0 = \frac{\partial \phi_{\sigma\mu}}{\partial q^\nu} + \frac{\partial \phi_{\mu\nu}}{\partial q^\sigma} + \frac{\partial \phi_{\nu\sigma}}{\partial q^\mu},$$
$$0 = \frac{\partial \psi_\sigma}{\partial q^\nu} - \frac{\partial \psi_\nu}{\partial q^\sigma} - \frac{\partial \phi_{\sigma\nu}}{\partial t},$$

which are conditions (5.30)–(5.32). □

Problem 5.4

Solution Suppose that the reference frame translates with the acceleration $\boldsymbol{a}$ and rotates with the angular velocity $\boldsymbol{\omega}$. Both these vectors are functions of time, in general. The complete expression including of all types of fictive (inertial) forces reads

$$\boldsymbol{F}^* = -m\boldsymbol{a} - 2\,m\boldsymbol{\omega} \times \boldsymbol{v} - m\boldsymbol{\omega} \times (\boldsymbol{\omega} \times \boldsymbol{r}) - m\boldsymbol{\varepsilon} \times \boldsymbol{r}, \tag{9.23}$$

where vectors $\boldsymbol{r} = (x, y, z)$ and $\boldsymbol{v} = (\dot{x}, \dot{y}, \dot{z})$ represent the position and velocity of the particle in the *non-inertial* reference frame, respectively, and $\boldsymbol{\varepsilon} = \dot{\boldsymbol{\omega}}$ is the angular acceleration of the reference frame. Individual summands in expression (9.23) are the *translational force*, *Coriolis force*, *centrifugal force* and *Euler force*. It is evident that the Coriolis force represents the term $\boldsymbol{v} \times \boldsymbol{B}$ in the general expression (5.34) for variational forces. So $\boldsymbol{B} = 2m\boldsymbol{\omega}$, i.e. rot $\boldsymbol{A} = 2m\boldsymbol{\omega}$. Suppose for simplicity that $\boldsymbol{\omega}$ is directed along the z-axis (measured by an observer in an *inertial* reference frame), i.e. $\boldsymbol{\omega} = (0, 0, \omega)$, $\omega = \omega(t)$. Then the vector $\boldsymbol{A}$ can be chosen as

$$\boldsymbol{A} = (-m\omega y, m\omega x, 0) + \text{grad}\, f$$

because the equation rot $\boldsymbol{A} = -2m\boldsymbol{\omega}$ allows the *calibration uncertainty* grad f of $\boldsymbol{A}$, where $f = f(t, \boldsymbol{r})$ is a function. Then $-\frac{\partial \boldsymbol{A}}{\partial t}$ is the Euler force together with the translational force, if we put

$$f = m\boldsymbol{a}\boldsymbol{r}.$$

Indeed, it holds

$$-\frac{\partial \boldsymbol{A}}{\partial t} = (m\dot{\omega}y - ma^1, -m\dot{\omega}x - ma^2, -ma^3) = (m\varepsilon y, -m\varepsilon x, 0)$$
$$= -m\boldsymbol{\varepsilon} \times \boldsymbol{r} - m\boldsymbol{a}.$$

The centrifugal force

$$-m\boldsymbol{\omega} \times (\boldsymbol{\omega} \times \boldsymbol{r}) = m(\omega^2 x, \omega^2 y, 0) = \text{grad}\left(\frac{1}{2}m\omega^2(x^2 + y^2)\right) + \text{const.},$$
$$V = -\frac{1}{2}\omega^2(x^2 + y^2) + \text{const.}$$

We have proved that all fictive forces occurring in non-inertial reference frame are variational. This means that for variationality of equations of motion of a particle or a system of particles the variationality of *real* forces is decisive. □

Problems of Chapter 6

Problem 6.1

Hint Note that a first-order trivial Lagrangian has the form $\lambda_0 = L_0 \, \mathrm{d}t$ where L_0 is the total derivative of a function $f = f(t, q^\sigma)$ with respect to the variable t and calculate the matrix $\mathcal{M}$ for $\tilde{L} = L + L_0$. □

Problem 6.2

Hint Using the chain rule for derivatives of composed functions calculate the derivatives

$$\frac{\partial H(t, q^\nu, p_\nu)}{\partial q^\sigma} = -\frac{\partial L\left(t, q^\nu, \dot{q}^\nu(t, q^\mu, p_\mu)\right)}{\partial q^\sigma} + \frac{p_\nu \dot{q}^\nu(t, q^\mu, p_\mu)}{\partial q^\sigma},$$
$$\frac{\partial H(t, q^\nu, p_\nu)}{\partial p_\sigma} = -\frac{\partial L\left(t, q^\nu, \dot{q}^\nu(t, q^\mu, p_\mu).\right)}{\partial p_\sigma} + \frac{p_\nu \dot{q}^\nu(t, q^\mu, p_\mu)}{\partial p_\sigma}.$$

□

Problem 6.3

Solution

$$\theta_\lambda = \sqrt{t^2 + x^2}\sqrt{1 + \dot{x}^2}\,\mathrm{d}t + \sqrt{t^2 + x^2}\frac{\dot{x}}{\sqrt{1 + \dot{x}^2}}(\mathrm{d}x - \dot{x}\,\mathrm{d}t),$$

$$E_\lambda = \sqrt{1 + \dot{x}^2}\frac{x}{\sqrt{t^2 + x^2}} - \frac{\mathrm{d}}{\mathrm{d}t}\left(\sqrt{t^2 + x^2}\frac{\dot{x}}{\sqrt{1 + \dot{x}^2}}\right)\mathrm{d}x \wedge \mathrm{d}t = \dots,$$

$$\frac{\partial^2 L}{\partial \dot{x}^2} = \frac{(t^2 + x^2)^{1/2}}{(1 + \dot{x}^2)^{3/2}} \neq 0,$$

$$p = \frac{\partial L}{\partial \dot{x}} = \sqrt{t^2 + x^2}\frac{\dot{x}}{\sqrt{1 + \dot{x}^2}} \implies \dot{x} = \frac{p}{\sqrt{t^2 + x^2 - p^2}}, \quad H = -\sqrt{t^2 + x^2 - p^2},$$

$$\frac{\mathrm{d}x}{\mathrm{d}t} = \frac{p}{\sqrt{t^2 + x^2 - p^2}}, \quad \frac{\mathrm{d}p}{\mathrm{d}t} = \frac{x}{\sqrt{t^2 + x^2 - p^2}} \implies \frac{\mathrm{d}}{\mathrm{d}t}(x^2 - p^2) = 0,$$

$$x^2 - p^2 = \text{const.}, \quad \text{denote this constant as } \pm K^2.$$

The last equation is the equation of the phase trajectory. Putting this result into Hamilton equations we obtain the system of first-order equations for functions $x(t)$ and $p(t)$ and then the second-order equation for each of these functions separately:

$$\dot{x} = \frac{p}{\sqrt{t^2 \pm K^2}}, \quad \dot{p} = \frac{x}{\sqrt{t^2 \pm K^2}},$$

$$\ddot{x}(t^2 \pm K^2) + t\dot{x} - x = 0, \quad \ddot{p}(t^2 \pm K^2) + t\dot{p} - p = 0.$$

□

Problem 6.7

Hint Use relations

$$\tilde{h}\, \mathrm{d}q_0^\sigma = q_{0,1}^\sigma\, \mathrm{d}t, \quad \tilde{h}\, \mathrm{d}p_\sigma^0 = p_\sigma^{0,1}\, \mathrm{d}t.$$

□

Problem 6.8

Hint The problem (a) is solved by Helmholtz conditions and by the fact that $E = E_\lambda$ for a certain first-order Lagrangian λ. For solving (b) take into account that $E = p_1\alpha$. For (c) use the fact that $\Delta_\alpha^0 = \{\mathrm{span}\, i_\zeta\alpha\}$ where ζ is a an arbitrary π-vertical vector field on Y. For a vector field ξ generating Δ_α it holds $i_\xi(i_\zeta\alpha) = 0$. □

Problem 6.12

Hint Take into account that the annihilator of the dynamical distribution in the case of the first-order Euler-Lagrange form is given by relation (6.26). Consider a vector field $\zeta = \zeta^0 \frac{\partial}{\partial t} + \zeta^\sigma \frac{\partial}{\partial q_0^\sigma}$ and the definition relation $i_\zeta\eta = 0$ for $\eta \in \Delta_\alpha^0$. Remember that the matrix $(B_{\sigma\nu})$, $1 \leq \sigma, \nu \leq m$ is regular. □

Solution Denoting as $(\beta^{\sigma\nu}) = (B_{\sigma\nu})^{-1}$ we obtain

$$\Delta_\alpha = \mathrm{span}\left\{\frac{\partial}{\partial t} - \beta^{\sigma\nu} A_\nu \frac{\partial}{\partial q_0^\sigma}\right\}.$$

□

Problem 6.13

Hint For both parts of the problem use direct calculations in charts. □

Problem 6.14

Solution The Euler-Lagrange form of the given Lagrangian is

$$E_\lambda = (-m\ddot{x} - b\dot{y} - kx)\, \mathrm{d}x \wedge \mathrm{d}t + (-m\ddot{y} + b\dot{x} - ky)\, \mathrm{d}y \wedge \mathrm{d}t.$$

For solving the Euler-Lagrange equations

$$m\ddot{x} + b\dot{y} + kx = 0, \quad m\ddot{y} - b\dot{x} + ky = 0$$

we chose the substitution $u = x + \mathrm{i}y$ and obtain

$$m\ddot{u} - \mathrm{i}b\dot{u} + ku = 0 \implies u(t) = C_1 \exp\left[\mathrm{i}\left(\delta + \sqrt{\delta^2 + \omega^2}\right)t\right] + C_2 \exp\left[\mathrm{i}\left(\delta - \sqrt{\delta^2 + \omega^2}\right)t\right],$$

where $\delta = \frac{b}{2m}$, $\omega^2 = \frac{k}{m}$ and C_1 and C_2 are complex constants depending on initial conditions, e.g. $x(0) = x_0$, $y(0) = y_0$, $\dot{x}(0) = v_{0x}$, $\dot{y}(0) = v_{0y}$.

Let us now solve this problem within the framework of Hamilton theory. The Lagrange function L (Example 6.6) has the form

$$\lambda = L(t, x, y, \dot{x}, \dot{y})\,\mathrm{d}t, \quad L = \frac{m}{2}(\dot{x}^2 + \dot{y}^2) + \frac{b}{2}(\dot{x}y - x\dot{y}) - \frac{k}{2}(x^2 + y^2),$$

with positive constants m, b and k. (The corresponding first- order Lagrange structure is regular according to (6.1).) The standard Legendre transformation then reads

$$\mathcal{L} : (t, x, y, \dot{x}, \dot{y}) \longrightarrow (t, x, y, p_x, p_y).$$

Then the generalized momenta and the Hamilton function are

$$p_x = \frac{\partial L}{\partial \dot{x}} = m\dot{x} + \frac{b}{2}y, \quad p_y = \frac{\partial L}{\partial \dot{y}} = m\dot{y} - \frac{b}{2}x$$

$$\Longrightarrow \dot{x} = \frac{p_x}{m} - \frac{b}{2m}y, \quad \dot{y} = \frac{p_y}{m} + \frac{b}{2m}x,$$

$$H = -L + p_x\dot{x} + p_y\dot{y} = \frac{1}{2m}(p_x^2 + p_y^2) + \left(\frac{b^2}{8m} + \frac{k}{2}\right)(x^2 + y^2)$$
$$+ \frac{b}{m}(xp_y - yp_x).$$

The Hamilton equations are

$$\frac{\mathrm{d}x}{\mathrm{d}t} = \frac{\partial H}{\partial p_x} \Longrightarrow \frac{\mathrm{d}x}{\mathrm{d}t} - \frac{p_x}{m} + \frac{b}{m}y = 0,$$
$$\frac{\mathrm{d}y}{\mathrm{d}t} = \frac{\partial H}{\partial p_y} \Longrightarrow \frac{\mathrm{d}y}{\mathrm{d}t} - \frac{p_y}{m} - \frac{b}{m}x = 0,$$
$$\frac{\mathrm{d}p_x}{\mathrm{d}t} = -\frac{\partial H}{\partial x} \Longrightarrow \frac{\mathrm{d}p_x}{\mathrm{d}t} + \frac{b}{2m}p_y + \left(\frac{b^2}{4m} + k\right)x = 0,$$
$$\frac{\mathrm{d}p_y}{\mathrm{d}t} = -\frac{\partial H}{\partial y} \Longrightarrow \frac{\mathrm{d}p_y}{\mathrm{d}t} - \frac{b}{2m}x + \left(\frac{b^2}{4m} + k\right)y = 0.$$

We have obtained Hamilton equations—four first-order differential equations for functions $x(t)$, $y(t)$, $p_x(t)$ and $p_y(t)$ instead of Euler-Lagrange equations—two second-order equations for $x(t)$ and $y(t)$. For solving the obtained Hamilton equations we can use, e.g. the substitution $u = x + \mathrm{i}y$, $v = p_x + \mathrm{i}p_y$. We leave to the reader the complete solution of Hamilton equations and comparison of the result with the solution of Euler-Lagrange equations. □

Problem 6.16

Result

$$\frac{\partial S}{\partial t} + \frac{1}{2m}\left(\frac{\partial S}{\partial x}\right)^2 + \frac{k}{2}x^2 = 0.$$

□

Problem 6.17

Hint With regard to the form of both sides of the equation, assume the function $S = S(t, x, u)$ to be a quadratic form $S = c_1 t^2 + c_2 tx + c_3 x^2$ with unknown coefficients c_1, c_2 and c_3 and determine them by putting them into the equation. □

Problems of Chapter 7

Problem 7.1

Hint Prove that $I(\mathrm{d}\eta) = I(p_1\, \mathrm{d}\eta)$ by the direct calculation. For general argumentation use the properties of the variational sequence expressed i.a. by diagrams on Figs. 7.2 and 7.3. □

Problem 7.2

Hint Express the form η in coordinates adapted to the contact structure and then proceed by direct calculation using Definition 7.4 of the interior Euler operator. □

Problems of Chapter 8

Problem 8.2

Hint For (a) use the proof by contradiction: Suppose that there exist two projections α_{01} and α_{02} of α. Using (8.5) prove that $\alpha_{01} = \alpha_{02}$. For (b) and (c) use again relation (8.5) for arbitrarily chosen points. For example, for (b) $y_1, y_2 \in Y$, such that $x = \pi(y_1) = \pi(y_2)$, express $\bar{y}_1 = \alpha(y_1)$, $\bar{y}_2 = \alpha(y_2)$, and with help of (8.5) prove that

$$\bar{\pi}(\bar{y}_1) = \bar{\pi}(\bar{y}_2) = \bar{x} = \alpha_0(x).$$

□

Problem 8.3

Hint Choose charts (V, ψ), $\psi = (x^i, y^\sigma)$, $(\bar{V}, \bar{\psi})$, $\bar{\psi} = (\bar{x}^i, \bar{y}^\sigma)$, $V \cap \bar{V} \neq \emptyset$, on Y and corresponding associated charts (U, φ), $\varphi = (x^i)$, $(\bar{U}, \bar{\varphi})$, $\bar{\varphi} = (\bar{x}^i)$, on X, $1 \le i \le n$, $1 \le \sigma \le m$. For a section $\gamma \in \Gamma_{U \cap \bar{U}}(\pi)$, and $x \in U \cap \bar{U}$, where

$$\begin{aligned}\gamma(x) &= \left(x^1, \dots, x^n, y^1\gamma(x^1, \dots, x^n), \dots, y^m\gamma(x^1, \dots, x^n)\right)\\ &= \left(\bar{x}^1, \dots, \bar{x}^n, \bar{y}^1\gamma(\bar{x}^1, \dots, \bar{x}^n), \dots, \bar{y}^m\gamma(\bar{x}^1, \dots, \bar{x}^n)\right),\end{aligned}$$

shortly

$$\gamma(x) = \left(x^i, y^\sigma \gamma(x^i)\right) = \left(\bar{x}^i, \bar{y}^\sigma \gamma(\bar{x}^i)\right),$$

express by the direct calculation (chain rule for derivatives) the derivatives

$$\frac{\partial \bar{y}^\sigma \gamma(\bar{x}^1, \ldots, \bar{x}^n)}{\partial \bar{x}^j} \quad \text{through the derivatives} \quad \frac{\partial y^\nu \gamma(x^1, \ldots, x^n)}{\partial x^\ell}$$

and transformation relations. Show that if

$$\frac{\partial y^\sigma \gamma_1(x^i)}{\partial x^j} = \frac{\partial y^\sigma \gamma_2(x^i)}{\partial x^j}$$

for all σ, $1 \le \sigma \le m$, and all j, $1 \le j \le n$, then the same holds in charts $(\bar{V}, \bar{\psi})$, $(\bar{U}, \bar{\varphi})$ □

Problem 8.4

Hint Suppose that $\gamma \in \Gamma_U(\pi)$. Using the definition of $J^r\gamma$ prove that $\pi_r \circ J^r\gamma = \mathrm{id}_U, U \subset X$. □

Problem 8.6

Hint Let (V, ψ), $\psi = (x^i, y^\sigma)$, be a fibred chart on Y such that $W \subset V$, (U, φ), $U = \pi(V)$, $\varphi = (x^i)$, the associated chart on X and (V_r, ψ_r), $\psi_r = (x^i, y^\sigma, y^\sigma_j, \ldots, y^\sigma_{j_1 \ldots j_r})$ the associated fibred chart on J^rY, $1 \le i, j, j_1 \ldots, j_r \le n$, $1 \le \sigma \le m$, and modify the calculations in the proof of Proposition 2.2. □

Results Chart mapping of $(J^r\alpha, \alpha_0)$ for $r = 2$:

$$\begin{aligned}
&\psi_2 \circ J^2\alpha \circ \psi_2^{-1} : \psi_2\left(\pi_{2,0}^{-1}(W)\right) \ni (x^j, y^\nu, y^\nu_k, y^\sigma_{k\ell}) \\
&\quad \longrightarrow \psi_2 \circ J^2\alpha \circ \psi_2^{-1}(x^j, y^\nu, y^\nu_k, y^\sigma_{k\ell}) \\
&= \Big(x^i(\psi_2 \circ J^2\alpha \circ \psi_2^{-1})(x^j, y^\nu, y^\nu_k, y^\nu_{k\ell}), y^\sigma(\psi_2 \circ J^2\alpha \circ \psi_2^{-1})(x^j, y^\nu, y^\nu_k, y^\nu_{k\ell}), \\
&\quad y^\sigma_p(\psi_2 \circ J^2\alpha \circ \psi_2^{-1})(x^j, y^\nu, y^\nu_k, y^\nu_{k\ell}), y^\sigma_{pq}(\psi_2 \circ J^2\alpha \circ \psi_2^{-1})(x^j, y^\nu, y^\nu_k, y^\nu_{k\ell}),\Big).
\end{aligned}$$

□

Problem 8.7

Hint Follow the procedure similar to that presented in Problem 2.3.

- Define a smooth parametrized curves as the mappings

$$\begin{aligned}
&(-\varepsilon, \varepsilon) \ni s \to C_x(s) \in X \text{ for } x \in X,\ (-\varepsilon, \varepsilon) \ni s \\
&\quad \to C_y(s) \in Y \text{ for } y \in Y, \text{ etc.}
\end{aligned}$$

- For curves, e.g. $C_{x,1}$ and $C_{x,2}$ going through the point $x \in X$ define their equivalence:

$$\frac{\mathrm{d}x^i\varphi\big(C_{x,1}(s)\big)}{\mathrm{d}s} = \frac{\mathrm{d}x^i\varphi\big(C_{x,2}(s)\big)}{\mathrm{d}s}.$$

  Similarly for Y, etc.
- On the set of classes, e.g. $TY = \cup_{y\in Y} T_y Y$ of equivalent curves on Y introduce the vector space structure, i.e. define $\big[\xi(y)\big] + \big[\zeta(y)\big]$ and $\alpha\big[\xi(y)\big] \in T_y Y, \alpha \in \mathbf{R}$.
- Show that (TY, Π, Y), $\Pi : TY \supset T_y Y \ni \big[\xi(y)\big] \to y \in Y$ is the fibred manifold. (Similarly for X, J^1Y, etc.)

□

Problem 8.8

Hint For a vector field ξ on Y, $\xi = \xi^i(x^j, y^\nu)\frac{\partial}{\partial x^i} + \Xi^\sigma(x^j, y^\nu)\frac{\partial}{\partial y^\sigma}$ defined on an open set $W \subset Y$ verify the condition $\Pi \circ \xi = \mathrm{id}_W$. □

Problem 8.10

Solution Let (V, ψ), $\psi = (x^j, y^\nu)$, be a fibred chart on Y such that $W \subset V$, $\big(\pi(V), \varphi\big)$, $\varphi = (x^j)$, the associated chart on X and $\big(\pi^{-1}(V), \psi_1\big)$, $\psi_1 = (x^j, y^\nu, y^\nu_j)$, $1 \le j \le n$, $1 \le \nu \le m$, the associated fibred chart on J^1Y. Following (8.10) we obtain

$$J^1\xi = \xi^i\frac{\partial}{\partial x^i} + \Xi^\sigma\frac{\partial}{\partial y^\sigma} + \left(\left.\frac{\mathrm{d}y^\sigma_i J^1\alpha_u}{du}\right|_{u=0}\right)\frac{\partial}{\partial y^\sigma_i},$$

$$\Xi^\sigma_i = \left.\frac{\mathrm{d}y^\sigma_i(\psi_1 \circ J^1\alpha_u \circ \psi_1^{-1})(x^j, y^\nu)}{du}\right|_{u=0}.$$

Using (8.11) we have

$$\begin{aligned}\Xi^\sigma_1(J^1_x\gamma) &= \frac{\mathrm{d}}{du}\left[y^\sigma_j J^1_{\alpha_{0u}(x)}(\alpha_u\gamma\alpha_{0u}^{-1})\right]_{u=0}\\ &= \frac{\mathrm{d}}{du}\left[y^\sigma_j\left(J^1\alpha_u \circ J^1\gamma \circ \alpha_{0u}^{-1}\right)\left(\alpha_{0u}(x)\right)\right]_{u=0}.\end{aligned}$$

Now use the chain rule for derivatives. Denote for brevity $x = (x^1, \dots x^n)$, $x^i_u = x^i\alpha_{0u}(x)$, $1 \le i \le n$, i.e. $x_u = (x^1_u, \dots, x^n_u)$, and $x^i\alpha_{0u}^{-1}(x^1_u, \dots, x^n_u) = x^i_0 = x^i$. For $1 \le j \le n$ calculate the coordinate y^σ_j of the mapping $J^1\alpha_u \circ J^1\gamma \circ \alpha_{0u}^{-1}$ at the point

$\alpha_{0u}(x)$, i.e. the expression $y^\sigma_j\left(J^1\alpha_u\circ J^1\gamma\circ\alpha_{0u}^{-1}\right)(\alpha_{0u}(x))$.

$$
\begin{aligned}
y^\sigma_j\left(J^1\alpha_u\circ J^1\gamma\circ\alpha_{0u}^{-1}\right)(\alpha_{0u}(x)) &= y^\sigma_j\left(J^1\alpha_u\circ J^1\gamma\circ\alpha_{0u}^{-1}\right)(x_u)\\
&= \frac{\mathrm{d}}{\mathrm{d}x^j_u}\left[y^\sigma\left(J^1\alpha_u\circ J^1\gamma\circ\alpha_{0u}^{-1}\right)\right]_{\alpha_{0u}(x)}\\
&= \left.\frac{\mathrm{d}y^\sigma(J^1\alpha_u\circ J^1\gamma)(x)}{\mathrm{d}x^\ell}\right|_x\cdot\left.\frac{\mathrm{d}x^\ell\alpha_{0u}^{-1}(x_u)}{\mathrm{d}x^j_u}\right|_{\alpha_{0u}(x)}\\
&= \left.\frac{\mathrm{d}y^\sigma(\alpha_u\circ\gamma)}{\mathrm{d}x^\ell}\right|_x\cdot\left.\frac{\mathrm{d}x^\ell\alpha_{0u}^{-1}(x_u)}{\mathrm{d}x^j_u}\right|_{\alpha_{0u}(x)}.
\end{aligned}
$$

(We used the fact that $y^\sigma(J^1\alpha_u\circ J^1\gamma)=y^\sigma(\alpha_u\circ\gamma)$.) The above obtained expression for $y^\sigma_j\left(J^1\alpha_u\circ J^1\gamma\circ\alpha_{0u}^{-1}\right)(\alpha_{0u}(x))$ now derive with respect to u and put $u=0$, taking into account the following facts:

- derivatives with respect to u a x^j are changeable, i.e. $\partial/\partial u\partial x^j=\partial/\partial x^j\partial u$,
- the section γ is arbitrary, i.e. the calculation is valid for every section γ for which $z=J^1_x\gamma\in J^1Y$,
- it holds $\alpha_{0u}^{-1}=\alpha_{(0,-u)}$ (one of properties of the one-parameter group),
- $\left.\frac{\mathrm{d}x^j\alpha_{0u}^{-1}(x)}{\mathrm{d}u}\right|_{u=0}=\left.\frac{\mathrm{d}x^j\alpha_{(0,-u)}(x)}{\mathrm{d}u}\right|_{u=0}=-\left.\frac{\mathrm{d}x^j\alpha_{0u}(x)}{\mathrm{d}u}\right|_{u=0}=-\xi^0(x)$,
- $\left.\frac{\mathrm{d}y^\sigma(\alpha_u\gamma)(x)}{\mathrm{d}u}\right|_{u=0}=\Xi^\sigma(\gamma(x))$,

We finally obtain

$$
\begin{aligned}
\Xi^\sigma_1 &= \left[\frac{\mathrm{d}}{\mathrm{d}u}\left(\left.\frac{\mathrm{d}y^\sigma(\alpha_u\gamma)(x)}{\mathrm{d}x^\ell}\right|_x\cdot\left.\frac{\mathrm{d}x^\ell\alpha_{0u}^{-1}(x_u)}{\mathrm{d}x^j_u}\right|_{\alpha_{0u}(x)}\right)\right]_{u=0}\\
&= \left[\frac{\mathrm{d}}{\mathrm{d}u}\frac{\mathrm{d}y^\sigma(\alpha_u\gamma)(x)}{\mathrm{d}x^\ell}\right]_{u=0}\cdot\left[\left.\frac{\mathrm{d}x^\ell\alpha_{0u}^{-1}}{\mathrm{d}x^j}\right|_{\alpha_{0u}(x)}\right]_{u=0}\\
&\quad+\left[\frac{\mathrm{d}y^\sigma(\alpha_u\gamma)(x)}{\mathrm{d}x^\ell}\right]_{u=0}\cdot\left[\frac{\mathrm{d}}{\mathrm{d}u}\frac{\mathrm{d}x^\ell\alpha_{0u}^{-1}(x_u)}{\mathrm{d}x^j}\right]_{u=0}\\
&= \frac{\mathrm{d}}{\mathrm{d}x^j}\left[\frac{\mathrm{d}y^\sigma(\alpha_u\gamma)(x)}{\mathrm{d}u}\right]_{u=0}-y^\sigma_\ell\gamma(x)\cdot\frac{\mathrm{d}}{\mathrm{d}x^\ell}\left[\frac{\mathrm{d}x^j\alpha_{0u}(x)}{\mathrm{d}u}\right]_{u=0}.
\end{aligned}
$$

The obtained result is valid for an arbitrary section γ, i.e.

$$\Xi_j^\sigma = \frac{\mathrm{d}\Xi^\sigma}{\mathrm{d}x^j} - y_\ell^\sigma \frac{\mathrm{d}\xi^\ell}{\mathrm{d}x^j}.$$

□

Problem 8.17

Hint Express all 1-forms of the type $\mathrm{d}y_J^\sigma$ with help of the basis (8.17) of 1-forms adapted to the contact structure and calculate the form $\pi_{r+1,r}^*$. □

Problem 8.18

Hint Express the form $\pi_{r+1,r}^* \mathrm{d}f$ in the basis of 1 forms adapted to the contact structure, Eq. (8.17). □

Problem 8.19

Hint For a general term

$$A_{\sigma_1 \ldots \sigma_p, i_{p+1} \ldots i_k}^{J_1 \ldots j_p} \mathrm{d}y_{J_1}^{\sigma_1} \wedge \ldots \wedge \mathrm{d}y_{J_p}^{\sigma_p} \wedge \mathrm{d}x^{i_{p+1}} \wedge \ldots \wedge \mathrm{d}x^{i_k}$$

of the chart expression (8.14) calculate the pullback by the mapping χ and for $\chi^* \mathrm{d}x^{i_\ell}$ and $\chi^* \mathrm{d}y_{J_s}^{\sigma_s}$ put expressions (8.21). □

Problem 8.20

Solution Taking into account the mapping χ, defined generally by relation (8.20), for the situation specified in the problem we have

$$\begin{aligned} \varrho = \mathcal{A}\eta &= y^1 \left(\int_{u=0}^{u=1} u(y_{22}^1 - 2y_1^2)\,\mathrm{d}u \right) \omega_0 + y^2 \left(\int_{u=0}^{u=1} u(y_{22}^2 + 2y_1^1)\,\mathrm{d}u \right) \omega_0 \\ &= \left[\frac{1}{2}(y^1 y_{22}^1 + y^2 y_{22}^2) - (y^1 y_1^2 - y^2 y_1^1) \right] \omega_0. \end{aligned}$$

Calculate the form $\tilde{\eta}$ for completeness (even though this form is not relevant for application of Poincaré lemma). □

Problem 8.21

Solution On (nonempty) intersections of charts it holds $\bar{L}\bar{\omega}_0 = L\omega_0$, i.e.

$$\bar{L}(\bar{x}^i, \bar{y}^\sigma, \ldots, \bar{y}_{j_1 \ldots j_r}^\sigma)\,\mathrm{d}\bar{x}^1 \wedge \ldots \wedge \mathrm{d}\bar{x}^n = L(x^j, y^\nu, \ldots, y_{\ell_1 \ldots \ell_r}^\nu)\,\mathrm{d}x^1 \wedge \ldots \wedge \mathrm{d}x^n,$$

$$\omega_0 = \mathrm{d}x^1 \wedge \ldots \wedge \mathrm{d}x^n = \frac{\partial x^1}{\partial \bar{x}^{j_1}} \mathrm{d}\bar{x}^{j_1} \wedge \ldots \wedge \frac{\partial x^n}{\partial \bar{x}^{j_n}} \mathrm{d}\bar{x}^{j_n} = \det D(\varphi \circ \bar{\varphi}^{-1}) \bar{\omega}_0$$

$$\Longrightarrow \bar{L} = L \det D(\varphi \circ \bar{\varphi}^{-1}).$$

□

Problem 8.23

Hint The difference of given Lagrangians is

$$\begin{aligned} \tilde{\lambda} - \pi^*_{2,1}\lambda = \Bigg\{ & \left[\frac{1}{2}(y^1 y^1_{22} + y^2 y^2_{22}) - (y^1 y^2_1 - y^2 y^1_1) \right] \\ & - \frac{1}{2}\left[(y^1_1)^2 + (y^2_1)^2 \right] - (y^1 y^2_1 - y^2 y^1_1) \Bigg\} \, \mathrm{d}x^1 \wedge \mathrm{d}x^2. \end{aligned}$$

□

Problem 8.24

Solution Consider the structures specified in Theorem 8.4. The basis for the proof of a) is given by the first variational Formula (8.27) and the definition of extremal (8.30). So, for a π-vertical vector field ξ we have

$$\int_\Omega J^{2r-1}\gamma^* i_{J^{2r-1}\xi} \, \mathrm{d}\varrho = 0 \text{ for every admissible } \Omega \implies J^{2r-1}\gamma^* i_{J^{2r-1}\xi} \, \mathrm{d}\varrho = 0.$$

As a immediate consequence we obtain

$$J^{2r}\gamma^* i_{J^{2r}\xi} E_\lambda = 0$$

for every π-vertical vector field ξ. Now suppose that

$$\xi = \xi^i \frac{\partial}{\partial x^i} + \Xi^\sigma \frac{\partial}{\partial y^\sigma}$$

is an arbitrary π- projectable vector field on Y, i.e. without the request of verticality. We obtain step by step

$$\begin{aligned} i_{J^{2r}\xi} E_\lambda &= E_\sigma (\Xi^\sigma - y^\sigma_i \xi^i) \, \omega_0 - (E_\sigma \xi^i) \omega^\sigma \wedge \omega_i, \\ J^{2r}\gamma^* \, i_{J^{2r}\xi} E_\lambda &= \left[E_\sigma (\Xi^\sigma - y^\sigma_i \xi^i) \right] \circ J^{2r}\gamma \; J^{2r}\gamma^* \omega_0 \\ &\quad - \left[(E_\sigma \xi^i) \circ J^{2r}\gamma \right] J^{2r}\gamma^* \omega^\sigma \wedge J^{2r}\gamma^* \omega_i \\ &= \left[E_\sigma (\Xi^\sigma - y^\sigma_i \xi^i) \right] \circ J^{2r}\gamma \; \omega_0. \end{aligned}$$

As it holds $J^{2r}\gamma^* i_{J^{2r}\xi} E_\lambda = 0$ for every π-vertical vector field ξ, then $E_\sigma \circ J^{2r}\gamma = 0$ for $1 \leq \sigma \leq m$ (components of the Euler-Lagrange form vanish along the prolonged extremal regardless of the requirements for the vector field $\tilde{\xi}$). Then $\tilde{\xi}$ does not have to be limited to a π-projectable vector field. It can be an arbitrary vector field on $J^{2r}Y$. □

Problem 8.25

Hint Performing the successive calculations of the Lie derivative

$$\partial_{J^2r\xi}(E_\sigma\omega^\sigma \wedge \omega_0) = i_{J^{2r}\xi}\,\mathrm{d}(E_\sigma\omega^\sigma \wedge \omega_0) + \mathrm{d}\left[i_{J^{2r}\xi}(E_\sigma\omega^\sigma \wedge \omega_0)\right]$$

remember that

$$\mathrm{d}\omega_0 = 0, \quad \mathrm{d}\omega_i = 0, \quad \mathrm{d}x^i \wedge \omega_0 = 0, \quad \mathrm{d}x^i \wedge \omega_i = \omega_0.$$

$$\mathrm{d}\omega^\sigma = -\mathrm{d}y_i^\sigma \wedge \mathrm{d}x^i \implies i_{J^{2r}\xi}\,\mathrm{d}\omega^\sigma = -\Xi_i^\sigma\,\mathrm{d}x^i + \xi^i\,\mathrm{d}y_i^\sigma.$$

Thanks to these relationships, a number of terms disappear in the detailed calculation. □

Problem 8.27

Solution The Euler-Lagrange form is given by the left-hand side of the wave equation. (This equation was obtained directly from the Lagrangian presented in Example 8.9, without additional modifications.) The underlying fibred manifold is $(Y, \pi, X) = (\mathbf{R}^4 \times \mathbf{R}, p, \mathbf{R}^4)$, The Euler-Lagrange form is of the second order, affine in higher coordinates on fibres,

$$\begin{aligned} E_\lambda &= -\left(y_{11}^1 - y_{22}^1 - y_{33}^1 - y_{44}^1\right)\omega^1 \wedge \omega_0, \\ \mathrm{d}E_1 &= -\mathrm{d}y_{11}^1 - \mathrm{d}y_{22}^1 - \mathrm{d}y_{33}^1 - \mathrm{d}y_{44}^1, \\ i_{J^2\xi}\,\mathrm{d}E_1 &= -\Xi_{11}^1 - \Xi_{22}^1 - \Xi_{33}^1 - \Xi_{44}^1, \end{aligned}$$

and Eq. (8.39) obtains the following form:

$$\left(\Xi_{11}^1 - \Xi_{22}^1 - \Xi_{33}^1 - \Xi_{44}^1\right) + \left(y_{11}^1 - y_{22}^1 - y_{33}^1 - y_{44}^1\right)\left(\frac{\partial \Xi^1}{\partial y^1} + \sum_{i=1}^{4}\frac{\mathrm{d}\xi^i}{\mathrm{d}x^i}\right) = 0.$$

For components Ξ_{ii}^1 we can use relations (8.12),

$$\Xi_{ii}^1 = \frac{\mathrm{d}\Xi_i^1}{\mathrm{d}x^i} - y_{ij}^1\frac{\partial \xi^j}{\partial x^i} = \frac{\partial \Xi_i^1}{\partial x^i} + \frac{\partial \Xi_i^1}{\partial y^1}y_i^1 + \frac{\partial \Xi^1}{\partial y_j^1}y_{ji}^1 - y_{ij}^1\frac{\partial \xi^j}{\partial x^i},$$

(without summation over the index i), where

$$\Xi_i^1 = \frac{\mathrm{d}\Xi^1}{\mathrm{d}x^i} - y_j^1\frac{\partial \xi^j}{\partial x^i} = \frac{\partial \Xi^1}{\partial x^i} + \frac{\partial \Xi^1}{\partial y^1}y_i^1 - y_j^1\frac{\partial \xi^j}{\partial x^i}.$$

It is clear that the complete solution of symmetries of the Euler-Lagrange form is difficult even in such a simple situation as get the problem of propagation of waves. (In Chap. 9 only the problem of Noether symmetries and currents will be solved.) □

Problem 8.28

Result

$$\frac{\partial E_\sigma}{\partial q_l^\nu} - (-1)^l \frac{\partial E_\nu}{\partial q_l^\sigma} - \sum_{j=l+1}^{s} (-1)^j \binom{j}{l} \frac{\mathrm{d}^{j-l}}{\mathrm{d}t^{j-l}} \left(\frac{\partial E_\nu}{\partial q_j^\sigma} \right) = 0, \quad 0 \le l \le s.$$

□

Problem 8.29

Solution The Helmholtz conditions for its variationality read:

$$\frac{\partial E_\sigma}{\partial y_{ij}^\nu} - \frac{\partial E_\nu}{\partial y_{ij}^\sigma} = 0,$$

$$\frac{\partial E_\sigma}{\partial y_i^\nu} + \frac{\partial E_\nu}{\partial y_i^\sigma} - 2\mathrm{d}_j \frac{\partial E_\nu}{\partial y_{ij}^\sigma} = 0,$$

$$\frac{\partial E_\sigma}{\partial y^\nu} - \frac{\partial E_\nu}{\partial y^\sigma} + \mathrm{d}_j \frac{\partial E_\nu}{\partial y_j^\sigma} - \mathrm{d}_i\, \mathrm{d}_j \frac{\partial E_\nu}{\partial y_{ij}^\sigma} = 0.$$

The Vainberg-Tonti Lagrangian is

$$\Lambda = \left(y^\sigma \int_0^1 E_\sigma(x^i, uy^\nu, uy_j^\nu, uy_{ij}^\nu)\, \mathrm{d}u \right) \omega_0, \quad 1 \le i, j \le n, \quad 1 \le \sigma, \nu \le m.$$

□

Problems of Chapter 9

Problem 9.1

Hint Choose an appropriate fibred space for the description of the problem, write the corresponding dynamical form and verify the Helmholtz conditions. Write the mapping χ and the integral leading to the Vainberg-Tonti Lagrangian. □

Problem 9.2

Hint In Problem 9.1 we obtained the minimal Lagrangian of the oscillator, $\lambda = \left(\frac{1}{2}m\ddot{x} - \frac{1}{2}kx^2\right) \mathrm{d}t$. For a π-projectable vector field

$$\xi = \xi^0(t)\frac{\partial}{\partial t} + \xi(t,x)\frac{\partial}{\partial x}, \quad J^1\xi = \xi^0(t)\frac{\partial}{\partial t} + \Xi(t,x)\frac{\partial}{\partial x} + \Xi_1(t,x,\dot{x})\frac{\partial}{\partial \dot{x}}$$

(express explicitly Ξ_1 with help of $\xi^0(t)$ and $\Xi(t,x)$) write the Noether equation $\partial_{J^1\xi}\lambda = 0$ using the well-known general relation

$$\partial_\zeta \eta = i_\zeta \, \mathrm{d}\eta + \mathrm{d}i_\zeta \eta$$

and obtain the Noether equation in the form

$$i_{J^1\xi} \, \mathrm{d}L + L\frac{\mathrm{d}\xi^0}{\mathrm{d}t} = 0.$$

Putting the vector field $J^1\xi$ into this equation find the solution. (Separate terms standing by $\dot{x}$, $\dot{x}^2$ and the term depending only on t and x.) □

Results Noether symmetries and currents have the expected form (time translations and conservation of mechanical energy along solutions of equation of motion):

$$\xi = C\frac{\partial}{\partial t}, \quad \Phi(\xi) = -C\left(\frac{1}{2}m\dot{x}^2 + \frac{1}{2}kx^2\right).$$

□

Problem 9.3

Hint For a π-projectable vector field

$$\xi = \xi^0(t)\frac{\partial}{\partial t} + \xi(t,x)\frac{\partial}{\partial x}, \quad J^1\xi = \xi^0(t)\frac{\partial}{\partial t} + \Xi(t,x)\frac{\partial}{\partial x} + \Xi_1(t,x,\dot{x})\frac{\partial}{\partial \dot{x}},$$

$$J^2\xi = \xi^0(t)\frac{\partial}{\partial t} + \Xi(t,x)\frac{\partial}{\partial x} + \Xi_1(t,x,\dot{x})\frac{\partial}{\partial \dot{x}} + \Xi_2(t,x,\dot{x},\ddot{x})\frac{\partial}{\partial \ddot{x}}$$

(express Ξ_1 and Ξ_2 with help of relations given in Chap. 3) write the NBH equation in the form

$$\left(\xi^0\frac{\partial E}{\partial t} + \Xi\frac{\partial E}{\partial x} + \Xi_1\frac{\partial E}{\partial \dot{x}} + \Xi_2\frac{\partial E}{\partial \ddot{x}}\right) + E\frac{\partial \Xi}{\partial x} + E\frac{\mathrm{d}\xi^0}{\mathrm{d}t} = 0,$$

put the function E into this equation, separate the relevant terms and solve obtained equations. □

Results The independent resulting equations for components $\xi^0(t)$ and $\Xi(t,x)$ of symmetries of the Euler-Lagrange form are

$$\frac{\mathrm{d}\xi^0}{\mathrm{d}t} - 2\frac{\partial \Xi}{\partial x} = 0,$$

$$k\Xi + m\frac{\partial^2 \Xi}{\partial t^2} + kx\frac{\partial \Xi}{\partial x} + kx\frac{\mathrm{d}\xi^0}{\mathrm{d}t} = 0.$$

Solution of these equations is

$$\Xi = A(t)x + B(t), \quad \text{where} \quad \ddot{A} + 4\frac{k}{m}A = 0, \quad \ddot{B} + \frac{k}{m}B = 0, \quad \xi^0 = \int 2A\,\mathrm{d}t.$$

Denoting, as it is usual, $\omega = \sqrt{k/m}$ and solving equations for A and B (again "equations of motion" of a linear harmonic oscillator) we finally obtain

$$\xi^0(t) = -\mathrm{i}C\omega^{-1}\mathrm{e}^{2\mathrm{i}\omega t} + \mathrm{i}D\omega^{-1}\mathrm{e}^{-2\mathrm{i}\omega t} + K,$$
$$\Xi(t,x) = x\left(C\mathrm{e}^{2\mathrm{i}\omega t} + D\mathrm{e}^{-2\mathrm{i}\omega t}\right) + c\mathrm{e}^{\mathrm{i}\omega t} + d\mathrm{e}^{-\mathrm{i}\omega t},$$

where C, D, K, c and d are constants. By the most simple choose of constants (always one equals one, the others zero) we obtain 5 vector fields—symmetries of equations of motion of a linear harmonic oscillator (see also [1]).

$$\xi_1 = \frac{\partial}{\partial t}, \quad \xi_2 = \mathrm{e}^{-2\mathrm{i}\omega t}\left(\frac{\mathrm{i}}{\omega}\frac{\partial}{\partial t} + x\frac{\partial}{\partial x}\right), \quad \xi_3 = \mathrm{e}^{2\mathrm{i}\omega t}\left(-\frac{\mathrm{i}}{\omega}\frac{\partial}{\partial t} + x\frac{\partial}{\partial x}\right),$$

$$\xi_4 = \mathrm{e}^{-\mathrm{i}\omega t}\frac{\partial}{\partial x}, \quad \xi_5 = \mathrm{e}^{\mathrm{i}\omega t}\frac{\partial}{\partial x}.$$

For completeness note that there are two triples of these vector fields generating algebras. There are (ξ_1, ξ_2, ξ_3) and (ξ_1, ξ_4, ξ_5). □

Problem 9.4

Hint For the Lagrangian $\lambda = \left(\frac{1}{2}m\dot{x}^2 - \frac{1}{2}kx^2\right)\mathrm{d}t$ calculate $\partial_{J^1\xi_j}\lambda$ for all ξ_j and verify that the obtained forms are trivial Lagrangians (their corresponding dynamical forms are identically zero). □

Problem 9.5

Hint The more general form of the wave equation described in the formulation of the problem is given by the (second order) dynamical form

$$E = E_1\omega^1 \wedge \omega_0 = g^{ij}u_{ij}\omega^1 \wedge \omega_0, \quad \omega^1 = \mathrm{d}u - u_j\,\mathrm{d}x^j,$$

$1 \le i, j \le n$, the corresponding form of the wave equation is

$$\left(\frac{\partial^2 u}{\partial (x^0)^2} - \sum_{i=1}^{n-1}\frac{\partial^2 u}{\partial (x^i)^2}\right) \circ J^2\gamma = 0.$$

(Here x^0 represents the time variable, usually $x^0 = ct$.) The underlying fibred manifold is $(Y, \pi, X) = (\mathbf{R}^n \times \mathbf{R}, p, \mathbf{R}^n)$, coordinates on (Y, π, X), (J^1Y, π_1, X), (J^2Y, π_2, X)

$$(x^i, u), \quad (x^i, u, u_i), \quad (x^i, u, u_i, u_{ij}), \quad u_{ij} = u_{ji},$$

$1 \le i, j \le n$. Verification of Helmholtz conditions of variationality (9.4) is simple, taking into account $m = 1$ and $y^1 = u$, $y^1_i = u_i$, $y^1_{ij} = u_{ij}$. They are fulfilled trivially. The mapping χ is

$$\chi : (s, x^i, u, u_i, u_{ij}) \longrightarrow (x^i, su, su_i, su_{ij}), \quad s \in [0, 1]$$

and the Vainberg-Tonti Lagrangian and minimal Lagrangian (derive it by subtracting a trivial Lagrangian) are

$$\Lambda = \left(u \int_0^1 s g^{ij} u_{ij} \, \mathrm{d}s \right) \omega_0 = \left(\frac{1}{2} g^{ij} u_{ij} \right) \omega_0 , \quad \lambda = \left(\frac{1}{2} g^{ij} u_i u_j \right) \omega_0 .$$

The Noether equation for a π-projectable vector field

$$\xi = \xi^i(x^j) \frac{\partial}{\partial x^i} + \Xi(x^j, u) \frac{\partial}{\partial u}$$

(express explicitly its first prolongation) reads (derive)

$$\begin{aligned} \partial_{j^1\xi} \lambda &= i_{J^1\xi} \, \mathrm{d}\lambda + \mathrm{d} i_{J^1\xi} \lambda = 0 \\ &\Longrightarrow u_i u_j \left(g^{ij} \frac{\partial \Xi}{\partial u} - \frac{1}{2} g^{ik} \frac{\partial \xi^j}{\partial x^k} - \frac{1}{2} g^{jk} \frac{\partial \xi^i}{\partial x^k} + \frac{1}{2} g^{ij} \frac{\partial \xi^k}{\partial x^k} \right) \\ &\quad + u_i \left(g^{ij} \frac{\partial \Xi}{\partial x^j} \right) = 0. \end{aligned}$$

So, for Noether symmetries we have equations

$$g^{ij} \frac{\partial \Xi}{\partial u} - \frac{1}{2} g^{ik} \frac{\partial \xi^j}{\partial x^k} - \frac{1}{2} g^{jk} \frac{\partial \xi^i}{\partial x^k} + \frac{1}{2} g^{ij} \frac{\partial \xi^k}{\partial x^k} = 0 , \quad g^{ij} \frac{\partial \Xi}{\partial x^j} = 0.$$

Solution of obtained conditions for Noether symmetries is laborious and can be found in [1]. □

Problem 9.6

Hint Derive the Noether equation putting the Lagrangian (9.7) into the condition

$$\partial_{J\xi} \lambda = 0 \quad \Longrightarrow \quad i_{J^1\xi} \, \mathrm{d}L + L \frac{\mathrm{d}\xi^i}{\mathrm{d}x^i} = 0, \quad 0 \le i \le 3,$$

$$\xi = \xi^i \frac{\partial}{\partial x^i} + \Xi^j \frac{\partial}{\partial A_j} , \quad J^1\xi = \xi^i \frac{\partial}{\partial x^i} + \Xi^i \frac{\partial}{\partial A_i} + \Xi^i_j \frac{\partial}{\partial A_{i,j}} .$$

□

Table A.1 Generators of Noether symmeries—quantum free particle, one-dimensional motion

	P	Q	a	b	c	e	Generators $\xi_1, \xi_2, \xi_3, \xi_4, \xi_5$ on Y
1	1	0	0	0	0	0	$\frac{\partial}{\partial t}$
2	0	1	0	0	0	0	$\frac{\partial}{\partial x}$
3	0	0	1	0	0	0	$-2t^2 \frac{\partial}{\partial t} - 2xt \frac{\partial}{\partial x} +$ $\left(tw + \frac{m}{\hbar} x^2 v\right) \frac{\partial}{\partial v} + \left(tw - \frac{m}{\hbar} x^2 v\right) \frac{\partial}{\partial w}$
4	0	0	0	1	0	0	$-4t \frac{\partial}{\partial t} - 2t \frac{\partial}{\partial x} + v \frac{\partial}{\partial v} + w \frac{\partial}{\partial w}$
5	0	0	0	0	1	0	$-w \frac{\partial}{\partial v} + v \frac{\partial}{\partial w}$
6	0	0	0	0	0	1	$t \frac{\partial}{\partial x} - \frac{m}{\hbar} xw \frac{\partial}{\partial v} + \frac{m}{\hbar} xv \frac{\partial}{\partial w}$

Results The equation for Noether symmetries reads

$$i_{J^1\xi} \left(\frac{\partial L}{\partial x^i} \, \mathrm{d}x^i + \frac{\partial L}{\partial A_i} \, \mathrm{d}A_i + \frac{\partial L}{\partial A_{i,j}} \, \mathrm{d}A_{i,j} \right) + L \frac{\mathrm{d}\xi^i}{\mathrm{d}x^i} = 0.$$

The Noether current for the symmetry $\xi = \xi^i \frac{\partial}{\partial x^i}$ is

$$\Phi(\xi) = \left[L\xi^i + \frac{\partial L}{\partial A_{\ell,i}} \left(-A_{\ell,j} \xi^j \right) \right] \omega_i,$$

and putting $\frac{\partial L}{\partial A_{\ell,j}} = \text{(calculate)} = -\frac{1}{c\mu_0} F^{j\ell}$

$$\Phi(\xi) = -\frac{1}{c\mu_0} \left[\frac{1}{4} F_{k\ell} F^{k\ell} \xi^i - A_{\ell,j} F^{i\ell} \xi^j \right] \omega_i.$$

Especially for $\frac{\partial}{\partial x^j}$ we have

$$\Phi\left(\frac{\partial}{\partial x^j} \right) = -\frac{1}{c\mu_0} \left(\frac{1}{4} F_{k\ell} F^{k\ell} \delta^i_j - A_{\ell,j} F^{i\ell} \right) \omega_i.$$

The components of this differential form rewritten to the contravariant form

$$T^{ij} = -\frac{1}{c\mu_0} \left(\frac{1}{4} g^{ij} F_{k\ell} F^{k\ell} - g^{js} g_{r\ell} A^{\ell}_{s} F^{ir} \right)$$

are the components of the well-known energy-momentum tensor. (See [2], more detailed calculations in [3,4].) □

Problem 9.7

Results Generators of Noether symmetries are summarized in Table A.1.

□

Problem 9.8

Hint Put the Lagrangian (9.9) into the general (first order) relation for Noether currents

$$\Phi(\xi) = \left[L\xi^i + \frac{\partial L}{\partial y_i^\sigma} \left(\Xi^\sigma - y_j^\sigma \xi^j \right) \right] \omega_i,$$

which for our case gives

$$\begin{aligned}\Phi(\xi) = &\left[L\xi^t + \frac{\partial L}{\partial v_t} \left(\Xi^v - v_t \xi^t - v_x \xi^x \right) + \frac{\partial L}{\partial w_t} \left(\Xi^w - w_t \xi^t - w_x \xi^x \right) \right] \mathrm{d}x \\ &- \left[L\xi^x + \frac{\partial L}{\partial v_x} \left(\Xi^v - v_t \xi^t - v_x \xi^x \right) + \frac{\partial L}{\partial w_x} \left(\Xi^w - w_t \xi^t - w_x \xi^x \right) \right] \mathrm{d}t.\end{aligned}$$

□

Results Put the generators $\xi_1 = \partial/\partial t$, $\xi_2 = \partial/\partial x$, and

$$\xi_5 = -w\frac{\partial}{\partial v} + v\frac{\partial}{\partial w}, \quad \text{here } \xi^t = \xi^x = 0, \ \Xi^v = -w, \ \Xi^w = v,$$

and Lagrange function

$$L = -\frac{\hbar^2}{4m}(v_x^2 + w_x^2) - \frac{\hbar}{2}(vw_t - wv_t)$$

into the relation for $\Phi(\xi)$ and obtain the corresponding currents:

$$\begin{aligned}\Phi(\xi_1) &= \Phi\left(\frac{\partial}{\partial t}\right) = \left[-\frac{\hbar^2}{4m}(v_x^2 + w_x^2) \right] \mathrm{d}x + \left[-\frac{\hbar^2}{2m} (w_x w_t + v_t v_x) \right] \mathrm{d}t, \\ \Phi(\xi_2) &= \Phi\left(\frac{\partial}{\partial x}\right) = \frac{\hbar}{2} (vw_x - wv_x) \, \mathrm{d}x \\ &\quad + \left[-\frac{\hbar^2}{4m}(v_x^2 + w_x^2) + \frac{\hbar}{2} (vw_t - wv_t) \right] \mathrm{d}t, \\ \Phi(\xi_5) &= -\frac{\hbar}{2}(v^2 + w^2) \, \mathrm{d}x - \frac{\hbar^2}{2m} (wv_x - vw_x) \, \mathrm{d}t.\end{aligned}$$

These forms are constant along extremals γ of the Schrödinger equation. For completeness calculate

$$\mathrm{d}\pi_{2,1}^* \Phi(\xi_1) \circ J^2\gamma, \ \ \mathrm{d}\pi_{2,1}^* \Phi(\xi_2) \circ J^2\gamma, \ \ \mathrm{d}\pi_{2,1}^* \Phi(\xi_5) \circ J^2\gamma),$$

and prove that these expressions are zero. Moreover calculate Noether currents for remaining generators. □

Problem 9.9

Solution Express $\Phi(\xi_5)$ with help of the wave function:

$$v^2 + w^2 = \psi\psi^* = |\psi|^2,$$

$$wv_x - vw_x = \frac{\psi - \psi^*}{2\mathrm{i}} \cdot \frac{\psi_x + \psi_x^*}{2} - \frac{\psi + \psi^*}{2} \cdot \frac{\psi_x - \psi_x^*}{2\mathrm{i}} = -\frac{\mathrm{i}}{2}\left(\psi\psi_x^* - \psi^*\psi_x\right).$$

The quantities

$$j = \frac{\mathrm{i}\hbar}{2m}\left(\psi\psi_x^* - \psi^*\psi_x\right), \quad |\psi\psi^*|$$

are the density of the flow of probability and density of probability, respectively. The Noether current $\Phi(\xi_5)$ is then the differential form

$$\Phi(\xi_5) = -\frac{\hbar}{2}\left(|\psi|^2\,\mathrm{d}x + j\,\mathrm{d}t\right).$$

We have verified that this form is closed along extremals of the dynamical form, i.e.

$$\mathrm{d}\Phi(\xi_5) = 0 \implies \frac{\partial j}{\partial x} + \frac{\partial|\psi|^2}{\partial t} = 0.$$

This condition has the form of equation of continuity and it represents the conservation law for density of probability. Note that the vector field ξ_5 is the generator of rotations on fibres. Such rotations lead only to change of the phase of the wave function. □

Problem 9.11

Hint The Schrödinger equation with the potential energy, and thus the corresponding dynamical form has the form

$$\begin{aligned}
\mathrm{i}\hbar\frac{\partial\psi}{\partial t} &= -\frac{\hbar^2}{2m}\frac{\partial^2\psi}{\partial x^2} + V\psi, \quad V = V(t,x),\\
E &= \left(\frac{\hbar^2}{2m}\frac{\partial^2 v}{\partial x^2} - \hbar\frac{\partial w}{\partial t} - Vv\right)\mathrm{d}v \wedge \mathrm{d}t \wedge \mathrm{d}x\\
&+ \left(\frac{\hbar^2}{2m}\frac{\partial^2 w}{\partial x^2} + \hbar\frac{\partial v}{\partial t} - Vw\right)\mathrm{d}w \wedge \mathrm{d}t \wedge \mathrm{d}x.
\end{aligned}$$

In the expression of the Vainberg-Tonti Lagrangian the additional term occurs

$$-v\int_0^1\left(uvV(t,x)\,\mathrm{d}u - w\int_0^1 uwV(t,x)\,\mathrm{d}u\right) = -\frac{1}{2}(v^2 + w^2)V(t,x).$$

Complete the calculation and prove the resulting additional term $\left(-V(t,x)|\psi(t,x)|^2\right)$. □

References

1. E. Búš, Symmetries and conservation laws in variational theories. Thesis, Masaryk University, Brno 2023, 86 pp. (Supervisor J. Musilová) (In Slovak)
2. J. Musilová, S. Hronek, The calculus of variations on jet bundles as a universal approach for a variational formulation of fundamental physical theories. Commun. Math. **24**(2), 173–193 (2016). https://doi.org/10.1515/cm-2016-0012
3. S. Hronek, Differential forms and variational theories. Thesis. Masaryk University, Brno 2017, 71 pp. Supervisor J. Musilová) (In Czech)
4. J. Musilová, P. Musilová, *Mathematics for understanding and praxis III* (VUTIUM, Brno 2017), 1068 pp. (In Czech)

Index

A
Acceleration
  generalized, 132
Action, 237
Action function, 80
  variation, 82
Angular momentum, 265

B
Base of forms
  adapted to contact structure, 54
Basis of 1-forms, 229
Body
  rigid, 259
Boson string, 281
Boundary conditions
  natural, 89
Brachistochrone, 79, 93

C
Chart
  associated, 15
  associated fibred, 25
  fibred, 15
    associated, 216
Christoffel symbols, 95, 259, 302
Classical particle
  Euler-Lagrange form, 259
  Lagrangian, 259
Cohomology, 207
Conservation law, 99
  generalized, 109
  mechanical energy, 264
  momentum component, 264
Constraint, 258
  non-holonomic, 261
Contact
  basis, 230
  derivative, 53
  form, 53
  ideal, 53, 226
  symmetry, 55, 233
Contactization, 53, 232
Coordinate
  cyclic, 97, 264
  generalized, 132
Current
  Noether, 248, 282
Curve
  parametrized, 31
Cyclic coordinate, 97

D
Decomposition of forms, 230
Degrees of freedom, 258
De Rham sequence, 189
Derivative
  contact, 53
  horizontal, 52
  Lie, 83, 293
  operator
    contact, 184
    horizontal, 184
  total, 52
    form, 182
  variational, 82, 238
Derivative operators
  properties, 184

J. Musilová et al., *Calculus of Variations on Fibred Manifolds and Variational Physics*, Lecture Notes in Physics 1033, https://doi.org/10.1007/978-3-031-77408-9

Differential k-form, 223
global, 48
local, 48
Differential form
π-projectable, 225
π_r-horizontal, 225
π_r-projectable, 225
$\pi_{r,s}$-horizontal, 225
$\pi_{r,s}$-projectable, 225
0-contact, 52
1-contact, 52
at least q-contact, 232
at most q-contact, 232
contact, 52, 226
dynamical, 50
Lepagean, 238
projection, 225
q-contact, 52, 226
strongly contact, 56, 232
Distribution
annihilator, 66, 236
Cartan, 67
characteristic, 69
codimension, 66
constant rank, 223
corank, 236
dynamical, 69, 73
Euler-Lagrange, 153
higher order, 166
global, 64
integral section, 66, 236
local, 63
of constant rank, 64
on $J^r Y$, 223
rank, 64, 223
tangent, 64
weakly horizontal, 67
Dynamical form, 235
affine in accelerations, 236
locally variational, 249
variational, 249
variational globally, 119
variational locally, 119
Dynamical (mechanical) system, 69
regular, 69

E
Element
volume, 224
Emmy Noether
theorem generalized, 108
Energy
mechanical, 97
potential
scalar, 260
vector, 260
Equation
Euler-Lagrange, 88, 242
Hamilton, 137, 152
canonical, higher order, 168
first integrals, 142, 173
Hamilton-Jacobi, 173
Klein-Gordon, 280
Maxwell, 133, 267, 273
four-dimensional formulation, 275
minimal Lagrangian, 277
Vainberg-Tonti Lagrangian, 276
motion, 115, 132
Noether, 98, 246
Noether-Bessel-Hagen, 107, 246
of motion
Newton, 96
Schrödinger, 277
minimal Lagrangian, 278
Noether symmetries, 280
Vainberg-Tonti Lagrangian, 278
wave, 270
generalization, 285
Equations of motion
Newton, 96
Euclidean group, 100, 265
Euler-Lagrange
form, 85
mapping, 85, 190
Euler-Lagrange form, 239
invariance transformation, 245
point symmetry, 246
regular, 157
Euler-Lagrange mapping, 240
Exact sequence, 189
short, 189
Exact subsequence, 189
Extremal, 87, 241
Hamilton, 143, 145, 148
higher order, regular, 164
regular, 151

F
Fibration, 13, 212
Fibre, 12
type, 13
Fibred chart
associated, 216
Fibred manifold, 12
first prolongation, 25
multidimensional base, 212
r-jet prolongation, 215
second prolongation, 25

Field
vector
π_r-vertical, 220
projectable, 219
First integrals, 142
First variational formula, 86
infinitesimal form, 86, 241
integral form, 86, 241
Flow, 99
Noether, 99
Force
fictive, 305
fictive centrifugal, 305
fictive Coriolis, 305
fictive Euler, 305
fictive translational, 305
generalized, 132
inertial
centrifugal, 268
Coriolis, 268
Euler, 268
translational, 268
inertial, fictional, 267
Lorentz, 133
Lorentzian, 267
variational, 96, 132, 267
Form
decomposition, 230
differential, 48
2-form, 73
at least q-contact, 232
at most q-contact, 232
contact, 52, 226
dynamical, 50
Lepagean, 238
projection, 225
q-contact, 226
strongly contact, 56, 232
dynamical, 167, 235
affine in accelerations, 236
Lepage equivalent, 206
variational, 119, 249
Euler-Lagrange, 85, 239
regular, 157
Hamilton, 151
Helmholtz-Sonin, 200
Lepage, 69, 201
generalized, 205
Poincaré-Cartan, 239
total derivative, 182
variational dynamical
Helmholtz conditions, 123
Formula
first variational, 83, 86
integral form, 241
Four-vector
four-potential, 261
Function
action, 80, 237
Hamilton, 139
higher order, 165
Lagrange, 103
Functional
Nambu-Goto, 281, 282
Polyakov, 282

G
Generalized momenta, 158
higher order, 165
Generator
space rotation, 263
space translation, 262
time translation, 262
Geodesics, 95
Group
Euclidean, 265

H
Hamilton equations, 140, 152
canonical, higher order, 168
first integrals, 142, 143, 173
higher order, 166
Hamilton extremal, 143, 148
higher order, 164
higher order, regular, 164
regular, 151
Hamilton form, 151
Hamilton function, 139, 158
higher order, 165
Hamilton-Jacobi
equations, 173
Hamilton vector field
higher order, 165
Helmholtz-Sonin
mapping, 190
Homogeneity
space, 103
time, 103
Homomorphism
of fibred manifolds, 20, 214
projection, 214
projection of, 20
Horizontal derivative, 52
Horizontalization, 52, 232

I
Image sheaf, 188

Integral
variational, 35, 80, 237
Interior Euler operator, 195
general order, 200
higher order, 200
Invariance-transformation
generalized, 106
Inverse problem, 116
strong, 128, 249
weak, 128
Isomorphism
first prolongation, 28
local, 28
of fibred manifolds, 214
second prolongation, 28
Isomporphism
local, 217

K
k-form
π_r-horizontal, 50
π_r-projectable, 49
$\pi_{r,s}$-horizontal, 50
$\pi_{r,s}$-projectable, 50
chart expression, 49
differential, 223

L
Lagrange structure
r-th order, 80, 117, 237
first order, 80
point symmetry, 245
regular, 85, 138
singular, 97
symmetry, 245
trivial, 96
variationally trivial, 243
Lagrangian, 83
r-th order, 50, 225, 237
Cawley, 112
Lepage equivalent, 85, 238
minimal, 116
minimal order, 128, 251
regular, 157, 167
rigid body, 259
singular, 97
trivial, 96, 243
Vainberg-Tonti, 129, 251
variationally equivalent, 243
variationally trivial, 243
Law
conservation, 99, 100
generalized, 109
Legendre transformation, 139
generalized, 158
generalized, higher order, 168
Lemma
Poincaré, 58, 73, 234
Lepage
2-form associated with E, 69
equivalent of E, 69
Lepage equivalent
of dynamical form, 206
of EL form, 121
of Lagrangian, 85, 238
principal, 240
Lepage form, 201
generalized, 205
Lepage mapping
second type, 121
Lift
of a function, 37
of a vector field, 38

M
Manifold
fibred, 11, 12, 80
prolonged, 22
two-dimensional, 16
Mapping
Euler-Lagrange, 190, 192, 240
Helmholtz-Sonin, 190, 192
Mechanical system
particles, 258
Metric tensor, 259
Möbius strip, 16
Momenta
generalized, 138
higher order, 165
Momentum
angular, 265

N
Noether
equation, 98
symmetry, 98
generators, 320
theorem, 99, 247
Noether-Bessel-Hagen (NBH)
equation, 107
Noether flow, 99

O
1-form
bases, 49
basis, 229
chart expressions, 49

Operator
 $\mathcal{A}$
  Poincaré lemma, 60
 contact derivative, 53
 derivative
  contact, 184
  horizontal, 184
 Euler
  interior, 195
  interior, general order, 200
  interior, higher order, 200
 exterior derivative, 189
 total derivative, 23, 217
Operator $\mathcal{A}$
 Poincaré lemma, 60
Oscillator
 damped, 127
 harmonic, 127
 linear
  harmonic, 284

P
Particle
 classical
  Euler-Lagrange form, 259
  Lagrangian, 259
 relativistic, 260
  Lagrangian, 260
Pendulum
 conical, 18
 simple, 18
 spherical, 18, 258
Permeability, 275
Permittivity, 275
Poincaré-Cartan form, 239
Poincaré lemma, 58, 234
 operator $\mathcal{A}$, 60
Point
 stationary, 80
Poisson bracket, 142
Potential
 scalar, 274
 vector
  magnetic, 274
Potential energy
 scalar, 260
 vector, 260
Problem
 brachistochrone, 79, 93
 inverse, 181, 266
  strong, 249
 singular, 96
 trivial, 96, 181
 variational, 80, 92
  trivial, 243
Projection, 12
 differential form, 225
 vector field, 219

Q
q-contactization, 232

R
Reference frame
 associated with a body, 259
Relativistic particle, 260
 Lagrangian, 260
 with constraint, 260
Representation sequence, 197
Rigid body, 259
 Lagrangian, 259
 rotation, 259
 translation, 259
Rotation
 generator
  space, 263
 rigid body, 259
 space, 101
r-th order Lagrangian
 standard regularity, 163

S
Section, 11
 0-equivalent, 22
 critical, 87, 241
 extremal, 87
 global, 17, 213
 holonomic, 25, 216
 jet prolongation, 11, 22
 local, 17, 213
 of 0-th order contact, 22
 prolongations of, 25
 r-equivalent, 22, 214
 r-jet, 22, 215
 r-th order contact, 22, 214
 smooth, 214
 stationary point, 241
 variation, deformation of, 237
Sequence
 de Rham, 189
 exact, 189
 representation, 197
 variational, 201
  first order, 190
  representation mapping, 192
Sheaf
 direct image, 187

of forms, 187
Space
configuration, 11
fibred, 11
phase, 11
total, 12
Standard model, 281
String theory, 281
Subsequence
exact, 189
Symmetry
contact, 233
generalized, 106
Lagrange structure, 245
Noether, 98
point, 98
space, 103
time, 103
time-space, 261
System
mechanical, 258

T
Tangent
bundle, 31
space, 31
vector, 31
Tensor
electromagnetic field, 275
of inertia, 259
principal axes, 259
Tensor of
inertia
diagonal, 259
Theorem
Emmy Noether, 247
generalized, 108
Noether, 99
Stokes, 83, 109
Theory
Hamilton-Jacobi, 170
Time-space
symmetries, 261
space homogeneity, 261
space isotropy, 262
time homogeneity, 261
Total derivative, 52
of forms, 183
Trajectory
stationary, 35
Transformation
coordinate, 161
invariance, 98
Euler-Lagrange form, 245
Legendre, 139
generalized, 158
generalized, higher order, 168
relations, 15
Translation
generator
space, 262
time, 262
rigid body, 259
space, 101
time, 100

V
Variational derivative, 238
Variational dynamical form
Helmholtz conditions, 123, 124
Variational integral, 80, 237
free ends, 170
Variationality
global, 249
local, 249
Variational problem
singular, 96
trivial, 96, 243
Variational sequence
first order, 190
representation mapping, 192
Variation, 80, 81, 237
first of the action, 82
fixed ends, 237
with fixed endpoints, 87
Variation (deformation
of a section), 81
Vector field
1-st prolongation, 43
2-nd prolongation, 43
π-projectable, 81
π_r-projectable, 38
π_r-projection of, 38
π_r-vertical, 40, 220
π_s-projectable, 219
π_s-projection, 219
$\pi_{r,s}$-projectable, 38, 219
$\pi_{r,s}$-projection, 219
$\pi_{r,s}$-projection of, 38
$\pi_{r,s}$-vertical, 40, 220
r-th prolongation, 221
chart expressions, 36
global, 36
Hamilton, 153
higher order, 165
Killing, 106
local, 36, 218
prolongations-components, 46

smooth, 36
total derivative, 183
Velocity
generalized, 132
Volume element, 224

www.ingramcontent.com/pod-product-compliance
Lightning Source LLC
Chambersburg PA
CBHW072337020325
22834CB00004B/64
9783031774072